Quantum Field Theory

A Tourist Guide for Mathematicians

Mathematical
Surveys
and
Monographs

Volume 149

Quantum Field Theory

A Tourist Guide for Mathematicians

Gerald B. Folland

American Mathematical Society
Providence, Rhode Island

2000 *Mathematics Subject Classification.* Primary 81-01; Secondary 81T13, 81T15, 81U20, 81V10.

For additional information and updates on this book, visit
www.ams.org/bookpages/surv-149

Library of Congress Cataloging-in-Publication Data

Folland, G. B.
Quantum field theory : a tourist guide for mathematicians / Gerald B. Folland.
p. cm. — (Mathematical surveys and monographs ; v. 149)
Includes bibliographical references and index.
ISBN 978-0-8218-4705-3 (alk. paper)
1. Quantum electrodynamics–Mathematics. 2. Quantum field theory–Mathematics. I. Title.

QC680.F65 2008
530.14′30151–dc22　　2008021019

∞ The paper used in this book is acid-free and falls within the guidelines
established to ensure permanence and durability.
Visit the AMS home page at http://www.ams.org/

10 9 8 7 6 5 4 3 2 1　　13 12 11 10 09 08

Contents

Preface

This book is an attempt to present the rudiments of quantum field theory in general and quantum electrodynamics in particular, *as actually practiced by physicists for the purpose of understanding the behavior of subatomic particles*, in a way that will be comprehensible to mathematicians.

It is, therefore, *not* an attempt to develop quantum field theory in a mathematically rigorous fashion. Sixty years after the growth of quantum electrodynamics (QED) and forty years after the discovery of the other gauge field theories on which the current understanding of the fundamental interactions of physics is based, putting these theories on a sound mathematical foundation remains an outstanding open problem — one of the Millennium prize problems, in fact (see [**66**]). I have no idea how to solve this problem. In this book, then, I give mathematically precise definitions and arguments when they are available and proceed on a more informal level when they are not, taking some care to be honest about where the problems lie. Moreover, I do not hesitate to use the informal language of distributions, with its blurring of the distinction between functions and generalized functions, when that is the easiest and clearest way to present the ideas (as it often is).

So: why would a self-respecting mathematician risk the scorn of his peers by undertaking a project of such dubious propriety, and why would he expect any of them to read the result?

In spite of its mathematical incompleteness, quantum field theory has been an enormous success for physics. It has yielded profound advances in our understanding of how the universe works at the submicroscopic level, and QED in particular has stood up to extremely stringent experimental tests of its validity. Anyone with an interest in the physical sciences must be curious about these achievements, and it is not hard to obtain information about them at the level of, say, *Scientific American* articles. In such popular accounts, one finds that (1) interaction processes are described pictorially by diagrams that represent particles colliding, being emitted and absorbed, and being created and destroyed, although the relevance of these diagrams to actual computations is usually not explained; (2) some of the lines in these diagrams represent real particles, but others represent some shadowy entities called "virtual particles" that cannot be observed although their effects can be measured; (3) quantum field theories are plagued with infinities that must be systematically subtracted off to yield meaningful answers; (4) in spite of the impression given by (1)–(3) that one has blundered into some sort of twilight zone, these ingredients can be combined to yield precise answers that agree exquisitely with experiment. (For example, the theoretical and experimental values of the magnetic moment of the electron agree to within one part in 10^{10}, which is like determining the distance from the Empire State Building to the Eiffel Tower to within a millimeter.)

People with mathematical training are entitled to ask for a deeper and more quantitative understanding of what is going on here. They may feel optimistic about attaining it from their experience with the older areas of fundamental physics that have proved very congenial to mathematical study: the differential equations of classical mechanics, the geometry of Hamiltonian mechanics, and the functional analysis of quantum mechanics. But when they attempt to learn quantum field theory, they are likely to feel that they have run up against a solid wall. There are several reasons for this.

In the first place, quantum field theory is *hard*. A mathematician is no more likely to be able to pick up a text on quantum fields such as Peskin and Schroeder [**88**] and understand its contents on a first reading than a physicist hoping to do the same with, say, Hartshorne's *Algebraic Geometry*. At the deep conceptual level, the absence of firm mathematical foundations gives a warning that some struggle is to be expected. Moreover, quantum field theory draws on ideas and techniques from many different areas of physics and mathematics. (Despite the fact that subatomic particles behave in ways that seem completely bizarre from the human perspective, our understanding of that behavior is built to a remarkable extent on classical physics!) At the more pedestrian level, the fact that the universe seems to be made out of vectors and spinors rather than scalars means that even the simplest calculations tend to involve a certain amount of algebraic messiness that increases the effort needed to understand the essential points. And at the mosquito-bite level of annoyance, there are numerous factors of -1, i, and 2π that are easy to misplace, as well as numerous disagreements among different authors as to how to arrange various normalization constants.

But there is another difficulty of a more cultural and linguistic nature: physics texts are usually written by physicists for physicists. They speak a different dialect, use different notation, emphasize different points, and worry about different things than mathematicians do, and this makes their books hard for mathematicians to read. (Physicists have exactly the same complaint about mathematics books!) In the mathematically better established areas of physics, there are books written from a more mathematical perspective that help to solve this problem, but the lack of a completely rigorous theory has largely prevented such books from being written about quantum field theory.

There have been some attempts at cross-cultural communication. Mathematical interest in theoretical physics was rekindled in the 1980s, after a period in which the long marriage of the two subjects seemed to be disintegrating, when ideas from gauge field theory turned out to have striking applications in differential geometry. But the gauge fields of interest to the geometers are not quantum fields at all, but rather their "classical" (unquantized) analogues, so the mathematicians were not forced to come to grips with quantum issues. More recently, motivated by the development of string theory, in 1996–97 a special year in quantum field theory at the Institute for Advanced Study brought together a group of eminent mathematicians and physicists to learn from each other, and it resulted in the two-volume collection of expository essays *Quantum Fields and Strings* [**21**]. These books contain a lot of interesting material, but as an introduction to quantum fields for ordinary mortals they leave a lot to be desired. One drawback is that the multiple authorships do not lead to a consistent and cohesively structured development of the subject. Another is that the physics is mostly on a rather formal and abstract level; the down-to-earth calculations that lead to experimentally verifiable results are given scant attention. Actually, I would suggest that the reader might study *Quantum Fields and Strings* more profitably *after* reading the present book, as the real focus there is on more advanced topics.

There is another book about quantum fields written by a mathematician, Ticciati's *Quantum Field Theory for Mathematicians* [**119**]. In its general purpose it has some similarity to the present book, but in its organization, scope, and style it is quite different. It turned out not to be the book I needed in order to understand the subject, but it may be a useful reference for others.

The foregoing paragraphs should explain why I thought there was a gap in the literature that needed filling. Now I shall say a few words about what this book does to fill it.

First of all, what are the prerequisites? On the mathematical side, the reader needs to be familiar with the basics of Fourier analysis, distributions (generalized functions), and linear operators on Hilbert spaces, together with a couple of more advanced results in the latter subject — most notably, the spectral theorem. This material can all be found in the union of Folland [**47**] and Reed and Simon [**93**], for example. In addition, a little Lie theory is needed now and then, mostly in the context of the specific groups of space-time symmetries, but in a more general way in the last chapter; Hall [**61**] is a good reference for this. The language of differential geometry is employed only in a few places that can safely be skimmed by readers who are not fluent in it. On the physical side, the reader should have some familiarity with the Hamiltonian and Lagrangian versions of classical mechanics, as well as special relativity, the Maxwell theory of electromagnetism, and basic quantum mechanics. The relevant material is summarized in Chapters 2 and 3, but these brief accounts are meant for review and reference rather than as texts for the novice.

As I mentioned earlier, quantum field theory is built on a very broad base of earlier physics, so the first four chapters of this book are devoted to setting the stage. Chapter 5 introduces free fields, which are already mathematically quite nontrivial although physically uninteresting. The aim here is not only to present the rigorous mathematical construction but also to introduce the more informal way of treating such objects that is common in the physics literature, which offers both practical and conceptual advantages once one gets used to it. The plunge into the deep waters of interacting field theory takes place in Chapter 6, which along with Chapter 7 on renormalization contains most of the really hard work in the book. I use some imagery derived from the Faust legend to describe the necessary departures from mathematical rectitude; its significance is meant to be purely literary rather than theological. Chapter 8 sketches the attractive alternative approach to quantum fields through Feynman's sum-over-histories view of quantum mechanics, and the final chapter presents the rudiments of gauge field theory, skirting most of the quantum issues but managing to derive some very interesting physics nonetheless.

There are several ways to get from the starting line to the goal of calculating quantities with direct physical meaning such as scattering cross-sections. The path I follow here is to start with free fields, apply perturbation theory to arrive at the integrals associated to Feynman diagrams, and renormalize as necessary. This has the advantages of directness and of minimizing the amount of time spent dealing with mathematically ill-defined objects. Its drawback is that it tethers one to perturbation theory, whereas nonperturbative arguments would be more satisfying in some situations. Physicists may also object to it on the grounds that free fields, although mathematically meaningful, are physically fictitious.

The problem with interacting fields, on the other hand, is exactly the reverse. Hence, although some might prefer to give them a more prominent role, I sequester them in the last section of Chapter 6, where the mathematical soundness of the narrative reaches its nadir, and do not use them at all in Chapter 7 except for a couple of passing mentions. Their credibility is somewhat enhanced, however, by the arguments in Chapter 8 using functional integrals, which are also mathematically ill-defined but intuitively more accessible and seductively close to honest mathematics. Some physicists like to use functional integrals as the principal route to the main results, but despite their appeal, I find them a bit too much like sorcery to be relied on until one already knows where one is going.

This book is meant to be only an introduction to quantum field theory, and it focuses on the goal of explaining actual physical phenomena rather than studying formal structures for their own sake. This means that I have largely (though not entirely) resisted the temptation to pursue mathematical issues when they do not add to the illumination of the physics, and also that I have nothing to say about the more speculative areas of present-day theoretical physics such as supersymmetry and string theory. Even within these restrictions, there are many important topics that are mentioned only briefly or omitted entirely — most notably, the renormalization group. My hope is that this book will better prepare those who wish to go further to tackle the physics literature. References to sources where further information can be obtained on various topics are scattered throughout the book. Here, however, I wish to draw the reader's attention to three physics books whose quality of writing I find exceptional.

First, everyone with any interest in quantum electrodynamics should treat themselves to a perusal of Feynman's *QED* [**37**], an amazingly fine piece of popular exposition. On a much more sophisticated level, but still with a high ratio of physical insight to technical detail, Zee's *Quantum Field Theory in a Nutshell* [**136**] makes very good reading. (Both of these books adopt the functional integral approach.) And finally, for a full-dress treatment of the subject, Weinberg's *The Quantum Theory of Fields* [**129**], [**130**], [**131**] is the sort of book for which the overworked adjective "magisterial" is truly appropriate. Weinberg does not aim for a mathematician's level of rigor, but he has a mathematician's respect for careful reasoning and for appropriate levels of generality, and his approach has influenced mine considerably. I will warn the reader, however, that Weinberg's notation is at variance with standard usage in some respects. Most notably, he takes the Lorentz metric (which he denotes by $\eta^{\mu\nu}$) to have signature $-+++$ rather than the usual $+---$, and since he wants his Dirac matrices γ^μ to satisfy $\{\gamma^\mu, \gamma^\nu\} = 2\eta^{\mu\nu}$, what he calls γ^μ is what most people call $-i\gamma^\mu$.[1]

I call this book a tourist guide for mathematicians. This is meant to give the impression not that it is easy reading (it's not) but that the intended audience consists of people who approach physics as tourists approach a foreign country, as a place to enjoy and learn from but not to settle in permanently. It is also meant to free me and my readers from guilt about omitting various important but technical

[1] There is yet a third convention for defining Dirac matrices, found in Sakurai [**102**] among other places.

topics, viewing others from a point of view that physicists may find perverse, failing to acquire a scholarly knowledge of the literature, and skipping the gruesome details of certain necessary but boring calculations.

I wish to state emphatically that I am a tourist in the realm of physics myself. I hope that my foreigner's perceptions do not do violence to the native culture and that my lack of expertise has not led to the perpetration of many outright falsehoods. Given what usually happens when physicists write about mathematics, however, I dare not hope that there are none. Corrections will be gratefully received at `folland@math.washington.edu` and recorded on a web page accessible from `www.math.washington.edu/~folland/Homepage/index.html`. The American Mathematical Society will also host a web page for this book, the URL for which can be found on the back cover above the barcode.

Acknowledgments. I am grateful to the students and colleagues who sat through the course I offered in 2001 in which I made my rather inept first attempt to put this material together. Several physicists, particularly David Boulware, have patiently answered many questions for me, and they are not to blame if their answers have become distorted in passing through my brain. Finally, an unnamed referee provided several helpful suggestions and useful references.

The Feynman diagrams in this book were created with JaxoDraw, available at `jaxodraw.sourceforge.net/sitemap.html`.

Gerald B. Folland
Seattle, April 2008

CHAPTER 1

Prologue

> Men can do nothing without the make-believe of a beginning. Even Science, the strict measurer, is obliged to start with a make-believe unit, and must fix on a point in the stars' unceasing journey when his sidereal clock shall pretend time is at Nought. His less accurate grandmother Poetry has always been understood to start in the middle, but on reflection it appears that her proceeding is not very different from his; since Science, too, reckons backwards as well as forwards, divides his unit into billions, and with his clock-finger at Nought really sets off *in medias res*. No retrospect will take us to the true beginning; and whether our prologue be in heaven or on earth, it is but a fraction of the all-presupposing fact with which our story sets out.
> —George Eliot, *Daniel Deronda* (epigraph of Chapter 1)

The purpose of this preliminary chapter is to present some notation and terminology, discuss a few matters on which mathematicians' and physics usages differ, and provide some basic definitions and formulas from physics and Lie theory that will be needed throughout the book.

1.1. Linguistic prologue: notation and terminology

Hilbert space and its operators. Quantum mechanicians and mathematical analysts both spend a lot of time in Hilbert space, but they have different ways of speaking about it. In this book we generally follow physicists' conventions, so it will be well to explain how to translate from one dialect to the other. (The physicists' dialect is largely due to Dirac, and its ubiquity is a testimony to the profound influence that his book **[22]** had on several generations of physicists.)

To begin with, complex conjugates are denoted by asterisks rather than overlines:

$$x - iy = (x + iy)^* \text{ rather than } \overline{x + iy}.$$

The inner product on a Hilbert space is denoted by $\langle\cdot|\cdot\rangle$ and is taken to be linear in the second variable and conjugate-linear in the first:

$$\langle u|av + bw\rangle = a\langle u|v\rangle + b\langle u|w\rangle, \qquad \langle v|u\rangle = \langle u|v\rangle^*.$$

The Hermitian adjoint (conjugate transpose) of a matrix or an operator is denoted by a dagger rather than an asterisk:

$$\text{if } A = (x_{jk} + iy_{jk}),\text{ then } (x_{kj} - iy_{kj}) = A^\dagger \text{ rather than } A^*.$$

(More about adjoints below.) These shifted uses of * and † may seem annoying at first, but one gets used to them — and there is another consideration. The overline is employed in physics not for conjugation but to denote the "Dirac adjoint" of a Dirac spinor, as we shall explain in §4.1. This usage turns up sufficiently frequently that any attempt to permute the physicists' meanings of *, †, and $^-$ would threaten mass confusion.

We shall, however, employ the standard mathematical notation for the norm on the Hilbert space: $\|u\| = \langle u|u\rangle^{1/2}$. We also denote the transpose of a complex matrix A by A^T; for real matrices we shall generally use $A^\dagger$ instead.

Now, some more subtle matters. In mathematicians' dialect, $\langle u|v\rangle$ is the inner product of two vectors u and v in the Hilbert space $\mathcal{H}$. Moreover, if $u \in \mathcal{H}$, the map $\phi_u(v) = \langle u|v\rangle$ is a bounded linear functional on $\mathcal{H}$; the correspondence $u \leftrightarrow \phi_u$ gives a conjugate-linear identification of $\mathcal{H}$ with its dual $\mathcal{H}'$, which mathematicians generally take for granted without employing any special notation for it. Physicists, on the other hand, distinguish between elements of $\mathcal{H}$ and elements of $\mathcal{H}'$, which they respectively call *ket* vectors and *bra* vectors and denote by symbols of the form $|u\rangle$ and $\langle u|$. If $|u\rangle$ is a ket vector (what mathematicians might call simply u), the corresponding bra vector $\langle u|$ is the linear functional denoted by ϕ_u above. The action of the bra vector (linear functional) $\langle u|$ on the ket vector (element of $\mathcal{H}$) $|v\rangle$ is the inner product (i.e., "bracket" or "bra-ket") $\langle u|v\rangle$.

In this system "u" can be any sort of convenient label to identify the vector: either a name for the vector itself, like the mathematicians' u, or an index or set of indices that specify the vector within an indexed family. For example, if one is working with an operator with a set of multiplicity-one eigenvalues, one might denote a unit eigenvector for the eigenvalue λ by $|\lambda\rangle$, or an eigenvector for the nth eigenvalue simply by $|n\rangle$. Likewise, joint eigenvectors for a pair of operators with eigenvalues λ_n and μ_j might be denoted by $|n, j\rangle$; and so forth. (In most situations the ambiguity of a scalar factor of modulus one in the choice of eigenvector is of no importance, for reasons that will become clear in Chapter 3.) Mathematicians may find this convention uncomfortably informal, but its virtue lies in its flexibility and its ability to strip away inessential symbols. Mathematicians would typically denote the nth eigenvector by something like u_n, but the symbol u is just a placeholder; it is the n that carries the useful information, and the physicists' notation gives it the starring role it deserves.

Next, let A be a linear operator[1] on $\mathcal{H}$: in mathematicians' notation, A maps a vector v to Av; and in physicists' notation, it maps $|v\rangle$ to $A|v\rangle$: no surprises here. But the physicists' notation for the scalar product of $A|v\rangle$ with the bra vector $\langle u|$ is $\langle u|A|v\rangle$:

$$\text{physicists' } \langle u|A|v\rangle = \text{mathematicians' } \langle u|Av\rangle.$$

Now, $\langle u|A|v\rangle$ can also be considered as the scalar product of a bra $\langle u|A$ with the ket $|v\rangle$, and in this way A defines a linear operator on bra vectors. *This operator is what the mathematicians call the adjoint of A* when $\mathcal{H}'$ is identified with $\mathcal{H}$ — a point that can lead to some confusion if one does not remain alert. Recalling that the adjoint of A is denoted by $A^\dagger$ and the conjugate of $a \in \mathbb{C}$ is denoted by a^*, we have

$$\langle u|A|v\rangle = \langle v|A^\dagger|u\rangle^*.$$

On the theory that flexibility and concision are more important than consistency, we shall feel free to denote elements of the Hilbert space by either u or $|u\rangle$, and we shall write either $\langle u|A|v\rangle$ or $\langle u|Av\rangle$ as convenience dictates.

[1] A need not be bounded. The notion of adjoint for unbounded operators involves some technicalities, but they are beside the point here.

By the way, we will often write a scalar multiple of the identity operator, λI, simply as λ. This commonly used bit of shorthand will cause no confusion if readers will keep in mind that when they encounter a scalar where an operator or a matrix seems to be needed, they should tacitly insert an I.

Vectors, tensors, and derivatives. In this book we reserve boldface type almost exclusively to denote 3-dimensional vectors related to $\mathbb{R}^3$ as a model for the physical space in which we live. (Exception: In §4.5 it is used to denote elements of tensor products of Hilbert spaces.) Elements of $\mathbb{R}^n$ for general n are usually denoted by lower-case italic letters (x, p, ...). We denote the n-tuple of partial derivatives $(\partial_1, \ldots, \partial_n)$ on $\mathbb{R}^n$ by ∇, and we denote the Laplacian $\nabla \cdot \nabla = \sum \partial_j^2$ by ∇^2 (rather than Δ, for which we will have other uses).

When $n = 4$, a variant of this notation will be used for calculations arising from relativistic mechanics. Four-dimensional space-time is taken to be $\mathbb{R}^4$ with coordinates

$$x^0 = ct, \quad x^1 = x, \quad x^2 = y, \quad x^3 = z$$

on $\mathbb{R}^4$, where t represents time, c the speed of light, and (x, y, z) a set of Cartesian coordinates on physical space $\mathbb{R}^3$. Thus, a point $x \in \mathbb{R}^4$ may be written as $(x^0, \mathbf{x})$ when it is important to separate the space and time components. The *Lorentz inner product* on $\mathbb{R}^4$ is the bilinear form Λ defined by

$$\Lambda(x, y) = x^0y^0 - x^1y^1 - x^2y^2 - x^3y^3,$$

and $\mathbb{R}^4$ equipped with this form is called *Minkowski space.* It is common in the physics literature to denote the vector whose components are $x^0, \ldots, x^3$ by x^μ rather than x. This is the same sort of harmless abuse of language that is involved in speaking of "the sequence a_n" or "the function $f(x)$"; we shall use it when it seems convenient.

We shall generally use classical tensor notation for vectors and tensors associated to Minkowski space. In particular, the Lorentz form is defined by the matrix

$$g_{\mu\nu} = g^{\mu\nu} = \operatorname{diag}(1, -1, -1, -1). \tag{1.1}$$

We employ the *Einstein summation convention*: in any product of vectors and tensors in which an index appears once as a subscript and once as a superscript, that index is to be summed from 0 to 3. Thus, for example,

$$\Lambda(x, y) = g_{\mu\nu}x^\mu y^\nu. \tag{1.2}$$

We shall also adopt the convention of using the matrix g to "raise or lower indices":

$$x_\mu = g_{\mu\nu}x^\nu, \qquad p^\mu = g^{\mu\nu}p_\nu,$$

the practical effect of which is to change the sign of the last three components. (Strictly speaking, vectors whose components are denoted by subscripts should be construed as elements of the dual space $(\mathbb{R}^4)^*$, and the map $x^\mu \mapsto x_\mu$ is the isomorphism of $\mathbb{R}^4$ with $(\mathbb{R}^4)^*$ induced by the Lorentz form. But we shall not attempt to distinguish between $\mathbb{R}^4$ and its dual.) Formula (1.2) can then be written as

$$\Lambda(x, y) = x^\mu y_\mu = x_\mu y^\mu.$$

The Lorentz inner product of a vector x with itself is denoted by x^2 and its Euclidean norm by $|x|$:

$$x^2 = x_\mu x^\mu, \qquad |x|^2 = x_0^2 + x_1^2 + x_2^2 + x_3^2.$$

Vectors x such that $x^2 > 0$, $x^2 = 0$, or $x^2 < 0$ are called *timelike*, *lightlike*, or *spacelike*, respectively.

In this framework, the notation for derivatives on $\mathbb{R}^4$ is

$$\partial_\mu = \frac{\partial}{\partial x^\mu},$$

so that, with respect to traditional space-time coordinates $\mathbf{x}$ and t,

$$(\partial_0, \dots, \partial_3) = (c^{-1}\partial_t, \nabla_{\mathbf{x}}), \qquad (\partial^0, \dots, \partial^3) = (c^{-1}\partial_t, -\nabla_{\mathbf{x}}), \tag{1.3}$$

and ∂^2 is the wave operator or d'Alembertian:

$$\partial^2 = \partial_0^2 - \partial_1^2 - \partial_2^2 - \partial_3^2 = c^{-2}\partial_t^2 - \nabla_{\mathbf{x}}^2.$$

Integrals. Physicists like to indicate the dimensionality of their integrals explicitly by writing the volume element on $\mathbb{R}^n$ as $d^n x$. This convention is occasionally an aid to clarity, and we shall generally follow it. Another convention for integrals more commonly used in physics than in mathematics is to put the differential next to the integral sign: $\int dx\, f(x)$ rather than $\int f(x)\, dx$. This can also be an aid to clarity, particularly in multiple integrals where writing $\int_a^b dx \int_c^d dy\, f(x,y)$ in preference to $\int_a^b \int_c^d f(x,y)\, dy\, dx$ makes it easier to see which variables go with which limits of integration. However, for reasons of inertia more than anything else, we shall not adopt this convention.

Fourier transforms. If we are considering functions on the space $\mathbb{R}^n$ equipped with the Euclidean inner product $(x,y) \mapsto x \cdot y$, the Fourier transform $f \mapsto \widehat{f}$ and its inverse $f \mapsto f^\vee$ will be defined by

$$\widehat{f}(y) = \int e^{-ix\cdot y} f(x)\, d^n x, \qquad f^\vee(x) = \int e^{ix\cdot y} f(y) \frac{d^n y}{(2\pi)^n}.$$

The Parseval identity is then

$$\int |f(x)|^2\, d^n x = \int |\widehat{f}(y)|^2 \frac{d^n y}{(2\pi)^n}. \tag{1.4}$$

However, most of the time the Fourier transform will pertain to functions on $\mathbb{R}^4$ equipped with the Lorentz metric, in which case we define

$$\widehat{f}(p) = \int e^{ip_\mu x^\mu} f(x)\, d^4 x, \qquad f^\vee(x) = \int e^{-ip_\mu x^\mu} f(p) \frac{d^4 p}{(2\pi)^4}.$$

(The Parseval identity is again (1.4), with $n = 4$.) Thus the Lorentz sign convention in the exponent agrees with the Euclidean one as far as the space variables are concerned, but it is reversed for the time variable. In this situation, generally x is in "position space" and p is in "momentum space," in which case the $p_\mu x^\mu$ in the exponent really needs to be divided by a normalization factor so that the argument of the exponent is a pure number, independent of the units of measurement. In quantum mechanics this factor is Planck's constant $\hbar$, but it will usually be suppressed since we will be using units in which $\hbar = 1$.

Lord Kelvin once quipped that a mathematician is someone to whom it is obvious that $\int_{-\infty}^{\infty} e^{-x^2}\, dx = \sqrt{\pi}$. In the same spirit, I submit that a mathematical physicist is someone to whom it is obvious that

$$\delta(x) = \int e^{ix\cdot y} \frac{d^n y}{(2\pi)^n}, \tag{1.5}$$

where δ denotes the delta-function (point mass at the origin) on $\mathbb{R}^n$. This is the Fourier inversion formula, stated in the informal language of distributions. Readers who do not yet qualify as mathematical physicists by this criterion are advised to spend some time playing with this formula until they understand how to make sense of it and are convinced of its validity, for it will be used on many occasions in this book. (The necessary tools from distribution theory can be found, for example, in Folland [**47**].)

It is useful to remember that the factors of 2π always go with the measure on momentum space; that is, the standard measure on momentum space (or, more abstractly, "Fourier space" as opposed to "configuration space") $\mathbb{R}^n$ is $d^n p/(2\pi)^n$. A corollary of this is that delta-functions on momentum space are normally accompanied by a factor of $(2\pi)^n$ so that their integral against this measure is still 1.

Symbolic homonyms. It is an unfortunate fact of life that some letters of the alphabet have more than one conventional meaning, while others are used sometimes conventionally and sometimes just as a convenient label for a variable. In such cases one must rely on context for clarification. For example, all of us have surely seen uses of the letter π that have nothing to do with the constant $3.14\ldots$; there are some in this book. Other notable examples: (1) e is both $\exp(1)$ and the electric charge that functions as the coupling constant in quantum electrodynamics (typically the charge of the electron or its absolute value); occasionally it is also the symbol for the electron. (Fortunately, we never need it to be the eccentricity of an ellipse!) (2) q is sometimes an electric charge and sometimes a momentum. (3) α^j is a type of Dirac matrix, but α is the fine structure constant. (4) γ^μ is another type of Dirac matrix, but γ is the Euler-Mascheroni constant and occasionally the symbol for the photon. (5) Z is sometimes the number of protons in a nucleus, sometimes a field renormalization constant, sometimes the generating functional in the functional integral approach to field theory, and sometimes the symbol for the neutral vector Boson of the weak interaction.

Caveat lector.

1.2. Physical prologue: dimensions, units, constants, and particles

The fundamental aspects of the universe to which all physical measurements relate are mass $[m]$, length or distance $[l]$, and time $[t]$. All quantities in physics come with "dimensions" that can be expressed in terms of these three basic ones. For example, velocity $\mathbf{v} = d\mathbf{x}/dt$ has dimensions of distance per unit time, or $[lt^{-1}]$. The dimensions of some of the basic quantities of mechanics are summarized in Table 1.1. The fact that angular momentum and action both have dimensions $[ml^2t^{-1}]$ seems a mere coincidence in classical mechanics, but it acquires a deeper significance in quantum mechanics, because $[ml^2t^{-1}]$ is the dimensions of Planck's constant.

Units for electromagnetic quantities are related to those of mechanics by taking the constant of proportionality in some basic law to be equal to one. For example, the ampere is the unit of current defined as follows: if two infinite, straight, perfectly conducting wires are placed parallel to each other one meter apart and a current of one ampere flows in each of them, the (magnetic) force between them per unit length is one newton per meter. The other everyday units of electromagnetism (coulomb, volt, ohm, etc.) are then defined in terms of the ampere and the MKS

Quantity	Dimensions	Defining relation
Momentum	$[mlt^{-1}]$	$\mathbf{p}=m\mathbf{v}$
Angular momentum	$[ml^2t^{-1}]$	$\mathbf{l}=\mathbf{x}\times\mathbf{p}$
Force	$[mlt^{-2}]$	$\mathbf{F}=m(d\mathbf{v}/dt)$
Energy	$[ml^2t^{-2}]$	$E=\frac{1}{2}m\lvert\mathbf{v}\rvert^2$ or $\int\mathbf{F}\cdot d\mathbf{x}$ or mc^2 or ...
Action	$[ml^2t^{-1}]$	$S=\int(E_{\text{kinetic}}-E_{\text{potential}})\,dt$

TABLE 1.1. Some basic quantities of mechanics.

mechanical units. However, for purposes of fundamental physics it is better to take electric charge as the basic quantity. Naively, the simplest procedure ("Gaussian units") is to define the unit of charge so that the constant in Coulomb's law is equal to one. That is, having chosen a system of units for mechanics, one sets the unit of charge so that the force between two point charges is *equal* in magnitude to the product of the charges divided by the square of the distance between them. From a slightly more sophisticated point of view, however, a better procedure ("Heaviside-Lorentz units") is to take the constant in Coulomb's law to be $1/4\pi$, essentially because $-1/4\pi|\mathbf{x}|$ is the fundamental solution of the Laplacian ($\nabla^2[-1/4\pi|\mathbf{x}|]=\delta$), and this is the procedure we shall follow. Its practical advantage is that it makes all the 4π's disappear from Maxwell's equations. Either way, the dimension $[q]$ of charge is related to mechanical dimensions by $[q^2l^{-2}]=[\text{force}]=[mlt^{-2}]$, or $[q]=[m^{1/2}l^{3/2}t^{-1}]$.

The idea of setting proportionality constants equal to one can be carried further. In relativistic physics it is natural to relate the units of length and time so that the speed of light c is equal to one; doing this makes $[l]$ equivalent to $[t]$. Moreover, in quantum mechanics it is natural to choose units so that Planck's constant $\hbar$ is equal to one. Since $\hbar$ has dimensions $[ml^2t^{-1}]$, in conjunction with the condition $c=1$ this makes $[m]$ equivalent to $[l^{-1}]$ or $[t^{-1}]$. We have $c=299792458$ m/s (an *exact* equality according to the current official definition of the meter) and $\hbar=1.054589\times10^{-34}$ kg $\cdot$ m^2/s, so the equivalences of the basic MKS units are as follows:

$$1\text{ second}\cong 299792458\text{ meter}\cong 8.522668\times10^{50}\text{ kilogram}^{-1}.$$

These large numbers make the MKS system awkward for the world of particle physics. Seconds and centimeters (accompanied by large negative powers of 10) are still generally used for times and lengths, but the other commonly used unit is the *electronvolt* (eV), which is the amount of energy gained by an electron when passing through an electrostatic potential of one volt, or its larger relatives the *mega-electronvolt* (MeV) or *giga-electronvolt* (GeV):

$$1\text{ eV}=1.602176\times10^{-19}\text{ kg}\cdot\text{m}^2/\text{s}^2,\qquad 1\text{ MeV}=10^6\text{ eV},\qquad 1\text{ GeV}=10^9\text{ eV}.$$

The eV is a unit of energy, but setting $c=1$ makes Einstein's $E=mc^2$ into an *equality* of mass and energy, and in the subatomic world this is not just a formal equivalence but an everyday fact of life. The eV, or more commonly the MeV or GeV, is therefore used as a unit of *mass*:

$$1\text{ MeV}=1.782661\times10^{-27}\text{ gram}.$$

The equivalence of units of mass, length, and time can be restated as follows:

$$(1.6)\quad 1 \text{ MeV} \cong (6.581966 \times 10^{-22} \text{ second})^{-1} \cong (1.973224 \times 10^{-11} \text{ centimeter})^{-1}.$$

Again, in the subatomic world these are more than formal equivalences. The precise constants are usually not important, but the orders of magnitude express relations among the characteristic energies, distances, and times of elementary particle interactions.

When we first discuss relativity in Chapter 2 and quantum theory in Chapter 3, we will include the factors of c and $\hbar$ explicitly for the sake of clarity, but starting with Chapter 4 we will always use "natural" units in which c and $\hbar$ are both equal to 1. We can still quote lengths in centimeters, times in seconds, and masses in MeV, but having chosen one of these units, we must relate the other two to it by (1.6).

Once one has made $[l] = [t] = [m^{-1}]$, the dimensions $[m^{1/2}l^{3/2}t^{-1}]$ of electric charge cancel out, so that charge is *dimensionless*, and the unit of charge in the Heaviside-Lorentz system is simply the number 1. The basic physical constant is then the fundamental unit of charge, the charge of a proton or the absolute value of the charge of an electron, which we denote here by e. The related quantity $e^2/4\pi$ (or $e^2/4\pi\hbar c$ if one does not use "natural" units) is the *fine structure constant* α, which is nearly equal to 1/137 and is often quoted that way:

$$\alpha = \frac{e^2}{4\pi} = 0.0072973525 = \frac{1}{137.035},$$

which gives

$$e = 0.30282212.$$

(In some parts of this book, the letter e may denote the charge of whatever particle is in question at the time; this is always an integer multiple [$\frac{1}{3}$-integer in the case of quarks] of the e here.)

One can carry the reduction of different dimensions one final step by including gravity. Having set $\hbar$ and c equal to one, one can set the coefficient G in Newton's law of gravity $F = Gm_1m_2/r^2$ equal to one too. This yields an *absolute* scale of length, time, and mass, called the *Planck scale.* Since $G = 6.674 \times 10^{-11}\ \text{m}^3/\text{kg}\cdot\text{s}^2$, we have

$$\begin{aligned} \text{Planck length} &= 1.616 \times 10^{-33} \text{ centimeter}, \\ \text{Planck time} &= 5.390 \times 10^{-44} \text{ second}, \\ \text{Planck mass} &= 2.176 \times 10^{-5} \text{ gram} = 1.221 \times 10^{19} \text{ GeV}. \end{aligned}$$

The Planck length and time are ridiculously small, and the Planck mass ridiculously large, on the scale of ordinary particle physics. There is much speculation about what particle physics on the Planck scale might look like, none of which will be discussed in this book.

Here is a quick review of the terminology concerning subatomic particles. To begin with, there are two basic dichotomies: all particles are either *Bosons*, with integer spin, or *Fermions*, with half-integer spin; and all particles are either *hadrons*, which participate in the strong interaction, or non-hadrons. The fundamental Fermions are *quarks*, which are hadrons, and *leptons*, which are not. Quarks combine in triplets to make *baryons*, of which the most familiar are the proton and neutron, and in quark-antiquark pairs to make *mesons*, of which the most important are the pions; baryons are Fermions, whereas mesons are Bosons. The leptons comprise electrons, muons,[2] tauons, and their associated neutrinos. The fundamental Bosons are the mediating particles of the various interactions: photons (electromagnetism), $W^{\pm}$ and Z particles (weak interaction), and gluons (strong interaction). There is also presumably a graviton for gravity, as well as the Higgs Boson demanded by currently accepted "standard model," but neither of them has been observed as of this writing. For each of the particles mentioned above there is also an antiparticle; in some cases (photons, gravitons, Higgs particles, Z particles, and some gluons and mesons) the particle and antiparticle coincide. Griffiths [**58**] is a good reference for the physics of elementary particles.

The rest masses of the particles that we will encounter most frequently in this book are as follows:

$$m_\gamma = 0, \quad m_e = 0.511 \text{ MeV}, \quad m_p = 938.280 \text{ MeV}, \quad m_n = 939.573 \text{ MeV},$$
$$m_\mu = 105.659 \text{ MeV}, \quad m_{\pi^\pm} = 139.569 \text{ MeV}, \quad m_{\pi^0} = 134.964 \text{ MeV},$$

where the subscripts denote photons (γ), electrons, protons, neutrons, muons, charged pions, and neutral pions. The mass of a proton in grams is 1.672635×10^{-24}; the reciprocal of this number, approximately 6×10^{23}, is essentially Avogadro's number, which to this one-significant-figure accuracy can be considered as the number of nucleons (protons or neutrons) in one gram of matter.[3]

A few characteristic lengths: The diameter of a proton is about 10^{-13} cm. Diameters of atoms are in the range 1–5×10^{-8} cm. (Heavy atoms about the same size as light ones, because although the former have more electrons, the increased charge of the nucleus causes them to be packed in more tightly.) Electrons, as far as we know, are point particles, but the *Compton radius* or *classical radius* of an electron is the number r_0 such that the mass of the electron is equal to the electrostatic energy of a solid ball of radius r_0 with a uniform charge distribution of total charge e, which in natural units is $e^2/4\pi m_e$:

$$r_0 = \frac{e^2}{4\pi m_e} = 2.8 \times 10^{-13} \text{ cm.} \tag{1.7}$$

1.3. Mathematical prologue: some Lie groups and Lie algebras

In this section we review the basic facts and terminology concerning the Lie groups and Lie algebras that play a central role in quantum mechanics and relativity.

The Lorentz and orthogonal groups and their Lie algebras. The *Lorentz group* $O(1,3)$ is the group of all linear transformations of $\mathbb{R}^4$ (or 4×4 real matrices) that preserve the Lorentz inner product:[4]

$$A \in O(1,3) \iff (Ax)_\mu (Ay)^\mu = x_\mu y^\mu \text{ for all } x, y \iff A^\dagger g A = g.$$

We employ the notation discussed in §1.1; in particular, g is the matrix (1.1).

[2]Muons were originally called "μ-mesons." This is a misuse of the word "meson" according to modern usage.

[3]The actual mass of an atom is generally less than the sum of the masses of its nucleons, even after adding in the mass of its electrons. One has to subtract the binding energy of the nucleus.

[4]$O(1,3)$ is more commonly called $O(3,1)$, but I prefer to keep the arguments of $O(\cdot,\cdot)$ in the same order as the coordinates on $\mathbb{R}^4$.

The group $O(1,3)$ has four connected components, which are determined by the values of two homomorphisms from $O(1,3)$ onto the two-element group $\{\pm 1\}$:

$$A \mapsto \det A \quad \text{and} \quad A \mapsto \operatorname{sgn}(A^0_0) \text{ (the sign of the } (0,0) \text{ entry of } A).$$

The kernel of the first of these is the *special Lorentz group* $SO(1,3)$, and the kernel of the second one is the *orthochronous Lorentz group* $O^\uparrow(1,3)$, the subgroup of $O(1,3)$ that preserves the direction of time. The intersection

$$SO^\uparrow(1,3) = SO(1,3) \cap O^\uparrow(1,3),$$

sometimes called the *restricted Lorentz group* or *proper Lorentz group*, is the connected component of the identity in $O(1,3)$. The linear isometry group or *orthogonal group* $O(3)$ of $\mathbb{R}^3$ sits inside $O^\uparrow(1,3)$ as the subgroup that fixes the point $(1,0,0,0)$, and the *rotation group* $SO(3)$ is $O(3) \cap SO^\uparrow(1,3)$.

The Lie algebra $\mathfrak{so}(1,3)$ of the Lorentz group consists of the 4×4 real matrices X that satisfy

$$X^\dagger g + gX = 0.$$

A convenient basis for it may be described as follows. Let $e_{\mu\nu}$ be the matrix whose (μ,ν) entry is 1 and whose other entries are 0, and define

$$X_{jk} = e_{kj} - e_{jk}, \quad X_{k0} = -X_{0k} = e_{k0} + e_{0k} \qquad (j,k = 1,2,3).$$

Then $\{X_{\mu\nu} : \mu < \nu\}$ is a basis for $\mathfrak{so}(1,3)$, and of course one can replace any $X_{\mu\nu}$ by $X_{\nu\mu} = -X_{\mu\nu}$. The commutation relations are given by

$$(1.8) \qquad \begin{aligned} &[X_{\mu\nu}, X_{\rho\sigma}] = 0 \text{ if } \{\mu,\nu\} = \{\rho,\sigma\} \text{ or } \{\mu,\nu\} \cap \{\rho,\sigma\} = \varnothing, \\ &[X_{\mu\nu}, X_{\nu\rho}] = g_{\nu\nu} X_{\mu\rho}. \end{aligned}$$

The Lie algebra $\mathfrak{so}(3)$ of the spatial rotation group is the span of $\{X_{jk} : j,k > 0\}$.

In detail, the relation between these basis elements and the Lie group is as follows. If (j,k,l) is a cyclic permutation of $(1,2,3)$, $\exp(sX_{jk})$ is the rotation through angle s about the x^l-axis (counterclockwise, as viewed from the positive x^l-axis). On the other hand, $\exp(sX_{0k})$ is a so-called *boost* along the x^k-axis that changes from the initial reference frame "at rest" to one moving with velocity $\tanh s$ along the x^k-axis. Thus, for example,

$$\exp(sX_{23}) = \begin{pmatrix} I & 0 \\ 0 & R(s) \end{pmatrix}, \qquad R(s) = \begin{pmatrix} \cos s & -\sin s \\ \sin s & \cos s \end{pmatrix},$$

and

$$\exp(sX_{01}) = \begin{pmatrix} B(s) & 0 \\ 0 & I \end{pmatrix}, \qquad B(s) = \begin{pmatrix} \cosh s & \sinh s \\ \sinh s & \cosh s \end{pmatrix}.$$

Orbits and invariant measures. We need to consider the geometry of the action of $O(1,3)$ on $\mathbb{R}^4$. Actually, there are two natural actions of $O(1,3)$ on $\mathbb{R}^4$: the identity representation of $O(1,3)$ and its contragredient $A \mapsto A^{\dagger -1}$. Strictly speaking, the latter is the action of $O(1,3)$ on the dual space $(\mathbb{R}^4)'$; physically, it is the action on momentum space rather than position space. Since the map $A \mapsto A^{\dagger -1}$ is an automorphism of $O(1,3)$, however, the orbits in $\mathbb{R}^4$ under the two actions are identical, and we need not distinguish them. In what follows we shall think of $\mathbb{R}^4$ as momentum space, since this is the context in which the orbits usually appear naturally.

The orbits of the restricted Lorentz group and the orthochronous Lorentz group are identical. Parametrized by a nonnegative real number m that generally has a physical interpretation as a mass, they are as follows:

$$\begin{gathered} X_m^+ = \{p : p^2 = m^2,\ p_0 > 0\}, \qquad X_m^- = \{p : p^2 = m^2,\ p_0 < 0\}, \\ Y_m = \{p : p^2 = -m^2\} \quad (m > 0), \\ \{0\}. \end{gathered} \tag{1.9}$$

(Recall that $p^2 = p_\mu p^\mu$.) The orbits X_m^+ (and sometimes also X_m^-) are known as *mass shells*, and the orbits X_0^+ and X_0^- are called the *forward* and *backward light cones*. (Under the action of the full group $O(1,3)$, the two orbits $X_m^\pm$ coalesce into one.)

Observe that

$$p \in X_m^+ \iff p = (\omega_{\mathbf{p}}, \mathbf{p}), \text{ where } \omega_{\mathbf{p}} = \sqrt{m^2 + |\mathbf{p}|^2}. \tag{1.10}$$

The symbol $\omega_{\mathbf{p}}$ will carry this meaning throughout the book.

Each of the orbits (1.9) has an $O^\uparrow(1,3)$-invariant measure, which is known on abstract grounds to be unique up to scalar multiples. The invariant measure on the mass shell X_m^+ with $m > 0$ may be derived as follows. Let $V = \bigcup_{m>0} X_m^+ = \{p : p^2 > 0,\ p_0 > 0\}$ be the region inside the forward light cone. Then V is $O^\uparrow(1,3)$-invariant, and since $|\det T| = 1$ for $T \in O^\uparrow(1,3)$, the restriction of Lebesgue measure d^4p to V is an $O^\uparrow(1,3)$-invariant measure on V; hence so is $f(p^2)d^4p$ for any nonnegative continuous f with support in $(0,\infty)$. One obtains the invariant measure on X_m^+ by letting f turn into a delta-function with pole at m^2: $\delta(p^2 - m^2)d^4p$. The result often appears in precisely this way in the physics literature, but in order to avoid possible pitfalls in using delta-functions with nonlinear arguments it is best to take a little more care.

To wit, consider the map $\phi : (0,\infty) \times \mathbb{R}^3 \to V$ defined by

$$\phi(y, \mathbf{p}) = (\sqrt{y + |\mathbf{p}|^2}, \mathbf{p}),$$

so that $q^2 = y$ when $q = \phi(y, \mathbf{p})$. ϕ is a diffeomorphism, and its Jacobian is $1/2\sqrt{y + \mathbf{p}|^2}$, so for $f \in C_c(0,\infty)$ we have

$$f(p^2)d^4p = \frac{f(y)\,dy\,d^3\mathbf{p}}{2\sqrt{y + |\mathbf{p}|^2}}.$$

Now let f approach the delta-function with pole at $y = m^2$: if we write points in X_m^+ as $(\omega_{\mathbf{p}}, \mathbf{p})$ where $\omega_{\mathbf{p}} = \sqrt{m^2 + |\mathbf{p}|^2}$, we obtain the invariant measure

$$\frac{d^3\mathbf{p}}{2\sqrt{m^2 + |\mathbf{p}|^2}} = \frac{d^3\mathbf{p}}{2\omega_{\mathbf{p}}}. \tag{1.11}$$

A limiting argument then shows that $d^3\mathbf{p}/2|\mathbf{p}|$ is an invariant measure on X_0^+. The invariant measure on X_m^-, of course, is also given by (1.11). We leave the calculation of the invariant measure on Y_m, for which we shall have no use, to the reader.

$SL(2,\mathbb{C})$, $SU(2)$, and the Pauli matrices. The three *Pauli matrices* are

$$\sigma_1 = \begin{pmatrix} 0 & 1 \\ 1 & 0 \end{pmatrix}, \qquad \sigma_2 = \begin{pmatrix} 0 & -i \\ i & 0 \end{pmatrix}, \qquad \sigma_3 = \begin{pmatrix} 1 & 0 \\ 0 & -1 \end{pmatrix}. \tag{1.12}$$

We shall denote the triple $(\sigma_1, \sigma_2, \sigma_3)$ by $\boldsymbol{\sigma}$. For any cyclic permutation (j,k,l) of $(1,2,3)$, we have

$$\sigma_j\sigma_k = -\sigma_k\sigma_j = i\sigma_l, \quad \text{and hence} \quad [\sigma_j,\sigma_k] = 2i\sigma_l. \tag{1.13}$$

$\frac{1}{2i}\sigma_1$, $\frac{1}{2i}\sigma_2$, and $\frac{1}{2i}\sigma_3$ are a basis for the Lie algebra $\mathfrak{su}(2)$ of skew-Hermitian 2×2 matrices of trace zero, and together with $\frac{1}{2}\sigma_1$, $\frac{1}{2}\sigma_2$, and $\frac{1}{2}\sigma_3$, they are a basis (over $\mathbb{R}$) for the Lie algebra $\mathfrak{sl}(2,\mathbb{C})$ of all 2×2 complex matrices of trace zero. In view of (1.8) and (1.13), the linear map $\kappa' : \mathfrak{sl}(2,\mathbb{C}) \to \mathfrak{so}(1,3)$ defined on this basis by

$$\kappa'(\tfrac{1}{2i}\sigma_j) = X_{kl} \text{ for } (j,k,l) \text{ a cyclic permutation of } (1,2,3), \qquad \kappa'(\tfrac{1}{2}\sigma_k) = X_{0k} \tag{1.14}$$

is an isomorphism of Lie algebras, and its restriction to $\mathfrak{su}(2)$ is an isomorphism from $\mathfrak{su}(2)$ to $\mathfrak{so}(3)$.

The corresponding homomorphism on the group level may be described as follows. Let us add the "fourth Pauli matrix"

$$\sigma_0 = I = \begin{pmatrix} 1 & 0 \\ 0 & 1 \end{pmatrix}$$

to obtain a basis $\sigma_0, \ldots, \sigma_3$ for the space H of 2×2 Hermitian matrices. We then identify $\mathbb{R}^4$ with H by the correspondence

$$x \in \mathbb{R}^4 \longleftrightarrow M(x) = x^\mu \sigma_\mu = \begin{pmatrix} x^0 + x^3 & x^1 - ix^2 \\ x^1 + ix^2 & x^0 - x^3 \end{pmatrix}. \tag{1.15}$$

The crucial feature of this correspondence is that

$$\det M(x) = (x^0)^2 - (x^1)^2 - (x^2)^2 - (x^3)^2 = x^2. \tag{1.16}$$

Every $A \in SL(2,\mathbb{C})$ acts on H by the map $X \mapsto AXA^\dagger$, and we denote the corresponding action on $\mathbb{R}^4$ by $\kappa(A)$:

$$M[\kappa(A)x] = AM(x)A^\dagger. \tag{1.17}$$

Since $\det A = 1$, we have $\det AM(x)A^\dagger = \det M(x)$, and by (1.16) this means that κ maps $SL(2,\mathbb{C})$ into $O(1,3)$. It is easily verified that the differential of κ at the identity, which takes $S \in \mathfrak{sl}(2,\mathbb{C})$ to the map $M(x) \mapsto SM(x)+M(x)S^\dagger$, is precisely the map κ' defined in (1.14). Thus κ is a local isomorphism, and since $SL(2,\mathbb{C})$ is connected, its image is the connected component of the identity in $O(1,3)$, namely, $SO^\uparrow(1,3)$. Finally, it is easy to verify that the kernel of κ is $\pm I$. In short, we have proved:

The map κ is a double covering of $SO^\uparrow(1,3)$ by $SL(2,\mathbb{C})$.

As we observed earlier, $SO(3)$ can be identified with the subgroup of $SO^\uparrow(1,3)$ that fixes the point $(1,0,0,0)$. The inverse image of $SO(3)$ in $SL(2,\mathbb{C})$ is therefore the set of all $A \in SL(2,\mathbb{C})$ that fix the point $I = M(1,0,0,0)$, i.e., that satisfy $AA^\dagger = I$. This is the group $SU(2)$ of 2×2 *unitary* matrices. It is easily verified that $SU(2)$ is the set of matrices of the form $\left(\begin{smallmatrix} a & -\bar{b} \\ b & \bar{a} \end{smallmatrix}\right)$ where $|a|^2 + |b|^2 = 1$, so that $SU(2)$ is homeomorphic to the unit sphere in $\mathbb{C}^2$ and in particular is simply connected. Hence:

$SU(2)$ is the universal cover of $SO(3)$, and the covering map is the restriction of κ to $SU(2)$.

There is another way to look at this. The map $\mathbf{x} \mapsto iM(0,\mathbf{x}) = i\mathbf{x}\cdot\boldsymbol{\sigma}$ is an isomorphism from $\mathbb{R}^3$ to the space of 2×2 skew-Hermitian matrices of trace zero, which is the Lie algebra $\mathfrak{su}(2)$. Since $A^\dagger = A^{-1}$ for $A \in SU(2)$, the action

$A \mapsto \kappa(A)|_{\{0\}\times\mathbb{R}^3}$ of $SU(2)$ on $\mathbb{R}^3$ is essentially the adjoint action of $SU(2)$ on its Lie algebra.

The map κ has one more important property that is not quite obvious: it respects adjoints, i.e.,

$$\kappa(A^\dagger) = \kappa(A)^\dagger.$$

This is most easily seen on the Lie algebra level: κ' takes the Hermitian matrices $\frac{1}{2}\sigma_j$ to the symmetric matrices X_{j0} and the skew-Hermitian matrices $\frac{1}{2i}\sigma_j$ to the skew-symmetric matrices X_{kl}. Thus $\kappa'(X^\dagger) = \kappa'(X)^\dagger$, so since $\kappa(\exp X) = \exp \kappa'(X)$ and exp preserves adjoints, the same is true of κ.

One can form a group that doubly covers the whole Lorentz group $O(1,3)$: it is a semidirect product of $SL(2,\mathbb{C})$ with the group $\mathbb{Z}_2 \times \mathbb{Z}_2$. We leave the details to the reader.

The Poincaré group. The *Poincaré group* or *inhomogeneous Lorentz group* $\mathcal{P}$ is the group of transformations of $\mathbb{R}^4$ generated by $O(1,3)$ and the group of translations (isomorphic to $\mathbb{R}^4$ itself). That is, $\mathcal{P}$ is the semi-direct product of $\mathbb{R}^4$ and $O(1,3)$,

$$\mathcal{P} = \mathbb{R}^4 \ltimes O(1,3),$$

whose underlying set is $\mathbb{R}^4 \times O(1,3)$ with group law given by

$$(a,S)(b,T) = (a + Sb,\, ST), \qquad (a,S)^{-1} = (-S^{-1}a, S^{-1}).$$

The action of $(a,A) \in \mathcal{P}$ on $\mathbb{R}^4$ is

$$(a,S)x = Sx + a.$$

Like $O(1,3)$, $\mathcal{P}$ has four connected components, and the component of the identity is $\mathcal{P}_0 = \mathbb{R}^4 \ltimes SO^\uparrow(3,1)$. The covering map κ of $O(1,3)$ by $SL(2,\mathbb{C})$ induces a double covering $(a,A) \mapsto (a,\kappa(A))$ of $\mathcal{P}_0$ by the group $\mathbb{R}^4 \ltimes SL(2,\mathbb{C})$, whose group law is

$$(a,A)(b,B) = (a + \kappa(A)b,\, AB).$$

CHAPTER 2

Review of Pre-quantum Physics

> I was lucky enough to attend a few lectures of S. S. Chern just before he retired from Berkeley in which he said that the cotangent bundle (differential forms) is the feminine side of analysis on manifolds, and the tangent bundle (vector fields) is the masculine side. From this perspective, Hamiltonian mechanics is the feminine side of classical physics, [and] its masculine side is Lagrangian mechanics.
> —Richard Montgomery ([**83**], p. 352)

This chapter is devoted to a brief review of classical mechanics, special relativity, and electromagnetic theory. Among the many available texts on these subjects, the Feynman Lectures [**41**] are excellent for Newtonian mechanics, relativity, and electromagnetism, and Purcell [**90**] perhaps even better for the latter. The classic physics text on Hamiltonian and Lagrangian mechanics is Goldstein [**55**]; Abraham and Marsden [**1**] and Arnold [**4**] are good treatments of this material by and for mathematicians.

2.1. Mechanics according to Newton and Hamilton

We may as well begin at the beginning with Newton's law:

$$F = ma. \tag{2.1}$$

Familiar as this is, it needs some explanation. We consider a system of k particles with fixed masses $m_1, \dots, m_k$, located at positions $\mathbf{x}_1, \dots, \mathbf{x}_k \in \mathbb{R}^3$ at time $t \in \mathbb{R}$. The jth particle is acted upon by a force $\mathbf{F}_j$ that depends on the positions $\mathbf{x}_1, \dots, \mathbf{x}_k$ and the time t (and perhaps on other parameters such as the masses m_j). We then concatenate the positions and forces into $3k$-vectors,

$$F = (\mathbf{F}_1, \dots, \mathbf{F}_k) \in \mathbb{R}^{3k}, \qquad x = (\mathbf{x}_1, \dots, \mathbf{x}_k) \in \mathbb{R}^{3k}$$

and the masses into a $3k \times 3k$ matrix,

$$m = \begin{pmatrix} m_1 I_3 & 0 & \dots & 0 \\ 0 & m_2 I_3 & \dots & 0 \\ \vdots & \vdots & & \vdots \\ 0 & 0 & \dots & m_k I_3 \end{pmatrix} \qquad (I_3 = 3 \times 3 \text{ identity matrix}),$$

and set

$$v = \frac{dx}{dt}, \qquad a = \frac{dv}{dt}.$$

Then (2.1) is a second-order ordinary differential equation for x when the force F is known, and a solution is determined by initial values $x_0 = x(t_0)$ and $v_0 = v(t_0)$.

Mathematicians look at physical laws such as (2.1) with the expectation that they will contain some universal truth about the world. They do, but not in the absolute way that the formula $A = \pi r^2$ is a universal truth about circles; they are always burdened with a certain amount of fine print. The law (2.1) is valid in all places at all times, as far as we know, but it has to be modified in relativistic situations where the masses cannot be treated as constants, and it fades into the background in the study of the submicroscopic world where the whole concept of force loses much of its utility. Moreover, (2.1) is valid only in inertial coordinate systems, so it gives the right answers about what goes on in a laboratory on the surface of the earth only to the extent that the rotation of the earth and its motion about the sun can be neglected. It is important to keep in mind that all physical theories have their limitations; the quantum field theories toward which we are heading are no exception.

We shall consider only *autonomous* and *conservative* forces, that is, those F that depend only on the positions x (and not explicitly on t) and for which the line integral $\int_C F \cdot dx$ vanishes for every closed curve C in $\mathbb{R}^{3n}$. In this case, F is the gradient of a function on $\mathbb{R}^{3n}$ denoted by $-V$ and called the *potential energy.* We also have the *kinetic energy* T of the system,

$$T = \tfrac{1}{2}mv \cdot v = \sum \tfrac{1}{2}m_j|\mathbf{v}_j|^2,$$

and the *total energy,*

$$E = T + V.$$

It is to be noted that V, and hence E, is well defined only up to an additive constant. Total energy is conserved:

$$\frac{dE}{dt} = mv \cdot \frac{dv}{dt} + \nabla V \cdot \frac{dx}{dt} = (ma - F) \cdot v = 0.$$

The Hamiltonian reformulation of Newtonian mechanics allows more flexibility and reveals some important mathematical structures. For reasons that will become clearer below, it takes the *momentum*

$$p = mv$$

instead of the velocity v as a primary object. The total energy, considered as a function of position and momentum, is called the *Hamiltonian* and is denoted by H:

$$H(x,p) = \tfrac{1}{2}m^{-1}p \cdot p + V(x).$$

We can rewrite Newton's law $F = ma$ as a first-order system in the variables x and p:

$$\frac{dx}{dt} = v = m^{-1}p, \qquad \frac{dp}{dt} = ma = -\nabla V(x).$$

The key point is that the quantities on the right are derivatives of the Hamiltonian:

$$\frac{dx}{dt} = \nabla_p H, \qquad \frac{dp}{dt} = -\nabla_x H. \tag{2.2}$$

These are *Hamilton's equations,* which tell how position and momentum evolve with time. From them one easily obtains the evolution equations for any function of position and momentum, $f(x,p)$:

$$\frac{df}{dt} = \nabla_x f \cdot \frac{dx}{dt} + \nabla_p f \cdot \frac{dp}{dt} = \{f, H\}, \tag{2.3}$$

where the *Poisson bracket* $\{f, g\}$ of any functions f and g is defined by

$$\{f, g\} = \nabla_x f \cdot \nabla_p g - \nabla_p f \cdot \nabla_x g.$$

Of particular importance are the Poisson brackets of the coordinate functions themselves:

(2.4) $$\{x_i, p_j\} = -\{p_j, x_i\} = \delta_{ij}, \qquad \{x_i, x_j\} = \{p_i, p_j\} = 0.$$

At this point a fundamental mathematical structure comes clearly into view (from the modern perspective): that of a symplectic manifold.

Let us pause for a brief review of this concept. A *symplectic manifold* is a C^∞ manifold M equipped with a differential 2-form Ω that is closed ($d\Omega = 0$) and pointwise nondegenerate, i.e., for each $a \in M$ and $v \in T_aM$, $\Omega_a(v, w) = 0$ for all $w \in T_aM$ only when $w = 0$. Nondegeneracy forces the dimension of M to be even, $\dim M = 2n$, and it is equivalent to the condition that the nth exterior power of Ω is everywhere nonvanishing.

The form Ω gives an identification of 1-forms with vector fields. Namely, if ϕ is a 1-form, the corresponding vector field X_ϕ is defined by

$$\Omega(X_\phi, Y) = \phi(Y)$$

for all vector fields Y. In particular, if f is a smooth function on M, X_{df} is a smooth vector field on M called the *Hamiltonian vector field* of f; by a small abuse of notation, we denote it by X_f rather than X_{df}. Thus, for any vector field Y,

$$\Omega(X_f, Y) = df(Y) = Yf.$$

The *Poisson bracket* of two smooth functions f and g is

$$\{f, g\} = \Omega(X_f, X_g) = X_g f = -X_f g.$$

The Poisson bracket makes $C^\infty(M)$ into a Lie algebra; the Jacobi identity is a consequence of the fact that $d\Omega = 0$ and the formula for the action of the exterior derivative of a form on vector fields. Moreover, the correspondence $f \mapsto X_f$ is a Lie algebra homomorphism:

$$\begin{aligned} X_{\{f,g\}}h &= -\{h, \{f, g\}\} = \{g, \{h, f\}\} + \{f, \{g, h\}\} \\ &= -X_g X_f h + X_f X_g h = [X_f, X_g]h. \end{aligned}$$

A celebrated theorem of Darboux states that for any point a in a symplectic manifold M there is a system of local coordinates $x_1, \ldots, x_n, p_1 \ldots, p_n$ on a neighborhood of a such that $\Omega = \sum dx_j \wedge dp_j$. A coordinate system with this property is called *canonical*, and the coordinates x_j and p_j are said to be *canonically conjugate.* In canonical coordinates, Hamiltonian vector fields and Poisson brackets are given by

$$X_f = \sum \left(\frac{\partial f}{\partial p_j}\frac{\partial}{\partial x_j} - \frac{\partial f}{\partial x_j}\frac{\partial}{\partial p_j} \right), \qquad \{f, g\} = \sum \left(\frac{\partial f}{\partial x_j}\frac{\partial g}{\partial p_j} - \frac{\partial f}{\partial p_j}\frac{\partial g}{\partial x_j} \right).$$

A diffeomorphism $\Phi : M \to M$ that preserves the symplectic structure is called a *canonical transformation* (by physicists) or a *symplectomorphism* (by geometers). Suppose $\{\Phi_t : t \in \mathbb{R}\}$ is a one-parameter group of canonical transformations; then its infinitesimal generator, the vector field X defined by $Xf = (d/dt)f \circ \Phi_t|_{t=0}$, satisfies $L_X\Omega = 0$, where L denotes the Lie derivative. Conversely, if X is a vector field such that $L_X\Omega = 0$, the flow it generates consists of (local) canonical transformations. But $L_X\Omega = i_X(d\Omega) + d(i_X\Omega) = d(i_X\Omega)$, where i_X denotes contraction with X, and $i_X\Omega$ is the 1-form associated to X by Ω, so X is an infinitesimal canonical transformation if and only if its associated 1-form is closed. But this means that $i_X\Omega = df$ for some function f (perhaps well-defined only on a covering space of M if M is not simply connected), or in other words, that $X = X_f$. Thus, with the understanding that functions and transformations may be only locally well-defined, *Hamiltonian vector fields are precisely the infinitesimal generators of canonical transformations.*

For Hamiltonian mechanics, the symplectic manifolds of primary importance are cotangent bundles. Let N be an n-dimensional manifold, T^*N its cotangent bundle, and $\pi : T^*N \to N$ the natural projection. There is a canonical 1-form ω on T^*N, defined by $\omega(v) = \phi(\pi_* v)$ for $\phi \in T^*N$ and $v \in T_\phi(T^*N)$, and $\Omega = -d\omega$ is a symplectic form on T^*N. (The minus sign is inserted to make this consistent with the preceding discussion and some standard conventions in the literature.) Indeed, any system $\{x_1, \ldots, x_n\}$ of local coordinates on an open set $U \subset N$ induces a frame $\{dx_1, \ldots, dx_n\}$ on T^*U and hence a system of linear coordinates $\{p_1, \ldots, p_n\}$ on each fiber of T^*U (that is, if ϕ is a 1-form on U, $\phi = \sum p_j(\phi) dx_j$), and $\{x_1, \ldots, x_n, p_1, \ldots, p_n\}$ is a system of local coordinates on T^*U. In these coordinates the canonical 1-form is given by $\omega = \sum p_j\, dx_j$, so $\Omega = \sum dx_j \wedge dp_j$. Thus Ω is indeed a symplectic form, and the coordinates x_j, p_j are canonical.

We return to physics. According to Newton's law, once the forces are given, the motion of the system is completely determined by (i) the position x and (ii) the velocity v or the momentum p at an initial time t_0. We therefore take this data as a complete description of the *state* of the system. To build a general mathematical framework for dealing with these matters, we start with a *configuration space* N, which is taken to be a manifold and is supposed to be a description of the possible "positions" of the system. There is quite a lot of flexibility here. For example, if the system consists of k particles moving in $\mathbb{R}^3$ as discussed previously, N will be $\mathbb{R}^{3k}$, or perhaps $\{(\mathbf{x}_1, \ldots, \mathbf{x}_n) \in \mathbb{R}^{3k} : \mathbf{x}_i \neq \mathbf{x}_j \text{ for } i \neq j\}$. If the motion is subject to some constraints, N could be a submanifold of $\mathbb{R}^{3k}$ instead. For an asymmetric rigid body, however, the appropriate configuration space is $\mathbb{R}^3 \times SO(3)$: three linear coordinates to give the location of the center of mass (or some other convenient point in the body), and three angular coordinates to give the body's orientation in space.

Velocities are taken to be tangent vectors to the configuration space, so the position-velocity state space is TN. On the other hand, the appearance of Poisson brackets in the Hamiltonian formalism leads us to take the position-momentum state space to be the symplectic manifold T^*N. That is, momenta should be considered as cotangent vectors, and in the relation $p = mv$ the mass matrix m should be interpreted as a Riemannian metric on N that mediates between vectors and covectors. Thus in the setting of $N = \mathbb{R}^{3k}$ considered above, the inner product on tangent vectors is given by $\langle v, w\rangle_m = mv \cdot w$; the corresponding inner product on cotangent vectors is given by $\langle p, q\rangle^m = m^{-1}p \cdot q$, and the cotangent vector corresponding to the tangent vector v is $p = mv$. T^*N is traditionally called the *phase space* of the system. (The origin of the name lies in statistical mechanics.)

As a function on T^*N, the Hamiltonian is $H(\phi) = \frac{1}{2}\langle \phi, \phi\rangle^m + V(\pi(\phi))$, where $\langle \cdot, \cdot\rangle^m$ is the Riemannian inner product just described and $\pi : T^*N \to N$ is the natural projection. Hamilton's equations (2.2) say that the Hamiltonian vector field X_H is the infinitesimal generator of the one-parameter group Φ_t of time translations of the system. (If (x, p) is the state at time t_0, $\Phi_t(x, p)$ is the state at time $t + t_0$. Actually Φ_t may be only a flow rather than a full one-parameter group, as particles might collide or escape to infinity in a finite time.) Consequently, the time translations are canonical transformations: the symplectic structure is an invariant of the dynamics.

In this setting it is easy to derive the connection between symmetries and conserved quantities generally known as *Noether's theorem.* To wit, suppose $\{\Psi_s : s \in \mathbb{R}\}$ is a one-parameter group of canonical transformations of phase space, and suppose the Hamiltonian H is invariant under these transformations. The infinitesimal generator of Ψ_s is a Hamiltonian vector field X_f, and invariance means that $X_f H = 0$. But $X_f H = \{H, f\} = -\{f, H\} = -df/dt$, so f is invariant under time translations, i.e., f is a conserved quantity. For example, if $N = \mathbb{R}^{3k}$ and Ψ_s is spatial translation in the direction of a unit vector $\mathbf{u} \in \mathbb{R}^3$, i.e., $\Psi_s(x,p) = (\mathbf{x}_1 + s\mathbf{u}, \ldots, \mathbf{x}_n + s\mathbf{u}, p)$, then f is the $\mathbf{u}$-component of the total linear momentum, $f(x,p) = (\mathbf{p}_1 + \ldots + \mathbf{p}_n) \cdot \mathbf{u}$. If $N = \mathbb{R}^3$ and Ψ_s is rotation through the angle s about the $\mathbf{u}$ axis, then f is the $\mathbf{u}$-component of the angular momentum, $f(\mathbf{x}, \mathbf{p}) = (\mathbf{x} \times \mathbf{p}) \cdot \mathbf{u}$.

One of the advantages of the Hamiltonian formulation of mechanics is that it allows the possibility of simplifying the problem by performing arbitrary canonical transformations, including ones that mix up the position and momentum variables. As an illustration, we consider a simple but important problem whose quantum analogue will be of fundamental importance later on: the one-dimensional harmonic oscillator. This is a single particle of mass m, moving along the real line subject to a linear restoring force $F(x) = -\kappa x$ $(\kappa > 0)$. The associated potential is $V(x) = \frac{1}{2}\kappa x^2$. Of course, Newton's equation $mx'' = -\kappa x$ is easy to solve directly, but the Hamiltonian method allows us to transform this simple differential equation into a completely trivial one.

Let us first observe that canonical transformations on $\mathbb{R}^2$ are precisely those transformations that preserve orientation and area, as the symplectic form $dx \wedge dp$ is just the element of oriented area.

Let $\omega = \sqrt{\kappa/m}$. The Hamiltonian is

$$H = \frac{1}{2m}p^2 + \frac{\kappa}{2}x^2 = \frac{\omega}{2}\left(\frac{1}{\sqrt{\kappa m}}p^2 + \sqrt{\kappa m}\, x^2\right).$$

As a first step, we make the canonical transformation $\widetilde{x} = (\kappa m)^{1/4}x$, $\widetilde{p} = p/(\kappa m)^{1/4}$, which makes $H = (\omega/2)(\widetilde{x}^2 + \widetilde{p}^2)$. Now, the polar-coordinate map $(\widetilde{x}, \widetilde{p}) \mapsto (r, \theta)$ is not area-preserving, but its close relative $(\widetilde{x}, \widetilde{p}) \mapsto (\frac{1}{2}r^2, \theta)$ is, since $r\,dr = d(\frac{1}{2}r^2)$. Thus, the map $(\widetilde{x}, \widetilde{p}) \mapsto (s, \theta)$ is canonical if we take $s = \frac{1}{2}(\widetilde{x}^2 + \widetilde{p}^2)$ and $\theta = \arctan(p/x)$, and in the new coordinates, H is simply ωs. Hamilton's equations therefore boil down to

$$\frac{ds}{dt} = \frac{\partial H}{\partial \theta} = 0, \qquad \frac{d\theta}{dt} = -\frac{\partial H}{\partial s} = -\omega.$$

Thus s is a constant — namely, $s = E/\omega$ where E is the total energy of the system — and $\theta = \theta_0 - \omega t$. Finally, $\widetilde{x} = \sqrt{2s}\cos\theta = \sqrt{2E/\omega}\cos(\omega t - \theta_0)$, so

$$x = \sqrt{\frac{2E}{\kappa}}\cos(\omega t - \theta_0).$$

The constant θ_0 can be chosen to make the initial values $x(t_0)$ and $p(t_0)$ whatever we wish, subject to the condition that $H(x_0, p_0) = E$.

Another canonical transformation that will be useful later is the one that simplifies the two-body problem. Suppose that two particles with masses m_1, m_2, positions $\mathbf{x}_1, \mathbf{x}_2$, and momenta $\mathbf{p}_1, \mathbf{p}_2$ are subject to a conservative force with potential V that depends only on their relative displacement $\mathbf{x}_1 - \mathbf{x}_2$, so the Hamiltonian is

$$H = \frac{|\mathbf{p}_1|^2}{2m_1} + \frac{|\mathbf{p}_2|^2}{2m_2} + V(\mathbf{x}_1 - \mathbf{x}_2).$$

Let

$$M = m_1 + m_2, \qquad m = \frac{m_1 m_2}{m_1 + m_2}, \qquad \mathbf{X} = \frac{m_1\mathbf{x}_1 + m_2\mathbf{x}_2}{m_1 + m_2}, \qquad \mathbf{x} = \mathbf{x}_1 - \mathbf{x}_2,$$

$$\mathbf{P} = M\frac{d\mathbf{X}}{dt} = \mathbf{p}_1 + \mathbf{p}_2, \qquad \mathbf{p} = m\frac{d\mathbf{x}}{dt} = \frac{m_2\mathbf{p}_1 - m_1\mathbf{p}_2}{m_1 + m_2}.$$

The transformation $(\mathbf{x}_1, \mathbf{x}_2, \mathbf{p}_1, \mathbf{p}_2) \mapsto (\mathbf{X}, \mathbf{x}, \mathbf{P}, \mathbf{p})$ is canonical, and in the new coordinates the Hamiltonian becomes

$$H = \frac{|\mathbf{P}|^2}{2M} + \frac{|\mathbf{p}|^2}{2m} + V(\mathbf{x}),$$

in which the coordinates are decoupled. Hamilton's equations then give $d\mathbf{X}/dt = \mathbf{P}/M$, $d\mathbf{P}/dt = 0$, so that $\mathbf{X}$, the center of mass of the two-body system, moves with constant velocity; and $d\mathbf{x}/dt = \mathbf{p}/m$, $d\mathbf{p}/dt = -\nabla V(\mathbf{x})$, so that $m\mathbf{x}''(t) = -\nabla V(\mathbf{x})$. In short, after removing the uniform motion of the center of mass, the problem reduces to that of a single particle of mass m, the *reduced mass* of the system, moving in the potential $V(\mathbf{x})$.

2.2. Mechanics according to Lagrange

We now turn to the other major reworking of Newtonian mechanics, the Lagrangian formulation. On the simplest level of a system of particles moving in a potential V, whose Hamiltonian is $H(x,p) = \frac{1}{2}m^{-1}p \cdot p + V(x)$, the *Lagrangian* is the function of position and velocity given by

$$L(x,v) = mv \cdot v - H(x, mv) = \tfrac{1}{2}mv \cdot v - V(x). \tag{2.5}$$

Stated in a more intrinsic fashion, if we regard H as a function on the cotangent bundle T^*N, the Lagrangian is the function on the tangent bundle TN given by $L(\xi) = \langle \jmath_m\xi, \xi\rangle - H(\jmath_m\xi)$, where $\jmath_m : TN \to T^*N$ is the isomorphism provided by the inner product m, and $\langle\cdot,\cdot\rangle$ is the pairing of cotangent vectors with tangent vectors. Returning to (2.5), we have

$$\nabla_x L = -\nabla_x H = \nabla V, \qquad \nabla_v L = mv,$$

so Newton's law becomes *Lagrange's equation*

$$\frac{d}{dt}\nabla_v L = \nabla_x L.$$

The significance of this is that it is the Euler-Lagrange equation for a certain problem in the calculus of variations. Namely, given a path $t \mapsto x(t)$, $t_0 \le t \le t_1$, in the configuration space N, we consider the *action*

$$S = \int_{t_0}^{t_1} L(x(t), x'(t))\, dt,$$

and the problem is to *minimize the action over all paths $x(t)$ that begin and end at two given points $x_0 = x(t_0)$ and $x_1 = x(t_1)$.* Indeed, suppose we replace $x(t)$ by a slightly different path $x(t)+\delta x(t)$, where $\delta x(t_0)=\delta x(t_1)=0$. To first order in δx, the change in the Lagrangian is

$$\delta L = L(x+\delta x,\, x'+\delta x') - L(x,x') = \nabla_x L\cdot \delta x + \nabla_v L \cdot \delta x',$$

so the change in the action is

$$\delta S = \int_{t_0}^{t_1} \delta L\, dt = \int_{t_0}^{t_1} \left[\nabla_x L - \frac{d}{dt}\nabla_v L\right]\cdot \delta x\, dt,$$

where the endpoint terms in the integration by parts vanish since δx vanishes at the endpoints. At a minimum, δS must vanish, and δx is arbitrary, so Lagrange's equation must hold.

This result is commonly known as the *principle of least action*; it will come back to haunt us in Chapter 8. Of course, this name is somewhat inaccurate: a solution of Newton's equations is a path that is a critical point, not necessarily a minimizer, of the action.

The calculation that led us from Hamiltonian mechanics to Lagrangian mechanics is easily reversible. Indeed, suppose we start with the Lagrangian $L(x,v)=\frac{1}{2}mv\cdot v - V(x)$. We define the Hamiltonian H by

$$H(x,p) = m^{-1}p\cdot p - L(x, m^{-1}p) = \tfrac{1}{2}m^{-1}p\cdot p + V(x).$$

Then $\nabla_p H = m^{-1}p = v = dx/dt$, and $\nabla_v L = mv = p$, so by Lagrange's equation, $-\nabla_x H = \nabla_x L = (d/dt)(\nabla_v L) = dp/dt$, and we have Hamilton's equations.

The Lagrangian and Hamiltonian machinery can be used in many situations other than the simple mechanical systems we have mentioned. The most common paradigm is to formulate a physical problem as a variational problem for a Lagrangian L defined on the tangent bundle of some configuration space N. The general procedure for converting such a problem to Hamiltonian form is as follows.

Let $x_1,\dots,x_n$ be local coordinates on N, and let $v_1,\dots,v_n$ be the corresponding linear coordinates on the fibers of TN; thus L is a function of x and v. For the moment, we fix a point $a\in N$ and think of L as a function on T_aN. Its differential $dL=\sum(\partial L/\partial v_j)dv_j$ is a 1-form on T_aN. If we identify the vector space T_aN with its own tangent spaces, we can think of dL as a map that assigns to each $v\in T_aN$ a linear functional on T_aN, i.e., a map from T_aN to T_a^*N. Now letting a vary too, we obtain a map $W: TN\to T^*N$.

Let us write this out in terms of local coordinates. Given coordinates $x_1,\dots,x_n$ on $U\subset N$, we identify TU and T^*U both with $U\times\mathbb{R}^n$; then L is a function of x and the fiber coordinates v on TU, and $W(x,v)=(x,\nabla_v L)$. The component $p_j=\partial L/\partial v_j$ of $\nabla_v L$ is called the *conjugate momentum* to the position variable x_j with respect to L.

Now suppose that W is invertible. From the Lagrangian L on TN we construct the Hamiltonian H on T^*N by

$$H(\xi) = \langle \xi, W^{-1}\xi\rangle - L(W^{-1}\xi),$$

where the first term on the right is the pairing of a covector with a vector. A calculation similar to the one we performed above then shows that the Euler-Lagrange equation for L is equivalent to Hamilton's equations for H. Exactly the same recipe can be used to go from H back to L; it is called the *Legendre transformation.*

In the general situation, this procedure for going from H to L and back may be problematic because of the question of the invertibility of W. However, it clearly coincides with the transformation we considered earlier for the case where $L(x,v) = \frac{1}{2}mv \cdot v - V(x)$. More generally, suppose that for each $a \in N$ the restriction of L to T_aN is a quadratic function whose pure second-order part is positive definite. Then W is indeed invertible: it is an affine-linear isomorphism of T_aN and T_a^*N for each a. This is the most important case in practice, and the only one that will concern us in the sequel.

The Lagrangian formulation of Noether's theorem about symmetries and conserved quantities (which is the original version) is as follows: Suppose $\{\psi_s : s \in \mathbb{R}\}$ is a one-parameter group of diffeomorphisms of the configuration space N that preserve the Lagrangian:

$$L(\psi_s(x), (\psi_s)_*(v)) = L(x,v).$$

Then the quantity $I(x,v) = \nabla_v L(x,v) \cdot d\psi_s(x)/ds$ is a constant of the motion. (See Arnold [4] or Goldstein [**55**].)

Our real interest is not in mechanical systems with finitely many degrees of freedom but in fields, which can be regarded as continuum mechanical systems with infinitely many degrees of freedom. Let us illustrate the transition to this situation with a simple example.

Consider a long elastic rod that is subject to longitudinal vibrations (or, if you prefer, a long air column such as an organ pipe). As a simplified model, we chop the rod up into small bits of equal size, consider each bit as a point particle, and replace the elastic forces within the rod by ideal springs connecting the particles. More precisely, we take each particle to have mass Δm and the separation between adjacent particles when the system is at rest to be Δx. Let u_j be the displacement of the jth particle from its position at rest. Then Newton's law gives the following system of equations for the u_j's:

$$\Delta m \frac{d^2 u_j}{dt^2} = k(u_{j+1} - u_j) - k(u_j - u_{j-1}) = k(u_{j+1} - 2u_j + u_{j-1}), \tag{2.6}$$

where k is the common spring constant, and the Lagrangian is

$$L = T - V = \tfrac{1}{2}\sum \Delta m \left(\frac{du_j}{dt}\right)^2 - \tfrac{1}{2}\sum k(u_{j+1} - u_j)^2. \tag{2.7}$$

Let us divide (2.6) by the equilibrium separation Δx,

$$\frac{\Delta m}{\Delta x}\frac{d^2 u_j}{dt^2} = (k\Delta x)\frac{u_{j+1} - 2u_j + u_{j-1}}{\Delta x^2},$$

and rewrite (2.7) as

$$L = \tfrac{1}{2}\sum \left[\frac{\Delta m}{\Delta x}\left(\frac{du_j}{dt}\right)^2 - (k\Delta x)\left(\frac{u_{j+1} - u_j}{\Delta x}\right)^2\right]\Delta x.$$

Now pass to the continuum limit: $\Delta m/\Delta x$ becomes the mass density μ, $k\Delta x$ becomes the Young's modulus Y, the displacements u_j become a displacement function $u(x)$, and we obtain

$$\mu \frac{\partial^2 u}{\partial t^2} = Y \frac{\partial^2 u}{\partial x^2} \tag{2.8}$$

from (2.6) and

$$L(u, \partial_t u, \partial_x u) = \tfrac{1}{2} \int_a^b \left[\mu \left(\frac{\partial u}{\partial t} \right)^2 - Y \left(\frac{\partial u}{\partial x} \right)^2 \right] dx \tag{2.9}$$

from (2.7), where $[a, b]$ is the interval occupied by the rod. Equation (2.8) is the familiar *wave equation* which is indeed the standard model for elastic vibrations, and (2.9) expresses the Lagrangian as the integral of the *Lagrangian density*

$$\mathfrak{L}(u, \partial_t u, \partial_x u) = \tfrac{1}{2} \left[\mu \left(\frac{\partial u}{\partial t} \right)^2 - Y \left(\frac{\partial u}{\partial x} \right)^2 \right]. \tag{2.10}$$

The wave equation (2.8) can be derived from the Lagrangian (2.9) by the calculus of variations just as before. That is, one considers the action

$$S(u) = \int_{t_0}^{t_1} L\, dt = \int_{t_0}^{t_1} \int_a^b \mathfrak{L}(u, \partial_x u, \partial_t u)\, dx\, dt,$$

and requires $\delta S = S(u+\delta u) - S(u)$ to vanish to first order in δu for appropriate δu. To make this precise, one must consider what happens at the ends of the rod. One possibility is to consider an infinitely long rod, $[a, b] = \mathbb{R}$. The vibrations $u(x, t)$ are then assumed to have compact support in x or vanish rapidly as $x \to \pm\infty$, so that the Lagrangian makes sense, and one imposes the same requirement on δu, together with $\delta u = 0$ at $t = t_0, t_1$. Likewise, for a rod with fixed ends or an air column with closed ends, one requires u and δu to vanish at $x = a, b$.

A little more interesting is the case of a rod with free ends or an air column with open ends. Here there is no *a priori* restriction on u and δu at the endpoints; let us see what happens. With L given by (2.9), to first order in δu we have

$$\delta S = \int_{t_0}^{t_1} \int_a^b \left[\mu \frac{\partial u}{\partial t} \frac{\partial \delta u}{\partial t} - Y \frac{\partial u}{\partial x} \frac{\partial \delta u}{\partial x} \right] dx\, dt.$$

Integration by parts yields

$$\delta S = \int_{t_0}^{t_1} \int_a^b \left[-m \frac{\partial^2 u}{\partial t^2} + Y \frac{\partial^2 u}{\partial x^2} \right] \delta u\, dx\, dt - Y \int_{t_0}^{t_1} \left[\frac{\partial u}{\partial x} \delta u \right]_{x=a}^{b} dt.$$

(We always assume δu vanishes at $t = t_0, t_1$, so the terms with t-derivatives contribute no boundary terms.) Thus if we want $\delta S = 0$ for arbitrary δu, u must satisfy not only the wave equation but the boundary conditions $\partial u / \partial x = 0$ at $x = a, b$. These are in fact the physically correct boundary conditions for a rod with free ends or an air column with open ends, and they yield a well-posed boundary value problem for the wave equation.

Let us be clear about the subtle shift of notation that has taken place here. For a system with finitely many degrees of freedom, we had position variables x_j and velocity variables v_j. Here the index j has become the continuous variable x, x_j has become $u(x, t)$, and v_j has become $\partial_t u$. It is not worthwhile to try to maintain the geometric language of tangent and cotangent vectors at this point, but we can still speak of the *canonically conjugate momentum density* to the field u,

$$p = \frac{\partial \mathfrak{L}}{\partial(\partial_t u)} = \mu \frac{\partial u}{\partial t},$$

and the *Hamiltonian density*,

$$\mathcal{H} = p\partial_t u - \mathcal{L} = \tfrac{1}{2}\left[\left(\frac{\partial u}{\partial t}\right)^2 + Y\left(\frac{\partial u}{\partial x}\right)^2\right].$$

The *Hamiltonian* $H = \int \mathcal{H}\,dx$ still represents the total energy, and it is a conserved quantity.

Similar ideas pertain to the wave equation in higher dimensions. Beginning with the Lagrangian density

$$\mathcal{L} = \tfrac{1}{2}\left[c^{-2}(\partial_t u)^2 - |\nabla_x u|^2\right] \tag{2.11}$$

for a function $u(x,t)$ with $x \in \Omega$, an open subset of $\mathbb{R}^n$, and $t \in \mathbb{R}$, one forms the action integral

$$S = \int_{t_0}^{t_1}\int_\Omega \mathcal{L}\,dx\,dt.$$

Setting the first variation of S equal to zero yields the wave equation

$$c^{-2}\partial_t^2 u - \nabla^2 u = 0.$$

If Ω is a bounded domain and one imposes no restrictions on the perturbation δu on the boundary $\partial\Omega$, one obtains the Neumann boundary conditions $\partial u/\partial n = 0$ for free. Alternatively, one can take $\Omega = \mathbb{R}^n$, $t_0 = -\infty$, and $t_1 = +\infty$, with some implicit or explicit assumptions on the vanishing of the fields at infinity, and obtain the wave equation on $\mathbb{R}^n$.

This last point leads to a glimpse of the importance of Lagrangians for quantum field theory. As we shall see, the transition from classical mechanics to quantum mechanics proceeds most easily by adopting the Hamiltonian point of view. However, since this approach singles out the time axis for special attention, it lacks relativistic invariance. A Lagrangian density, on the other hand, or the action which is its integral over space-time, employs the space and time variables on an equal footing, and it may have relativistic invariance built in — as (2.11) does, for example. It therefore serves as a useful starting point for building relativistic field theories.

When we return to these ideas in later chapters, we shall often be cavalier about dropping the word "density"; that is, we shall often refer to $\mathcal{L}$ as the Lagrangian and $\mathcal{H}$ as the Hamiltonian; the meaning will always be clear from the context.

2.3. Special relativity

It is taken as axiomatic that the fundamental laws of physics should be invariant under the Euclidean group of translations and rotations of space, as well as under translations of time. Newtonian physics and its immediate descendants are also invariant under the *Galilean transformations*

$$(t, \mathbf{x}) \mapsto (t,\, \mathbf{x} + t\mathbf{v}),$$

for any $\mathbf{v} \in \mathbb{R}^3$. The fundamental equations governing electromagnetism, however, are not, as we shall see in the next section. Rather, they are invariant under translations and under the Lorentz group $O(1,3)$ (see §1.3). For the time being we adopt notation that displays the speed of light c explicitly, so $O(1,3)$ is the group of linear transformations of $\mathbb{R}^4$ that preserve the Lorentz bilinear form

$$\Lambda((t,\mathbf{x}),\,(t',\mathbf{x}')) = c^2tt' - \mathbf{x}\cdot\mathbf{x}' = c^2tt' - xx' - yy' - zz'.$$

Einstein's great insight to realize that the laws of mechanics should be modified so as to be invariant under this group too.

A typical Lorentz transformation that mixes space and time variables is the boost along the x-axis with velocity v ($|v| < c$):

$$(t, x, y, z) \mapsto \left(\frac{t + (vx/c^2)}{\sqrt{1 - (v/c)^2}}, \frac{x + vt}{\sqrt{1 - (v/c)^2}}, y, z \right). \tag{2.12}$$

(This description of the boost is related to the one in §1.3 by the substitution $v/c = \tanh s$.) The factors of $\sqrt{1 - (v/c)^2}$ in (2.12) are what account for the "Fitzgerald contraction" and "time dilation" effects in relativistic motion. (The fact that the word "contraction" is used in one case and "dilation" is used in the other is purely a matter of psychology. The factor $\sqrt{1 - (v/c)^2}$ is in the denominator for both x and t.) The feature of these phenomena that is perplexing to the intuition is that the inverse transformation of (2.12) is of exactly the same form, with v replaced by $-v$, so the *same* factors of $\sqrt{1 - (v/c)^2}$ appear in the inverse, not their reciprocals. Nonetheless, everything works out consistently.

For example, suppose an observer A fires a bomb at an object O that is at rest with respect to A, and an observer B rides along with the bomb like Slim Pickens in *Dr. Strangelove*. A and B synchronize their clocks so that the bomb is launched at time 0. Suppose the bomb is set to go off 1 second after launch, the speed of the bomb is $0.6c$, and the distance from A to O is 2.25×10^8 meters (= 0.75 light-second). Result: The bomb goes off when it reaches O. From A's point of view, the bomb takes 1.25 seconds to reach O, and it takes that long to go off because the clock on board runs slow by a factor of $\sqrt{1 - (v/c)^2} = 0.8$. From B's point of view, the bomb takes 1 second to reach O, but the distance from A to O is only 1.8×10^8 meters because it is shortened by a factor of 0.8. (This is why muons created by cosmic rays in the upper atmosphere, around 10^{-4} light-second from earth, manage to reach the earth's surface by traveling at almost the speed of light, even though the half-life of a muon is only about 10^{-6} second. We think the muons' clocks run slow; they think the distance is short.) On the other hand, from B's point of view, A's clock runs slow, so (according to B) A's clock only reads 0.8 second when the bomb goes off. The point here is that the events of the explosion and the reading of 0.8 second on A's clock are simultaneous for B but not for A. This is possible because the two events happen at different places: the transformation (2.12) does not take equal times to equal times when $x \neq 0$. In constrast, A and B agree that B's clock reads 1 second when the bomb goes off; these two events happen at the same place, namely O, so their simultaneity is independent of the reference frame.

An elementary derivation of the relativistic formulas for momentum, energy, etc., can be found in Feynman [**41**]. Here we take a different approach, via Lagrangian mechanics, as a simple exercise in the sort of thought processes that go into finding the laws of quantum field theory.

We consider a free particle and wish to derive its laws of motion from an action functional, $S = \int_{t_0}^{t_1} L(\mathbf{x}, \mathbf{v})\, dt$, where $\mathbf{x} = \mathbf{x}(t)$ is a path with given endpoints $\mathbf{x}(t_0) = \mathbf{x}_0$ and $\mathbf{x}(t_1) = \mathbf{x}_1$, and $\mathbf{v}(t) = \mathbf{x}'(t)$. (We implicitly assume that $|\mathbf{v}(t)| < c$ and that $c^2(t_1 - t_0)^2 > |\mathbf{x}_1 - \mathbf{x}_0|^2$.) We wish the action to be Lorentz invariant, i.e., to be unchanged if $(\mathbf{x}_0, t_0)$ and $(\mathbf{x}_1, t_1)$ are subjected to the same Lorentz transformation, and we also wish it to yield Newtonian mechanics for a free particle in the limit of small velocities. The first requirement narrows down the possibilities enormously; the simplest one is to take the integrand to be a constant multiple of the Lorentz analogue of the differential of arc length,

$$a\,ds = a\sqrt{c^2dt^2 - dx^2 - dy^2 - dz^2}, \tag{2.13}$$

which in a given reference frame becomes

$$a\sqrt{1 - \frac{|\mathbf{v}|^2}{c^2}}\,c\,dt.$$

Thus we try the Lagrangian

$$L(\mathbf{x}, \mathbf{v}) = L(\mathbf{v}) = ac\sqrt{1 - \frac{|\mathbf{v}|^2}{c^2}}.$$

To figure out the right value of a, observe that when $|\mathbf{v}| \ll c$ we have

$$L(\mathbf{v}) \approx ac - \frac{a|\mathbf{v}|^2}{2c}.$$

Adding or subtracting a constant such as ac to the Lagrangian does not affect the equations of motion, and the second term $-a|\mathbf{v}|^2/2c$ becomes the classical (nonrelativistic) Lagrangian $m|\mathbf{v}|^2/2$ for a free particle of mass m if we take $a = -mc$. Thus we are led to the Lagrangian

$$L = -mc^2\sqrt{1 - \frac{|\mathbf{v}|^2}{c^2}}. \tag{2.14}$$

The corresponding conjugate momentum is

$$\mathbf{p} = \nabla_{\mathbf{v}} L = \frac{m\mathbf{v}}{\sqrt{1 - |\mathbf{v}^2|/c^2}}.$$

Again, this is approximately the classical momentum $m\mathbf{v}$ when $\mathbf{v}$ is small. In general, it is $M\mathbf{v}$ where

$$M = M(\mathbf{v}) = \frac{m}{\sqrt{1 - |\mathbf{v}|^2/c^2}}.$$

We are led to the conclusion that *the effective mass of a particle of (rest) mass m moving with velocity $\mathbf{v}$ is $M(\mathbf{v})$*. Thus mass is subject to the same dilation effect as time and distance.

The Euler-Lagrange equation for the Lagrangian (2.14) is

$$0 = \frac{d}{dt}\nabla_{\mathbf{v}} L - \nabla_{\mathbf{x}} L = \frac{d\mathbf{p}}{dt},$$

which gives motion in a straight line with constant speed as expected. (In general, if forces are present, Newton's law remains valid if "$m\mathbf{a}$" is reinterpreted as $d\mathbf{p}/dt$. But the question of a proper relativistic interpretation of forces is in general problematic.)

Next, the quantity

$$E = \mathbf{p} \cdot \mathbf{v} - L,$$

considered as a function of $\mathbf{v}$, is a constant of the motion that constitutes the total energy in nonrelativistic mechanics. In our situation,

$$E = \frac{m|\mathbf{v}|^2}{\sqrt{1 - |\mathbf{v}|^2/c^2}} - mc^2\sqrt{1 - |\mathbf{v}|^2/c^2} = \frac{mc^2}{\sqrt{1 - |\mathbf{v}|^2/c^2}} = Mc^2,$$

which is Einstein's formula for the energy of a free particle in relativistic mechanics. When $\mathbf{v}$ is small,

$$E \approx mc^2 + \tfrac{1}{2}m|\mathbf{v}|^2,$$

the sum of the "rest energy" mc^2 and the classical kinetic energy. We observe that

$$\frac{E^2}{c^2} - |\mathbf{p}|^2 = \frac{m^2c^2}{1 - |\mathbf{v}|^2/c^2} - \frac{m^2|\mathbf{v}|^2}{1 - |\mathbf{v}|^2/c^2} = m^2c^2, \tag{2.15}$$

so that

$$E = c\sqrt{|\mathbf{p}|^2 + m^2c^2}. \tag{2.16}$$

Expressed in this way as a function of $\mathbf{p}$, E is the *Hamiltonian* of the free particle. Again, for small velocities we have $|\mathbf{p}| \ll mc$ and hence

$$E \approx mc^2 + \frac{|\mathbf{p}|^2}{2m},$$

the rest energy plus the classical Hamiltonian.

The momentum $\mathbf{p}$ and the energy E are different in different frames of reference related by Lorentz transformations. However, it is not hard to check that $\mathbf{p}$ and E/c transform as the space and time components of a 4-vector, called the (relativistic) 4-momentum or energy-momentum vector. The relation (2.15) expresses the fact that the Lorentz inner product of this vector with itself is m^2c^2. It is important to note that in the tensor notation explained in §1.1, this vector is p^μ rather than p_μ; otherwise the sign of $\mathbf{p}$ comes out wrong.

2.4. Electromagnetism

Classically, the electric field $\mathbf{E}$ and the magnetic field $\mathbf{B}$ are vector-valued functions of position $\mathbf{x} \in \mathbb{R}^3$ and time t that may be operationally defined by the *Lorentz force law*: The force on a particle with charge q moving with velocity $\mathbf{v}$ is

$$\mathbf{F} = q\mathbf{E} + \frac{q}{c}\mathbf{v} \times \mathbf{B}, \tag{2.17}$$

where c is the speed of light, and $\mathbf{E}$ and $\mathbf{B}$ are evaluated at the location of the particle.[1]

The behavior of $\mathbf{E}$ and $\mathbf{B}$ is governed by *Maxwell's equations*. Using the Heaviside-Lorentz convention on the scale of electric charge (see §1.1), they are[2]

$$\operatorname{div} \mathbf{E} = \rho \tag{2.18}$$

$$\operatorname{div} \mathbf{B} = 0 \tag{2.19}$$

$$\operatorname{curl} \mathbf{E} + \frac{1}{c}\frac{\partial \mathbf{B}}{\partial t} = 0 \tag{2.20}$$

$$\operatorname{curl} \mathbf{B} - \frac{1}{c}\frac{\partial \mathbf{E}}{\partial t} = \frac{1}{c}\mathbf{j} \tag{2.21}$$

Here ρ is the charge density and $\mathbf{j}$ is the current density, both of which are functions of position and time. Everything here may be interpreted in the sense of

---

[1] In some texts the factor of c is omitted, which amounts to a redefinition of $\mathbf{B}$. The point of including it is to make $\mathbf{E}$ and $\mathbf{B}$ have the same dimensions, namely, force per unit charge.

[2] Some texts write Maxwell's equations in a way that involves two additional vector fields called $\mathbf{D}$ and $\mathbf{H}$. This is appropriate for the study of electromagnetism in bulk matter, but on the level of fundamental physics, $\mathbf{D} = \mathbf{E}$ and $\mathbf{H} = \mathbf{B}$.

distributions, so that one may allow point charges, charge distributions on curves or surfaces, etc.

The quantities ρ and $\mathbf{j}$ are not independent; indeed, by (2.18) and (2.21),

$$\frac{\partial \rho}{\partial t} = \operatorname{div} \frac{\partial \mathbf{E}}{\partial t} = c \operatorname{div}(\operatorname{curl} \mathbf{B}) - \operatorname{div} \mathbf{j} = -\operatorname{div} \mathbf{j},$$

since $\operatorname{div}\operatorname{curl} = 0$. This so-called *continuity equation*,

$$\frac{\partial \rho}{\partial t} + \operatorname{div} \mathbf{j} = 0, \tag{2.22}$$

is a strong form of the law of conservation of charge. More precisely, the total charge in a region $R \subset \mathbb{R}^3$ is the integral of ρ over R, and by the divergence theorem,

$$\frac{d}{dt} \iiint_R \rho \, dV = -\iiint \operatorname{div} \mathbf{j} \, dV = -\iint_{\partial R} \mathbf{j} \cdot \mathbf{n} \, dS,$$

where $\mathbf{n}$ is the unit outward normal to ∂R. Thus the total charge in R can change only to the extent that charge enters and leaves through ∂R.

On the one hand, electromagnetic fields are *produced by* charges and currents, in accordance with Maxwell's equations; on the other, they *act on* charged bodies in accordance with the Lorentz force law. If one combines these things with a specification of the relation between charge and mass distributions, one obtains from (2.17)–(2.21) a system of differential equations that, in principle, completely describes the evolution of the system. It is nonlinear because of the products in (2.17). However, in practice one often considers electromagnetic fields produced by a given system of charges and currents or the motion of some charged particles induced by a given electromagnetic field (produced by some laboratory apparatus, say) without worrying about feedback.

If ρ and $\mathbf{j}$ are taken as given (subject to (2.22)), Maxwell's equations alone provide a *linear* system of differential equations for the fields. The two equations (2.20) and (2.21) that involve time derivatives form a *symmetric hyperbolic system*,

$$\frac{\partial \mathbf{E}}{\partial t} = c \operatorname{curl} \mathbf{B} - \mathbf{j},$$
$$\frac{\partial \mathbf{B}}{\partial t} = -c \operatorname{curl} \mathbf{E}.$$

The mathematical theory of such systems is well understood. In particular, the Cauchy problem is well posed: there is a family of theorems that say that if $\mathbf{j}$ is in some nice space $\mathcal{X}$ of functions or distributions on $\mathbb{R}^4$ and the initial data $\mathbf{E}(t_0, \cdot)$ and $\mathbf{B}(t_0, \cdot)$ are in some related space $\mathcal{Y}$ of functions or distributions on $\mathbb{R}^3$, there is a unique solution $(\mathbf{E}, \mathbf{B})$ of this system in another related space $\mathcal{Z}$. Versions of this result can be found, for example, in Taylor [**115**], §6.5, and Treves [**120**], §15. As for the other two Maxwell equations, observe that

$$\frac{\partial}{\partial t} \operatorname{div} \mathbf{B} = \operatorname{div} \frac{\partial \mathbf{B}}{\partial t} = -c \operatorname{div}\operatorname{curl} \mathbf{E} = 0,$$
$$\frac{\partial}{\partial t} \operatorname{div} \mathbf{E} = \operatorname{div} \frac{\partial \mathbf{E}}{\partial t} = c \operatorname{div}\operatorname{curl} \mathbf{B} - \operatorname{div} \mathbf{j} = \frac{\partial \rho}{\partial t}.$$

Hence if the equations (2.18) and (2.19) are satisfied at the initial time t_0, they are satisfied at all other times too, so they are merely restrictions on the initial values of $\mathbf{E}$ and $\mathbf{B}$.

In what follows we assume that the quantities in Maxwell's equations are defined on all of space-time $\mathbb{R}^4$, although they may have singularities, in which case the equations are to be interpreted in the sense of distributions. (However, one should *not* think of the spatial component $\mathbb{R}^3$ as representing the entire universe. Rather, one should regard it as a model for some region of which the phenomena in which one is interested are largely concentrated in some compact subregion. Depending on the phenomena one is studying, the actual size of such a region could be anything from a fraction of a meter to many light-years. But one almost always has in mind that the fields and charge distributions vanish at infinity.)

By (2.19), $\mathbf{B}$ is the curl of a vector field $\mathbf{A}$ known as the *vector potential.* The equation (2.20) then can be rewritten as $\operatorname{curl}(\mathbf{E} + c^{-1}\partial_t\mathbf{A}) = 0$, so $\mathbf{E} + c^{-1}\partial_t\mathbf{A}$ is the gradient of a function $-\phi$, the *scalar potential.* Of course $\mathbf{A}$ and ϕ are far from unique: given any (smooth) function χ on $\mathbb{R}^4$, one can replace $\mathbf{A}$ by $\mathbf{A} - \nabla_{\mathbf{x}}\chi$ and ϕ by $\phi + c^{-1}\partial_t\chi$ without changing $\mathbf{E}$ and $\mathbf{B}$. Such adjustments to $\mathbf{A}$ and ϕ are called *gauge transformations* (for reasons that will be explained in §9.1), and the choice of a particular χ to make $\mathbf{A}$ and ϕ satisfy some desired condition is called a *choice of gauge.*

One of the commonly imposed gauge conditions is

$$\operatorname{div}\mathbf{A} + \frac{1}{c}\frac{\partial\phi}{\partial t} = 0, \tag{2.23}$$

which can be achieved by starting with any $\mathbf{A}$ and ϕ and replacing them with $\mathbf{A} - \nabla\chi$ and $\phi + c^{-1}\partial_t\chi$ where χ satisfies the inhomogeneous wave equation $\nabla^2\chi - c^{-2}\partial_t^2\chi = \operatorname{div}\mathbf{A} + c^{-1}\partial_t\phi$. Potentials $\mathbf{A}$ and ϕ that satisfy (2.23) are said to be in the *Lorentz gauge* or *Landau gauge.* In terms of such potentials, Maxwell's equations take a particularly appealing form. The equations (2.19) and (2.20) are already embodied in the definition of $\mathbf{A}$ and ϕ. For (2.18), we have

$$\rho = \operatorname{div}\mathbf{E} = -\nabla^2\phi - \frac{1}{c}\frac{\partial}{\partial t}\operatorname{div}\mathbf{A} = -\nabla^2\phi + \frac{1}{c^2}\frac{\partial^2\phi}{\partial t^2},$$

and for (2.21), since $\operatorname{curl}(\operatorname{curl}\mathbf{A}) = \nabla(\operatorname{div}\mathbf{A}) - \nabla^2\mathbf{A}$ we have

$$\begin{aligned}\frac{1}{c}\mathbf{j} &= \operatorname{curl}(\operatorname{curl}\mathbf{A}) - \frac{1}{c}\frac{\partial}{\partial t}\left(-\nabla\phi - \frac{1}{c}\frac{\partial\mathbf{A}}{\partial t}\right)\\ &= \frac{1}{c^2}\frac{\partial^2\mathbf{A}}{\partial t^2} - \nabla^2\mathbf{A}.\end{aligned}$$

In short, denoting the wave operator or d'Alembertian by $\square$,

$$\square = \frac{1}{c^2}\frac{\partial^2}{\partial t^2} - \nabla^2,$$

we see that Maxwell's equations boil down to

$$\square\phi = \rho, \qquad \square\mathbf{A} = \mathbf{j} \tag{2.24}$$

in the presence of the Lorentz gauge condition (2.23).

The situation here is best understood by thinking relativistically and putting space and time on an equal footing. Henceforth we adopt the space-time coordinates

$$x^0 = ct, \quad (x^1, x^2, x^3) = \mathbf{x}$$

to get rid of the factors of c, and we adopt the tensor notation explained in §1.1. We define the *electromagnetic 4-potential* by

(2.25)
$$A^\mu = (\phi, \mathbf{A}), \quad \text{or} \quad A_\mu = (\phi, -\mathbf{A}).$$

Gauge transformations now take the form

$$A^\mu \mapsto A^\mu + \partial^\mu \chi, \quad \text{or} \quad A_\mu \mapsto A_\mu + \partial_\mu \chi.$$

The electric and magnetic fields are combined into the *electromagnetic field tensor*

(2.26)
$$F_{\mu\nu} = \partial_\mu A_\nu - \partial_\nu A_\mu, \quad \text{or} \quad F^{\mu\nu} = \partial^\mu A^\nu - \partial^\nu A^\mu,$$

that is,

$$F_{\mu\nu} = \begin{pmatrix} 0 & E_x & E_y & E_z \\ -E_x & 0 & -B_z & B_y \\ -E_y & B_z & 0 & -B_x \\ -E_z & -B_y & B_x & 0 \end{pmatrix}, \qquad F^{\mu\nu} = \begin{pmatrix} 0 & -E_x & -E_y & -E_z \\ E_x & 0 & -B_z & B_y \\ E_y & B_z & 0 & -B_x \\ E_z & -B_y & B_x & 0 \end{pmatrix}.$$

The Maxwell equations (2.19) and (2.20) become

(2.27)
$$\partial_\kappa F_{\mu\nu} + \partial_\mu F_{\nu\kappa} + \partial_\nu F_{\kappa\mu} = 0,$$

which is an immediate consequence of the form (2.25) of $F_{\mu\nu}$, and the equations (2.18) and (2.21) become

(2.28)
$$\partial_\mu F^{\mu\nu} = j^\nu,$$

where the *4-current density* j^μ is defined by

$$j^\mu = (\rho, \mathbf{j}).$$

The Lorentz gauge condition (2.23) becomes

$$\partial_\mu A^\mu = 0,$$

and since

$$\partial_\mu F^{\mu\nu} = (\partial_\mu \partial^\mu) A^\nu - \partial^\nu (\partial_\mu A^\mu),$$

the Lorentz condition together with (2.28) yields the wave equations (2.24):

$$\partial^2 A^\mu = j^\mu.$$

Moreover, the continuity equation (2.22) becomes

(2.29)
$$\partial_\mu j^\mu = 0.$$

Fans of the language of differential geometry may prefer to see this restated in terms of differential forms. The potential and current density are expressed as 1-forms:

$$A = A_\mu \, dx^\mu = \phi \, dt - A_x \, dx - A_y \, dy - A_z \, dz, \quad j = j_\mu \, dx^\mu = \rho \, dt - j_x \, dx - j_y \, dy - j_z \, dz$$

(with t measured in units so that $c = 1$). The electromagnetic field tensor F is the negative of the exterior derivative of A:

$$F = -dA = (E_x \, dx + E_y \, dy + E_z \, dz) \wedge dt + B_x \, dy \wedge dz + B_y \, dz \wedge dx + B_z \, dx \wedge dy.$$

We have

$$dF = -ddA = 0,$$

which is equivalent to (2.27). The other Maxwell equation (2.28) and the continuity equation (2.29) become

$$*d{*}F = j, \qquad d{*}j = 0,$$

where $*$ is the Hodge star operator relative to the Lorentz form. (It is defined by

$$\alpha \wedge *\beta = \Lambda(\alpha, \beta) dt \wedge dx \wedge dy \wedge dz,$$

where $\Lambda(\alpha, \beta)$ denotes the bilinear functional on differential forms induced by the Lorentz form.)

In terms of the potential A, we have

$$-{*d*dA} = j.$$

In the presence of the Lorentz gauge condition

$$d{*A} = 0$$

this is equivalent to

$$-({*d*dA} + d{*d*})A = j.$$

But $-({*d*d} + d{*d*})$ is the "Lorentz Laplacian" — that is, the d'Alembertian ∂^2, acting componentwise; hence we recover (2.24) in the form

$$\partial^2 A = j \qquad (d{*A} = 0).$$

We now turn to the Lagrangian formulation of the laws of electromagnetism. We first derive the Lagrangian for a charged particle with rest mass m and charge[3] q, moving in a given electromagnetic field with potential $A^\mu = (\phi, \mathbf{A})$. Relativistic invariance severly restricts the possibilities for a Lorentz-invariant action functional, and the simplest one is

$$S = \int_{(t_0, \mathbf{x}_0)}^{(t_1, \mathbf{x}_1)} \left(-m\, ds - qA_\mu\, dx^\mu\right), \tag{2.30}$$

where ds is as in (2.13) and the integral is taken over a path in spacetime with the given endpoints. The first term gives the action for a free particle; the second one is meant to represent the interaction with the electromagnetic field; and as with the coefficient $-m$ for the free particle, the coefficient $-q$ for the interaction is dictated by the need to obtain the correct equations of motion. Notice that the gauge ambiguity in A_μ is irrelevant: replacing A_μ by $A_\mu + \partial_\mu \chi$ subtracts the exact differential $d\chi$ from the integrand and the constant $\chi(t_1, \mathbf{x}_1) - \chi(t_0, \mathbf{x}_0)$ from the action, which does not affect the dynamics.

Choosing a particular reference frame with coordinates $(t, \mathbf{x})$ (and units chosen so that $c = 1$) and rewriting the action as an integral in t with $\mathbf{v} = d\mathbf{x}/dt$ gives

$$S = \int_{t_0}^{t_1} \left(-m\sqrt{1 - |\mathbf{v}|^2} + q\mathbf{A} \cdot \mathbf{v} - q\phi\right) dt,$$

so the Lagrangian is

$$L = -m\sqrt{1 - |\mathbf{v}|^2} + q\mathbf{A} \cdot \mathbf{v} - q\phi.$$

At this point there are two different momenta to be considered. We shall refer to the ordinary mass-times-velocity momentum $m\mathbf{v}/\sqrt{1 - |\mathbf{v}|^2}$ as the *kinematic* or *mechanical* momentum and denote it by $\mathbf{p}_\kappa$. On the other hand, $\mathbf{p}$ will denote the *canonical* momentum associated to the Lagrangian L:

$$\mathbf{p} = \nabla_{\mathbf{v}} L = \frac{m\mathbf{v}}{\sqrt{1 - |\mathbf{v}|^2}} + q\mathbf{A} = \mathbf{p}_\kappa + q\mathbf{A}.$$

[3] It is an experimental fact that charge, unlike mass, does not depend on velocity.

(This is the momentum that needs to be considered when we pass to quantum mechanics.) The Euler-Lagrange equation is

$$\frac{d\mathbf{p}}{dt} = \nabla_{\mathbf{x}} L = q\nabla_{\mathbf{x}}(\mathbf{A}\cdot\mathbf{v}) - q\nabla\phi.$$

Here L is considered as a function of the *independent* variables $\mathbf{x}$ and $\mathbf{v}$, so $\mathbf{v}$ is treated as a constant in the expression $\nabla_{\mathbf{x}}(\mathbf{A}\cdot\mathbf{v})$; hence,

$$\nabla_{\mathbf{x}}(\mathbf{A}\cdot\mathbf{v}) = (\mathbf{v}\cdot\nabla_{\mathbf{x}})\mathbf{A} + \mathbf{v}\times\operatorname{curl}\mathbf{A}.$$

Moreover,

$$\frac{d\mathbf{A}}{dt} = \frac{\partial\mathbf{A}}{\partial t} + (\mathbf{v}\cdot\nabla_{\mathbf{x}})\mathbf{A}.$$

Hence the Euler-Lagrange equation is equivalent to

$$\begin{aligned}\frac{d\mathbf{p}_\kappa}{dt} = \frac{d}{dt}(\mathbf{p} - q\mathbf{A}) &= q\left(-\nabla\phi - \frac{\partial\mathbf{A}}{\partial t}\right) + q\mathbf{v}\times\operatorname{curl}\mathbf{A}\\ &= q\mathbf{E} + q\mathbf{v}\times\mathbf{B}.\end{aligned}$$

This is the Lorentz force law (2.17) (with $c = 1$), so our choice (2.30) for the action is indeed correct.

The corresponding Hamiltonian $H = \mathbf{p}\cdot\mathbf{v} - L$ is, in terms of the velocity $\mathbf{v}$,

$$H = \frac{m|\mathbf{v}|^2}{\sqrt{1-|\mathbf{v}|^2}}\cdot\mathbf{v} + m\sqrt{1-|\mathbf{v}|^2} + q\phi = \frac{m}{\sqrt{1-|\mathbf{v}|^2}} + q\phi,$$

the sum of the free energy and the Coulomb potential energy. By the same calculation as in the case of a free particle, we find that $(H - q\phi)^2 - |\mathbf{p}_\kappa|^2 = m^2$, so in terms of the canonical momentum $\mathbf{p}$,

$$H = \sqrt{m^2 + |\mathbf{p} - q\mathbf{A}|^2} + q\phi. \tag{2.31}$$

When $\mathbf{v}$ is small we have

$$L \approx -m + \tfrac{1}{2}m|\mathbf{v}|^2 + q\mathbf{A}\cdot\mathbf{v} - q\phi, \qquad H \approx m + \frac{1}{2m}|\mathbf{p} - q\mathbf{A}|^2 + q\phi,$$

which yield the Lagrangian and Hamiltonian for nonrelativistic motion of a particle in an electromagnetic field after subtraction of the rest mass m.

We observed earlier that the energy and momentum $(E, \mathbf{p})$ of a free particle make up a 4-vector p^μ. In the presence of an electromagnetic potential A^μ, so do the total energy given by (2.31) and the *canonical* momentum $\mathbf{p}$, and we still denote this 4-vector by p^μ. Equation (2.31) implies that

$$(E - q\phi)^2 = m^2 + |\mathbf{p} - q\mathbf{A}|^2,$$

which says that the Lorentz inner product of $p^\mu - qA^\mu$ with itself is m^2: $(p - qA)^2 = m^2$. Comparing this with the formula (2.16) for the free energy, we obtain a rule for incorporating the electromagnetic field that will be of basic importance in the quantum theory: *To obtain the energy-momentum vector for a particle with charge q in an electromagnetic field A^μ from that for a free particle, simply replace p^μ by $p^\mu - qA^\mu$.*

Next we consider the Lagrangian form of the field equations for electromagnetism. Here, since we are dealing with fields instead of particles, the action will be a 4-dimensional integral of a Lagrangian density that is a function of the fields and their derivatives. Moreover, the basic field is taken to be the *potential* A^μ rather than the electromagnetic field $F^{\mu\nu}$; the justification for this is that it is the only thing that works at the quantum level, in spite of the ambiguity in the definition of A^μ. Moreover, there is an implicit assumption that all fields and charges vanish at spatial infinity in such a way that these 4-dimensional integrals converge and all boundary terms in spatial integrations by parts vanish. In the preceding paragraphs we developed the Lagrangian for a particle moving in a given electromagnetic field; it was the sum of the Lagrangian for a free particle and a Lagrangian for the interaction. Similarly, we now develop the Lagrangian for a field in the presence of a given system of charges and currents; it will be the sum of a Lagrangian involving only the field and a Lagrangian for the interaction.

The form of the interaction term is suggested by the $-qA_\mu dx^\mu/dt = -q\phi + q\mathbf{A}\cdot\mathbf{v}$ that we found before. That was for a single charge q; if instead we consider a continuous distribution of charge with charge density ρ and current density $\mathbf{j}$, the obvious analogue is $-A_\mu j^\mu = -\rho\phi + \mathbf{A}\cdot\mathbf{j}$ where $j^\mu = (\rho, \mathbf{j})$ is the 4-current density. Thus, the interaction part of the action functional will be

$$S_{\text{int}} = -\int_{\mathbb{R}^4} A_\mu j^\mu \, d^4x. \tag{2.32}$$

As before, the ambiguity in A^μ is irrelevant: If we change A_μ by adding $\partial_\mu \chi$, where χ vanishes at infinity, the change in the action is

$$-\int (\partial_\mu \chi) j^\mu \, d^4x = \int \chi \partial_\mu j^\mu \, d^4x = 0$$

because of (2.29).

Now, what about the free-field Lagrangian? It should be Lorentz invariant; it should be unaffected by gauge transformations; it should be quadratic so that the field equations will turn out to be linear (Maxwell's equations). The only possibility is a constant multiple of $F_{\mu\nu}F^{\mu\nu}$ (recall that $F_{\mu\nu} = \partial_\mu A_\nu - \partial_\nu A_\mu$), and the correct constant turns out to be $-\frac{1}{4}$. Thus the full action functional is

$$S = S_{\text{field}} + S_{\text{int}} = \int_{\mathbb{R}^4} \left(-\tfrac{1}{4} F_{\mu\nu}F^{\mu\nu} - A_\mu j^\mu\right) d^4x. \tag{2.33}$$

(In the classical notation, $-\frac{1}{4}F_{\mu\nu}F^{\mu\nu} = \frac{1}{2}(|\mathbf{E}|^2 - |\mathbf{B}|^2)$.)

To confirm the validity of (2.33), let us derive the field equations. If we add a perturbation δA^μ that vanishes at (spatial and temporal) infinity to A^μ, the change in S to first order is

$$\begin{aligned}
\delta S &= \int \left(-\tfrac{1}{2}F^{\mu\nu}\delta F_{\mu\nu} - j^\nu \delta A_\nu\right) d^4x \\
&= \int \left(-\tfrac{1}{2}F^{\mu\nu}[\partial_\mu(\delta A_\nu) - \partial_\nu(\delta A_\mu)] - j^\nu \delta A_\nu\right) d^4x \\
&= \int \left(-F^{\mu\nu}\partial_\mu(\delta A_\nu) - j^\nu \delta A_\nu\right) d^4x \\
&= \int \left(\partial_\mu F^{\mu\nu} - j^\nu\right) \delta A_\nu \, d^4x.
\end{aligned}$$

In the first line used the fact that $F_{\mu\nu}\delta F^{\mu\nu} = F^{\mu\nu}\delta F_{\mu\nu}$, and in passing from the second to the third we used the fact that $-F^{\mu\nu}\partial_\nu \delta A_\mu = -F^{\nu\mu}\partial_\mu \delta A_\nu = F^{\mu\nu}\partial_\mu \delta A_\nu$ (by relabeling indices and using the skew symmetry of $F^{\mu\nu}$). Thus the condition $\delta S = 0$ yields the Maxwell equation (2.28). (Recall that the other equation (2.27) is a consequence of the formula $F_{\mu\nu} = \partial_\mu A_\nu - \partial_\nu A_\mu$.)

We have developed a Lagrangian formulation of the evolution equations for a charged particle moving in a given electromagnetic field and for an electromagnetic field in the presence of a given charge-current distribution. An attempt to combine these into a relativistically correct Lagrangian that describes a system of charges and fields interacting with each other is problematic, largely because of the question of finding a suitable form for the pure-matter term (the analogue of $-\int m\,ds$ for a single particle); see Goldstein [**55**]. When we come to quantum electrodynamics, however, this problem will evaporate. The charge-current distribution will be derived from an electron *field*, and the pure-matter term will be its free-field Lagrangian.

CHAPTER 3

Basic Quantum Mechanics

Raffiniert ist der Herrgott, aber boshaft ist er nicht. [God is subtle, but he is not malicious.]
—Albert Einstein

This chapter is a brief exposition of the fundamentals of quantum theory from a mathematical point of view. Our development of the underlying mathematical structure follows Mackey [**78**]; see also Jauch [**67**] and Varadarajan [**123**]. For quantum mechanics proper, there are many good texts available, so I will just mention three old classics with that I have found valuable: the introductory account in the Feynman Lectures [**41**] and the comprehensive treatises of Messiah [**82**] and Landau and Lifshitz [**76**]; the latter is particularly good from a mathematical point of view. Reed and Simon [**93**] is an excellent source for the relevant functional analysis; all the results about Hilbert spaces and their operators that are quoted in this chapter without references can be found there. We also need a few facts about Lie groups, Lie algebras, and their representations; we refer the reader to Hall [**61**] for more background.

3.1. The mathematical framework

The fundamental objects in a mathematical description of a physical system are *states* and *observables*. *Observables* are the physical quantities such as position, momentum, energy, etc., that one is interested in studying. In this discussion we shall consider only *real-valued* observables. This is not much of a restriction: for vector-valued observables one can consider their components with respect to a basis, and one can generally assign numerical labels to the values of other observables. In particular, one can think of true-false statements about the system as observables whose only values are 1 and 0. (Mackey [**78**] calls such observables "questions," but we prefer to think of them as assertions.) The *state* of a system is supposed to be a specification of the condition of the system from which one can read off all the available information about the observables of interest; the set of all possible states is called the *state space*.

In Hamiltonian mechanics, the state space is a symplectic manifold M, the observables are (Borel measurable) real-valued functions on M, and the true-false observables can be identified with the (Borel) subsets of M (the observable $f : M \to \{0, 1\}$ corresponds to $\{x \in M : f(x) = 1\}$). Thus the set of true-false statements on M is identified with the Borel σ-algebra on M, and the basic logical connectives "or," "and," and "not" correspond to the Boolean operations of union, intersection, and complement.

General observables can be analyzed in terms of true-false statements. A Borel function $f : M \to \mathbb{R}$ is completely specified by its level sets $f^{-1}(\{a\})$, $a \in \mathbb{R}$, or more generally by the sets $f^{-1}(E)$, E a Borel subset of $\mathbb{R}$. That is, an observable f is specified by the sets on which the statements "$f(x) = a$" or "$f(x) \in E$" are true.

The striking difference between classical and quantum mechanics is that *in a quantum system, the the true-false statements do not form a Boolean algebra.* Specifically, what fails is the distributive law:

$$(A \text{ or } B) \text{ and } C \quad \Longleftrightarrow \quad (A \text{ and } C) \text{ or } (B \text{ and } C).$$

The classic counterexample is the double-slit experiment in which a beam of electrons is directed at a barrier with two slits in it and the electrons that pass through the slits arrive at a screen where they are detected. Thus, a given electron passes through one of the two slits (A or B) and arrives at the screen (C). The resulting pattern of arrivals at the screen exhibits the sort of interference oscillations one would expect if the electrons were waves rather than particles. But if one puts detectors at the slits to see which slit an electron goes through (and thus changes the process to (A and C) or (B and C)), the interference pattern disappears. See Feynman [**41**], Chapter 3 of vol. 3, for a detailed discussion.

Thus one must find a new model for states and observables in quantum mechanics, and what turns out to work is the following. The *state space* of a quantum system is a projective Hilbert space $\mathbb{P}\mathcal{H}$, that is, the set of nonzero vectors in a Hilbert space $\mathcal{H}$ modulo the equivalence relation that $u \sim v$ if and only if v is a scalar multiple of u. In practice, one usually thinks of $\mathcal{H}$ itself as the state space, with the understanding that two equivalent vectors define the same state. It is almost always desirable to take the vectors representing states to be *unit* vectors; one speaks of *normalized* states.

At this point we refer the reader back to §1.1 for a discussion of the physicists' notation and terminology that we will be using from now on.

The true-false statements about the quantum system correspond not to subsets of $\mathcal{H}$ but to *closed linear subspaces* of $\mathcal{H}$. The logical "and" still corresponds to intersection, but now "or" and "not" correspond to *closed linear span* and *orthogonal complement*, respectively. Thus the lattice of subspaces takes the place of the Boolean algebra of subsets. However, to develop the theory further, it is more convenient to identify a closed linear subspace with the orthogonal projection onto that subspace. Thus, from now on, true-false statements will be identified with orthogonal projection operators.

If the system is in the state represented by the unit vector u, and a true-false statement is represented by the projection P, the statement is definitely true if u is in the range $\mathcal{R}(P)$ and definitely false if u is in the nullspace $\mathcal{N}(P)$. In general, we write $u = u_0 + u_1$ where $u_0 \in \mathcal{N}(P)$ and $u_1 \in \mathcal{R}(P)$. Then $\|u_0\|^2 + \|u_1\|^2 = 1$, and we interpret $\|u_0\|^2$ and $\|u_1\|^2$ as the *probabilities* that the statement is false and true, respectively. (In experimental terms, probability is interpreted in terms of statistical frequency: If the system is prepared N times in state u where N is large, the statement will be true about $N\|u_1\|^2$ times.) This probabilistic interpretation is an essential feature of quantum mechanics, not just the result of an incomplete description of the system.

The state u can be identified with the true-false statement "The system is in the state u." Taking u to be a unit vector, the corresponding projection is the orthogonal projection P onto the line spanned by u, $Pv = \langle u|v\rangle u$, and for any unit vector v, $|\langle u|v\rangle|^2 = |\langle u|P|v\rangle|^2$ is the *transition probability* that a system known to be in the state u will also be found to be in the state v (or vice versa: $|\langle u|v\rangle|^2$ is symmetric in u and v). The fact that a system that is definitely in one state may also be in a different state is another peculiarity of quantum systems.

By the way, physicists refer to the inner product $\langle u|v\rangle$ as a probability *amplitude*. This usage of the word "amplitude" is in conflict with the classical one according to which the amplitude of a sinusoidal wave is the coefficient a (always assumed positive) in the expression $a\sin(\omega t + c)$. This is just one of those semantic peculiarities that one has to learn to live with.

The usual device for combining simpler quantum systems into more complicated ones is the tensor product of Hilbert spaces, a concept which we now briefly review. Given two Hilbert spaces $\mathcal{H}_1$ and $\mathcal{H}_2$, one way to define their (Hilbert-space) tensor product $\mathcal{H}_1 \otimes \mathcal{H}_2$ is to begin with their algebraic tensor product (the set of finite linear combinations of products $u_1 \otimes u_2$ where $u_j \in \mathcal{H}_j$). There is a unique inner product on this space such that

$$\langle u_1 \otimes u_2 | v_1 \otimes v_2 \rangle = \langle u_1|v_1\rangle\langle u_2|v_2\rangle,$$

and $\mathcal{H}_1 \otimes \mathcal{H}_2$ is its completion with respect to the associated norm. Alternatively, one can define $\mathcal{H}_1 \otimes \mathcal{H}_2$ to be the space of all bounded conjugate-linear maps $A : \mathcal{H}_1 \to \mathcal{H}_2$ such that $\sum \|Ae_j\|^2 < \infty$ for some (and hence any) orthonormal basis $\{e_j\}$ for $\mathcal{H}_1$, equipped with the inner product $\langle A|B\rangle = \sum \langle Ae_j|Be_j\rangle$. (In this picture $u_1 \otimes u_2$ is the map $v \mapsto \langle v|u_1\rangle u_2$.) Either way, if $\{e_j\}$ and $\{f_k\}$ are orthonormal bases for $\mathcal{H}_1$ and $\mathcal{H}_2$, $\{e_j \otimes f_k\}$ is an orthonormal basis for $\mathcal{H}_1 \otimes \mathcal{H}_2$. Other variations on this theme are possible; see Reed and Simon [**93**], §II.4, or Folland [**44**], §7.3 for more details.

Here are two typical uses of tensor products in quantum mechanics. First, if $\mathcal{H}_1$ and $\mathcal{H}_2$ are the quantum state spaces for two particles p_1 and p_2, the state space for the two-particle system is $\mathcal{H}_1 \otimes \mathcal{H}_2$, with $u \otimes v$ representing the state in which p_1 is in state u and p_2 is in state v.[1] (In this situation it is often appropriate to take $\mathcal{H}_1 = \mathcal{H}_2 = L^2(\mathbb{R}^3)$, in which case $\mathcal{H}_1 \otimes \mathcal{H}_2$ can be identified with $L^2(\mathbb{R}^6)$ by taking $f_1 \otimes f_2$ to be the function $(x_1, x_2) \mapsto f_1(x_1)f_2(x_2)$.) Second, just as the motion of a classical rigid body may be analyzed on one level by considering it as a featureless "particle" and then in more detail by considering its rotations about its center of mass, a quantum particle may have one set of "states" (i.e., one Hilbert space $\mathcal{H}_1$) to describe its motion in space and another one $\mathcal{H}_2$ to describe its "internal degrees of freedom" such as spin; then the full state space for the particle will be $\mathcal{H}_1 \otimes \mathcal{H}_2$.

Now, how do we model observables? Given a real-valued physical quantity O that we want to be an observable of our system, we follow the hint indicated earlier and analyze it in terms of the true-false statements

$$S(E)\text{: The value of } O \text{ is in } E,$$

where E a Borel subset of $\mathbb{R}$. Let $P(E)$ be the projection corresponding to $S(E)$. Obviously $S(\varnothing)$ is always false and $S(\mathbb{R})$ is always true, so

$$P(\varnothing) = 0 \quad \text{and} \quad P(\mathbb{R}) = I. \tag{3.1}$$

If E_1 and E_2 are disjoint, $S(E_1)$ is definitely false whenever $S(E_2)$ is definitely true, and vice versa, so the ranges of $P(E_1)$ and $P(E_2)$ are mutually orthogonal;

---

[1] If p_1 and p_2 are identical — two electrons, for example — this picture needs to be modified; see §4.5.

equivalently,

$$P(E_1)P(E_2) = 0 \text{ whenever } E_1 \cap E_2 = \varnothing. \tag{3.2}$$

In this case, the closed linear span of the ranges is their orthogonal sum, and the projection onto the latter space is $P(E_1) + P(E_2)$. But that projection corresponds to the statement "O is in E_1 or E_2," which is the same as $S(E_1 \cup E_2)$ and hence corresponds to the projection $P(E_1 \cup E_2)$. It then follows by induction that $P(\bigcup_1^n E_j) = \sum_1^n P(E_j)$ whenever $E_1, \ldots, E_n$ are disjoint, and it is eminently reasonable to assume that the same relation holds for countable unions:

$$P\left(\bigcup_1^\infty E_j\right) = \sum_1^\infty P(E_j) \text{ whenever } E_i \cap E_j = \varnothing \text{ for } i \neq j. \tag{3.3}$$

(The convergence of the sum on the right is in the strong operator topology, i.e., the topology of pointwise convergence in norm.) A map $E \mapsto P(E)$ from the Borel sets in $\mathbb{R}$ to the orthogonal projections on $\mathcal{H}$ that satisfies (3.1)–(3.3) is called a *projection-valued measure* on $\mathbb{R}$.

It is worth observing that projection-valued measures behave in the natural way with respect to complements and intersections. Namely, since $P(\mathbb{R} \setminus E) + P(E) = P(\mathbb{R}) = I$, $P(\mathbb{R} \setminus E)$ is the projection whose range and nullspace are the nullspace and range of $P(E)$, respectively. Moreover, for any E_1 and E_2, the sets $E_1 \setminus E_2$, $E_1 \cap E_2$, and $E_2 \setminus E_1$ are disjoint, so by (3.2) and (3.3),

$$\begin{aligned} P(E_1)P(E_2) &= \big[P(E_1 \setminus E_2) + P(E_1 \cap E_2)\big]\big[P(E_1 \cap E_2) + P(E_2 \setminus E_1)\big] \\ &= P(E_1 \cap E_2)^2 = P(E_1 \cap E_2). \end{aligned}$$

If P is any projection-valued measure on $\mathbb{R}$ and u is a unit vector, the map $P_u : E \mapsto \langle u|P(E)|u\rangle$ is an ordinary probability measure on $\mathbb{R}$. It is interpreted as the probability distribution of the observable P in the state u. This is consistent with the probabilistic interpretation of single projections mentioned earlier. The states (if any) in which the observable P has a definite value a are those in the range of $P(\{a\})$.

Now comes the leap into analysis: By the spectral theorem, *projection-valued measures are in one-to-one correspondence with self-adjoint operators.* To wit, if A is a self-adjoint operator, the spectral functional calculus yields a self-adjoint operator $f(A)$ for any Borel function $f : \mathbb{R} \to \mathbb{R}$, and the operators $P(E) = \chi_E(A)$ ($\chi =$ characteristic function) form a projection-valued measure. On the other hand, if P is a projection-valued measure, there is a unique self-adjoint operator A, whose domain $\mathcal{D}(A)$ is the set of all $u \in \mathcal{H}$ with $\int_{\mathbb{R}} t^2\, dP_u(t) < \infty$, such that $\langle u|A|u\rangle = \int_{\mathbb{R}} t dP_u(t)$ for all $u \in \mathcal{D}(A)$. (The diagonal matrix elements $\langle u|A|u\rangle$ determine all other matrix elements $\langle u|A|v\rangle$, and hence A itself, by polarization.) Moreover, these correspondences $A \mapsto P$ and $P \mapsto A$ are mutually inverse. In short, *observables may be identified with self-adjoint operators*, and we shall do so henceforth.

The diagonal matrix elements $\langle u|A|u\rangle$ have an immediate physical interpretation: $\langle u|A|u\rangle = \int_{\mathbb{R}} t\, dP_u(t)$ is the expected value of the probability distribution P_u, and it therefore represents the expected value — or, as physicists usually say, the *expectation value* — of the observable A in the state u. Moreover, the states in which the observable A has a definite value a are simply the eigenvectors of A with eigenvalue a.

Incidentally, this is the appropriate point to mention a bit of physicists' argot that the reader may encounter occasionally. In the early days of quantum mechanics when the idea that observables are represented by noncommuting operators was still new and strange, some people spoke of quantum observables as "quantities whose values are q-numbers" — the notion of "q-number" being meant to suggest noncommutativity — as opposed to "quantities whose values are c-numbers," i.e., ordinary complex-valued quantities whose algebra is commutative. One still finds the terms *q-number* (rarely) and *c-number* (more frequently) in the physics literature; in particular, to say that an operator is a c-number is to say that it is a scalar multiple of the identity. (E.g., "The commutator of A and B is just a c-number.") That usage is echoed in this book by our tendency to write just λ instead of λI for a scalar multiple of the identity.

The reader will notice that we have entered the perilous world of unbounded operators. *Bounded* self-adjoint operators correspond to projection-valued measures P such that $P(E) = I$ for some *bounded* set E. But many important observables can potentially have values anywhere on the real line, so the corresponding operators will be unbounded. In this situation the word "self-adjoint" has to be understood in its precise technical sense. Although these technicalities play a remarkably small role in quantum field theory, we had better take a little time to set them out clearly.

To recall the precise definitions: let A be an operator defined on the domain $\mathcal{D}(A) \subset \mathcal{H}$. If $\mathcal{D}(A)$ is dense in $\mathcal{H}$, the *adjoint* of A is the operator $A^\dagger$ on the domain of all $u \in \mathcal{H}$ such that the map $v \mapsto \langle u|Av\rangle$ ($v \in \mathcal{D}(A)$) extends to a bounded linear functional on all of $\mathcal{H}$ (uniquely, since $\mathcal{D}(A)$ is dense); $A^\dagger u$ is the element of $\mathcal{H}$ corresponding to this functional, so that $\langle A^\dagger u|v\rangle = \langle u|Av\rangle$. A is called *symmetric* if $\langle Au|v\rangle = \langle u|Av\rangle$ for all $u, v \in \mathcal{D}(A)$. This means that $A^\dagger$ is an extension of A, i.e. $\mathcal{D}(A^\dagger) \supset \mathcal{D}(A)$ and $A^\dagger|\mathcal{D}(A) = A$. A is called *self-adjoint* if $A^\dagger = A$. If A is symmetric, it may be possible to extend A to a larger domain to make it self-adjoint, perhaps in many different ways. A is called *essentially self-adjoint* if it has a unique self-adjoint extension, or equivalently, if the graph of $A^\dagger$ in $\mathcal{H} \times \mathcal{H}$ is the closure of the graph of A. The principal reason for insisting upon these distinctions is that *among the symmetric operators, the self-adjoint ones are the only ones to which the spectral theorem and its consequences apply.*

In the mathematics literature the word *Hermitian* is usually a synonym for *symmetric*. In the physics literature, too, if one looks for a definition of "Hermitian" one will usually find it stated that A is Hermitian if $\langle u|A|v\rangle = \langle v|A|u\rangle^*$. However, when physicists *speak* of Hermitian operators, they often *want* self-adjoint ones, for those are the ones with a good spectral theory and the ones that really correspond to observables. Generally this is not a serious problem; the diligent mathematician can usually figure out what the appropriate domain is with a certain amount of work. In particular, when the Hilbert space is something like $L^2(\mathbb{R}^n)$, it often happens that a symmetric operator is essentially self-adjoint when considered on an obvious domain of "nice" functions, and then there is no problem. (The standard position and momentum observables defined by (3.12) below are good examples. It takes some thought to come up with a dense subspace of their domains on which they are *not* essentially self-adjoint.) On the other hand, when the operator in question is a differential operator on a region in $\mathbb{R}^n$, the question of specifying an appropriate domain usually amounts to choosing appropriate boundary conditions, and these conditions are usually dictated by physical considerations.

In any case, we shall be careful to use the term "self-adjoint" when it is really an issue, and we shall generally follow the physicists in using the term "Hermitian" when it is not.

For those who are not on easy terms with these ideas, it may be useful to consider an example. Let $\mathcal{H} = L^2([0,1])$, and let $A = d^2/dx^2$ on the domain of all $f \in C^1([0,1])$ such that f' is absolutely continuous and f'' (defined a.e.) is in L^2. The fundamental formula arises from integration by parts:

$$\begin{aligned}(3.4)\quad &\langle Af|g\rangle - \langle f|Ag\rangle \\ &= \int_0^1 [f''(x)^*g(x) - f(x)^*g''(x)]\,dx = \big[f'(x)^*g(x) - f(x)^*g'(x)\big]_0^1.\end{aligned}$$

(Remember that * denotes complex conjugation!) Thus A is not symmetric. However, the restriction A_0 of A to $\{f \in \mathcal{D}(A) : f(0) = f'(0) = f(1) = f'(1) = 0\}$ is symmetric, and it is not hard to show that $A = A_0^\dagger$. Note that $\mathcal{D}(A_0)$ is defined by the imposition of four boundary conditions; it is too small to give a self-adjoint operator. The self-adjoint extensions of A_0 are defined by restricting A to a domain defined by *two* boundary conditions in such a way that the boundary terms in (3.4) vanish for all f and g satisfying the conditions. There is, in fact, a four-parameter family of such conditions (naturally parametrized by the unitary group $U(2)$), and the resulting operators all have different spectral resolutions. Three cases will suffice to indicate the variety of possibilities:

i. A_1 is the restriction of A to $\{f \in \mathcal{D}(A) : f(0) = f(1) = 0\}$. The eigenvalues of A_1 are $-n^2\pi^2$, $n \geq 1$, each with multiplicity 1, and the corresponding eigenfunctions are $\sin n\pi x$.
ii. A_2 is the restriction of A to $\{f \in \mathcal{D}(A) : f(0) = f'(1) = 0\}$. The eigenvalues of A_2 are $-(n - \frac{1}{2})^2\pi^2$, $n \geq 1$, each with multiplicity 1, and the corresponding eigenfunctions are $\sin(n - \frac{1}{2})\pi x$.
iii. A_3 is the restriction of A to $\{f \in \mathcal{D}(A) : f(0) = f(1) \text{ and } f'(0) = f'(1)\}$. The eigenvalues are $-4n^2\pi^2$, $n \geq 0$; the eigenvalue 0 has multiplicity 1 and all others have multiplicity 2, and the corresponding eigenfunctions are linear combinations of $\cos 2n\pi x$ and $\sin 2n\pi x$.

In each of these cases the operator has an orthonormal eigenbasis, and each of these bases has a nice interpretation in terms of the classical physics of one-dimensional vibrations. The first one models the normal modes of a string fixed at both ends; the second models the normal modes of a cylindrical pipe (an organ pipe or clarinet, say) that is closed at one end and open at the other; and the third models the normal modes of a vibrating circular hoop. So the physicists may not talk about domains of unbounded operators, but they have no trouble distinguishing these cases! On the other hand, to bring home the point about good spectral theory, let us add A and A_0 to the list.

iv. Every $\lambda \in \mathbb{C}$ is an eigenvalue of A with multiplicity 2; the eigenfunctions are linear combinations of $e^{\sqrt{-\lambda}x}$ and $e^{-\sqrt{-\lambda}x}$ for $\lambda \neq 0$ and of 1 and x for $\lambda = 0$. There is no canonical eigenfunction expansion associated to A.
v. The operator A_0 has no eigenvalues at all. However, its spectrum, the set of all $\lambda \in \mathbb{C}$ such that $A_0 - \lambda$ is not invertible, is the whole complex plane because $A_0 - \lambda$ is never surjective; in fact, its range isn't even dense. (The orthogonal complement of its range consists of the eigenfunctions of A with eigenvalue λ^*.) Again there is no good spectral resolution.

If one replaces the condition "$f \in \mathcal{D}(A)$" with "$f \in C^2([0,1])$" or "$f \in C^\infty([0,1])$" in the definition of A_1, A_2, or A_3, the resulting operator is essentially self-adjoint. But adding an extra boundary condition, or taking one away, really destroys the self-adjointness.

Next we consider symmetries of the quantum system. These are modeled by automorphisms of the projective Hilbert space $\mathbb{P}\mathcal{H}$, that is, the bijections from $\mathbb{P}\mathcal{H}$ to itself that preserve the projectively invariant part of the inner product,

$$(u,v) = \frac{|\langle u|v\rangle|}{\|u\|\,\|v\|} \qquad (u, v \neq 0).$$

(Geometrically, (u,v) is the cosine of the angle between the complex lines spanned by u and v. Physically, $(u,v)^2$ is the transition probability between the states u and v.) Clearly any unitary operator on $\mathcal{H}$ induces an automorphism of $\mathbb{P}\mathcal{H}$, as does any anti-unitary operator (a conjugate-linear operator V such that $\langle u|V|w\rangle = \langle w|u\rangle$). A theorem of Wigner (see Bargmann [**7**] or Jauch [**67**]) assures us that these are the *only* automorphisms of $\mathbb{P}\mathcal{H}$. Thus the automorphism group can be identified with the group of unitary or antiunitary operators on $\mathcal{H}$ modulo scalar multiples of the identity operator (which induce the identity transformation on $\mathbb{P}\mathcal{H}$). We equip it with the topology inherited from the strong operator topology; the unitary and antiunitary parts are then its connected components. (The group of unitary operators is connected because, by spectral theory, every unitary operator belongs to a continuous one-parameter group of unitary operators and so is connected to the identity by a continuous path.) We denote the component of the identity, $\mathcal{U}(\mathcal{H})/\{cI : |c| = 1\}$, by $\mathbb{P}\mathcal{U}(\mathcal{H})$.

Let us pass to continuous groups of symmetries. If G is a connected topological group, a *projective representation* or *ray representation* of G is a continuous homomorphism $\rho : G \to \mathbb{P}\mathcal{U}(\mathcal{H})$. The question that immediately arises is whether such a ρ can be lifted to a continuous homomorphism $\widetilde{\rho} : G \to \mathcal{U}(\mathcal{H})$, i.e., a unitary representation of G. There are two possible obstructions, one topological and one cohomological. The topological problem is that if G is not simply connected, it may not even be possible to lift ρ to a continuous map from G into $\mathcal{U}(\mathcal{H})$, let alone a homomorphism. (Examples are provided by the double-valued representations of the rotation group; see §3.5.) The solution to this problem is to pass to the universal cover of G. Assuming then that G is simply connected, a continuous lifting $\widehat{\rho} : G \to \mathcal{U}(\mathcal{H})$ always exists, and it satisfies $\widehat{\rho}(xy) = \omega(x,y)\widehat{\rho}(x)\widehat{\rho}(y)$ where $\omega(x,y)$ is a numerical factor of absolute value 1. Moreover, the associative law for G implies that ω satisfies the cocycle condition $\omega(xy,z)\omega(x,y) = \omega(x,yz)\omega(y,z)$. If ω is a coboundary, that is, if $\omega(x,y) = \lambda(xy)/\lambda(x)\lambda(y)$ for some continuous $\lambda : G \to \mathbf{T}^1$, then $\widetilde{\rho}(x) = \lambda(x)\widehat{\rho}(x)$ is the desired homomorphic lift of ρ. (Those familiar with group cohomology will recognize the general framework into which the words "cocycle" and "coboundary" should be placed; others need not worry about it.) If ω is not a coboundary, it determines a group $\widetilde{G}$ that fits into an exact sequence $1 \to U(1) \to \widetilde{G} \to G \to 1$ (that is, $\widetilde{G}$ is an extension of G by the circle group $U(1)$), and the best one can do is to find a unitary representation $\widetilde{\rho}$ of $\widetilde{G}$ such that $\pi_{\mathcal{U}} \circ \widetilde{\rho} = \rho \circ \pi_G$, where $\pi_{\mathcal{U}}$ and π_G are the projections from $\mathcal{U}(\mathcal{H})$ to $\mathbb{P}\mathcal{U}(\mathcal{H})$ and from $\widetilde{G}$ to G.

When G is a simply connected Lie group, the problem in group cohomology adumbrated above can be reduced to a problem in Lie algebra cohomology, which is more readily computable. The relevant cohomology group is $H^2(\mathfrak{g}, \mathbb{R})$, which is the quotient of the space of all bilinear maps $\phi : \mathfrak{g} \times \mathfrak{g} \to \mathbb{R}$ that satisfy the cocycle condition $\phi(X, [Y, Z]) + \phi(Y, [Z, X]) + \phi(Z, [X, Y]) = 0$ by the subspace of all such ϕ of the form $\phi(X, Y) = f([X, Y])$ for some linear functional f on $\mathfrak{g}$. The upshot is *Bargmann's theorem* [**6**] (see also Simms [**110**] and Varadarajan [**123**]):

Let G be a simply connected Lie group with Lie algebra $\mathfrak{g}$. Every projective representation ρ of G determines an element $\beta_\rho \in H^2(\mathfrak{g}, \mathbb{R})$, and ρ can be lifted to a unitary representation of G if and only if $\beta_\rho = 0$. In particular, if $H^2(\mathfrak{g}, \mathbb{R}) = 0$, every projective representation of G can be lifted to a unitary representation of G.

The cohomology group $H^2(\mathfrak{g}, \mathbb{R})$ vanishes whenever G is semisimple, and also when $G = \mathbb{R}$ and when G is the Poincaré group. (It does not vanish when $G = \mathbb{R}^n$ for $n > 1$, however. Examples of projective representations of $\mathbb{R}^n$ for even n that cannot be lifted to unitary representations are provided by the representations of the canonical commutation relations that we shall discuss in §3.2.) For the time being, the essential case for us is $G = \mathbb{R}$: *Every one-parameter subgroup of $\mathbb{PU}(\mathcal{H})$ comes from a one-parameter group of unitary operators on $\mathcal{H}$.*

We are now in a position to apply another of the big theorems of functional analysis, the Stone representation theorem for (strongly continuous) one-parameter unitary groups. Recall that an operator S is called *skew-adjoint* if $S^\dagger = -S$. If S is any skew-adjoint operator, then $U(t) = e^{tS}$ (defined by the spectral functional calculus) is a strongly continuous one-parameter group of unitary operators, and Stone's theorem asserts that *every* strongly continuous one-parameter group of unitary operators is of this form. Moreover, skew-adjoint and self-adjoint operators are almost the same thing: multiplication by a purely imaginary scalar converts one into the other. Thus, if one fixes a nonzero real constant b, the map $A \mapsto ibA$ defines a bijection from the self-adjoint to the skew-adjoint operators. (Why not take $b = 1$? There are good reasons: more about this in the next section.) Thus there is a correspondence between observables (self-adjoint operators) and one-parameter groups of symmetries of a quantum system: $A \leftrightarrow e^{ibtA}$. This is the quantum version of Noether's theorem.

Let us explore this a little further. A one-parameter group e^{ibtA} can be considered as transforming the states by $u \mapsto u_t = e^{ibtA}u$ while the observables remain fixed, or as transforming the observables by $B \mapsto B_t = e^{-ibtA}Be^{ibtA}$ while the states remain fixed. These are known as the *Schrödinger* and *Heisenberg pictures*, respectively. If P is the projection-valued measure associated to B, then $P_t(\cdot) = e^{-ibtA}P(\cdot)e^{ibtA}$ is the projection-valued measure associated to B_t, and the probability distribution of the observable B in the state u, after the action of e^{ibtA}, is $E \mapsto \langle u_t|P(E)|u_t\rangle = \langle u|P_t(E)|u\rangle$.

In the Schrödinger picture, the differential equation governing the evolution of u_t is

$$\frac{du_t}{dt} = ibAu_t,$$

the *(abstract) Schrödinger equation.* (More precisely, this equation is valid when u is in the domain of A; this domain is invariant under the group e^{ibtA}.) In the Heisenberg picture, the differential equation governing the evolution of B_t is

$$\frac{dB_t}{dt} = -ibAB_t + ibB_tA = ib[B_t, A] = ib[B, A]_t.$$

(More precisely, this equation is valid as an operator identity on the domain of all $u \in \mathcal{D}(B_t) \cap \mathcal{D}(A)$ such that $B_t u \in \mathcal{D}(A)$ and $Au \in \mathcal{D}(B_t)$ for all t.)

The resemblance to Hamilton's equation (2.3) is notable, and it leads us to think about the analogy between Poisson brackets and commutators. Both operations are Lie products, in the sense of being skew-symmetric and satisfying the Jacobi identity. The set of C^∞ functions on a symplectic manifold forms a Lie algebra under Poisson brackets, and the set of bounded skew-adjoint operators on a Hilbert space forms a Lie algebra under commutators. The correspondence $A \leftrightarrow ibA$ between self-adjoint and skew-adjoint operators then induces a Lie algebra structure on the set of bounded self-adjoint operators by

$$(A_1, A_2) \mapsto \frac{[ibA_1, ibA_2]}{ib} = ib[A_1, A_2].$$

Alas, one sometimes needs to consider classical observables that aren't C^∞ and quantum observables that aren't bounded, so the Lie algebra structures aren't perfect. The classical problem is not serious; one just has to remember that the Poisson bracket of two C^k functions is only C^{k-1}. On the quantum side, if A_1 and A_2 are self-adjoint operators, not both bounded, $ic[A_1, A_2]$ is at least a symmetric operator on the domain of all $u \in \mathcal{D}(A_1) \cap \mathcal{D}(A_2)$ such that $A_1 u \in \mathcal{D}(A_2)$ and $A_2 u \in \mathcal{D}(A_1)$, but in general that domain will be too small for $[A_1, A_2]$ to be self-adjoint. (If one is fortunate, it will be essentially self-adjoint. There are substantial technical issues to be dealt with here, but they do not concern us now.)

Let us pause to take stock of the analogies between the formalisms of Hamiltonian and quantum mechanics.

- In Hamiltonian mechanics, the state space is phase space, a symplectic manifold M; the observables are real-valued functions on M, and the symmetries are canonical transformations of M. Every (smooth) observable defines a Hamiltonian vector field, which generates a one-parameter group (or at least a flow) of canonical transformations, and every such flow arises from an observable in this way (at least if M is simply connected). Moreover, the (smooth) observables form a Lie algebra under the Poisson bracket. If f is an observable whose Hamiltonian vector field generates the flow Φ_t, the evolution of any other observable g under the flow is given by $(d/dt)(g \circ \Phi_t) = \{g, f\} \circ \Phi_t$.
- In quantum mechanics, the state space is a projective Hilbert space $\mathbb{P}\mathcal{H}$, the observables are self-adjoint operators on $\mathcal{H}$, and the symmetries are unitary or anti-unitary operators on $\mathcal{H}$. A choice of a nonzero real constant b (to be specified shortly) sets up a bijection between self-adjoint and skew-adjoint operators, $A \leftrightarrow ibA$. Via this correspondence, observables are precisely the infinitesimal generators of one-parameter groups of symmetries: $U(t) = e^{ibtA}$. Moreover, self-adjoint operators form a Lie algebra under the Lie product $(A_1, A_2) \mapsto ib[A_1, A_2]$ (except for technical difficulties concerning the domains of unbounded operators). If A is an observable, the evolution of any other observable B under the one-parameter group e^{ibtA} is given by $(d/dt)(e^{-ibtA} B e^{ibtA}) = e^{-ibtA}(ib[B, A])e^{ibtA}$.

These analogies are a little ragged around the edges because of the differentiability assumptions, the local nature of most flows on the classical side, and the hangups with unbounded operators on the quantum side, but they are sufficiently striking to assure us that we are onto something good.

3.2. Quantization

So far we have only constructed a skeleton of a quantum theory. To put some flesh on the bones, we need a way to find the quantum analogues of familiar classical observables such as position, momentum, and energy. (Once we understand the observables, the nature of the states will become clearer.) We shall take the Hamiltonian formulation of classical mechanics as a starting point so as to take advantage of the analogies outlined above, and the resulting quantum theory will therefore be *nonrelativistic*. Building a relativistic theory will require more work.

What we want is a way to assign to each classical observable (a real-valued function f on phase space) a quantum observable A_f (a self-adjoint operator on Hilbert space). We (optimistically) ask for this correspondence to enjoy the following properties:

- Constant functions correspond to constant multiples of the identity:

$$A_\alpha = \alpha I \qquad (\alpha \in \mathbb{R}). \tag{3.5}$$

- If $\phi : \mathbb{R} \to \mathbb{R}$ is a Borel function, and $E \subset \mathbb{R}$ is a Borel set, we have $(\phi \circ f)(x) \in E$ if and only if $f(x) \in \phi^{-1}(E)$. Let P_f be the projection-valued measure corresponding to the operator A_f. Recalling the way in which projection-valued measures are to be interpreted as observables, we infer that the projection-valued measure corresponding to the classical observable $\phi \circ f$ should be $E \mapsto P_f(\phi^{-1}(E))$. But the self-adjoint operator corresponding to this measure is just $\phi(A_f)$, defined by the spectral functional calculus. In short,

$$A_{\phi\circ f} = \phi(A_f). \tag{3.6}$$

- If f and g are classical observables, the expected value of the quantum observable A_{f+g} in any state u should be the sum of the expected values of A_f and A_g, that is, $\langle u|A_{f+g}|u\rangle = \langle u|A_f|u\rangle + \langle u|A_g|u\rangle$. But the diagonal matrix elements of an operator completely determine the operator by polarization, so

$$A_{f+g} = A_f + A_g. \tag{3.7}$$

- Poisson brackets should correspond to commutators, as modified by the correspondence between self-adjoint and skew-adjoint operators:

$$A_{\{f,g\}} = ib[A_f, A_g]. \tag{3.8}$$

  (We ignore questions about domains here.)

It turns out that this is too much to ask for; quantum mechanics isn't *that* similar to classical mechanics. (See Folland [**43**] for further discussion of this point.) But we are only looking for guidelines, not a functor that will magically transform classical systems into quantum ones, and we shall use this wish-list in that spirit. In particular, we expect only that (3.8) will hold in an approximate or asymptotic sense in situations where the classical and quantum pictures can be compared directly.

It is time to explore the meaning of the mysterious constant b. The reason for not simply taking $b = 1$ is that its value depends on the units of measurement being used. Differentiation with respect to a position or momentum variable changes the dimensions by adding a factor of $[l^{-1}]$ or $[m^{-1}l^{-1}t]$, respectively. Thus, if f and g have dimensions $[d_f]$ and $[d_g]$, their Poisson bracket will have dimensions $[d_f d_g m^{-1} l^{-2} t]$. On the quantum side, the "dimensions" of a self-adjoint operator

A are the dimensions of the variable λ in its spectral resolution $A = \int \lambda\, dP(\lambda)$ (or equivalently, if A has discrete spectrum, the dimensions of its eigenvalues). (This point may seem less than perfectly clear in the present abstract setting, but it is always obvious in the concrete models actually used in quantum theory, where the dimensions of an operator are manifest from its form and the coefficients occuring in it.) If A and B are self-adjoint operators with dimensions $[d_A]$ and $[d_B]$, the operator $i[A,B] = i(AB - BA)$ will have dimensions $[d_A d_B]$. But this means that *if the correspondence* (3.8) *is to be valid, even in an approximate sense, the constant b must have dimensions $[m^{-1}l^{-2}t]$. Hence, its value depends on the system of units being used, and its value in a given system must be determined empirically.* Moreover, in order to make the correspondences between symmetries and observables in classical and quantum mechanics parallel to one another, it turns out to be necessary to take b to be *negative.* (Alternatively, one can take b positive but replace the i in the correspondence $A \leftrightarrow ibA$ with $1/i$.)

We are now ready to reveal the secret: the constant $-1/b$ is called *Planck's constant* and is denoted by $\hbar$; in MKS units it is approximately 1.054×10^{-34} kg $\cdot$ m^2/s. (Planck's original constant was $h = 2\pi\hbar$, but now everyone uses $\hbar$. Basically, it is a question of whether one wants to measure frequency in cycles per second or radians per second.) The dimensions $[ml^2t^{-1}]$ of $\hbar$ can be interpreted as $[ml^2t^{-2} \times t]$ (energy times time) or as $[l \times mlt^{-1}]$ (moment arm times momentum), and they are therefore the same as the dimensions of action, the basic quantity in Lagrangian mechanics, and angular momentum. As we shall see, Planck's constant functions as a basic unit of both action and angular momentum in quantum theory.

The correspondences between observables and one-parameter symmetry groups and between Poisson brackets and commutators can now be rewritten as

$$A \leftrightarrow e^{tA/i\hbar}, \qquad A_{\{f,g\}} \leftrightarrow \frac{1}{i\hbar}[A_f, A_g]. \tag{3.9}$$

While we are thinking of dimensions, let us point out that the dimensions of the t in the one-parameter group must be related to those of A so that $tA/\hbar$ is dimensionless (independent of the system of units). In particular, if $e^{-itA/\hbar}$ is the group that describes the time evolution of the system, so that t is time, then A must have the dimensions of energy, in accordance with the analogy with Hamiltonian mechanics.

We now work toward the quantum description of a system of moving particles with classical position and momentum coordinates $x_1, \dots, x_n$ and $p_1, \dots, p_n$. (That is, the phase space is $T^*\mathbb{R}^n \cong \mathbb{R}^{2n}$, where $n = 3k$ if there are k particles moving in $\mathbb{R}^3$.) As a starting point, we try to quantize these basic observables in such a way that the correspondence between Poisson brackets and commutators in (3.9) is an exact equality. Since

$$\{x_j, p_k\} = \delta_{jk}, \qquad \{x_j, x_k\} = \{p_j, p_k\} = 0,$$

we are looking for self-adjoint operators $X_1, \dots, X_n$ and $P_1, \dots, P_n$ on a Hilbert space $\mathcal{H}$ that satisfy

$$[X_j, P_k] = i\hbar\delta_{jk}I, \qquad [X_j, X_k] = [P_j, P_k] = 0. \tag{3.10}$$

These are called the *canonical commutation relations*, and P_j is said to be *canonically conjugate to* X_j when they are satisfied.

One must immediately face the fact that the operators X_j and P_j in (3.10) are necessarily unbounded — not just because they are supposed to correspond to unbounded classical observables but for purely mathematical reasons. (Suppose X

and P are bounded self-adjoint operators satisfying $[X, P] = i\hbar I$. An easy induction shows that $[X, P^n] = ni\hbar P^{n-1}$, so $n\hbar\|P^{n-1}\| \le \|XP^n - P^nX\| \le 2\|X\|\,\|P^n\|$. The self-adjointness of P implies that $\|P^k\| = \|P\|^k$ for any k, so $n\hbar \le 2\|X\|\,\|P\|$ for all n, which is absurd.) Consequently, the interpretation of the relation $[X_j, P_k] = i\hbar\delta_{jk}I$ is problematic: the operator on the right is defined on all of $\mathcal{H}$, but the operator on the left is not. To take the extreme case, there exist self-adjoint operators X and P whose domains contain only the zero vector in common, and then $[X, P] = i\hbar I$ is valid on $\mathcal{D}([X, P])$ for the stupid reason that $\mathcal{D}([X, P]) = \{0\}$. (Less stupid pathologies can occur too, as we shall point out later.)

The best resolution of this problem is as follows. The commutation relations (3.10) imply that the operators X_j, P_j, and iI span a Lie algebra. Let us consider this algebra in the abstract; that is, let $\mathfrak{h}_n$ be $\mathbb{R}^n \times \mathbb{R}^n \times \mathbb{R}$ equipped with the Lie bracket

$$[(v, w, t),\, (v', w', t')] = (0,\, 0,\, v \cdot w' - w \cdot v').$$

($\mathfrak{h}_n$ is called the *Heisenberg algebra* of dimension $2n + 1$.) Then the standard basis vectors

$$V_j = (e_j, 0, 0), \quad W_j = (0, e_j, 0), \quad T = (0, 0, 1),$$

where $\{e_1, \ldots, e_n\}$ is the standard basis for $\mathbb{R}^n$, satisfy

$$[V_j, W_k] = \delta_{jk}T, \qquad [V_j, V_k] = [W_j, W_k] = [V_j, T] = [W_j, T] = 0,$$

so the relations (3.10) mean that the map $(v, w, t) \mapsto \sum(v_jX_j + w_jP_j) + i\hbar tI$ is a Lie algebra homomorphism. The "good" interpretation of (3.10) is that this homomorphism should come from a unitary representation of the corresponding Lie group.

That is, let H_n (the *Heisenberg group*) be the simply connected Lie group with Lie algebra $\mathfrak{h}_n$. H_n can be identified with $\mathbb{R}^n \times \mathbb{R}^n \times \mathbb{R}$ with the group law

$$(v, w, t)(v', w', t') = (v + v',\, w + w',\, t + t' + \tfrac{1}{2}(v \cdot w' - w \cdot v'))$$

(see Folland [**43**]). If ρ is a unitary representation of H_n on $\mathcal{H}$ and $\Xi \in \mathfrak{h}_n$, let $d\rho(\Xi)$ be the infinitesimal generator of the one-parameter unitary group $t \mapsto \rho(\exp(t\Xi))$ (a skew-adjoint operator on $\mathcal{H}$). Then the domains of the operators $d\rho(\Xi)$, $\Xi \in \mathfrak{h}_n$, all contain the space $C^\infty(\rho)$ of C^∞ vectors for ρ and map $C^\infty(\rho)$ into itself; they are all essentially skew-adjoint on $C^\infty(\rho)$, and the relation

$$[d\rho(\Xi_1), d\rho(\Xi_2)] = d\rho([\Xi_1, \Xi_2])$$

holds when both sides are interpreted as operators on $C^\infty(\rho)$. Let $X_j = i\hbar d\rho(V_j)$ and $P_j = i\hbar d\rho(W_j)$; then the relations (3.10) hold on $C^\infty(\rho)$ provided that $d\rho(T) = I/i\hbar$, which means that $\rho(0, 0, t) = e^{t/i\hbar}I$.

Thus, we are looking for unitary representations ρ of H_n such that $\rho(0, 0, t) = e^{t/i\hbar}I$. One such representation is the *Schrödinger representation* σ on $L^2(\mathbb{R}^n)$, defined by

$$[\sigma(v, w, t)f](x) = \exp\left(\frac{v \cdot x + t - \frac{1}{2}v \cdot w}{i\hbar}\right) f(x - w), \tag{3.11}$$

which yields

$$[X_jf](x) = i\hbar d\sigma(V_j)f(x) = x_jf(x), \qquad P_jf = i\hbar d\sigma(W_j)f = \frac{\hbar}{i}\frac{\partial f}{\partial x_j}. \tag{3.12}$$

(The domains of X_j and P_j are simply the set of all $f \in L^2$ such that $x_j f$ and the distribution derivative $\partial_j f$ are in L^2, respectively; the space $C^\infty(\sigma)$ is the Schwartz class $\mathcal{S}(\mathbb{R}^n)$.)

The representation σ of H_n defined by (3.11) is easily seen to be irreducible. Indeed, suppose $\mathcal{M}$ is a nonzero closed invariant subspace of L^2, and pick $f \neq 0 \in \mathcal{M}$. If $g \in \mathcal{M}^\perp$, then $g \perp \sigma(v, w, 0)f$ for all $v, w \in \mathbb{R}^n$, that is,

$$0 = \int e^{(v\cdot x-(v\cdot w)/2)/i\hbar} f(x-w)g(x)^*\, dx = \int e^{v\cdot x/i\hbar} f(x - \tfrac{1}{2}w)g(x+\tfrac{1}{2}w)^*\, dx.$$

By Fourier uniqueness, $f(x-\frac{1}{2}w)g(x+\frac{1}{2}w)^* = 0$ for a.e. x and w. Since the linear map $(x, w) \mapsto (x - \frac{1}{2}w,\ x + \frac{1}{2}w)$ is invertible, we have $f(x)g(y)^* = 0$ for a.e. x and y, and so either $f = 0$ or $g = 0$ as elements of L^2. But $f \neq 0$ by assumption; hence $\mathcal{M}^\perp = \{0\}$.

Now, what about other representations ρ of H_n such that $\rho(0, 0, t) = e^{t/i\hbar}I$? The answer is provided by the *Stone-von Neumann theorem*:

Suppose ρ is a unitary representation of H_n on a Hilbert space $\mathcal{H}$ such that $\rho(0, 0, t) = e^{t/i\hbar}I$. Then $\mathcal{H}$ is a direct sum of mutually orthogonal subspaces $\mathcal{H}_\alpha$ that are invariant under ρ, such that the subrepresentation of ρ on $\mathcal{H}_\alpha$ is unitarily equivalent to σ for each α. In particular, if ρ is irreducible, then ρ is unitarily equivalent to σ.

There are two good ways to prove this theorem: by von Neumann's original argument and as a corollary of the Mackey imprimitivity theorem. See Folland [**43**], [**44**].

A few mathematical remarks are in order at this point.

i. Given a unitary representation ρ of H_n, let $\rho_1(v) = \rho(v, 0, 0)$ and $\rho_2(w) = \rho(0, w, 0)$. Then ρ_1 and ρ_2 are unitary representations of $\mathbb{R}^n$. If ρ satisfies $\rho(0, 0, t) = e^{t/i\hbar}I$, then ρ_1 and ρ_2 satisfy the *integrated form of the canonical commutation relations*,

$$\rho_2(w)\rho_1(v) = e^{v\cdot w/i\hbar}\rho_1(v)\rho_2(w). \tag{3.13}$$

Conversely if ρ_1 and ρ_2 are unitary representations of $\mathbb{R}^n$ that satisfy (3.13), then $\rho(v, w, t) = e^{(it-(v\cdot w/2))/i\hbar}\rho_1(v)\rho_2(w)$ is a representation of H_n that satisfies $\rho(0, 0, t) = e^{t/i\hbar}I$. The Stone-von Neumann theorem is often stated in terms of ρ_1 and ρ_2 rather than ρ.

ii. The map $(v, w) \mapsto \rho(v, w, 0)$ is a projective representation of $\mathbb{R}^{2n}$ that does not arise from a unitary representation of $\mathbb{R}^{2n}$.

iii. To appreciate the delicacy of passing from the commutation relations (3.10) to a representation of H_n, consider the following example. Let $\mathcal{H} = L^2([0, 1])$, $[Xf](x) = xf(x)$, and $Pf = (\hbar/i)f'$ on the domain $\mathcal{D}(P) = \{f \in AC : f' \in L^2 \text{ and } f(0) = f(1)\}$ (AC is the space of absolutely continuous functions on $[0, 1]$.) Then X and P are both self-adjoint operators on $\mathcal{H}$, and $[X, P] = i\hbar I$ on the domain of $[X, P]$, which is $\{f \in AC : f' \in L^2 \text{ and } f(0) = f(1) = 0\}$. *The representation $(u, v, t) \mapsto uX + vP + (t/i\hbar)I$ of the Heisenberg algebra $\mathfrak{h}_1$ does not come from a unitary representation of H_1.* Indeed, the reader may verify that the one-parameter groups generated by $X/i\hbar$ and $P/i\hbar$ (namely, $[e^{tX/i\hbar}f](x) = e^{tx/i\hbar}f(x)$ and $[e^{tP/i\hbar}f](x) = f(x - t)$, where $x - t$ is taken modulo 1) do not satisfy (3.13).

iv. The Schrödinger representation σ depends on the parameter $\hbar$. That is, for each nonzero $\hbar \in \mathbb{R}$ there is a Schrödinger representation $\sigma_\hbar$, and these representations are all unitarily inequivalent. (Indeed, they are already inequivalent on the center of H_n, namely, $Z = \{(0,0,t) : t \in \mathbb{R}\}$.) On the other hand, if ρ is any irreducible unitary representation of H_n, Schur's lemma implies that $\rho(0,0,t) = e^{iat}I$ for some $a \in \mathbb{R}$. If $a \neq 0$, then ρ is unitarily equivalent to $\sigma_{-1/a}$ by the Stone-von Neumann theorem. If $a = 0$, ρ factors through the quotient group $H_n/Z \cong \mathbb{R}^{2n}$, and hence ρ acts on a one-dimensional space by $\rho(v,w,t) = e^{i\alpha\cdot v + i\beta\cdot w}$ for some $\alpha, \beta \in \mathbb{R}^n$. Thus the Stone-von Neumann theorem yields a complete classification of the irreducible unitary representations of H_n.

Returning to physics, the upshot of all this is that we have a good candidate for the quantum description of a (nonrelativistic) particle moving in $\mathbb{R}^n$: the quantum state space is $L^2(\mathbb{R}^n)$, and the position and momentum observables are given by (3.12). It is clear that these operators have the right dimensions: X_j = multiplication by x_j, which carries dimensions of length; and $P_j = (\hbar/i)\partial/\partial x_j$, whose dimensions are $[ml^2t^{-1}]$ (from the $\hbar$) times $[l^{-1}]$ (from the $\partial/\partial x_j$), or $[mlt^{-1}]$, which is right for momentum.

The irreducibility of the Schrödinger representation is a desirable feature. By the Stone-von Neumann theorem, any other representation of the canonical commutation relations (suitably interpreted via the Heisenberg group) is equivalent to a direct sum of copies of the Schrödinger representation, and taking two or more copies would be like taking the union of two or more copies of of $\mathbb{R}^{2n}$ for the classical phase space.

The projection-valued measures $\mathcal{P}_j$ (so denoted in order to distinguish them from the momentum operators P_j) corresponding to the position observables X_j = multiplication by x_j are simply given by $\mathcal{P}_j(E)$ = multiplication by the characteristic function of $\{x : x_j \in E\}$. In fact, these operators have a common spectral resolution, namely, the projection-valued measure $\mathcal{P}$ on $\mathbb{R}^n$ defined by $\mathcal{P}(E)$ = multiplication by χ_E for E a Borel set in $\mathbb{R}^n$; the measures $\mathcal{P}_j$ on $\mathbb{R}$ are the push-forwards of $\mathcal{P}$ by the coordinate functions. This means that if f is a unit vector in $L^2(\mathbb{R}^n)$ and $E \subset \mathbb{R}^n$ is a Borel set, $\langle f|\mathcal{P}(E)|f\rangle = \int_E |f|^2$ is the probability that the particle will be located in E when the system is in the state f. Thus, in this model the normalized state vectors have the physical significance that *the probability density of the position of the particle in* $\mathbb{R}^n$ *in the state* f *is* $|f|^2$. (Incidentally, if $f \in L^2(\mathbb{R}^n)$ is a unit vector representing a state, the function $f(x)$ carries dimensions $[l^{-n/2}]$, so that $|f(x)|^2$ has the dimensions $[l^{-n}]$ appropriate to a probability per unit volume.)

A similar picture for the momentum variables is obtained by applying the Fourier transform. For this purpose we employ the following version of the Fourier transform that incorporates Planck's constant: if $f \in L^2(\mathbb{R}^n)$,

$$\widehat{f}(p) = \int e^{-ip\cdot x/\hbar} f(x)\, d^n x. \tag{3.14}$$

The inversion and Parseval formulas are

$$f(x) = \int e^{ip\cdot x/\hbar}\widehat{f}(p)\,\frac{d^n p}{(2\pi\hbar)^n}, \qquad \int |f(x)|^2\, d^n x = \int |\widehat{f}(p)|^2 \frac{d^n p}{(2\pi\hbar)^n}. \tag{3.15}$$

By integration by parts,

$$(P_j f)\widehat{\ }(p) = \frac{\hbar}{i}\int e^{-ip\cdot x/\hbar}\frac{\partial f}{\partial x_j}(x)\,dx = \int p_j e^{-ip\cdot x/\hbar} f(x)\,dx = p_j \widehat{f}(p).$$

Hence we arrive at the following picture: We think of the copy of $\mathbb{R}^n$ on which $\widehat{f}$ lives as "momentum space," just as the original $\mathbb{R}^n$ is "position space." (Thus the coordinates p_j have the dimensions of mometum, so that $p\cdot x/\hbar$ is dimensionless, as it should be.) In momentum space, P_j is just multiplication by the jth coordinate function, so the same analysis as in the preceding paragraph shows that $|\widehat{f}|^2$ *is the probability density of the momentum of the particle in the state* f with respect to the normalized Lebesgue measure $d^n p/(2\pi\hbar)^n$.

It should be noted that there are no states in which the particle has a definite position or momentum. The "generalized state" (or, as the physicists say, "non-normalizable state") in which the particle is definitely at the position $x_0 \in \mathbb{R}^n$ is represented by the delta-function $\delta(x - x_0)$, and the "generalized state" in which the particle definitely has momentum $p_0 \in \mathbb{R}^n$ is represented by the inverse Fourier transform of the normalized delta-function $(2\pi)^n\delta(p - p_0)$, i.e., the plane wave $e^{ip_0\cdot x/\hbar}$. It is frequently convenient to allow such idealized states into one's universe of discourse. This can, of course, be done with complete rigor by invoking the theory of distributions (or, on a more abstract level, rigged Hilbert spaces; see Gelfand and Vilenkin [**51**]), but we shall keep the discussion on an informal level as the physicists do and take the opportunity to introduce and explain some notation and formulas that may seem peculiar to mathematicians but are common in the physics literature.

The generalized state $\delta(x - x_0)$ where the particle is located at x_0 is the simultaneous eigenvector of the position operators X_j with eigenvalues x_{0j}, so it can conveniently be labeled by these eigenvalues and denoted $|x_0\rangle$. The corresponding bra $\langle x_0|$ is then nothing but the evaluation functional at x_0: $\langle x_0|f\rangle = f(x_0)$. The states $|x\rangle$, $x \in \mathbb{R}^n$, form a "generalized eigenbasis" for the position operators: just as $f = \sum\langle e_k|f\rangle e_k$ when we have a genuine orthonormal basis $\{e_k\}$, here we can write

$$|f\rangle = \int |x\rangle\langle x|f\rangle\,d^n x, \quad \text{or} \quad \int |x\rangle\langle x|\,d^n x = I,$$

which is just a restatement of the fact that $\int f(x)\delta(y - x)\,dx = f(y)$. Likewise, the generalized state where the particle has momentum p may be denoted by $|p\rangle$. (Here, of course, it must be understood that, contrary to common mathematical practice, the letters x and p are not just arbitrarily chosen symbols but carry some semantic content: x means position and p means momentum.) Here the scalar product $\langle p|f\rangle$ is $\widehat{f}(p)$, and the fact that the states $|p\rangle$ are again a "generalized orthonormal basis," that is,

$$|f\rangle = \int |p\rangle\langle p|f\rangle\,\frac{d^n p}{(2\pi\hbar)^n}, \quad \text{or} \quad \int |p\rangle\langle p|\,\frac{d^n p}{(2\pi\hbar)^n} = I,$$

is a restatement of the Fourier inversion formula (3.15); cf. (1.5).

This is a good place to bring up an intriguing difference in the way physicists and mathematicians regard mathematical objects. Mathematicians are trained to think that all mathematical objects should be realized in some specific way as sets. When we do calculus on the real line we may not wish to think of real numbers as Dedekind cuts or equivalence classes of Cauchy sequences of rationals, but it reassures us to know that we can do so. Physicists' thought processes, on the other hand, are anchored in the physical world rather than in set theory, and they see no need to tie themselves down to specific set-theoretic models. A physicist may speak of the state space $\mathcal{H}$ of a quantum system $\mathcal{Q}$; if a mathematician asks "What Hilbert space is $\mathcal{H}$?", the response is "I just told you, it's the state space of $\mathcal{Q}$." If one is studying a particular observable, one may wish to *represent* $\mathcal{H}$ in a way that makes the analysis more transparent. For example, if one is studying an operator with discrete spectrum — eigenvalues λ_n and eigenvectors $|\lambda_n\rangle$ — one may wish to represent $\mathcal{H}$ as l^2 by identifying the state $|f\rangle$ with the sequence $\{\langle\lambda_n|f\rangle\}$; if one is studying the position of a particle, one will represent $\mathcal{H}$ as $L^2(\mathbb{R}^n)$ in a particular way by using the generalized states $|x\rangle$ as a "basis" to form the wave function $f(x) = \langle x|f\rangle$; if one is studying momentum, one will represent $\mathcal{H}$ as $L^2(\mathbb{R}^n)$ in a different way by using the generalized states $|p\rangle$ instead; and so forth. But, from this point of view, choosing a particular representation is akin to choosing a particular coordinate system for doing analysis on a manifold, and it is not unreasonable to regard committing oneself to such a choice at the outset as a mistake.

(Similarly, it is clear from the way physicists talk about Lie algebras that they regard a Lie algebra as an algebraic structure unattached to any particular set, rather than as a set endowed with an algebraic structure. If you ask "What is $\mathfrak{so}(3)$?", a mathematician will answer "the set of skew-symmetric 3×3 real matrices, with the commutator as Lie bracket," but a physicist will answer something like "three generators X, Y, and Z satisfying $[X,Y] = Z$, $[Y,Z] = X$, and $[Z,X] = Y$," and he will say that X, Y, and Z "satisfy the Lie algebra $\mathfrak{so}(3)$.")

Nonetheless, as this book is being written by a mathematician for mathematicians, we shall continue to take state spaces for particular quantum systems to be particular concrete Hilbert spaces.

It remains to quantize more general functions of position and momentum. For functions of position alone or momentum alone, there is an obvious way to do this: If ϕ is a function on $\mathbb{R}^n$, the quantization of the classical observable $\phi(x)$ is multiplication by ϕ in position space, and the quantization of the classical observable $\phi(p)$ is multiplication by ϕ in momentum space:

$$[\phi(X)f](x) = \phi(x)f(x), \qquad [\phi(P)f]\widehat{\ }(p) = \phi(p)\widehat{f}(p).$$

However, one runs into difficulties in going further. For functions of the form $\phi(x) + \psi(p)$, the symmetric operator $\phi(X) + \psi(P)$ may not be self-adjoint on the domain $\mathcal{D}(\phi(X))\cap\mathcal{D}(\psi(P))$, and when one considers functions involving products of position and momentum variables it is impossible to satisfy (3.6) or (3.8) even on the formal level: see Folland [**43**], p. 17 and pp. 197–199. These no-go theorems simply reflect that fact that quantum mechanics is too different from classical mechanics for a quantization procedure to exist that has all the properties in the wish list (3.5)–(3.8). After all, classical mechanics is a limiting case of quantum mechanics, not the other way around, and there is no reason to expect that one can fully recover the quantum theory from the limiting case. From the physicists' point of view, quantization procedures are to be taken only as possible ways of arriving at a good quantum theory. Once one has constructed such a theory, the test of its correctness comes from working out its consequences and comparing them with experiment, not from any *a priori* considerations.

There is, however, some purely mathematical interest in quantization procedures. One reason is the desire for a useful functional calculus for the noncommuting operators X_j and P_j, motivated by problems in the theory of partial differential equations. Indeed, the calculus of pseudodifferential operators, which sets up a correspondence between "symbols" $f(x,p)$ and operators $f(X,P)$ (usually with $\hbar = 1$ or $\hbar = 1/2\pi$), is such a procedure. We refer to Folland [**43**] for an extensive discussion of it from a point of view similar to the one here.

The other mathematical motivation for studying quantization procedures comes from group representation theory. Suppose one wishes to find all the irreducible unitary representations of a connected Lie group G. In physical language, this can be considered as finding all quantum systems on which G acts irreducibly as a group of symmetries. The corresponding "classical" problem is to find all symplectic manifolds on which G acts transitively as a group of canonical transformations, i.e., all symplectic homogeneous G-spaces. The latter problem is actually quite explicitly solvable: The orbits of the co-adjoint action of G on $\mathfrak{g}^*$ are symplectic homogeneous G-spaces, as are the orbits of the coadjoint action of central extensions of G by $\mathbb{R}$ (essentially the same family of groups that intervenes in the issue of projective vs. unitary representations). Moreover, *all* symplectic homogeneous G-spaces are more or less of this form (the phrase "more or less" has to do with questions about covering spaces). Hence, if one has a good quantization procedure for such spaces, one has a hope of understanding the unitary representations of G. This idea works beautifully when G is nilpotent or when G is compact; the Kirillov theory for nilpotent groups and the Borel-Weil theory for compact groups can both be understood as special cases of quantization of coadjoint orbits. It also provides much illumination when G is solvable or semisimple, although these cases present additional technical problems. (See Kirillov [**74**] together with Vogan's review [**126**], and also Vogan [**125**].)

Back to physics: The time evolution of a quantum system is given by a one-parameter group $U(t)$ of unitary maps, and the observable corresponding to its infinitesimal generator is the quantum *Hamiltonian* H; that is, $U(t) = \exp(tH/i\hbar)$. We expect this operator to be a quantized version of the Hamiltonian in classical mechanics. Here we consider the case of particles moving in a potential V, for which the classical Hamiltonian is $\frac{1}{2}m^{-1}p \cdot p + V(x)$. (Recall that m is a diagonal matrix whose diagonal entries are the masses of the particles.) The obvious quantum analogue is the differential operator on $L^2(\mathbb{R}^n)$

$$H = -\tfrac{1}{2}\hbar^2 m^{-1}\nabla \cdot \nabla + V, \tag{3.16}$$

the *Schrödinger operator* with potential V, where V is considered as the operator $f \mapsto Vf$. The relation $U(t) = e^{tH/i\hbar}$ can be expressed as a differential equation for the evolution of a state vector f:

$$i\hbar\frac{\partial f}{\partial t} = -\frac{\hbar^2}{2}m^{-1}\nabla \cdot \nabla f + Vf \qquad \big(f(\cdot,t) = U(t)f\big). \tag{3.17}$$

This is the classic *Schrödinger equation* with potential V.

The missing ingredient here is a specification of H as a self-adjoint operator on L^2. (In the study of particles confined to a bounded region, this will involve imposition of boundary conditions, but we consider only particles moving in $\mathbb{R}^n$ here.) The operator $m^{-1}\nabla \cdot \nabla$, a variant of the Laplacian, is self-adjoint on the natural domain of all $f \in L^2$ such that $m^{-1}\nabla \cdot \nabla \in L^2$ (i.e., the Sobolev space of order 2), and V is self-adjoint on the domain of all $f \in L^2$ such that $Vf \in L^2$. Hence, H is at least a symmetric operator on the intersection of these two domains, and in most important cases it turns out to be essentially self-adjoint. There is a considerable body of theorems relating to this question. Here are a couple of them, with inessential constants omitted, in which ∇^2 denotes the Laplacian on $\mathbb{R}^n$:

i. If $V \in L^2_{\text{loc}}$ and V is bounded below, then $-\nabla^2 + V$ is essentially self-adjoint on C_c^∞.
ii. Suppose V is the sum of an L^p function and an L^∞ function, where $p = 2$ if $n \leq 3$, $p > 2$ if $n = 4$, and $p = \frac{1}{2}n$ if $n \geq 5$. Then $-\nabla^2 + V$ is essentially self-adjoint on C_c^∞.
iii. (Kato) Suppose $V = \sum_{j=1}^k V_j \circ \pi_j$, where each π_j is a partial isometry of $\mathbb{R}^n$ onto $\mathbb{R}^3$ and each V_j is the sum of an L^2 function and an L^∞ function on $\mathbb{R}^3$. Then $-\nabla^2 + V$ is essentially self-adjoint on C_c^∞.

In (iii), a "partial isometry" is a linear map that is an isometry on the orthogonal complement of its nullspace. In (ii) and (iii), $-\nabla^2 + V$ is actually self-adjoint on $\mathcal{D}(\nabla^2)$, the Sobolev space of order 2. In (ii), the essential point is that by the Sobolev imbedding theorems, $\mathcal{D}(\nabla^2)$ is contained in C_0 (continuous functions vanishing at infinity) when $n \leq 3$, in $\bigcap_{2 \leq q < \infty} L^q$ when $n = 4$, and in $L^{2n/(n-4)}$ when $n \geq 5$, so that the product of a function in $\mathcal{D}(\nabla^2)$ and an L^p function (with p as in (ii)) is in L^2. (iii) is a consequence of (ii) and a little Fourier analysis. Proofs, as well as other related theorems, can be found in Reed and Simon [**94**], Sections X.2–X.4.

The result (iii) applies, in particular, to the Hamiltonian for an atom with atomic number Z when the nucleus is considered as fixed at the origin and the moving particles are taken to be the electrons:

$$H = -\frac{\hbar^2}{2m}\nabla^2 - \sum_{j=1}^{Z} \frac{Ze^2}{4\pi|\mathbf{x}_j|} + \sum_{1 \leq j < k \leq Z} \frac{e^2}{4\pi|\mathbf{x}_j - \mathbf{x}_k|}.$$

(Here m and $-e$ are the mass and charge of an electron.) More generally, it applies to any system of N particles attracting or repelling each other by inverse-square-law forces; here $n = 3N$, each π_j is of the form $\pi_j(\mathbf{x}_1, \ldots, \mathbf{x}_m) = (\mathbf{x}_{i_1} - \mathbf{x}_{i_2})/\sqrt{2}$ for $\mathbf{x}_1, \ldots, \mathbf{x}_m \in \mathbb{R}^3$, and $V_j(\mathbf{x}) = c_j|\mathbf{x}|^{-1}$ for $\mathbf{x} \in \mathbb{R}^3$. The analysis of such Hamiltonians is complicated when N is large, and it is remarkable that interesting things can be deduced from them about assemblages of particles of macroscopic or even astronomical size. The most famous result along these lines is the "stability of matter" theorem, proved originally by Dyson and Lenard with subsequent improvements and extensions by various other people. Lieb [**77**] gives a very interesting account of this and related results.

One can also imagine physically unrealistic but conceptually reasonable situations in which essential self-adjointness fails. For example, consider a particle of mass 1 moving on a line subject to a potential $V(x) = -|x|^a$, with $a > 1$. (Thus the particle is subject to a force $(\operatorname{sgn} x)a|x|^{a-1}$ that *repels* it from the origin, and the force increases in strength along with the distance from the origin.) The Hamiltonian $H = -\frac{1}{2}(d/dx)^2 + V(x)$ is esentially self-adjoint on C_c^∞ when $a \leq 2$, but not when $a > 2$. (See Dunford and Schwartz [**24**], Corollary XIII.6.22.) When $a > 2$, H has a four-parameter family of self-adjoint extensions, which correspond to imposing "boundary conditions at $\pm\infty$." This phenomenon is actually reflected in the classical physics, where the position of the particle evolves by the differential equation $x''(t) = (\operatorname{sgn} x)a|x|^{a-1}$. This differential equation can be solved more or less explicitly, and the upshot is that the position of the particle (starting at a position $x_0 > 0$ with velocity $v_0 > 0$, say) behaves roughly like $t^{2/(2-a)}$ if $a < 2$ and like e^t if $a = 2$, but it reaches infinity in a finite time if $a > 2$. Classically, that's the end of the story; quantum mechanically, one can impose a boundary condition to tell the particle what to do when it gets to infinity, and the motion can then continue to further times.

At any rate, once the Hamiltonian has been properly defined as a self-adjoint operator, calculating the time-evolution group $e^{tH/i\hbar}$ amounts to solving the eigenvalue equation $H\phi = \lambda\phi$, $\lambda \in \mathbb{R}$, and determining how an arbitrary $f \in L^2$ can be expanded in terms of the solutions. When the spectrum of H is purely discrete, this just means finding an orthonormal basis of eigenvectors, but when H has continuous spectrum, the expansion of f will involve an integral as well as (perhaps) a sum. In §3.4 and §3.6 we shall work out two examples: the harmonic oscillator and the Coulomb potential.

3.3. Uncertainty inequalities

Again we consider the Schrödinger model for position and momentum with the state space $L^2(\mathbb{R}^n)$. Either the position or the momentum of a particle may be localized to as small a region as we please (i.e., for any nonempty open set $U \subset \mathbb{R}^n$ there are nonzero L^2 functions f such that either f or $\widehat{f}$ vanishes outside U), but there are limits on the extent to which position and momentum can be localized simultaneously; this is one aspect of the *uncertainty principle.*

If μ is a probability measure on $\mathbb{R}$, a natural measure of the "uncertainty" in μ is the standard deviation

$$\sigma(\mu) = \left[\inf_{a\in\mathbb{R}} \int (s-a)^2\, d\mu(s)\right]^{1/2}.$$

When μ is the probability distribution of an observable A in the state u (A a self-adjoint operator, u a unit vector), i.e., $\mu(E) = \langle u|P(E)|u\rangle$ where P is the projection-valued measure associated to A, we have

$$\sigma(\mu) = \left[\inf_{a\in\mathbb{R}} \int (s-a)^2\, \langle u|dP(s)|u\rangle\right]^{1/2} = \left[\inf_{a\in\mathbb{R}} \langle u|(A-a)^2|u\rangle\right]^{1/2} = \inf_{a\in\mathbb{R}} \|(A-a)u\|.$$

(When $\sigma(\mu) < \infty$, the infimum is achieved when a is the expectation value $\langle A\rangle = \langle u|A|u\rangle$.) This notion easily yields an uncertainty inequality for any two noncommuting observables (self-adjoint operators) A and B. Indeed, if u is in the domain of $[A,B]$, i.e., the set of all $u \in \mathcal{D}(A)\cap\mathcal{D}(B)$ such that $Au \in \mathcal{D}(B)$ and $Bu \in \mathcal{D}(A)$, we have

$$|\langle u|[A,B]|u\rangle| = |\langle Au|Bu\rangle - \langle Bu|Au\rangle| = 2|\operatorname{Im}\langle Au|Bu\rangle| \le 2\|Au\|\,\|Bu\|.$$

Replacing A by $A-a$ and B by $B-b$ for arbitrary $a, b \in \mathbb{R}$ (which does not affect $[A,B]$), we deduce that

$$\|(A-a)u\|\,\|(B-b)u\| \ge \tfrac{1}{2}|\langle u|[A,B]|u\rangle| \text{ for all } u \in \mathcal{D}([A,B]). \tag{3.18}$$

This result is not as strong as one might wish, because the domain of $[A,B]$ may be "too small." Indeed, if C is the closure of $[A,B]$, it is in general *false* that $\|Au\|\,\|Bu\| \ge \frac{1}{2}|\langle u, Cu\rangle|$ for all $u \in \mathcal{D}(A)\cap\mathcal{D}(B)\cap\mathcal{D}(C)$. (Counterexample: Let X and P be as in the third remark following the Stone-von Neumann theorem, and let

u be the constant function 1.) For the position and momentum operators defined by (3.12), however, we can do better.

We state the result in the one-dimensional case for simplicity. For $u \in L^2(\mathbb{R})$ with $\|u\|_2 = 1$, the standard deviations of X and P in the state u are

$$\Delta x = \left[\inf_{a\in\mathbb{R}} \int (x-a)^2 |u(x)|^2\,dx\right]^{1/2}, \qquad \Delta p = \left[\inf_{b\in\mathbb{R}} \int (p-b)^2 |\widehat{u}(p)|^2 \frac{dp}{2\pi\hbar}\right]^{1/2},$$

which make sense for *all* $u \in L^2$ with the understanding that they may equal $+\infty$, and the uncertainty inequality

$$\Delta x\,\Delta p \geq \tfrac{1}{2}\hbar \tag{3.19}$$

is valid with no restriction on u. Moreover, equality holds if and only if u is a Gaussian: $u(x) = (c/\pi)^{1/4}e^{ibx}e^{-c(x-a)^2}$ for some $a, b \in \mathbb{R}$ (namely, the points where the infimum is achieved in the formulas for Δx and Δp) and some $c > 0$. We refer to Folland and Sitaram [**49**] (where $\hbar$ is taken to be $1/2\pi$) for the proof, the generalizations to n dimensions, and a large assortment of related results.

There is another important uncertainty inequality that is of a somewhat different nature: that relating to time and energy, which is usually stated in the simple form $\Delta t\,\Delta E \geq \frac{1}{2}\hbar$. That there should be such a relation is strongly suggested by the classical theory of relativity, in which energy is the time component of a 4-vector of which the momenta are the space components. However, unlike position and momentum, time and energy are not observables that satisfy the canonical commutation relations. Indeed, there is no observable T that is canonically conjugate to the Hamiltonian (energy) H, for if there were, by the Stone-von Neumann theorem the spectrum of T and H would have to be the whole real line; but in any reasonable quantum system the energy must be bounded below.

To make sense of the time-energy uncertainty inequality, one can proceed as follows. Let H be the Hamiltonian of the system under consideration, and let $A = A(t)$ be a Heisenberg-picture observable; thus, $dA/dt = (1/i\hbar)[A, H]$. For any state $|u\rangle$ of the system, let $\langle A\rangle = \langle u|A|u\rangle$ and $\Delta A = \|(A - \langle A\rangle)u\|$ be the expectation and standard deviation of A (assumed finite) in the state $|u\rangle$, and let ΔE be the standard deviation of H in $|u\rangle$. By (3.18), we have

$$(\Delta A)(\Delta E) \geq \tfrac{1}{2}|\langle [A, H]\rangle| = \tfrac{1}{2}\hbar|\langle dA/dt\rangle| = \tfrac{1}{2}\hbar|d\langle A\rangle/dt|.$$

Now, the quantity

$$\Delta t_A = \left|\frac{\Delta A}{d\langle A\rangle/dt}\right|$$

represents an amount of time: it is the time required for the expectation of A to change by an amount equal to the standard deviation, and hence the time required for the statistics of A to change appreciably. The preceding inequality thus says that

$$(\Delta t)(\Delta E) \geq \tfrac{1}{2}\hbar, \quad \text{where} \quad \Delta t = \inf_A \Delta t_A. \tag{3.20}$$

This is the appropriate interpretation of the time-energy uncertainty inequality.

3.4. The harmonic oscillator

Harmonic oscillators are interesting and important in their own right, and as we shall see, they play a fundamental role in the modeling of quantum fields. The n-dimensional harmonic oscillator Hamiltonian, with a single scalar mass m, is

$$H = -(\hbar^2/2m)\nabla^2 + V$$

where V is a positive definite quadratic form on $\mathbb{R}^n$. By a suitable rotation of coordinates, one can assume that V is diagonal, in which case H decouples into a sum of one-dimensional operators. It therefore suffices to consider the one-dimensional case, as the n-dimensional eigenfunctions are just products of one-dimensional ones. For a particle of mass m in the potential $V(x) = \frac{1}{2}\kappa x^2$, the Hamiltonian is

$$H = -\frac{\hbar^2}{2m}\frac{d^2}{dx^2} + \frac{\kappa}{2}x^2. \tag{3.21}$$

We can get rid of the constants by rescaling: setting $Sf(x) = f((\hbar^2/m\kappa)^{1/4}x)$, a simple calculation shows that

$$SHS^{-1} = \frac{\hbar}{2\sqrt{m\kappa}}\left(-\frac{d^2}{dx^2} + x^2\right).$$

The factor of $\hbar/\sqrt{m\kappa}$ just comes along for the ride, but it is convenient to keep the factor of $\frac{1}{2}$ explicitly. Thus, we consider the rescaled Hamiltonian

$$H_0 = \frac{1}{2}\left(-\frac{d^2}{dx^2} + x^2\right). \tag{3.22}$$

To analyze this operator, we introduce the operators

$$A = \frac{1}{\sqrt{2}}\left(x + \frac{d}{dx}\right), \qquad A^\dagger = \frac{1}{\sqrt{2}}\left(x - \frac{d}{dx}\right). \tag{3.23}$$

(There is no need to worry about domains right now; all these operators can be considered as acting on the Schwartz space $\mathcal{S}(\mathbb{R})$. Observe that the notation is formally correct, as $\langle f|A^\dagger|g\rangle = \langle g|A|f\rangle^*$ for $f, g \in \mathcal{S}$.) Simple calculations yield the following formulas:

$$H_0 = AA^\dagger - \tfrac{1}{2}I = A^\dagger A + \tfrac{1}{2}I, \tag{3.24}$$

$$[A, A^\dagger] = I, \tag{3.25}$$

and more generally, by (3.25) and induction on k,

$$[A, (A^\dagger)^k] = k(A^\dagger)^{k-1}. \tag{3.26}$$

Let

$$\phi_0(x) = \pi^{-1/4}e^{-x^2/2}, \qquad \phi_k = \frac{1}{\sqrt{k!}}(A^\dagger)^k\phi_0 \text{ for } k = 1, 2, 3, \ldots \tag{3.27}$$

It is easy to verify that $\phi_k(x) = P_k(x)e^{-x^2/2}$ where P_k is a polynomial of degree k, odd or even according as k is odd or even. It is called the kth *normalized Hermite polynomial,* and ϕ_k is the kth *normalized Hermite function.* (The usual Hermite polynomials are $H_k = \pi^{1/4}\sqrt{2^k k!}\, P_k$.) We have

$$A^\dagger\phi_k = \sqrt{k+1}\,\phi_{k+1}, \tag{3.28}$$

and by (3.26) and the fact that $A\phi_0 = 0$,

$$A\phi_k = \frac{1}{\sqrt{k!}}\big([A, (A^\dagger)^k] + (A^\dagger)^k A\big)\phi_0 = \frac{k}{\sqrt{k!}}(A^\dagger)^{k-1}\phi_0 = \sqrt{k}\,\phi_{k-1}. \tag{3.29}$$

It follows that

$$A^\dagger A\phi_k = k\phi_k, \qquad AA^\dagger\phi_k = (k+1)\phi_k, \tag{3.30}$$

and hence, by (3.24),

$$H_0\phi_k = (k + \tfrac{1}{2})\phi_k. \tag{3.31}$$

We claim that $\{\phi_k\}_0^\infty$ is an orthonormal basis for L^2. We first observe that $\|\phi_0\|_2 = 1$ (that's why the $\pi^{-1/4}$ is there) and that $A\phi_0 = 0$. For $k > 0$, we have

$$\langle\phi_0|\phi_k\rangle = \frac{1}{\sqrt{k!}}\langle\phi_0|(A^\dagger)^k\phi_0\rangle = \frac{1}{\sqrt{k!}}\langle A\phi_0|(A^\dagger)^{k-1}\phi_0\rangle = 0.$$

Moreover, if $k \geq l > 0$, by (3.30) we have

$$\langle\phi_l|\phi_k\rangle = \frac{1}{\sqrt{kl}}\langle A^\dagger\phi_{l-1}|A^\dagger\phi_{k-1}\rangle = \frac{1}{\sqrt{kl}}\langle AA^\dagger\phi_l|\phi_k\rangle = \sqrt{\frac{k}{l}}\langle\phi_{l-1}|\phi_{k-1}\rangle.$$

By induction and the preceding calculation, it follows that

$$\langle\phi_l|\phi_k\rangle = \sqrt{\frac{k!}{l!}}\langle\phi_0|\phi_{k-l}\rangle = \sqrt{\frac{k!}{l!}}\delta_{k-l,0} = \delta_{kl},$$

and orthonormality is proved.

Finally, suppose $\langle\phi_k|f\rangle = 0$ for all k. It follows that f is orthogonal to $p(x)e^{-x^2/2}$ for every polynomial p, and hence

$$\int f(x)e^{i\xi x}e^{-x^2/2}\,dx = \sum_{k=0}^\infty \int f(x)\frac{(i\xi x)^k}{k!}e^{-x^2/2}\,dx = 0$$

for every ξ. (The interchange of summation and integration is justified since both f and $e^{|\xi x|}e^{-x^2/2}$ are in L^2, so their product is in L^1.) By Fourier uniqueness we conclude that $f(x)e^{-x^2/2} = 0$ for a.e. x, and hence $f = 0$ in L^2.

In short, we have shown that *$\{\phi_k\}_0^\infty$ is an orthonormal eigenbasis for H_0*. If we put the constants $\hbar$, m, and κ back in, we see that the Hamiltonian H given by (3.21) has eigenfunctions $\phi_k((m\kappa/\hbar^2)^{1/4}x)$, with eigenvalues $\hbar(k+\frac{1}{2})/\sqrt{m/\kappa}$.

Let us pause to see what this means for a *macroscopic* harmonic oscillator such as a weight hanging from a perfectly elastic spring, where m and κ are of the order of magnitude of unity in MKS units. The classical equation of motion is $mx'' = -\kappa x$, whose solutions are trig functions of $\sqrt{\kappa/m}\,t$. To be specific, let us assume the weight starts at rest at position $x = a$ at time $t = 0$; then the classical solution is $x(t) = a\cos(\sqrt{\kappa/m}\,t)$, and its energy and momentum are $\frac{1}{2}\kappa a^2$ and $mx'(t) = -a\sqrt{m\kappa}\sin(\sqrt{\kappa/m}\,t)$. Now, $\hbar \approx 10^{-34}$ in MKS units, so the quantum energy levels are extremely closely spaced, and there is no trouble finding one that is equal to $\frac{1}{2}\kappa a^2$ to reasonable accuracy. However, the quantum states of definite energy are *stationary* — if $H\phi = \lambda\phi$ then $e^{tH/i\hbar}\phi = e^{t\lambda/i\hbar}\phi$, which represents the same state as ϕ — and there is no sign of the classical oscillatory motion. What is going on here?

The trouble is that the classical initial value problem, in which the position and momentum at time $t = 0$ are specified exactly, makes no sense quantum mechanically because of the uncertainty principle. The best one can do is to specify an initial state in which the position and momentum are as well localized as the uncertainty principle permits; this will be a superposition of energy eigenstates, and the classical motion effectively results from interference between the oscillations $e^{t\lambda/i\hbar}$ for the different eigenvalues λ. In fact, if the initial state is a suitable uncertainty-minimizing Gaussian as described following (3.19), the quantum evolution turns out to be exactly the classical periodic motion. This little miracle is sufficiently entertaining that it is worth taking some space to work out in detail. As above, we work with the rescaled Hamiltonian H_0 and its eigenfunctions.

The essential ingredient in the calculation is a version of a classical generating function identity for Hermite polynomials. We begin by observing that

$$A^\dagger f(x) = \frac{1}{\sqrt{2}}[xf(x) - f'(x)] = -\frac{1}{\sqrt{2}}e^{x^2/2}\frac{d}{dx}[e^{-x^2/2}f(x)],$$

so that

$$(A^\dagger)^k f(x) = \frac{(-1)^k}{2^{k/2}}e^{x^2/2}\frac{d^k}{dx^k}[e^{-x^2/2}f(x)],$$

and hence

$$\begin{aligned}\phi_k(x) = \frac{1}{\pi^{1/4}\sqrt{k!}}(A^\dagger)^k e^{-x^2/2} &= \frac{(-1)^k}{\pi^{1/4}\sqrt{2^k k!}}e^{x^2/2}\frac{d^k}{dx^k}e^{-x^2} \\ &= \frac{1}{\pi^{1/4}\sqrt{2^k k!}}e^{x^2/2}\frac{d^k}{dw^k}e^{-(x-w)^2}\bigg|_{w=0}.\end{aligned}$$

Therefore, by Taylor's theorem, for any $w \in \mathbb{C}$ we have

$$\pi^{-1/4}e^{-(x-w)^2} = e^{-x^2/2}\sum_0^\infty \sqrt{\frac{2^k}{k!}}\phi_k(x)w^k.$$

Setting $w = \frac{1}{2}z$, so that $e^{-(x-w)^2} = e^{-x^2/2}e^{-(x-z)^2/2}e^{z^2/4}$, we obtain the result we want: For any $z \in \mathbb{C}$,

$$\pi^{-1/4}e^{-(x-z)^2/2} = e^{-z^2/4}\sum_0^\infty \frac{1}{\sqrt{2^k k!}}\phi_k(x)z^k. \tag{3.32}$$

Now, for $a, b \in \mathbb{R}$, let

$$\psi_{a,b}(x) = \pi^{-1/4}e^{-(x-a)^2/2}e^{ibx}.$$

Then the expected values of position and momentum in the state $\psi_{a,b}$ are a and b, respectively. We take $\psi_{a,0}$ as the initial state for the quantum oscillator corresponding to the classical initial values $x_0 = a$ and $p_0 = 0$. The expansion of $\psi_{a,0}$ in terms of the energy eigenfunctions is obtained by taking $z = a$ in (3.32). It is now a simple matter to compute its time evolution, using (3.31) and (3.32):

$$
\begin{aligned}
e^{-itH_0}\psi_{a,0}(x) &= e^{-a^2/4}\sum_0^\infty \frac{e^{-it(k+\frac{1}{2})}}{\sqrt{2^k k!}}\phi_k(x)a^k \\
&= e^{-a^2/4}e^{-it/2}\sum_0^\infty \frac{1}{\sqrt{2^k k!}}\phi_k(x)(ae^{-it})^k \\
&= e^{-a^2/4}e^{(ae^{-it})^2/4}e^{-it/2}\pi^{-1/4}e^{-(x-ae^{-it})^2/2},
\end{aligned}
$$

which simplifies to

$$
e^{i(a^2\sin t\cos t - t)/2}\pi^{-1/4}e^{-(x-a\cos t)^2/2}e^{-ixa\sin t} = e^{i(a^2\sin t\cos t-t)/2}\psi_{a\cos t,-a\sin t}(x).
$$

The factor $e^{i(a^2\sin t\cos t-t)/2}$ is just a scalar of modulus 1, so $e^{-itH_0}\psi_{a,0}$ is a Gaussian wave packet with expected position $a\cos t$ and momentum $-a\sin t$: the classical periodic motion! (Putting the constants $\hbar$, m, and κ back in changes these to $a\cos\sqrt{\kappa/m}\,t$ and $-a\sqrt{m\kappa}\sin\sqrt{\kappa/m}\,t$ as it should.) The states $\psi_{a,0}$, or more generally $\psi_{a,b}$, are called *coherent states* because of the way they maintain their shape under the time evolution.

Of course this exact quantum-classical correspondence is a highly unusual phenomenon, but one can show that the evolution of Gaussian wave packets under much more general potentials exhibits similar behavior in an approximate sense when $\hbar$ is taken sufficiently small; see Hagedorn [**60**].

To complete the picture, let us calculate the expected energy of the state $\psi_{a,0}$. Using the expansion of $\psi_{a,0}$ in terms of energy eigenfunctions given (3.32), we have

$$
\langle\psi_{a,0}|H_0|\psi_{a,0}\rangle = e^{-a^2/2}\sum_0^\infty (k+\tfrac{1}{2})\frac{a^{2k}}{2^k k!} = \frac{e^{-a^2/2}}{2}\frac{d}{da}\left[ae^{a^2/2}\right] = \frac{1}{2}(a^2+1).
$$

If the constants $\hbar$, m, and κ are reinserted appropriately, the expected energy turns out to be $\frac{1}{2}(\kappa a^2 + \hbar\sqrt{\kappa/m})$. The term $\frac{1}{2}\kappa a^2$ is the classical energy; the extra $\frac{1}{2}\hbar\sqrt{\kappa/m}$ is the ground state energy. (In the classical ground state, $x(t)\equiv 0$, the particle is absolutely at rest, but quantum mechanically this is not quite possible.)

3.5. Angular momentum and spin

Classically, the angular momentum (about the origin) of a particle in $\mathbb{R}^3$ with position $\mathbf{x}$ and momentum $\mathbf{p}$ is $\mathbf{l} = \mathbf{x}\times\mathbf{p}$ (cross product). The Hamiltonian vector fields corresponding to the three components of this vector observable are the infinitesimal generators of rotations about the three coordinate axes. In more detail, the first component of $\mathbf{l}$ is $l_1 = x_2p_3 - x_3p_2$, and its Hamiltonian vector field is

$$
X_{l_1} = x_2\frac{\partial}{\partial x_3} - x_3\frac{\partial}{\partial x_2} + p_2\frac{\partial}{\partial p_3} - p_3\frac{\partial}{\partial p_2},
$$

which generates the flow in phase space given by a rotation about the x_1-axis in position space and the same rotation about the p_1-axis in momentum space. Similarly for the other two components.

Quantization of $\mathbf{l}$ in the obvious way leads to the vector-valued quantum observable $\mathbf{L} = \mathbf{X}\times\mathbf{P}$, that is,

$$
L_1 = X_2P_3 - X_3P_2 = \frac{\hbar}{i}\left(x_2\frac{\partial}{\partial x_3} - x_3\frac{\partial}{\partial x_2}\right), \text{ etc.,}
$$

acting on $L^2(\mathbb{R}^3)$. (Note that there is no problem about ordering the factors here, because X_i and P_j commute when $i \neq j$.) The one-parameter groups generated by L_1, L_2, and L_3 are again the rotations about the coordinate axes. That is,

$$e^{tL_1/i\hbar}f(\mathbf{x}) = f(R_{t,1}^{-1}\mathbf{x}), \text{ where } R_{t,1} = \begin{pmatrix} 1 & 0 & 0 \\ 0 & \cos t & \sin t \\ 0 & -\sin t & \cos t \end{pmatrix},$$

and likewise, if $\mathbf{u}$ is any unit vector in $\mathbb{R}^3$, $e^{t\mathbf{u}\cdot\mathbf{L}/i\hbar}f(\mathbf{x}) = f(R_{t,\mathbf{u}}^{-1}\mathbf{x})$, where $R_{t,\mathbf{u}}$ is rotation through the angle t about the $\mathbf{u}$ axis.

In group-theoretic terms: The rotation group $SO(3)$ acts on $L^2(\mathbb{R}^3)$ by

$$[\rho(R)f](\mathbf{x}) = f(R^{-1}\mathbf{x}).$$

Its Lie algebra $\mathfrak{so}(3)$ can be identified with $\mathbb{R}^3$ with Lie bracket = cross product, the unit vector $\mathbf{u} \in \mathbb{R}^3$ corresponding to the generator of rotations about the $\mathbf{u}$ axis. The angular momentum operators are the range of the infinitesimal representation $d\rho$ of $\mathbb{R}^3 \cong \mathfrak{so}(3)$, that is, the angular momentum about the $\mathbf{u}$ axis is the operator

$$u_1L_1 + u_2L_2 + u_3L_3 = i\hbar\, d\rho(\mathbf{u}) = \frac{d}{dt}\rho(R_{t,\mathbf{u}})\big|_{t=0}.$$

The observables L_j are important in quantum mechanics, but they do not tell the whole story about angular momentum: most quantum particles[2] have an additional *intrinsic* angular momentum that has nothing to do with their state of motion, known as *spin*. Spin is a purely quantum phenomenon with no classical analogue, so there is no point in trying to motivate the way it works; we simply state the result. This discussion pertains to nonrelativistic particles of mass $m > 0$; we shall fit it into the relativistic framework in the next chapter, where we shall also discuss massless particles.

Until now we have been taking the state space for a quantum particle in $\mathbb{R}^3$ to be $L^2(\mathbb{R}^3)$. However, this usually does not provide a complete description of the state of the particle; there are additional observables that do not fit into this picture. Instead, the state space must be taken as $L^2(\mathbb{R}^3)\otimes\mathcal{H}$, where $\mathcal{H}$ is another Hilbert space that describes the "internal degrees of freedom" of the particle. We assume given a unitary (possibly projective) representation of the rotation group $SO(3)$ on this state space that describes how the state of the particle changes under spatial rotations. The (total) angular momentum operators will then be taken to be the image of $\mathfrak{so}(3)$ under the corresponding infintesimal representation.

Since we are interested only in angular momentum at this point, and not in other possible internal features of the particles, we take $\mathcal{H}$ to be the smallest Hilbert space that will encompass this phenomenon. Namely, $\mathcal{H}$ will be a finite-dimensional space equipped with an irreducible (possibly projective) representation π of $SO(3)$, and the action of $SO(3)$ on $L^2(\mathbb{R}^3) \otimes \mathcal{H}$ will be $\rho \otimes \pi$ where ρ is the natural representation on $L^2(\mathbb{R}^3)$ discussed above. We therefore need some facts about representations of $SO(3)$ and its universal cover $SU(2)$. We refer the reader back to §1.3 for the details of the relation between $SO(3)$ and $SU(2)$, and in particular for the calculation of the covering map κ.

Up to unitary equivalence, $SU(2)$ has exactly one irreducible unitary representation on $\mathbb{C}^n$ for each positive integer n; it may be realized as the action of $SU(2)$ on the homogeneous (holomorphic) polynomials of degree $n-1$ on $\mathbb{C}^2$. (See

[2]Particles can be elementary or composite here.

Folland [44] or Hall [61] for a proof.) For physical purposes it is customary to set $n = 2k+1$, where k is an integer or half integer; the irreducible representation on $\mathbb{C}^{2k+1}$ is denoted by π_k. π_0 is the trivial representation; $\pi_{1/2}$ is the defining representation of $SU(2)$ on $\mathbb{C}^2$, and π_1 is (equivalent to) the adjoint representation on the complexification of $\mathfrak{su}(2)$.

We have $\pi_k(-I) = (-1)^{2k}I$, so π_k descends to a representation of $SO(3) \cong SU(2)/\{\pm I\}$ precisely when k is an integer, and in any case it defines a projective representation of $SO(3)$. We shall denote these representations also by π_k. Thus, if $R \in SO(3)$, by $\pi_k(R)$ we really mean $\pi_k(\kappa^{-1}(R))$ where κ is the covering map defined in §1.3. This is defined only up to a factor of $\pm I$ when k is a half-integer; we say that π_k is a *double-valued* representation. The representations π_k exhaust the irreducible projective representations of $SO(3)$, and those with k integral exhaust the irreducible unitary ones, up to equivalence.

Let $Y_1, Y_2, Y_3 \in \mathfrak{so}(3)$ be the infinitesimal generators of the rotations about the three coordinate axes; thus $Y_j = \kappa(\frac{1}{2}i\sigma_j)$ in the notation of §1.3. Let π be *any* unitary representation of $SU(2)$ and $d\pi$ the corresponding representation of $\mathfrak{so}(3) \cong \mathfrak{su}(2)$. The operators

$$J_j = \frac{\hbar}{i}d\pi(Y_j) = \tfrac{1}{2}\hbar\, d\pi(\sigma_j) \qquad j = 1,2,3,$$

are the three components of the *angular momentum* $\mathbf{J}$ in the representation π. Further, let $\mathbf{Y}^2 = Y_1^2 + Y_2^2 + Y_3^2$, an element of the universal enveloping algebra of $\mathfrak{so}(3)$. The operator

$$\mathbf{J}^2 = J_1^2 + J_2^2 + J_3^2 = -\hbar^2 d\pi(\mathbf{Y}^2),$$

the *squared total angular momentum*, plays an important role. As is well known, $\mathbf{Y}^2$ is the so-called *Casimir element* of the universal enveloping algebra of $\mathfrak{so}(3)$, and it belongs to (in fact, generates) the center of that algebra. It follows that $\mathbf{J}^2$ commutes with the representation π; therefore, by Schur's lemma, it acts as a scalar on each irreducible subspace of the representation space of π.

Let us consider the angular momentum operators for the irreducible representations π_k, with units of measurement chosen so that $\hbar = 1$. Each operator J_j has the $2k+1$ distinct eigenvalues $-k, -k+1, \dots, k-1, k$. It is enough to verify this for J_3, as the J_j are all unitarily equivalent (rotations about all axes look the same). If we realize π_k as the action of $SU(2)$ on the homogeneous polynomials of degree $2k$ on $\mathbb{C}^2$, we have

$$\pi_k(e^{it\sigma_3/2})(z_1^m z_2^{2k-m}) = (e^{it/2}z_1)^m(e^{-it/2}z_2)^{2k-m} = e^{i(m-k)}z_1^m z_2^{2k-m},$$

and so

$$J_3(z_1^m z_2^{2k-m}) = \frac{1}{i}\frac{d}{dt}\pi_k(e^{it\sigma_3/2})(z_1^m z_2^{2k-m})\big|_{t=0} = (m-k)z_1^m z_2^{2k-m}.$$

Thus the eigenvalues of J_3 are $\{k-m : m = 0, \dots, 2k\}$, as advertised. Furthermore, since π_k is irreducible, $\mathbf{J}^2$ is a scalar multiple of the identity. To find the scalar, it suffices to compute the action of $\mathbf{J}^2$ on one vector, for example, the monomial z_1^{2k}:

$$\mathbf{J}^2(z_1^{2k}) = -\frac{d^2}{dt^2}\big[(e^{it/2}z_1)^{2k} + (z_1\cos\tfrac{1}{2}t + iz_2\sin\tfrac{1}{2}t)^{2k} + (z_1\cos\tfrac{1}{2}t - z_2\sin\tfrac{1}{2}t)^{2k}\big]_{t=0},$$

which the reader may verify to be $k(k+1)z_1^{2k}$; thus $\mathbf{J}^2 = k(k+1)I$.

Now, a quantum particle whose state space is $L^2(\mathbb{R}^3) \otimes \mathbb{C}^{2k+1}$, equipped with the representation $\rho \otimes \pi_k$ where ρ is the natural representation of $SO(3)$ on $L^2(\mathbb{R}^3)$, is said to have *spin* k. We can think of $L^2(\mathbb{R}^3)\otimes\mathbb{C}^{2k+1}$ as $L^2(\mathbb{R}^3,\mathbb{C}^{2k+1})$, the space of square-integrable $\mathbb{C}^{2k+1}$-valued functions on $\mathbb{R}^3$, and the representation $\rho\otimes\pi_k$ is then given by

$$[(\rho\otimes\pi_k)(R)f](\mathbf{x}) = \pi_k(R)[f(R^{-1}\mathbf{x})].$$

The infinitesimal representation of $\rho\otimes\pi_k$ is $d\rho\otimes I + I\otimes d\pi_k$, so the angular momentum operators are

$$\mathbf{J} = \mathbf{L}+\mathbf{S}, \text{ where } \mathbf{L} = \frac{\hbar}{i}d\rho(\mathbf{Y}) \text{ and } \mathbf{S} = \frac{\hbar}{i}d\pi_k(\mathbf{Y}).$$

$\mathbf{L}$ is just the angular momentum discussed at the beginning of this section, now called the *orbital angular momentum*, acting *componentwise* on the vector-valued functions in $L^2(\mathbb{R}^3,\mathbb{C}^{2k+1})$. $\mathbf{S}$ is the *spin angular momentum*, acting *on the components* of the functions.

We briefly describe the decomposition of the representation ρ into irreducible components; see Folland [**45**] for more details. The basic ingredients are the spaces of *spherical harmonics* $\mathcal{SH}_l$, $l = 0, 1, 2, \ldots$, whose elements are functions of the form $Y(\mathbf{x}) = P(\mathbf{x})|\mathbf{x}|^{-l}$ where P is a homogeneous harmonic polynomial of degree l. (Note that such functions are homogeneous of degree 0, i.e., are effectively functions of $\mathbf{x}/|\mathbf{x}|$.) The spaces $\mathcal{SH}_l$ are invariant under the natural action of the rotation group, and they are irreducible; the representation of $SO(3)$ on $\mathcal{SH}_l$ is equivalent to π_l. For each l, let

$$\mathcal{H}_l = \{f\in L^2(\mathbb{R}^3) : f(\mathbf{x}) = g(|\mathbf{x}|)Y(\mathbf{x}) : Y\in\mathcal{SH}_l\}. \tag{3.33}$$

The condition that f be in $L^2(\mathbb{R}^3)$ is equivalent to the condition that g be in $L^2((0,\infty), r^2\,dr)$. The rotation group acts trivially on the factor g and by the representation π_l on the factor Y, so a choice of orthonormal basis for $L^2((0,\infty), r^2\,dr)$ exhibits $\mathcal{H}_l$ as a direct sum of spaces isomorphic to $\mathcal{SH}_l$, and the representation $\rho|\mathcal{H}_l$ as a direct sum of copies of π_l. We have

$$L^2(\mathbb{R}^3) = \bigoplus_0^\infty \mathcal{H}_l,$$

yielding the decomposition of ρ into $SO(3)$-isotypic components. Observe that this can also be construed as the spectral decomposition of the squared orbital angular momentum operator $\mathbf{L}^2$: the space $\mathcal{H}_l$ is the eigenspace of $\mathbf{L}^2$ with eigenvalue $l(l+1)$.

From this one can obtain the irreducible decomposition of $\rho\otimes\pi_k$ for any k. The preceding analysis yields

$$L^2(\mathbb{R}^3,\mathbb{C}^{2k+1}) = \bigoplus_0^\infty \mathcal{H}_l\otimes\mathbb{C}^{2k+1},$$

and the representation of $SO(3)$ on $\mathcal{H}_l\otimes\mathbb{C}^{2k+1}$ is a direct sum of copies of $\pi_l\otimes\pi_k$. What remains, therefore, is to decompose $\pi_l\otimes\pi_k$ into its irreducible components. This problem is well understood, but the details of its solution will not concern us here. (See Landau and Lifshitz [**76**]; the key-words to look for are "Clebsch-Gordan coefficients.") For us it will suffice to describe the most important case, $k=\frac{1}{2}$.

Of course π_0 is the trivial one-dimensional representation, so $\pi_0\otimes\pi_{1/2}\cong\pi_{1/2}$. For $l>0$, consider an irreducible subspace of $\mathcal{H}_l$, say

$$\mathcal{X} = \{f : f(\mathbf{x}) = g(|\mathbf{x}|)Y(\mathbf{x}),\ Y\in\mathcal{SH}_l\}$$

for a fixed g, and consider the action of the z component of the total angular momentum, $J_3 = L_3 + S_3$, on $\mathcal{X} \otimes \mathbb{C}^2$. The operators L_3 and S_3, considered as acting on $\mathcal{X}$ and $\mathbb{C}^2$ respectively, have orthonormal eigenbases $\{\phi_m\}_{m=0}^{2l}$ and $\{v_+, v_-\}$ with $L_3\phi_m = (l-m)\phi_m$, $S_3 v_\pm = \pm\frac{1}{2}$. Then $\{\phi_m \otimes v_a : m = 0, \ldots, 2l;\ a = \pm\}$ is an orthonormal eigenbasis for J_3 with $J_3(\phi_m \otimes v_\pm) = l - m \pm \frac{1}{2}$. The eigenvalues $\pm(l + \frac{1}{2})$ occur with multiplicity one and the others occur with multiplicity two. The largest of them is $l + \frac{1}{2}$, which occurs only for the eigenvector $\phi_0 \otimes v_+$; hence the representation of $SO(3)$ on the invariant subspace generated by this vector is a copy of $\pi_{l+(1/2)}$. This accounts for one copy of each of the eigenvalues $l - m \pm \frac{1}{2}$. Its orthogonal complement in $\mathcal{X} \otimes \mathbb{C}^2$ is an $SO(3)$-invariant space on which J_3 has eigenvalues $l - \frac{1}{2}, l - \frac{3}{2}, \ldots, -l + \frac{1}{2}$, each with multiplicity one; the representation on it is therefore a copy of $\pi_{l-(1/2)}$. In short,

$$\pi_l \otimes \pi_{1/2} \cong \pi_{l+(1/2)} \oplus \pi_{l-(1/2)} \text{ for } l > 0, \qquad \pi_0 \otimes \pi_{1/2} \cong \pi_{1/2}. \tag{3.34}$$

3.6. The Coulomb potential

We now perform a spectral analysis of the Hamiltonian

$$H = \frac{\hbar^2}{2m}\nabla^2 - \frac{a}{|\mathbf{x}|}$$

for a particle of mass m moving in the potential $V(\mathbf{x}) = -a/|\mathbf{x}|$ ($a > 0$) corresponding to an attractive inverse-square-law force. The operator H acts on $L^2(\mathbb{R}^3, \mathbb{C}^{2k+1})$ if the particle has spin k, but the spin variables just come along for the ride here, so there is no loss of generality in taking $k = 0$. Our main task is to find and analyze the solutions of the differential equation $Hf = Ef$ for $E \in \mathbb{R}$ (E for energy). We sketch the main points but omit many of the details of the calculations. A more complete treatment can be found in Landau and Lifshitz [**76**].

The basic example is an electron with mass m_e and charge $-e$ moving in the Coulomb potential generated by an atomic nucleus with mass m_N and charge Ze where Z is the number of protons in the nucleus, so that $a = Ze^2/4\pi$ in Heaviside-Lorentz units. It is a pretty good approximation to pretend that $m_N = \infty$, so that the nucleus does not move; however, this two-body problem, just like the classical one discussed at the end of §2.1, is mathematically exactly equivalent to a single particle moving in the fixed potential $-a/|\mathbf{x}|$ if one takes its mass m to be the reduced mass $m_e m_N/(m_e + m_N)$.

First, we get rid of the constants by rescaling. The constant a has dimensions $[ml^3t^{-2}]$ since $a/|\mathbf{x}|$ represents an energy, so $\hbar^2/ma$ has dimensions of length and $ma^2/\hbar^2$ has dimensions of energy. Setting $\mathbf{x} = (\hbar^2/ma)\widetilde{\mathbf{x}}$ (so $\widetilde{\mathbf{x}}$ is dimensionless) makes

$$H = \frac{ma^2}{2\hbar^2}\left[-\nabla^2_{\widetilde{\mathbf{x}}} - \frac{2}{|\widetilde{\mathbf{x}}|}\right].$$

We now drop the tildes and choose the unit of energy to be $ma^2/2\hbar^2$, so

$$H = -\nabla^2 - \frac{2}{|\mathbf{x}|}.$$

(For a hydrogen atom consisting of one electron and one proton, with m taken to be the reduced mass $m_e m_p/(m_e + m_p)$, the resulting units of length and energy are $\hbar^2/ma = 5.29 \times 10^{-9}$ cm and $ma^2/2\hbar^2 = 13.6$ eV. The latter is known as the

Rydberg energy ; as we shall see, it is the absolute value of the energy of the electron in the ground state.)

Now, H is invariant under rotations, so it respects the decomposition $L^2(\mathbb{R}^3) = \bigoplus_0^\infty \mathcal{H}_l$, where $\mathcal{H}_l$ is as in (3.33). Moreover, the Laplacian is given in spherical coordinates by

$$\nabla^2 = \frac{d^2}{dr^2} + \frac{2}{r}\frac{d}{dr} - \frac{1}{r^2}\mathbf{L}^2 \qquad (r = |\mathbf{x}|),$$

where $\mathbf{L}^2$ is our friend the squared orbital angular momentum operator. On $\mathcal{H}_l$ we have $\mathbf{L}^2 = l(l+1)I$, so the restriction of H to $\mathcal{H}_l$ is given by

$$H(g(r)Y_l(\mathbf{x})) = \left[-g''(r) - \frac{2}{r}g'(r) + \frac{l(l+1)}{r^2}g(r) - \frac{2}{r}g(r)\right]Y_l(\mathbf{x}).$$

Hence the partial differential equation $Hf = Ef$ boils down to the ordinary differential equations

$$g''(r) + \frac{2}{r}g'(r) + \left(\frac{2}{r} - \frac{l(l+1)}{r^2} + E\right)g(r) = 0 \qquad (l = 0, 1, 2, \ldots). \tag{3.35}$$

As far as the discrete spectrum of H (i.e., the stationary states of the quantum system) goes, we are interested in solutions g of (3.35) such that the function $f(\mathbf{x}) = g(|\mathbf{x}|)Y_l(\mathbf{x})$ belongs to $L^2(\mathbb{R}^3)$; this requires that g be well-behaved at 0 and ∞. Moreover, we expect that the stationary states will occur when the particle is "trapped in the potential well," i.e., for energies $E < 0$. Thus we take the condition $E < 0$ as a working assumption for the time being.

The equation (3.35) has a regular singular point at the origin, where its characteristic exponents are l and $-l-1$; thus it has two solutions that are asymptotic to r^l and r^{-l-1} as $r \to 0$. The second solution must be discarded as the resulting function on $\mathbb{R}^3$ is not square-integrable.[3] Moreover, for large r the equation (3.35) is approximately $g''(r) + Eg(r) = 0$, so we expect to find solutions that are asymptotic to $e^{\pm\sqrt{-E}\,r}$ as $r \to \infty$, and we are only interested in the one with the negative exponent. (Remember that we are assuming $E < 0$.)

Motivated by these considerations, we make some changes of variables. First, to standardize the behavior at infinity (and, frankly, with some foreknowledge of the final result), we set

$$\nu = \frac{1}{\sqrt{-E}}, \qquad s = \frac{2r}{\nu}, \qquad \widetilde{g}(s) = g(r) = g(\nu s/2).$$

This turns (3.35) into

$$\widetilde{g}''(s) + \frac{2}{s}\widetilde{g}'(s) + \left(\frac{\nu}{s} - \frac{l(l+1)}{s^2} - \frac{1}{4}\right)\widetilde{g}(s) = 0. \tag{3.36}$$

The solutions are now expected to be asymptotic to $e^{\pm s/2}$ as $s \to \infty$ and to s^l as $s \to 0$. This suggests the further substitution

$$\widetilde{g}(s) = s^l e^{-s/2}h(s),$$

which turns (3.36) into

$$sh''(s) + (2l + 2 - s)h'(s) + (\nu - l - 1)h(s) = 0. \tag{3.37}$$

[3]When $l = 0$ this argument is insufficient, as the function $f(\mathbf{x}) = |\mathbf{x}|^{-1}$ is in $L^2(\mathbb{R}^3)$. However, it is not in the domain of H, viz., the Sobolev space of order 2. Indeed, $\nabla^2|\mathbf{x}|^{-1} = -4\pi\delta(\mathbf{x})$.

This, finally, is something in standard form: it is the confluent hypergeometric equation with parameters $a = l + 1 - \nu$ and $c = 2l + 2$, and the solutions that are regular at the origin are constant multiples of

$$F(l+1-\nu,\, 2l+2;\, s) = 1 + \sum_{k=1}^{\infty} \frac{(l+1-\nu)(l+2-\nu)\cdots(l+k-\nu)}{(2l+2)(2l+3)\cdots(2l+k+1)} \frac{s^k}{k!}$$

(The other solutions are asymptotic to multiples of s^{-2l-1} as $s \to 0$.) A wealth of information about these confluent hypergeometric functions is known; in particular, they grow like e^s as $s \to \infty$ except when $l + 1 - \nu$ is a nonpositive integer, in which case the series terminates and gives a polynomial of degree $\nu - l - 1$. In this case ν is itself an integer $\geq l + 1$, which (following standard practice) we denote by n, and the resulting polynomial is one of the (generalized) *Laguerre polynomials*:

$$F(l+1-n,\, 2l+2;\, s) = \frac{(2l+1)!(n-l-1)!}{(n+l)!} L_{n-l-1}^{2l+1}(s).$$

Undoing the changes of variables, we see that the numbers $E_n = -1/n^2$ are eigenvalues of $H|\mathcal{H}_l$ for every integer $n \geq l+1$, and the corresponding (unnormalized) eigenfunctions are

$$f_{n,Y}(\mathbf{x}) = r^l e^{-r/n} L_{n-l-1}^{2l+1}(2r/n) Y(\mathbf{x}), \quad Y \in \mathcal{SH}_l \qquad (r = |\mathbf{x}|).$$

For $E > 0$ one can make the same changes of variables to find that the solutions of (3.35) that are regular at the origin are multiples of $r^l e^{-r/\nu} F(l+1-\nu,\, 2l+2;\, 2r/\nu)$, but here $\nu = 1/\sqrt{-E}$ is pure imaginary. One can use contour integral techniques to figure out the asymptotic behavior of this function as $r \to \infty$; it turns out to be asymptotic to a constant times $r^{-1}\sin(\beta r + \beta^{-1}\log 2\beta r + C_l)$ with $\beta = \sqrt{E}$, which is not in $L^2((0,\infty), r^2\,dr)$. For $E = 0$, the equation (3.35) can be transformed into a Bessel equation, and the solution turns out to be $\sqrt{1/r}\, J_{2l+1}(\sqrt{8r})$, which also is not in L^2. Hence there is no discrete spectrum for $E \geq 0$. However, the corresponding eigenfunctions of the differential operator H,

$$f_{\beta,Y}(\mathbf{x}) = r^l e^{-i\beta r} F(l+1+(i/\beta),\, 2l+2;\, 2i\beta r) Y(\mathbf{x}), \qquad Y \in \mathcal{SH}_l,$$

are close enough to being in L^2 that they contribute infinitesimally to the spectral resolution of the Hilbert space operator H, just as $e^{i\omega x}$ contributes infinitesimally to the spectral resolution of id/dx on $L^2(\mathbb{R})$ (viz., the Fourier transform).

In short, the discrete spectrum of $H|\mathcal{H}_l$ is $\{-1/n^2 : n \geq l+1\}$ and the continuous spectrum is $[0, \infty)$. Let us choose a basis $Y_1, \ldots, Y_{2l+1}$ for $\mathcal{SH}_l$ (the physicists' standard choice is the eigenbasis of the z-angular momentum L_3). Then any $\phi \in \mathcal{H}_l$ can be written as $\phi(\mathbf{x}) = \sum_1^{2l+1} \phi_m(r) Y_m(\mathbf{x})$, and the expansion of ϕ in terms of eigenfunctions of H has the form

$$\phi(\mathbf{x}) = \sum_{m=1}^{2l+1} \left[\sum_{n=l+1}^{\infty} C_{mn} f_{n,Y_m}(\mathbf{x}) + \int_0^{\infty} \widetilde{\phi}_m(\beta) f_{\beta,Y_m}(\mathbf{x})\, d\beta \right].$$

The coefficients C_{mn} and $\widetilde{\phi}_m(\beta)$ are easily determined once the $f_{n,Y}$'s and $f_{\beta,Y}$'s are properly normalized.

Putting together all the $\mathcal{H}_l$'s, we see that the discrete spectrum of H is $\{-1/n^2 : n \geq 1\}$. The eigenvalue $-1/n^2$ occurs in $\mathcal{H}_l$ when $l \leq n-1$, and its multiplicity there is $\dim(\mathcal{SH}_l) = 2l+1$, so its total multiplicity is $\sum_0^{n-1}(2l+1) = n^2$. The continuous spectrum is $[0, \infty)$, and it occurs in each $\mathcal{H}_l$, so its total multiplicity is infinite. Finally, if the particle in the Coulomb potential has spin k, the Hamiltonian is the direct sum of $2k+1$ copies of the scalar Hamiltonian, so the multiplicities of all the eigenvalues must be multiplied by $2k+1$.

In particular, for an electron, we have $k = \frac{1}{2}$, and the multiplicity of the eigenvalue $-1/n^2$ is $2n^2$. In this case it is customary to label the joint eigenstates of the energy and the angular momentum by the integers n (the "principal quantum number"), l (the "orbital angular momentum quantum number"), and j (the "total angular momentum quantum number"). Here n and l are as in the preceding discussion, and j is the half-integer such that the eigenvalue of the total squared angular momentum $\mathbf{J}^2$ is $j(j+1)$. In view of (3.34), j is either $l+\frac{1}{2}$ or $l-\frac{1}{2}$, with $j = \frac{1}{2}$ being the only possibility when $l = 0$. Moreover, in deference to the old spectroscopists' terminology, for $l = 0, 1, 2, 3$ it is customary to denote the states with quantum numbers (n, l, j) by nL_j where $L = S, P, D, F$ according as $l = 0, 1, 2, 3$. Thus, for example, the states with quantum numbers $(1, 0, \frac{1}{2})$, $(2, 1, \frac{1}{2})$, and $(3, 2, \frac{5}{2})$ are denoted by $1S_{1/2}$, $2P_{1/2}$, and $3D_{5/2}$.

This description of an electron in a Coulomb potential, which accounts for the "electron shells" of the Bohr model as the eigenspaces of the Hamiltonian, was one of the early triumphs of the quantum theory. Its prediction of the electron energy levels in a hydrogen atom agrees very well with the spectroscopic data that were available when quantum mechanics was developed. Moreover, it provides a somewhat cruder but still very informative model for larger atoms when one neglects the electrical interaction of the electrons with each other except to note that the inner electrons tend to shield the outer electrons from the charge of the nucleus. The only further necessary ingredient is the Pauli exclusion principle, which we shall discuss in §5.4, §5.5, and §6.5; in the present context it says that the state vectors of the various electrons of the atom must all be orthogonal to each other. Thus there is room for 2 electrons in the lowest energy level ($n = 1$), 8 electrons in the next level ($n = 2$), and so forth. From this one can see the periodic table of elements taking shape before one's eyes.

The agreement with experiment is not perfect, however: the actual energy levels of electrons in the hydrogen atom depend to a small extent on the angular momentum quantum numbers too. The biggest of these effects, the so-called "fine splitting" between the states with the same n but different j, is explained by the Dirac model, which we shall discuss in the next chapter.

CHAPTER 4

Relativistic Quantum Mechanics

> When an equation is as successful as Dirac's, it is never simply a mistake. It may not be valid for the reason supposed by the author, it may break down in new contexts, and it may not even mean what its author thought it meant. ... But the great equations of modern physics are a permanent part of scientific knowledge, which may outlast even the beautiful cathedrals of earlier ages.
> —Steven Weinberg ([**30**], p. 257)

The historical development of relativistic quantum theory is too tangled a tale to recount in detail here. (An interesting brief account can be found in Weinberg [**129**], and Schweber [**105**] has produced a masterful comprehensive history. See also Schweber [**106**] and the references given there.) Initially the idea was to develop a relativistic theory for the motion of a single particle, or a fixed finite collection of particles, subject to electromagnetic forces. Eventually it was realized that this is the wrong goal, as high-energy interactions necessarily involve creation, annihilation, and transmutation of particles. (One should keep in mind that in the late 1920s, when the initial attempts at a relativistic theory were made, the only known subatomic particles were electrons, protons, and atomic nuclei, and the composition of the latter was still a matter of conjecture. The experimental discovery of neutrons and positrons, not to mention other species of particles, was still in the future.) The early attempts at a one-particle theory yielded differential equations, now known by the names of Klein, Gordon, and Dirac, that were meant to be relativistic analogues of the Schrödinger equation.[1] Interpreted as wave equations for single quantum particles, the Klein-Gordon equation was largely unsuccessful, and the Dirac equation was hugely successful in accounting for low-energy phenomena but presented serious problems in the high-energy regime. But you can't keep a good differential equation down: these equations turn out to be of fundamental importance for quantum field theory — but with a different interpretation that we shall describe briefly at the end of §4.3 and exploit in the following chapters. Before we get to that, however, it will be well to study the Klein-Gordon and Dirac equations as the early quantum physicists did.

From now on we shall adopt "natural" units in which Planck's constant $\hbar$ and the speed of light c are both equal to 1. If one keeps track of the dimensions of the quantities one is working with, reinserting the factors of $\hbar$ and c is always just a matter of getting equations and formulas to be dimensionally correct.

[1]The Klein-Gordon equation was proposed independently by Schrödinger, so it could equally well be named the relativistic Schrödinger equation, and it is so called by some people.

4.1. The Klein-Gordon and Dirac equations

The Schrödinger equation $i\partial\psi/\partial t = -(1/2m)\nabla^2\psi + V\psi$ is obviously unsatisfactory from the relativistic point of view, as it treats the time and space variables on a drastically different footing. But recall how we derived it: (i) The (classical, nonrelativistic) energy of a particle of mass m in a potential V is $H = |\mathbf{p}|^2/2m + V(\mathbf{x})$. (ii) H is the generator of time translations according to the rules of Hamiltonian mechanics. (iii) In quantum mechanics, $\mathbf{p}$ becomes $(1/i)\nabla$, and $V(\mathbf{x})$ acts as a multiplication operator. We might therefore try to obtain a relativistic equation by modifying (i) so as to obtain a relativistically correct formula for the energy and then proceeding as in (ii) and (iii). For this purpose we need to drop the potential V, which is not relativistically meaningful as it stands, and replace it with something else. The only Lorentz-covariant interaction we have at hand is the electromagnetic one, so we shall consider only the cases of a free particle and a particle in an electromagnetic field; and we shall not add the electromagnetic field to the recipe until the case of a free particle is well in hand.

Recall that in relativistic mechanics, the energy E and the momentum $\mathbf{p}$ of a free particle combine to give a 4-vector $p^\mu = (E, \mathbf{p})$ (in units such that $c = 1$) whose Lorentz inner product with itself is the rest mass of the particle. The quantum analogue of the momentum $\mathbf{p}$ is the operator $-i\nabla$, and the quantum analogue of the energy is the Hamiltonian operator H, which generates the time-translation group $e^{tH/i}$ and hence corresponds to $i\partial_t$. In short, in relativistic notation the recipe for quantizing energy and momentum is simply

$$p^\mu \to i\partial^\mu, \tag{4.1}$$

where ∂^μ is given by (1.3). (Note that this makes the signs come out right!) Applying (4.1) to the free-particle energy-momentum relation $p^2 = m^2$ yields our first relativistic quantum wave equation, the *Klein-Gordon equation*:

$$-\partial^2\phi = m^2\phi, \quad \text{or} \quad \partial_t^2\phi = \nabla^2\phi - m^2\phi. \tag{4.2}$$

The first thing that catches the eye is that (4.2) is second-order in t, so the properly posed initial value problem requires not only ψ but also $\partial_t\psi$ to be given at $t = 0$. What to do about this? Perhaps one should take the state vector to be not ψ but the pair $(\psi_1, \psi_2) = (\psi, \partial_t\psi)$. One can then rewrite (4.2) as a first-order system:

$$\frac{\partial}{\partial t}\begin{pmatrix}\psi_1\\ \psi_2\end{pmatrix} = \begin{pmatrix}\psi_2\\ \nabla^2\psi_1 - m^2\psi_1\end{pmatrix} = \begin{pmatrix}0 & 1\\ \nabla^2 - m^2 & 0\end{pmatrix}\begin{pmatrix}\psi_1\\ \psi_2\end{pmatrix}. \tag{4.3}$$

The trouble with (4.3) is that the operator on $L^2 \times L^2$ represented by the matrix on the right is not skew-adjoint, so it does not generate a one-parameter unitary group. One might also think of replacing (4.2) by the equation $i\partial_t\psi = A\psi$ where A is a self-adjoint operator whose square is the positive operator $-\nabla^2 + m^2$. This idea is not completely without merit, and it will resurface at the end of §4.3. But of course $i\partial_t\psi = A\psi$ is no longer a differential equation, and it was not considered seriously by the pioneers of quantum theory.

Another way of looking at the problem is to think of the unitarity of time evolution as a conservation law. That is, if $\psi(\mathbf{x}, t) = e^{-itH}\psi_0(\mathbf{x})$ is a solution of the ordinary Schrödinger equation $i\partial_t\psi = H\psi$, where H is (a self-adjoint form of) $-\nabla^2/2m + V$, then $\|\psi\|^2 = \int |\psi(\mathbf{x}, t)|^2\, d\mathbf{x}$ is a conserved quantity. We can express this conservation law as a continuity equation as we did for conservation of charge in §2.4. Indeed,

$$\begin{aligned}\frac{\partial}{\partial t}|\psi|^2 = \frac{\partial \psi}{\partial t}\psi^* + \psi\frac{\partial \psi^*}{\partial t} &= \frac{i}{2m}\left[(\nabla^2\psi)\psi^* - \psi(\nabla^2\psi^*)\right] \\ &= \frac{i}{2m}\operatorname{div}[\psi^*\nabla\psi - \psi\nabla\psi^*].\end{aligned}$$

Thus, taking $\|\psi\| = 1$ so that $\rho = |\psi|^2$ represents a probability density, if we define the "probability current density" to be $\mathbf{J} = -(i/2m)(\psi^*\nabla\psi - \psi\nabla\psi^*) = (1/m)\operatorname{Im}(\psi^*\nabla\psi)$, we have

$$\frac{\partial \rho}{\partial t} + \operatorname{div}\mathbf{J} = 0.$$

There is a similar continuity equation associated to (4.2). Namely, if ψ satisfies (4.2) and we set $j^\mu = (1/m)\operatorname{Im}(\psi^*\partial^\mu\psi)$, then $\partial_\mu j^\mu = 0$, i.e., $\partial_t\rho + \operatorname{div}\mathbf{j} = 0$ where $j^\mu = (j^0, \mathbf{j})$. Thus the conserved quantity for (4.2) is $\rho = (1/m)\int \operatorname{Im}\psi^*\partial_t\psi$. But $\rho = (1/m)\operatorname{Im}\psi^*\partial_t\psi$ (or its negative) cannot be interpreted as a probability density because, in general, it is not positive. (If ψ is real, as it might well be since (4.2) has real coefficients, ρ vanishes identically!) For this reason, among others, (4.2) was found to be unsatisfactory as a relativistic wave equation for a free particle. (See Weinberg [**129**] and Schweber [**105**] for more extensive discussions of this matter.)

Dirac had the insight that what is impossible with scalar-valued functions might be feasible with vector-valued functions. What is desired is a *Lorentz-covariant* Schrödinger-type equation $i\partial_t\psi = H\psi$ for a free particle of mass $m > 0$, where the wave function ψ takes values in $\mathbb{C}^n$ for some n to be determined. For this purpose the space and time variables must be on an equal footing, so H must be a first-order differential operator in the space variables with constant coefficients:

$$H = \frac{1}{i}(\alpha^1\partial_1 + \alpha^2\partial_2 + \alpha^3\partial_3) + m\beta.$$

Here $\alpha^1, \alpha^2, \alpha^3$ and β are $n \times n$ complex matrices that we require to be Hermitian so that H will be self-adjoint, and the factor of m is inserted for later convenience. Moreover, H represents the total energy of the particle, which in classical relativity satisfies $E^2 = |\mathbf{p}|^2 + m^2$, so we want $H^2 = -\nabla^2 + m^2$. But

$$H^2 = -\sum_1^3(\alpha^j)^2\partial_j^2 - \sum_{1\le j<k\le 3}(\alpha^j\alpha^k + \alpha^k\alpha^j)\partial_j\partial_k + \frac{m}{i}\sum_1^3(\alpha^j\beta + \beta\alpha^j)\partial_j + m^2\beta^2,$$

so to make $H^2 = -\nabla^2 + m^2$ we need

$$\alpha^j\alpha^k + \alpha^k\alpha^j = 2\delta_{jk}I, \qquad \alpha^j\beta + \beta\alpha^j = 0, \qquad \beta^2 = I. \tag{4.4}$$

We are now on familiar mathematical ground: the conditions (4.4) say that α^1, α^2, α^3, and β are the generators of a Clifford algebra. The smallest n for which one can find $n \times n$ matrices satisfying (4.4) is $n = 4$, and the reader may verify that we may take

$$\alpha^j = \begin{pmatrix} 0 & \sigma_j \\ \sigma_j & 0 \end{pmatrix}, \qquad \beta = \begin{pmatrix} I & 0 \\ 0 & -I \end{pmatrix}, \tag{4.5}$$

(written in 2×2 blocks) where the σ_j are the Pauli matrices (1.12).[2] Note that these matrices are all Hermitian, as desired. We have now arrived at the *Dirac*

[2] If $m = 0$ we need only the α's, not β, and in this case $n = 2$ suffices: we can take $\alpha^j = \sigma_j$.

equation for a free particle of mass m:

$$i\frac{\partial\psi}{\partial t} = \frac{1}{i}\sum_1^3 \alpha^j \frac{\partial\psi}{\partial x_j} + m\beta\psi. \tag{4.6}$$

There is still a slight asymmetry between the space and time variables that may be removed as follows. Define the *Dirac matrices* γ^μ by

$$\gamma^0 = \beta = \begin{pmatrix} I & 0 \\ 0 & -I \end{pmatrix}, \qquad \gamma^j = \beta\alpha^j = \begin{pmatrix} 0 & \sigma_j \\ -\sigma_j & 0 \end{pmatrix} \text{ for } j = 1, 2, 3. \tag{4.7}$$

Since $\beta^2 = I$, multiplying (4.6) on the left by β yields

$$i\gamma^\mu \partial_\mu \psi = m\psi, \tag{4.8}$$

the *covariant form* of the Dirac equation. The γ^μ are easily seen to satisfy the relations

$$\{\gamma^\mu, \gamma^\nu\} \equiv \gamma^\mu\gamma^\nu + \gamma^\nu\gamma^\mu = 2g^{\mu\nu} I \quad (\mu, \nu = 0, 1, 2, 3), \tag{4.9}$$

$$(\gamma^\mu)^\dagger = (\gamma^\mu)^{-1} = \gamma_\mu \ (= g_{\mu\nu}\gamma^\nu) \quad (\mu = 0, 1, 2, 3), \tag{4.10}$$

where $g^{\mu\nu} = g_{\mu\nu}$ is the Lorentz metric. (Here and in the sequel, we employ the relativistic tensor notation introducted in §1.1.)

We refer to the copy of $\mathbb{C}^4$ on which the Dirac matrices act as *spinor space*, and its elements are called *(Dirac) spinors*. (On the other hand, *Pauli spinors* are elements of the copy of $\mathbb{C}^2$ on which the Pauli matrices act.) To avoid confusion, it is important to remember that *spinor space has nothing to do with space-time*. The fact that the dimensions of these two spaces are both 4 is a coincidence.

It is not hard to verify that the 16 matrices I, γ^μ, $\gamma^\mu\gamma^\nu$ $(\mu < \nu)$, $\gamma^\mu\gamma^\nu\gamma^\rho$ $(\mu < \nu < \rho)$, and $\gamma^0\gamma^1\gamma^2\gamma^3$ are a basis for the space of all 4×4 complex matrices, so the Clifford algebra generated by the γ^μ is the full 4×4 matrix algebra. By the way, the last of these matrices occurs frequently enough to have its own name; multiplied by i so as to make it real, it is conventionally called γ^5:

$$\gamma^5 = i\gamma^0\gamma^1\gamma^2\gamma^3 = \begin{pmatrix} 0 & I \\ I & 0 \end{pmatrix}. \tag{4.11}$$

(Obviously this notation was introduced by someone who took the Lorentz index μ to run from 1 to 4 rather than 0 to 3. The convention changed but the name stuck.)

There is nothing sacred about the choice (4.7) for the γ^μ; any four matrices $\widetilde{\gamma}^\mu$ satisfying (4.9) and (4.10) would do just as well. One can obtain such sets of matrices by taking $\widetilde{\gamma}^\mu = U\gamma^\mu U^{-1}$ where γ^μ is given by (4.7) and U is any unitary 4×4 matrix; conversely, every such set $\widetilde{\gamma}^\mu$ is obtained in this way. (This follows from the well-known fact that the full matrix algebra has a unique irreducible representation; see Messiah [**82**], §XX.10, for a direct proof.) The choice (4.7) is called the *Dirac representation*. The other choice that we shall find useful is the *Weyl* or *chiral representation* of the Dirac matrices,

$$\gamma^0 = \begin{pmatrix} 0 & I \\ I & 0 \end{pmatrix}, \qquad \gamma^j = \begin{pmatrix} 0 & \sigma_j \\ -\sigma_j & 0 \end{pmatrix} \quad (j = 1, 2, 3), \qquad \gamma^5 = \begin{pmatrix} -I & 0 \\ 0 & I \end{pmatrix} \tag{4.12}$$

which is related to the Dirac representation by

$$\gamma^\mu_{\text{Weyl}} = U\gamma^\mu_{\text{Dirac}} U^{-1}, \qquad U = \frac{1}{\sqrt{2}}\begin{pmatrix} I & -I \\ I & I \end{pmatrix}. \tag{4.13}$$

Many calculations with Dirac matrices are representation-independent. When a particular representation is needed, for most purposes the Weyl representation is the convenient one, particularly when questions of parity (left- or right-handedness) arise. (We shall explain this at the beginning of §4.3.) In fact, in this book we shall use the Dirac representation *only* in the discussion of the nonrelativistic approximation and the Coulomb potential in §4.3 and in a brief reference to the latter in §7.11; elsewhere the reader should keep the Weyl representation in mind.

Unlike the Klein-Gordon equation, the Dirac equation yields a conservation-of-probability law; this was one of the main things that persuaded Dirac that his theory was a good one. To explain it, and for other purposes later on, we need to introduce a modified adjoint for Dirac wave functions. We recall that the matrices α^j and β in (4.6) are Hermitian. Thus $\gamma^0 = \beta$ is Hermitian, but $\gamma^j = \gamma^0\alpha^j$ ($j = 1, 2, 3$) is not; rather, $(\gamma^j)^\dagger = \gamma^0\gamma^j\gamma^0$, and this requires an extra twist in setting up certain quantities.

Suppose ψ is a solution of the Dirac equation $i\gamma^\mu\partial_\mu\psi - m\psi = 0$; here ψ is a 4×1 column vector of functions. Taking Hermitian adjoints yields $-i\partial_\mu\psi^\dagger(\gamma^\mu)^\dagger - m\psi^\dagger = 0$, $\psi^\dagger$ being the 1×4 row vector whose components are the conjugates of the components of ψ. To turn this into an equation involving γ^μ rather than $(\gamma^\mu)^\dagger$, we multiply it on the right by γ^0 and set

$$\overline{\psi} = \psi^\dagger\gamma^0, \tag{4.14}$$

obtaining

$$-i\partial_\mu\overline{\psi}\gamma^\mu - m\overline{\psi} = 0, \tag{4.15}$$

the *adjoint Dirac equation.*

$\overline{\psi} = \psi^\dagger\gamma^0$ is called the *(Dirac) adjoint* spinor of ψ; this notation will be employed throughout the sequel.

Now, if we multiply the original Dirac equation on the left by $\overline{\psi}$ and the adjoint equation on the right by ψ and subtract, we obtain

$$i\overline{\psi}\gamma^\mu\partial_\mu\psi + i\partial_\mu\overline{\psi}\gamma^\mu\psi = 0,$$

in other words,

$$\partial_\mu j^\mu = 0, \quad \text{where} \quad j^\mu = \overline{\psi}\gamma^\mu\psi.$$

This is the promised conservation law, for

$$j^0 = \overline{\psi}\gamma^0\psi = \psi^\dagger\gamma^0\gamma^0\psi = \psi^\dagger\psi,$$

which is the probability density for the position of the Dirac particle; the spatial components (j^1, j^2, j^3) therefore define the corresponding current density.

A related fact is that the Dirac equation can be derived from the Lagrangian density

$$\mathfrak{L} = i\overline{\psi}\gamma^\mu\partial_\mu\psi - m\overline{\psi}\psi,$$

in which $\overline{\psi}$ and ψ are to be regarded as independent dynamical variables. Variation of $\int \mathfrak{L}\, d^4x$ with respect to $\overline{\psi}$ gives (4.8), whereas variation with respect to ψ gives the adjoint equation (4.15). This Lagrangian is invariant under the transformations $\psi \mapsto e^{i\theta}\psi$ for $\theta \in \mathbb{R}$, and j^0 is the corresponding conserved quantity according to Noether's theorem.

This gives the clue for incorporating an (unquantized, external) electromagnetic field into the Dirac equation. As in the classical situation that we discussed in §2.4, the rule is simply to replace the energy-momentum vector p^μ by $p^\mu - qA^\mu$ where A^μ

is the electromagnetic potential and q is the charge of the particle. At this point, however, we make a shift in notation to avoid confusion in later sections where the letter q is needed to denote momenta: we denote the charge of the particle by e. (The letter e is conventionally used to denote the charge of an electron or its absolute value, and in most applications it will indeed have that meaning since most charged subatomic particles (except quarks) have charge $\pm e$. But for the present discussion, e can be anything.)

As noted at the beginning of this section, the quantized energy-momentum vector p^μ is $i\partial^\mu$, so the prescription for including electromagnetism in the Dirac equation is

$$i\partial_\mu \to i\partial_\mu - eA_\mu.$$

The Lagrangian density is now

$$\mathcal{L} = \overline{\psi}\gamma^\mu(i\partial_\mu - eA_\mu)\psi - m\overline{\psi}\psi, \tag{4.16}$$

which yields the Dirac equation and adjoint Dirac equation

$$\gamma^\mu(i\partial_\mu - eA_\mu)\psi - m\psi = 0, \tag{4.17}$$

$$(-i\partial_\mu - eA_\mu)\overline{\psi}\gamma^\mu - m\overline{\psi} = 0. \tag{4.18}$$

(In taking the adjoint, the i acquires a minus sign, but the A_μ does not, because it is real.) The probability current $j^\mu = \overline{\psi}\gamma^\mu\psi$ still works the same way, because when one multiplies (4.17) by $\overline{\psi}$ and (4.18) by ψ and subtracts, the A_μ's cancel out.

The covariant formulation of the Dirac equation may have obscured the initial idea of an evolution equation for vectors in a Hilbert space. Considered simply as a differential operator, the Dirac operator on the left of (4.17) acts on more-or-less arbitrary $\mathbb{C}^4$-valued functions on $\mathbb{R}^4$. But as an evolution equation for quantum states, its domain should be taken to be the space of differentiable functions on $\mathbb{R}$ (the time variable) with values in $L^2(\mathbb{R}^3, \mathbb{C}^4)$, and it should be rewritten in the form (4.6):

$$i\partial_t\psi = \left[\sum_1^3 \alpha^j(-i\partial_j - eA_j) + eA_0 + m\beta\right]\psi. \tag{4.19}$$

The operator in brackets is the *Dirac Hamiltonian*, which acts on $L^2(\mathbb{R}^3, \mathbb{C}^4)$. Of course, this formulation spoils the relativistic symmetry between space and time. To some extent this is inevitable when one singles out the time variable, but we shall see in §4.4 that it is possible to give a "Lorentz-friendly" description of the situation without losing sight of the Hilbert space.

If our derivation of (4.17)-(4.18) seems a bit breezy, one should keep in mind that quantization rules are only guidelines. The real justification for the Dirac equation (4.17) is the extent to which it *works* as a description of quantum phenomena. We shall present the fundamental evidence on this score in §4.3. First, though, we shall examine the behavior of the Dirac equation under relativistic changes of reference frame.

4.2. Invariance and covariance properties of the Dirac equation

The Dirac equation was proposed as a relativistically correct evolution equation for quantum particles. If it is to fulfill that requirement, it had better maintain its form under the Poincaré group of space-time symmetries. The key idea is that *when applying Lorentz transformations, the 4-tuple of matrices* $(\gamma^0, \gamma^1, \gamma^2, \gamma^3)$ *should be treated like an ordinary 4-vector.* (In particular, we can write $\gamma_\mu = g_{\mu\nu}\gamma^\nu$; we then have, for example, $\gamma_\mu\gamma^\mu = 4I$.)

To see how this works, suppose that $L \in O(1,3)$; we identify L with the matrix (L^μ_ν) so that $(Lx)^\mu = L^\mu_\nu x^\nu$. If we set

$$[L\gamma]^\mu = L^\mu_\nu \gamma^\nu,$$

the matrices $\widetilde{\gamma}^\mu = [L\gamma]^\mu$ still satisfy the anticommutation relations (4.9), for

$$\widetilde{\gamma}^\mu\widetilde{\gamma}^\nu + \widetilde{\gamma}^\nu\widetilde{\gamma}^\mu = L^\mu_\rho L^\nu_\sigma(\gamma^\rho\gamma^\sigma + \gamma^\sigma\gamma^\rho) = 2L^\mu_\rho g^{\rho\sigma} L^\nu_\rho = 2g^{\mu\nu},$$

by the defining property of $O(1,3)$. It follows on general grounds that there must be a matrix B_L such that $\widetilde{\gamma}^\mu = B_L\gamma^\mu B_L^{-1}$. Rather than presenting the general argument, we shall identify B_L explicitly.

For this purpose we employ the Weyl representation (4.12) of the Dirac matrices, which enables us to use the double covering map $\kappa : SL(2,\mathbb{C}) \to SO^\uparrow(1,3)$ defined by (1.17) in an efficient way. In fact, if $x \in \mathbb{R}^4$, by (4.12) we have

$$\gamma \cdot x = \begin{pmatrix} 0 & M(x) \\ M(Px) & 0 \end{pmatrix}, \tag{4.20}$$

where $\gamma \cdot x = \sum_1^4 \gamma^\mu x^\mu$ is the *Euclidean* scalar product of γ and x, $M(x)$ is defined by (1.15), and P is the spatial inversion or parity operator,

$$P(x^0, \mathbf{x}) = (x^0, -\mathbf{x}).$$

Now, if

$$L = \kappa(A) \in SO^\uparrow(1,3),$$

we have $M(Lx) = AM(x)A^\dagger$. Moreover, it is easily verified that if $M(x)$ is invertible then $M(Px) = (\det M(x))M(x)^{-1}$, so that

$$\begin{aligned} M(PLx) = (\det M(Lx))M(Lx)^{-1} &= (\det M(x))A^{\dagger -1}M(x)^{-1}A^{-1} \\ &= A^{\dagger -1}M(Px)A^{-1}. \end{aligned}$$

As the invertible matrices are dense in $M(\mathbb{R}^4)$ (the space of Hermitian 2×2 matrices), we have $M(PLx) = A^{\dagger -1}M(Px)A^{-1}$ in general. Therefore, since $L^\dagger = \kappa(A)^\dagger = \kappa(A^\dagger)$,

$$\begin{aligned} L\gamma \cdot x = \gamma \cdot L^\dagger x = \begin{pmatrix} 0 & M(L^\dagger x) \\ M(PL^\dagger x) & 0 \end{pmatrix} &= \begin{pmatrix} 0 & A^\dagger M(x) A \\ A^{-1}M(Px)A^{\dagger -1} & 0 \end{pmatrix} \\ &= \begin{pmatrix} A^\dagger & 0 \\ 0 & A^{-1} \end{pmatrix} \begin{pmatrix} 0 & M(x) \\ M(Px) & 0 \end{pmatrix} \begin{pmatrix} A^{\dagger -1} & 0 \\ 0 & A \end{pmatrix}. \end{aligned}$$

In short, if we define the homomorphism $\Phi : SL(2,\mathbb{C}) \to SL(4,\mathbb{C})$ by

$$\Phi(A) = \begin{pmatrix} A^{\dagger -1} & 0 \\ 0 & A \end{pmatrix}, \tag{4.21}$$

we have

$$L\gamma \cdot x = \Phi(A^{-1})(\gamma \cdot x)\Phi(A),$$

whence, by taking x to be one of the standard basis vectors,

$$L^\mu_\nu \gamma^\nu = \Phi(A^{-1})\gamma^\mu\Phi(A) \qquad (L = \kappa(A)). \tag{4.22}$$

At this point it will be convenient to sweep the double covering map under the rug and regard the correspondence $L \mapsto \Phi(A)$ as a double-valued representation of $SO^{\uparrow}(1,3)$. That is, we shall write

$$\Phi(L) = \Phi(\kappa^{-1}(L)),$$

with the understanding that $\Phi(L)$ is defined only up to a factor ± 1 (an ambiguity that will never cause any difficulty as long as the the same sign is used for both terms in (4.22)). Since κ preserves adjoints, (4.22) can then be rewritten as

$$L^{\mu}_{\nu}\gamma^{\nu} = \Phi(L^{-1})\gamma^{\mu}\Phi(L). \tag{4.23}$$

The representation Φ can be extended to the full Lorentz group $O(1,3)$. In fact, since the substitution $x \mapsto Px$ simply interchanges the two M's in (4.20), it is clear that for the parity operator P we can take

$$\Phi(P) = (\pm)\gamma^0 = (\pm)\begin{pmatrix} 0 & I \\ I & 0 \end{pmatrix}.$$

(A direct verification of (4.23) is also easy: $(\gamma^0)^3 = \gamma^0$ and $\gamma^0\gamma^j\gamma^0 = -\gamma^j$ for $j > 0$.) For the time inversion $T(x^0, \mathbf{x}) = (-x^0, \mathbf{x}) = -P(x^0, \mathbf{x})$ it is then readily verified that

$$\Phi(T) = (\pm)\gamma^1\gamma^2\gamma^3 = (\pm)\begin{pmatrix} 0 & -I \\ I & 0 \end{pmatrix}$$

does the job.[3]

We are now in a position to analyze the Lorentz covariance of the Dirac equation. First suppose that L is an *orthochronous* Lorentz transformation. The electromagnetic potential A transforms under coordinate changes as a covariant vector field, i.e., a differential 1-form. In other words, if two reference frames are related by the transformation L, the components A_μ with respect to the first frame are related to the components A^L_μ with respect to the second one by

$$A^L_\mu(x) = L^{\nu}_{\mu}A_\nu(Lx).$$

Now, if ψ is a $\mathbb{C}^4$-valued function on $\mathbb{R}^4$, let us set[4]

$$\psi^L(x) = \Phi(L^{-1})\psi(Lx). \tag{4.24}$$

Then since $\partial_\mu(f \circ L) = L^{\nu}_{\mu}(\partial_\nu f) \circ L$ (the chain rule), we have

$$\begin{aligned}
&[\gamma^\mu(i\partial_\mu - eA^L_\mu(x)) - m]\psi^L(x) \\
&\qquad = \gamma^\mu L^{\nu}_{\mu}\Phi(L^{-1})[i(\partial_\nu\psi)(Lx) - eA_\nu(Lx)\psi(Lx)] - m\Phi(L^{-1})\psi(Lx) \\
&\qquad = (\Phi(L^{-1})\gamma^\nu\Phi(L)\Phi(L^{-1})[i(\partial_\nu\psi)(Lx) - eA_\nu(Lx)\psi(Lx)] - m\Phi(L^{-1})\psi(Lx) \\
&\qquad = \Phi(L^{-1})[\gamma^\nu(i\partial_\nu\psi - eA_\nu\psi) - m\psi](Lx).
\end{aligned}$$

Thus, *if ψ satisfies the Dirac equation with potential A_μ, then ψ^L satisfies the Dirac equation with potential A^L_μ.*

---

[3]The restriction of Φ to the orthochronous Lorentz group comes from a genuine representation of its double cover. However, the ambiguity of sign when one adds in time inversion is unavoidable, because P and T commute whereas $\Phi(P)$ and $\Phi(T)$ anticommute.

[4]If one wants the map taking L to the transformation $\psi \mapsto \psi^L$ to be a group homomorphism, one should replace L by L^{-1} in (4.24).

For the time inversion $Tx = (-x^0, \mathbf{x})$ the transformation law for A_μ has an extra minus sign in it:

$$A_\mu^T(x) = -T_\mu^\nu A_\nu(Tx),$$

that is,

$$A_0^T(x^0, \mathbf{x}) = A_0(-x_0, \mathbf{x}) \text{ and } A_j^T(x^0, \mathbf{x}) = -A_j(-x_0, \mathbf{x}).$$

The reason is that $\nabla^2 A_0$ is the charge density, which transforms as a scalar function under time inversion, but $\nabla^2 \mathbf{A}$ is the current density, which acquires a minus sign because the motion of the charges is reversed. This necessitates an extra twist in the transformation law for ψ that makes it *antiunitary*. To wit, since the complex conjugate matrices $\gamma^{\mu *}$ satisfy the same anticommutation relations as γ^μ and are still unitary, there is a unitary matrix B, determined up to a phase factor, such that

$$\gamma^\mu B = B\gamma^{\mu *}.$$

(In the Dirac and Weyl representations, γ^0, γ^1, and γ^3 are real while γ^2 is imaginary, and it follows easily that $B = \gamma^0\gamma^1\gamma^3$ works.) If we define

$$\psi^T(x) = \gamma^0 B\psi(Tx)^*, \tag{4.25}$$

we have

$$\begin{aligned}
&[\gamma^\mu(i\partial_\mu - eA_\mu^T(x)) - m]\psi^T(x) \\
&= \gamma^0 B\Big[\gamma^{0*}(i\partial_0 - eA_0(Tx)) - \textstyle\sum_{j=1}^3 \gamma^{j*}(i\partial_j + eA_j(Tx)) - m\Big](\psi\circ T)(x)^* \\
&= \gamma^0 B\Big\{\Big[\gamma^0(-i\partial_0 - eA_0(Tx)) - \textstyle\sum_{j=1}^3 \gamma^j(-i\partial_j + eA_j(Tx)) - m\Big](\psi\circ T)(x)\Big\}^* \\
&= \gamma^0 B\Big[\gamma^\mu(i\partial_\mu - eA_\mu)\psi - m\psi\Big]^*(Tx).
\end{aligned}$$

(Here's what has happened: γ^0 commutes with itself and anticommutes with the γ^j, and $\gamma^\mu B = B\gamma^{\mu *}$, which gives the first equality. Putting the conjugation on the whole expression to the right of $\gamma^0 B$ changes the i's to $-i$'s, and finally $\partial_0(\psi\circ T) = -(\partial_0\psi)\circ T$.) In short, *if ψ satisfies the Dirac equation with potential A_μ, then ψ^T satisfies the Dirac equation with potential A_μ^T.*

The time-reversal transform $\psi \mapsto \psi^T$ is closely related to another basic antiunitary symmetry of the system, *charge conjugation*, which is defined by

$$\psi^C(x) = \gamma^5 B\psi(x)^* \tag{4.26}$$

where B is as above and γ^5 is defined by (4.11). A calculation much like the preceding one shows that *if ψ satisfies the Dirac equation $\gamma^\mu(i\partial_\mu - eA_\mu)\psi = m\psi$, then ψ^C satisfies $\gamma^m(i\partial_\mu + eA_\mu)\psi^C = m\psi^C$.* The point here is that γ^5 anticommutes with all the γ^μ; combining this with complex conjugation, which changes i to $-i$, the net effect is to change e to $-e$.

The transformation laws $A \mapsto A^L$, $\psi \mapsto \psi^L$ under Lorentz transformations are complemented by the simple transformation law $A_\mu^a(x) = A_\mu(x-a)$, $\psi^a(x) = \psi(x-a)$ under space-time translations. Thus the Dirac equation is covariant under the full Poincaré group.

There is one other crucial invariance property that must be noted. The electromagnetic potential A_μ is well-defined only modulo the gauge transformations $A_\mu \mapsto A_\mu + \partial_\mu\chi$, where χ is an arbitrary (smooth) real-valued function. The product rule readily shows that the compensating transformation of the wave function is $\psi \mapsto \exp(ie\chi)\psi$; that is, *if ψ and A_μ satisfy $\gamma^\mu(i\partial_\mu - eA_\mu)\psi = m\psi$, then*

$\psi' = \exp(ie\chi)\psi$ *and* $A'_\mu = A_\mu + \partial_\mu \chi$ *satsify* $\gamma^\mu(i\partial_\mu - eA'_\mu)\psi' = m\psi'$. We shall place this result into a broader context in Chapter 9.

There is much more that can be said about the mathematics of the Dirac equation; a comprehensive account can be found in Thaller [**116**].

4.3. Consequences of the Dirac equation

Spin and parity. Since $A^{\dagger -1} = A$ for $A \in SU(2)$, the restriction of the representation Φ defined by (4.21) to the rotation group $SO(3)$ is just the direct sum of two copies of the irreducible representation $\pi_{1/2}$ described in §3.5. It follows that *the solutions of the Dirac equation describe particles of spin* $\frac{1}{2}$. For Dirac the fact that spin is built into his equation in this way was an "unexpected bonus."

However, the fact that there are two copies of $\pi_{1/2}$ — that is, the shift from $\mathbb{C}^2$-valued wave functions in §3.5 to $\mathbb{C}^4$-valued wave functions here to describe particles of spin $\frac{1}{2}$ —requires further comment. The point is that the change from two to four components does *not* represent an additional two degrees of freedom. The easiest way to see this is to write out the Dirac equation using the Weyl representation (4.12) of the Dirac matrices: setting

$$\psi_L = \begin{pmatrix} \psi_1 \\ \psi_2 \end{pmatrix}, \qquad \psi_R = \begin{pmatrix} \psi_3 \\ \psi_4 \end{pmatrix} \tag{4.27}$$

(L and R stand, conventionally, for "left" and "right"), the Dirac equation (4.17) becomes

$$m\psi_L = (i\partial_t - eA_0)\psi_R - \sum_1^3 (i\partial_j - eA_j)\sigma_j \psi_R,$$

$$m\psi_R = (i\partial_t - eA_0)\psi_L + \sum_1^3 (i\partial_j - eA_j)\sigma_j \psi_L.$$

Thus, *either of the two-component functions* ψ_L *and* ψ_R *is completely determined once the other one is given.* The enlargement from two to four components is necessary in order to make room for the Dirac matrices, but the spin variables still represent only two degrees of freedom for the wave functions.

The reason for associating the labels "left" and "right" to ψ_L and ψ_R is that they are interchanged by the parity operator P. More precisely, since $\Phi(P) = \left(\begin{smallmatrix} 0 & I \\ I & 0 \end{smallmatrix}\right)$, by (4.24) we have $(\psi^P)_L(x) = \psi_R(Px)$ and $(\psi^P)_R(x) = \psi_L(Px)$. Note also that the left- and right-handed wave functions are ∓ 1-eigenvectors of $\gamma^5 = \left(\begin{smallmatrix} -I & 0 \\ 0 & I \end{smallmatrix}\right)$: if $\psi_R = 0$ then $\gamma^5\psi = -\psi$, and if $\psi_L = 0$ then $\gamma^5\psi = \psi$. This will be important in §9.4.

The nonrelativistic approximation. The Dirac model for a spin-$\frac{1}{2}$ particle, involving a first-order differential equation for a $\mathbb{C}^4$-valued wave function, does not seem to bear much resemblance to the nonrelativistic model of Chapter 3, involving a second-order differential equation for a $\mathbb{C}^2$-valued wave function. Nonetheless, the Dirac theory does reproduce the old theory in the low-energy limit. We now give a quick sketch of the way this works.

For this purpose we write the Dirac equation in the form with the time variable singled out:

$$(i\partial_t - eA_0)\psi = \sum_1^3 \alpha^j(-i\partial_j - eA_j)\psi + m\beta\psi,$$

where α^j and β are given by (4.5). We also set

$$\psi_l = \begin{pmatrix}\psi_1\\ \psi_2\end{pmatrix}, \qquad \psi_s = \begin{pmatrix}\psi_3\\ \psi_4\end{pmatrix} \tag{4.28}$$

where l and s now stand for "large" and "small," for reasons shortly to be explained. (These are *not* the ψ_L and ψ_R of (4.27), because there we were employing a basis for $\mathbb{C}^4$ that gives the Weyl representation (4.12) of the Dirac matrices, whereas here we are employing a basis that gives the Dirac representation (4.7). By (4.13), the two are related by $\psi_l = (\psi_L + \psi_R)/\sqrt{2}$, $\psi_s = (\psi_L - \psi_R)/\sqrt{2}$.) With the abbreviation

$$D = \sum_1^3 \sigma_j(-i\partial_j - eA_j),$$

the Dirac equation then becomes the pair of coupled equations

$$i\partial_t\psi_l = (m + eA_0)\psi_l + D\psi_s, \qquad i\partial_t\psi_s = (-m + eA_0)\psi_s + D\psi_l. \tag{4.29}$$

We shall give a quick and easy account of the nonrelativistic limit of (4.29) under the assumption that the electromagnetic field A is constant in time, so that we are in the usual setting of an equation in the form $i\partial_t\psi = H\psi$ where H is a time-independent operator. We proceed in the informal style of physicists by assuming at first that ψ is an eigenfunction of the Hamiltonian with eigenvalue E, so that $i\partial_t\psi = E\psi$. (Honest L^2 wave functions are continuous superpositions of these eigenfunctions; to rework the argument in terms of them would be more laborious but not more enlightening. For a more sophisticated and rigorous mathematical discussion, see Thaller [**116**].) The second equation in (4.29) then says that

$$\psi_s = \frac{1}{E + m - eA_0} D\psi_l. \tag{4.30}$$

We now make the nonrelativistic approximation, namely, that (1) the relativistic energy E consists almost entirely of the rest mass m, i.e., $E - m \ll m$, (2) the electromagnetic field is weak so that $|eA_\mu(\mathbf{x})| \ll m$, at least in the region where $\psi(\mathbf{x})$ is nonnegligible, (3) $\|\nabla\psi\|$, the root-mean-square momentum in the state ψ, is small in comparison with $m\|\psi\|$. (With the factors of c and $\hbar$ reinserted, these approximations are $E - mc^2 \ll mc^2$, $|eA_\mu| \ll mc^2$ and $\hbar\|\nabla\psi\| \leq mc^2\|\psi\|$.) Then $\|D\psi_l\| \ll m\|\psi\|$, whereas $E + m - eA_0 \approx 2m$, so *ψ_s is small in comparison with ψ*:

$$\|\psi_s\| \ll \|\psi\|.$$

In the nonrelativistic limit, ψ_s becomes negligible, and we are left with the two-component wave function ψ_l. The equation it satisfies is obtained by substituting (4.30) into the first equation in (4.29), again replacing $i\partial_t$ by E:

$$(E - m)\psi_l = eA_0\psi_l + D\frac{1}{E + m - eA_0}D\psi_l.$$

Again making the approximation $E + m - eA_0 \approx 2m$, we obtain

$$(E - m)\psi_l = eA_0\psi_l + \frac{1}{2m}D^2\psi_l.$$

A straightforward calculation shows that

$$D^2 = (-i\nabla - e\mathbf{A})^2 - \frac{e}{2m}\sum_1^3 B_j\sigma_j = (-i\nabla - e\mathbf{A})^2 - \frac{e}{m}\mathbf{B}\cdot\mathbf{S},$$

where $\mathbf{B} = \operatorname{curl}\mathbf{A}$ is the magnetic field and $\mathbf{S} = \frac{1}{2}\boldsymbol{\sigma}$ is the spin operator for the 2-component wave function ψ_l. Finally, we drop the assumption that ψ is an energy eigenstate and turn E back into $i\partial_t$. Setting $\psi_0 = e^{imt}\psi_l$ to compensate for the shift from $i\partial_t$ to $i\partial_t - m$ (which does not affect the state represented by ψ_l at time t), we obtain

$$i\partial_t\psi_0 = eA_0\psi_0 + \frac{1}{2m}(-i\nabla - e\mathbf{A})^2\psi_0 - \frac{e}{m}\mathbf{B}\cdot\mathbf{S}\psi_0. \tag{4.31}$$

This equation, with the final term omitted, is the classical Schrödinger equation for a particle in the electromagnetic potential A_μ, obtained from the free-particle equation by the canonical substitution $p_\mu \mapsto p_\mu - eA_\mu$ without taking spin into account. The final term represents the interaction of the spin with the magnetic field.

Let us consider the special case of a spin-$\frac{1}{2}$ charged particle moving in a *constant* magnetic field $\mathbf{B}$ (together with, perhaps, an electrostatic field such as a Coulomb field). We can then take $\mathbf{A} = \frac{1}{2}\mathbf{B}\times\mathbf{x}$, which gives, after a little algebraic manipulation,

$$(-i\nabla - e\mathbf{A})^2 = -\nabla^2 - e\mathbf{B}\cdot\mathbf{L} + e^2|\mathbf{B}\times\mathbf{x}|^2.$$

The Hamiltonian for ψ_0 is therefore

$$eA_0 - \frac{1}{2m}\nabla^2 - \frac{e}{2m}\mathbf{B}\cdot(\mathbf{L}+2\mathbf{S}) + \frac{e^2}{2m}|\mathbf{B}\times\mathbf{x}|^2. \tag{4.32}$$

(When the magnetic field is weak, the last term is generally negligible.) The striking feature here is the term $\mathbf{L}+2\mathbf{S}$: *the spin interacts with the magnetic field twice as strongly as the orbital angular momentum.* This phenomenon, for atomic electrons, was one of the outstanding puzzles of atomic physics in the 1920s, and Dirac's explanation of it was one of the triumphs of his theory.

Classically, a magnet in a constant magnetic field $\mathbf{B}$ experiences a force $\nabla(\boldsymbol{\mu}\cdot\mathbf{B})$ where $\boldsymbol{\mu}$ is the *magnetic moment* of the magnet, and the associated potential energy is $-\boldsymbol{\mu}\cdot\mathbf{B}$. The intrinsic magnetic moment of a spin-$\frac{1}{2}$ particle (i.e., the moment attributable to spin rather than the motion of the particle) has the form

$$\boldsymbol{\mu} = \frac{ge}{2m}\mathbf{S} = \frac{ge}{4m}\boldsymbol{\sigma},$$

where g, the *Landé g-factor*, is determined experimentally and depends on the particle in question. Thus, the preceding calculations show that the Dirac theory predicts $g = 2$. This is quite accurate for electrons, but the true value of g_{electron} is a bit bigger than 2. The discrepancy is known as the *anomalous magnetic moment*, and the theoretical calculation of it is one of the triumphs of quantum electrodynamics; we shall discuss it in §7.11.

On the other hand, the g-factor of a proton is about 5.59, which is nowhere near 2, and the magnetic moment of a neutron, which ought to be 0 since the neutron has no charge, is actually about $-2/3$ times that of a proton. These figures, which were a mystery for a long time, are now understood to be reflections of the fact that protons and neutrons are not truly elementary but are made up of quarks.

An approximate calculation of these magnetic moments in the quark model can be found in Griffiths [**58**], §5.10.

The Coulomb potential revisited. The Dirac equation yields a model for the bound states of electrons in atoms similar to the one we derived from the Schrödinger equation in §3.6, but more refined. To wit, it takes the wave function for an electron in an atom of atomic number Z (i.e., with Z protons in the nucleus) to be a solution of the Dirac equation with $e =$ the charge of the electron, $A_0 = Z|e|/4\pi r$, and $\mathbf{A} = \mathbf{0}$, where r is the distance to the origin. The assumption that the nucleus is infinitely heavy results in errors of the same order of magnitude as the fine-structure effects that the Dirac equation captures, so one should use a reference frame in which the center of mass of the electron-nucleus system is fixed and take the mass m in the Dirac equation to be the reduced mass $m_e m_N/(m_e + m_N)$, as in the classical two-body problem discussed at the end of §2.1. The assumption that the charge of the nucleus is all located precisely at the origin is also an idealization, but it gives a very good approximation for realistic values of Z. Since we are now taking $\hbar = c = 1$, the constant $e^2/4\pi$ in the coefficient of the potential $eA_0 = -Ze^2/4\pi r$ is just the fine structure constant α (see §1.2):

$$\alpha = \frac{e^2}{4\pi} \approx \frac{1}{137}.$$

In short, the mathematical problem to be solved is the spectral theory of the Dirac Hamiltonian

$$H = \frac{1}{i}\sum_1^3 \alpha^j \partial_j + m\beta - \frac{Z\alpha}{r} \qquad (Z \in \mathbb{Z}_+,\ \alpha \approx \tfrac{1}{137}), \tag{4.33}$$

particularly the eigenfunctions and eigenvalues of H that represent stationary states of the electron. We shall concentrate on the structural features rather than going through all the calculations in detail. Fuller discussions can be found in Thaller [**116**], Sakurai [**102**], or Messiah [**82**].

The first point to be made is that unlike the Schrödinger Hamiltonian with a $-Z\alpha/r$ potential, this H is scale-invariant, because both ∂_j and $1/r$ are homogeneous of degree -1 under dilations, and so is m in the natural units where $[m] = [l^{-1}]$. Thus, whereas in the Schrödinger theory we could make the coefficient of $1/r$ equal to 2 by choosing the units appropriately, this is not possible for the Dirac theory, and the size of the coefficient $Z\alpha$ is significant.

The spherical symmetry of the problem points to a consideration of the decomposition of the state space $L^2(\mathbb{R}^3, \mathbb{C}^4)$ into irreducible pieces under the action of the rotation group. We recall from (4.24) that the rotation $R \in SO(3)$ acts on the state vector ψ by

$$\psi^R(t, \mathbf{x}) = \Phi(R)\psi(t, R^{-1}\mathbf{x}). \tag{4.34}$$

This action is the tensor product of the natural representation $\rho(R)\phi(\mathbf{x}) = \phi(R^{-1}\mathbf{x})$ of $SO(3)$ on $L^2(\mathbb{R}^3)$ and the representation Φ on $\mathbb{C}^4$. The Schrödinger Hamiltonian of §3.6 commutes with each of these actions separately, but the Dirac Hamiltonian does not, so we need to take spin into account more carefully here than we did in §3.6.

As we showed in §3.6, ρ decomposes via spherical harmonics as $\rho \cong \bigoplus_0^\infty \pi_l$, where π_l is as in §3.5. Moreover, by the remarks on spin at the beginning of this section, $\Phi|SO(3)$ is the direct sum of two copies of $\pi_{1/2}$; but when $\rho\otimes\Phi$ is restricted to solutions of the Dirac equation, where the first two components of ψ in the Weyl representation determine the last two, there is only one copy. Hence, by (3.34), the representation of $SO(3)$ given by (4.34) decomposes as

$$\bigoplus_{l=0,1,2,\ldots} \pi_l \otimes \pi_{1/2} \cong \bigoplus_{j=\frac{1}{2},\frac{3}{2},\frac{5}{2},\ldots} 2\pi_j.$$

The 2 indicates multiplicity; one copy of π_j comes from $l = j - \frac{1}{2}$, the other from $l = j + \frac{1}{2}$.

One can think of this in terms of angular momentum. The angular momentum operators $\mathbf{J} = (J^1, J^2, J^3)$ are the infinitesimal generators of the action (4.34) of the rotations about the coordinate axes. Corresponding to the two occurrences of R on the right side of (4.34), we have

$$\mathbf{J} = \mathbf{L} + \mathbf{S},$$

where $\mathbf{L}$ is the orbital angular momentum and $\mathbf{S}$ is the spin,

$$\mathbf{S} = \frac{1}{2}\begin{pmatrix} \boldsymbol{\sigma} & 0 \\ 0 & \boldsymbol{\sigma} \end{pmatrix}.$$

The irreducible representation spaces of type π_j are eigenspaces of the total squared angular momentum $\mathbf{J}^2$ with eigenvalue $j(j+1)$, just as in §3.5. The Hamiltonian (4.33) is invariant under the action (4.34) of the rotations and hence commutes with $\mathbf{J}^2$, but it does not commute with $\mathbf{L}^2$.

We can resolve the multiplicity-two representations $2\pi_j$ by passing to the full orthogonal group $O(3)$, that is, by considering the parity or spatial inversion operator $P(t, \mathbf{x}) = (t, -\mathbf{x})$. Its action on the state vectors commutes with the action of $SO(3)$ and with the Dirac Hamitonian, and the space of state vectors on which $2\pi_j$ acts decomposes into the ± 1 eigenspaces of the action of P, that is, into the states of even and odd parity. Identifying "even" with 0 and "odd" with 1, we denote the parity of such an eigenstate by ϵ:

$$\psi^P = (-1)^\epsilon \psi.$$

At this point we need to be a little more concrete. We work in the Dirac representation, where the coefficients α^j and β in the Hamiltonian (4.33) are given by (4.5), and we write ψ_l and ψ_s for the first and second pair of components of ψ as in (4.28). We recall from §4.2 that $\Phi(P) = \gamma^0 = \beta$, which in the Dirac representation is $\left(\begin{smallmatrix} I & 0 \\ 0 & -I \end{smallmatrix}\right)$, so that

$$\psi^P(t, \mathbf{x}) = \begin{pmatrix} \psi_l(t, -\mathbf{x}) \\ -\psi_s(t, -\mathbf{x}) \end{pmatrix}.$$

Thus if ψ is a state of either even or odd parity, the components ψ_l and ψ_s must have opposite spatial parity: if $\psi^P = \psi$, then ψ_l is even and ψ_s is odd, and vice versa if $\psi^P = -\psi$.

When one works out the implications of these remarks in more detail, one finds that the eigenstates of $\mathbf{J}^2$ with eigenvalue $j(j+1)$ and definite parity ϵ are of the form

$$\psi = \begin{pmatrix} \psi_l \\ \psi_s \end{pmatrix} = \frac{1}{r}\begin{pmatrix} f(r)Y_l(\theta,\phi) \\ ig(r)Y_s(\theta,\phi) \end{pmatrix} \tag{4.35}$$

where Y_l and Y_s are $\mathbb{C}^2$-valued functions whose components are spherical harmonics of degree $j \pm \frac{1}{2}$. More precisely, if $j + \frac{1}{2} \equiv \epsilon \pmod 2$, then Y_l is of degree $j + \frac{1}{2}$ and

Y_s is of degree $j - \frac{1}{2}$; vice versa if $j - \frac{1}{2} \equiv \epsilon \pmod 2$. The requirement that ψ be an eigenfunction of the Hamiltonian H with eigenvalue E is then equivalent to the following pair of differential equations for f and g:

$$\begin{aligned} f'(r) + \frac{\kappa}{r} f(r) &= \left(E + m + \frac{Z\alpha}{r}\right) g(r), \\ -g'(r) + \frac{\kappa}{r} g(r) &= \left(E - m + \frac{Z\alpha}{r}\right) f(r), \end{aligned} \tag{4.36}$$

where

$$\kappa = (-1)^{\epsilon + j + (1/2)}(j + \tfrac{1}{2}). \tag{4.37}$$

The analysis up to this point is valid with $-Z\alpha/r$ replaced by any central potential $V(r)$.

The equations (4.36) are analytic with a regular singular point at the origin, so they yield readily to standard power series techniques. Beginning with the *Ansatz*

$$f(r) = a_0 r^\lambda + \text{higher order}, \qquad g(r) = b_0 r^\mu + \text{higher order},$$

one sees first that μ must equal λ. (The left side of the equation involving g' is $O(r^{\mu-1})$, whereas the right side is $\geq Cr^{\lambda-1}$ for r small; hence $\lambda \geq \mu$. Similarly, $\mu \leq \lambda$.) With $\mu = \lambda$, then, examination of the terms of order $\lambda - 1$ in (4.36) shows that

$$(\lambda + \kappa) a_0 = Z\alpha b_0, \qquad (-\lambda + \kappa) b_0 = Z\alpha a_0. \tag{4.38}$$

This pair of equations for a_0 and b_0 has nonzero solutions only when $\kappa^2 - \lambda^2 = (Z\alpha)^2$, that is, when $\lambda = \pm\sqrt{\kappa^2 - (Z\alpha)^2}$.

At this point we must address the square-integrability of the solutions at the origin. The factor of $1/r$ in (4.35) compensates for the factor of r^2 in the measure in spherical coordinates, so the ψ in (4.35) will be L^2 near the origin with respect to volume measure if and only if f and g are in L^2 near the origin with respect to linear measure; this happens precisely when $\operatorname{Re}\lambda > -\frac{1}{2}$. Thus, if $\kappa^2 - (Z\alpha)^2 > \frac{1}{4}$ we take the solution with $\lambda = \sqrt{\kappa^2 - (Z\alpha)^2}$ and discard the one with $\lambda = -\sqrt{\kappa^2 - (Z\alpha)^2}$. If $\kappa^2 - (Z\alpha)^2 < \frac{1}{4}$, however, *both* solutions are L^2 near the origin, and this is a disaster. It means that the differential operators in (4.36), and hence the Dirac Hamiltonian itself, do not determine a unique self-adjoint operator on L^2 and so do not have a well-defined spectral theory. One could specify a self-adjoint form by imposing "boundary conditions" at the origin, but their physical significance is unclear.

Fortunately, this is not a serious cause for worry. Recall that $|\kappa| = j + \frac{1}{2}$ is a positive integer, so we are always safe when $(Z\alpha)^2 < \frac{3}{4}$, i.e., $Z < \sqrt{3}/2\alpha \approx 118$.[5] All atomic nuclei with a more-than-ephemeral existence satisfy this condition with room to spare. Even for large Z, one avoids the mathematical catastrophe by recognizing that the charge of the nucleus is not really located at a single point. Taking the nucleus to be a solid ball of charge of radius $\sim 10^{-13}$ cm gives a more realistic model in which the potential has no singularity at the origin. (However, the resulting model for an atom is still physically bizarre for $Z > \sqrt{3}/2\alpha$, as the

[5] Some physics books say that the critical condition is $(Z/\alpha)^2 < 1$, i.e., $Z < 137$, because for $Z/\alpha > 1$ the solutions are all bounded (and wildly oscillatory) near the origin. But the real problem occurs earlier.

effective diameters of the electron orbits for small $|\kappa|$ are of the same order of magnitude as that of the nucleus.)

We henceforth assume that $Z < 118$. The next step is to solve the equations (4.36), with

$$\lambda = \sqrt{\kappa^2 - (Z\alpha)^2}.$$

For this purpose it is convenient to make a change of variables: keeping in mind that for bound states, the total energy E will be *less* than the rest mass m, we set $\rho = r\sqrt{m^2 - E^2}$ and take

$$f(r) = e^{-\rho}\rho^\lambda \sum_0^\infty a_n \rho^n, \qquad g(r) = e^{-\rho}\rho^\lambda \sum_0^\infty b_n \rho^n.$$

Plugging these formulas into (4.36) gives simple recursion formulas for a_n and b_n that can be solved explicitly; the resulting power series can be expressed in terms of confluent hypergeometric functions. From this one finds that f and g grow exponentially at infinity, and hence are not in L^2, except in those cases where the power series actually terminate; moreover, the series terminate with the ρ^N term $(N = 0, 1, 2, \ldots)$ precisely when $\sqrt{m^2 - E^2}(\lambda + N) = EZ\alpha$. Solving this equation for E gives

$$E = m\left[1 + \left(\frac{Z\alpha}{\lambda + N}\right)^2\right]^{1/2} = m\left[1 + \frac{(Z\alpha)^2}{(N + \sqrt{\kappa^2 - (Z\alpha)^2})^2}\right]^{1/2}.$$

Recall from (4.37) that $\kappa = \pm(j + \frac{1}{2})$. For $N > 0$ there is a solution of the differential equation for each of these values of κ, but for $N = 0$ there is a solution only for $\kappa < 0$. (In brief, the reason is that when $N = 0$, the recursion formula together with the fact that $a_1 = b_1 = 0$ implies that a_0 is a negative multiple of b_0; this is incompatible with the first equation in (4.38) if $\kappa > 0$.)

To put this formula into a more recognizable form, we assume that $Z\alpha \ll 1$ and expand the right side in powers of $Z\alpha$. Using the fact that $\kappa^2 = (j + \frac{1}{2})^2$ and setting

$$n = N + j + \tfrac{1}{2},$$

after some manipulation of Taylor polynomials we obtain

$$E = E(n, j) = m\left[1 - \frac{(Z\alpha)^2}{2n^2} + (Z\alpha)^4\left(\frac{3}{8n^4} - \frac{1}{n^3(2j + 1)}\right) + O((Z\alpha)^6)\right]. \tag{4.39}$$

Here $n = 1, 2, 3, \ldots$ and $j = \frac{1}{2}, \frac{3}{2}, \frac{5}{2}, \ldots$, subject to the relation $j < n$ (since $n - j - \frac{1}{2} = N \geq 0$). For each such n and j there are four independent eigenfunctions with eigenvalue $E(n, j)$ except when $n = j + \frac{1}{2}$, in which case there are two: more precisely, there are two for each allowable value of κ, which can be taken to be the states with spin up or down along some specified axis.

This is now amenable to comparision with the results of the Schrödinger equation. The first term on the right of (4.39) is, of course, just the rest mass. The next term, $-m(Z\alpha)^2/2n^2$, gives the energy levels predicted by the Schrödinger equation, n being the "principal quantum number" that tells which "shell" the electron lives in. (In the discussion in §3.6 we took the unit of energy to be $m(Z\alpha)^2/2$; here this factor appears explicitly.) The $(Z\alpha)^4$ term gives Dirac's corrections to the energy levels. The interesting feature here is the dependence on j, which is not present in the Schrödinger theory; this is the "fine splitting" of energy levels whose agreement with experiment is one of the major successes of the Dirac theory. In particular, the difference in the energies of the $2P_{1/2}$ and $2P_{3/2}$ states[6] for a hydrogen atom (i.e., $Z = 1$, $n = 2$, $j = \frac{1}{2}, \frac{3}{2}$) is about $m\alpha^4/32 \approx 4.5 \times 10^{-5}$ eV, which is about 10^{-5} times smaller than the main term $-m\alpha^2/8 \approx -3.4$ eV for their energies.

But as with the magnetic moment of the electron, this is not the end of the story. The Dirac theory gives the same energy to the $2S_{1/2}$ and $2P_{1/2}$ states, but experimentally there is a difference of about 4.4×10^{-6} eV (the "Lamb shift"). Explaining this is another job for quantum electrodynamics; we shall say more about it in §6.2 and §7.11.

There is one respect in which the predictions of the Dirac and Schrödinger equations for the electron in a hydrogen atom are both completely wrong. According to both of them, if the electron is in an energy eigenstate, it will remain there for all time unless some external force comes along to knock it out. But in fact, if the electron is in an eigenstate other than the ground state (an "excited state"), it eventually gets tired of being there and drops down to a lower energy level with the emission of a photon.[7] The problem is that the Schrödinger and Dirac models as presented here recognize only the Coulomb force generated by the nucleus and not the whole electromagnetic field associated to the nucleus and the electron together. The full analysis of this situation requires quantum field theory; we shall present a simplified model of it in §6.2.

Negative-energy states. A problem with the Dirac equation that was recognized from the beginning is the existence of states of negative energy — that is, the fact that the spectrum of the free Dirac Hamiltonian

$$H = -i\sum \alpha^j \partial_j + m\beta$$

is not bounded below. Indeed, on applying the Fourier transform on $\mathbb{R}^3$ and using the Dirac matrices (4.5), we see that

$$\widehat{H\psi}(\mathbf{p}) = (\mathbf{p}\cdot\boldsymbol{\alpha} + m\beta)\widehat{\psi}(\mathbf{p}) = \begin{pmatrix} mI & \mathbf{p}\cdot\boldsymbol{\sigma} \\ \mathbf{p}\cdot\boldsymbol{\sigma} & -mI \end{pmatrix}\widehat{\psi}(\mathbf{p}).$$

A straightforward calculation shows that for each $\mathbf{p}$, the 4×4 Hermitian matrix multiplying $\widehat{\psi}(\mathbf{p})$ has eigenvalues $\pm\sqrt{p^2+m^2}$ ($p = |\mathbf{p}|$), each with multiplicity 2, and that the unitary matrix

$$U(\mathbf{p}) = \frac{(m + \sqrt{p^2+m^2})I + \beta\mathbf{p}\cdot\boldsymbol{\alpha}}{[2m\sqrt{p^2+m^2} + 2(p^2+m^2)]^{1/2}}$$

diagonalizes it. Let $P_+(z_1, z_2, z_3, z_4) = (z_1, z_2)$ and $P_-(z_1, z_2, z_3, z_4) = (z_3, z_4)$; then the transformation V defined by

$$V\psi(\mathbf{p}) = \big(P_+U(\mathbf{p})\widehat{\psi}(\mathbf{p}),\, P_-U(\mathbf{p})\widehat{\psi}(\mathbf{p})\big)$$

is a unitary map from $L^2(\mathbb{R}^3, \mathbb{C}^4)$ to $L^2(\mathbb{R}^3, \mathbb{C}^2) \oplus L^2(\mathbb{R}^3, \mathbb{C}^2)$ that intertwines H with the operator M defined by

$$M(f_+, f_-)(\mathbf{p}) = \big(\sqrt{p^2+m^2}f_+(\mathbf{p}),\, -\sqrt{p^2+m^2}f_-(\mathbf{p})\big).$$

---

[6] Recall that the P in "$2P_{1/2}$" referred to the orbital angular momentum quantum number in the nonrelativistic theory. Here it is the orbital angular momentum quantum number of the large component ψ_l.

[7] "Eventually" is typically around 10^{-9} second.

Thus, if we denote the inverse images of the two summands in $L^2(\mathbb{R}^3,\mathbb{C}^2) \oplus L^2(\mathbb{R}^3,\mathbb{C}^2)$ under V by $\mathcal{H}_\pm$, we see that the restrictions of H to $\mathcal{H}_+$ and $\mathcal{H}_-$ are operators on these spaces with spectra $[m,\infty)$ and $(-\infty,-m]$, respectively.

The existence of states with arbitrarily large negative energy is physically very unsatisfactory. One might try to do away with them by assuming that the physical Hilbert space consists only of the subspace $\mathcal{H}_+$, but this assumption is not tenable, for various reasons. For example, $\mathcal{H}_+$ does not contain states that are highly localized in space, nor does it contain the ground state wave function for the electron in a hydrogen atom that we discussed earlier. Therefore, there must be some mechanism for forbidding a particle to release an infinite amount of energy by falling to lower and lower energy levels.

Dirac's remarkable (not to say fantastic) solution to this problem was to invoke the Pauli exclusion principle and to posit that almost all the negative-energy states are already occupied by a completely invisible "sea" of electrons. The only observable things are the electrons with positive energy and the unoccupied negative-energy states or "holes" in the negative-energy sea, which are perceived as particles with *positive* charge. A positive-energy electron could fall into one of these unoccupied states by emitting a photon; this would be perceived as mutual annihilation of the electron and the positively charged particle. Dirac originally proposed that the positively charged particles should be protons, but it quickly became clear that this is incompatible both with the discrepancy in mass between electrons and protons and the complete lack of observational evidence for electron-proton annihilation. Dirac was then forced to accept the possible existence of what we now call positrons, and he did so in a paper in 1931. The discovery of the positron by Anderson in 1932 was yet another triumph of the Dirac theory.

The "hole" theory, however, presents problems of its own. Aside from the epistemological difficulties associated to the "negative energy sea," it implies that the Dirac equation cannot really work as it was originally intended, as an evolution equation for a single electron, for the possibility of electron-positron creation or annihilation is always present. The interpretation of antiparticles as holes in a negative energy sea is also untenable for particles of integer spin, to which the exclusion principle does not apply. The idea that negative-energy states have *something* to do with antiparticles, however, is supported by the fact that the charge conjugation operator $\psi \mapsto \psi^C$ that we introduced in (4.26) intertwines H with $-H$ and hence interchanges $\mathcal{H}_+$ and $\mathcal{H}_-$. (The calculation that $H(\psi^C) = -(H\psi)^C$ is easy and left to the reader.)

We refer the reader to physics books for a fuller discussion of these matters, which took a fair amount of time for the physicists to sort out. The final upshot, however, was that a much more radical reinterpretation of the Dirac equation is necessary to produce a satisfactory theory, and that a similar reinterpretation of the Klein-Gordon equation gives it a new life as a fundamental theoretical tool. To wit, the Dirac and Klein-Gordon equations must be considered as wave equations not for wave functions of a quantum particle but for *classical fields*, even though these classical fields do not represent anything in classical physics. One must then *quantize* these classical fields to produce *quantum fields* whose quanta are the particles one wishes to study. (This reinterpretation is sometimes given the rather misleading name of "second quantization." There is only one quantization; it just takes place at a different level than one was originally expecting.)

4.4. Single-particle state spaces

So far we have taken the state space for a quantum particle to be $L^2(\mathbb{R}^3, V)$ where V is a suitable finite-dimensional vector space. The $\mathbb{R}^3$ here has to do with the position of the particle in space; the time variable appears only as a separate parameter. To do things on a more relativistically invariant footing, this picture needs to be revised. More precisely, the state of a particle is determined by various observations made on the particle from a particular inertial reference frame. Two observers in two different reference frames will assign different state vectors to the particle, but the underlying physics must be invariant — at least when the two frames are related by a transformation T in the connected component of the identity $\mathcal{P}_0$ of the Poincaré group $\mathcal{P}$. (Space and time inversion may present some problems that need to be considered separately.) The correspondence between the state vectors in the two reference frames must be given by a unitary map U_T of the state space, determined up to a phase factor, and clearly $U_{T_1T_2}$ must be $U_{T_1}U_{T_2}$ up to a phase factor. (Antiunitary maps are a possibility in principle, but since $\mathcal{P}_0$ is connected, continuity considerations force all the U_T to be unitary.)

In short, relativistic invariance implies that the state space $\mathcal{H}$ must come equipped with a projective unitary representation of the group $\mathcal{P}_0$. Moreover, the theorem of Bargmann [**6**] quoted in §3.1 guarantees that every projective unitary representation of $\mathcal{P}_0$ comes from a genuine unitary representation of its universal cover $\mathbb{R}^4 \ltimes SL(2,\mathbb{C})$. Finally, we shall assume that this representation is *irreducible,* which means that the particle is in some sense "elementary." (The particle may, in fact, have some additional structure, as long as that structure is irrelevant for the physical problems under consideration. Thus, for some purposes a proton, or even a whole atomic nucleus, can be considered "elementary." The situation is analogous to the problems in classical celestial mechanics in which the sun and the planets can be assumed to be point masses.)

We are therefore faced with the problem of describing the irreducible unitary representations of the group

$$G = \mathbb{R}^4 \ltimes SL(2,\mathbb{C}).$$

This was first accomplished by Wigner [**132**] and is now usually treated as a classic application of the "Mackey machine." We refer to Folland [**44**] and Varadarajan [**123**] for a detailed account of the mathematics and its physical implications; here we just present the results succinctly.

In this discussion we will be dealing with two copies of $\mathbb{R}^4$, "position space" and "momentum space," which should be considered to be in duality with each other via the Lorentz inner product $(p, x) \mapsto p_\mu x^\mu$. Since this notation with indices will become very cumbersome in places, we shall also write the Lorentz inner product as $\langle p, x\rangle$:

$$\langle p, x\rangle = p_\mu x^\mu.$$

The group $SL(2,\mathbb{C})$ acts as Lorentz transformations on position space, and hence also on momentum space by the dual or contragredient action. We denote the action of $A \in SL(2,\mathbb{C})$ on $x \in \mathbb{R}^4$ simply by Ax, and the contragredient action of A on $p \in (\mathbb{R}^4)^*$ by $A^{\dagger -1}p$. Thus, strictly speaking, $Ax = \kappa(A)x$ where $\kappa : SL(2,\mathbb{C}) \to SO^\uparrow(1,3)$ is the covering map defined by (1.17), and $A^{\dagger -1}p = \kappa(A)^{\dagger -1}p = \kappa(A^{\dagger -1})p$.

Given $p \in \mathbb{R}^4$, let H_p (the "little group" of p) be the subgroup of $SL(2, \mathbb{C})$ that fixes p, let $G_p = \mathbb{R}^4 \ltimes H_p$, and let σ be an irreducible unitary representation of H_p on a Hilbert space $\mathcal{H}_\sigma$. Then one can define an irreducible unitary representation $\rho_{p,\sigma}$ of G_p on $\mathcal{H}_\sigma$ by

$$\rho_{p,\sigma}(a, A) = \exp(-ip_\mu a^\mu)\sigma(A).$$

Finally, let $\pi_{p,\sigma}$ be the unitary representation of G induced from $\rho_{p,\sigma}$ (to be described in more detail later):

$$\pi_{p,\sigma} = \mathrm{ind}_{G_p}^G(\rho_{p,\sigma}).$$

Then $\pi_{p,\sigma}$ *is irreducible, and every irreducible unitary representation of* G *is equivalent to some* $\pi_{p,\sigma}$. *Moreover,* $\pi_{p,\sigma}$ *and* $\pi_{p',\sigma'}$ *are equivalent if and only if* p *and* p' *belong to the same* $SL(2, \mathbb{C})$*-orbit, say* $p' = Bp$*, and* $A \mapsto \sigma(A)$ *and* $A \mapsto \sigma'(BAB^{-1})$ *are equivalent representations of* H_p.

We recall from §1.3 that the $SL(2, \mathbb{C})$-orbits in $\mathbb{R}^4$ are as follows:

$$X_m^+ = \{p : p^2 = m^2,\ p_0 > 0\}, \text{ and } X_m^- = \{p : p^2 = m^2,\ p_0 < 0\}$$

for $m \in [0, \infty)$,

$$Y_m = \{p : p^2 = -m^2\}$$

for $m \in (0, \infty)$, and

$$\{0\}.$$

For $m > 0$, we may take the representative point in the orbit $X_m^\pm$ to be $p_m^\pm = (\pm m, 0, 0, 0)$ and the representative point in Y_m to be $q_m = (0, 0, m, 0)$; the corresponding little groups are $SU(2)$ (the double cover of $SO(3)$, acting in the spatial variables) and $SL(2, \mathbb{R})$ (the double cover of $SO^\uparrow(1, 2)$, acting in $p_0p_1p_3$-space). For $m = 0$, we may take the representative point in $X_0^\pm$ to be $p_0^\pm = (\pm 1, 0, 0, \pm 1)$; the little group is

$$H_{p_0^\pm} = \left\{ M_{\theta,b} = \begin{pmatrix} e^{i\theta} & b \\ 0 & e^{-i\theta} \end{pmatrix} : \theta \in \mathbb{R},\ b \in \mathbb{C} \right\}. \tag{4.40}$$

The reader may verify that this is a double cover of the group of rigid motions (translations and rotations) in the complex plane. Finally, the little group for the orbit $\{0\}$ is $SL(2, \mathbb{C})$ itself.

To proceed further we need to describe $\pi_{p,\sigma}$ more explicitly. In general, if G is a locally compact group and H is a closed subgroup such that G/H has a G-invariant measure μ, and σ is a unitary representation of H on $\mathcal{H}$, the induced representation $\mathrm{ind}_H^G(\sigma)$ may be described as follows. For $g \in G$, let $\overline{g}$ be the image of g in G/H, and let $\mathcal{F}_0$ be the space of continuous $\mathcal{H}$-valued functions f on G that satisfy

$$f(gh) = \sigma(h)^{-1} f(g) \qquad (g \in G,\ h \in H)$$

such that $\{\overline{g} : g \in \operatorname{supp} f\}$ is compact in G/H. Since σ is unitary, for $f \in \mathcal{F}_0$ we have $\|f(gh)\|_{\mathcal{H}} = \|f(g)\|_{\mathcal{H}}$, so $\|f(g)\|_{\mathcal{H}}$ depends only on $\overline{g}$. We define $\mathcal{F}$ to be the completion of $\mathcal{F}_0$ with respect to the Hilbert norm

$$\|f\|_{\mathcal{F}} = \left[\int_{G/H} \|f(g)\|_{\mathcal{H}}^2 \, d\mu(\overline{g}) \right]^{1/2}.$$

Then $\mathrm{ind}_H^G(\sigma)$ is the representation of G on $\mathcal{F}$ by left translation:

$$[\mathrm{ind}_H^G(\sigma)(g)]f(g') = f(g^{-1}g').$$

An alternative, more geometrically appealing description of $\operatorname{ind}_H^G(\sigma)$ is as the representation of G by left translations on the L^2 sections of the homogeneous vector bundle B determined by σ. B is the quotient of $G \times \mathcal{H}$ by the equivalence relation

$$(g, v) \sim (gh, \sigma(h)^{-1}v) \qquad (g \in G,\ h \in H,\ v \in \mathcal{H}),$$

and $f \in \mathcal{F}$ is identified with the section whose value at g is the image of $(g, f(g))$ under the equivalence relation.

In our case, the representation $\pi_{p,\sigma}$ of $G = \mathbb{R}^4 \ltimes SL(2, \mathbb{C})$ acts on $\mathcal{H}_\sigma$-valued functions f on G that satisfy

(4.41)
$$f((a, A)(b, B)) = e^{i\langle p,b\rangle}\sigma(B)^{-1}f(a, A) \qquad (a, b \in \mathbb{R}^4,\ A \in SL(2, \mathbb{C}),\ B \in H_p).$$

Since

$$(b, I)(a, A) = (a + b, A) = (a, A)(A^{-1}b, I)$$

the action of the space-time translation group is given by

$$\begin{aligned}[\pi_{p,\sigma}(b, I)f](a, A) &= f((-b, I)(a, A)) = f((a, A)(-A^{-1}b, I)) = e^{-i\langle p,A^{-1}b\rangle}f(a, A)\\ &= e^{-i\langle A^{\dagger-1}p,b\rangle}f(a, A).\end{aligned}$$

Hence, the energy and momentum operators — the infinitesimal generators of the time and space translations — are multiplication by the functions $(a, A) \mapsto (A^{\dagger-1}p)_\mu$. In particular, the operator corresponding to $E^2 - |\mathbf{p}|^2$, which is the square of the rest mass of a particle, is the scalar m^2 on $X_m^\pm$ and $-m^2$ on Y_m, and the energy operator is positive on X_m^+ and negative on X_m^-. It follows that *the representations associated to the orbits X_m^+ describe particles of mass m.* The orbits X_m^- correspond to the negative-energy solutions of the relativistic wave equations whose explanation requires quantum fields and antiparticles as discussed in the following chapter. The orbits Y_m would correspond to particles of imaginary mass, and the orbit $\{0\}$ would correspond to particles whose state is invariant under all space-time translations, neither of which exist in reality.[8]

For $m > 0$, the little group of the point $p_m^+ = (m, 0, 0, 0)$ (the energy and momentum in the rest frame of the particle) is $SU(2)$, and the action of $SU(2)$, the double cover of $SO(3)$, is what determines the spin of the particle. More precisely, *if we take $\sigma = \sigma_s$ to be the unique irreducible representation of $SU(2)$ of dimension $2s + 1$ ($s = 0, \frac{1}{2}, 1, \ldots$), the representation $\pi_{p_m^+,\sigma_s}$ describes a particle of mass m and spin s.* To nail this down completely, one should compute $\pi_{p_m^+,\sigma_s}(0, A)$ for $A \in SU(2)$ and show that its infinitesimal generators (the angular momentum operators) are the sum of the appropriate orbital angular momentum and spin operators. For this purpose it is better to give an alternate description of $\pi_{p_m^+,\sigma_s}$ as acting on a space of $\mathcal{H}_\sigma$-valued functions on the orbit X_m^+ rather than a space of functions on G. We shall work this out for $s = 0$ and $s = \frac{1}{2}$ below; see Varadarajan **[123]** or Simms **[110]** for the general case.

Henceforth we shall simplify the notation by writing

$$\pi_{m,s} = \pi_{p_m^+,\sigma_s}.$$

In the simplest case $s = 0$, where σ_s is the trivial one-dimensional representation of $SU(2)$, the space on which $\pi_{m,0}$ acts consists of functions on G that are

---

[8]However, the trivial one-dimensional representation, which is associated to the orbit $\{0\}$, may be said to represent the vacuum.

constant on the cosets of $G_{p_m^+}$ and hence can be considered as functions on the orbit $X_m^+ \cong G/G_{p_m^+}$. With this identification, the Hilbert space on which $\pi_{m,0}$ acts is $L^2(X_m^+, d^3\mathbf{p}/(2\pi)^3\omega_\mathbf{p})$, where

$$\omega_\mathbf{p} = \sqrt{|\mathbf{p}|^2 + m^2}$$

and $d^3\mathbf{p}/(2\pi)^3\omega_\mathbf{p}$ is the invariant measure on X_m^+. (See §1.3; we have chosen a different normalization of the invariant measure here.) For each fixed t, the inverse Fourier transform

$$\phi(t, \mathbf{x}) = \int \exp(-i\omega_\mathbf{p} t + \mathbf{p} \cdot \mathbf{x}) \frac{f(\omega_\mathbf{p}, \mathbf{p})}{\sqrt{\omega_\mathbf{p}}} \frac{d^3\mathbf{p}}{(2\pi)^3}$$

defines a unitary map $f \mapsto \phi(t, \cdot)$ from $L^2(X_m^+, d^3\mathbf{p}/(2\pi)^3\omega_\mathbf{p})$ onto $L^2(\mathbb{R}^3)$. The maps $f(\omega_\mathbf{p}, \mathbf{p}) \mapsto e^{-i\omega_\mathbf{p} t} f(\omega_\mathbf{p}, \mathbf{p})$ form a one-parameter unitary group on the former space whose infinitesimal generator is multiplication by $\omega_\mathbf{p} = \sqrt{|\mathbf{p}|^2 + m^2}$, and the inverse Fourier transform intertwines it with the time-translation group $\phi(s, \mathbf{x}) \mapsto \phi(s+t, \mathbf{x})$, whose infinitesimal generator is therefore $\sqrt{-\nabla^2 + m^2}$. Hence ϕ satisfies the Klein-Gordon equation, and we have recovered the description of the state space as position-space wave functions. A similar description is available for higher spin.

For $m = 0$, the little group of the point $p_0^+ = (1, 0, 0, 1)$ is given by (4.40). It has two sets of irreducible representations: the one-dimensional representations τ_n $(n \in \mathbb{Z})$ given by $\tau_n(M_{\theta,b}) = e^{in\theta}$, and a family of infinite-dimensional ones induced from the one-dimensional representations of the subgroup $\{M_{0,b} : b \in \mathbb{C}\}$. The latter do not correspond to any known physical particles, but *the representation* $\pi_{p_0^+, \tau_n}$ *describes a massless particle of spin* $\frac{1}{2}|n|$. Again we shall abbreviate: for $s = 0, \frac{1}{2}, 1, \frac{3}{2}, \ldots,$

$$\pi_{0,\pm s} = \pi_{p_0^+, \tau_{2s}}.$$

This requires some more explanation. For massive particles, the spin is the particle's angular momentum when the particle is at rest. It is a vector-valued operator, and for any unit vector $\mathbf{u} \in \mathbb{R}^3$ one can consider its component in the direction $\mathbf{u}$, i.e., the spin about the $\mathbf{u}$-axis. When one says that the particle has spin s, one means that s is the largest eigenvalue of one (and hence all) of these components. Massless particles, however, are never at rest. They always travel with the speed of light, and the only component of spin that makes sense is the one about the axis along which they are traveling. We are therefore dealing with irreducible representations not of $SO(3)$ but of $SO(2)$, and these are parametrized by integers. In particular, the momentum vector p_0^+ corresponds to travel along the x^3-axis, and rotation through an angle θ about this axis is the image of $M_{\theta/2,0}$ in $SO(3)$. Thus, in the representation induced from τ_n, the x^3-spin has the value $\frac{1}{2}n$. To distinguish this from the situation pertaining to massive particles, one often says that the particle has *helicity* $\frac{1}{2}n$.

Now, n can be either positive or negative, so one must ask what is the physical difference beteween $\pi_{0,s}$ and $\pi_{0,-s}$. To understand this, one needs to consider the effect of the parity transformation $(t, \mathbf{x}) \mapsto (t, -\mathbf{x})$, that is, to enlarge the symmetry group to the double cover $\widetilde{G}$ of the orthochronous Poincaré group. The representations $\pi_{m,s}$ with $m > 0$ all extend to representations of this larger group, but $\pi_{0,\pm s}$ does not. Rather, the direct sum $\pi_{0,s} \oplus \pi_{0,-s}$ extends to a representation of $\widetilde{G}$ in which the parity operator interchanges the two summands. Since parity is a symmetry of electromagnetism, gravity, and quantum chromodynamics, it is this direct sum that properly describes the associated massless particles: photons, gluons, and gravitons, which have spins 1, 1, and 2, respectively. However, parity is emphatically *not* a symmetry of the weak interaction, so in theories in which neutrinos are treated as massless, the two representations $\pi_{0,\pm 1/2}$ describe two *different* species of particle: $\pi_{0,-1/2}$ gives neutrinos and $\pi_{0,1/2}$ gives antineutrinos.[9]

How does the Dirac model for massive spin-$\frac{1}{2}$ particles, which uses functions with values in $\mathbb{C}^4$ rather than $\mathbb{C}^2$, fit into this picture? In terms of the representation $\pi_{m,1/2}$, the answer is quite simple. The direct sum of two copies of $\pi_{m,1/2}$ acts on certain functions on G with values in $\mathbb{C}^4 = \mathbb{C}^2 \times \mathbb{C}^2$, and the representation π_D that corresponds to the positive-energy solutions of the Dirac equation is the subrepresentation of $\pi_{m,1/2} \oplus \pi_{m,1/2}$ on the subspace of functions $f = (f_1, f_2, f_3, f_4)$ such that $(f_1, f_2) = (f_3, f_4)$. Here the little group is $SU(2)$, and $\sigma_{1/2}$ is simply the identity representation of $SU(2)$ on $\mathbb{C}^2$.

To see how this works, recall the representation Φ of $SL(2,\mathbb{C})$ defined by (4.21). Since $A^{\dagger -1} = A$ for $A \in SU(2)$, we can state the covariance condition (4.41) for the representation π_D in terms of Φ as $f((a,A)(b,B)) = e^{i\langle p_m^+, b\rangle}\Phi(B^{-1})f(a,A)$. That is, the Hilbert space for π_D is

(4.42)

$$\mathcal{H}_{\pi_D} = \Bigg\{ f : G \to \mathbb{C}^4 : f((a,A)(b,B)) = e^{i\langle p_m^+, b\rangle}\Phi(B^{-1})f(a,A) \text{ for } B \in SU(2),$$

$$(f_1, f_2) = (f_3, f_4), \quad \int_{X_m^+} |f(a,A)|^2 \frac{d^3\mathbf{p}}{(2\pi)^3 \omega_{\mathbf{p}}} < \infty \quad \left((\omega_{\mathbf{p}}, \mathbf{p}) = A^{\dagger -1} p_m^+\right) \Bigg\}.$$

The connection with the Dirac equation is as follows. Let V be the subbundle of the product bundle $X_m^+ \times \mathbb{C}^4$ over X_m^+ whose fiber at p is $V_p = \{v : p_\mu \gamma^\mu v = mv\}$, where the γ^μ are the Weyl form (4.12) of the Dirac matrices. Thus, the sections of V are the $\mathbb{C}^4$-valued functions f on X_m^+ that satisfy the Fourier-transformed Dirac equation

$$p_\mu \gamma^\mu f(p) = mf(p).$$

We define a Hilbert norm on the fiber V_p by

$$\|v\|_p^2 = \frac{m|v|^2}{p_0}, \tag{4.43}$$

where $|v|$ is the Euclidean norm on $\mathbb{C}^4$. The crucial point is that *if* $A \in SL(2,\mathbb{C})$, *the operator* $\Phi(A)$ *is an isometry from* V_p *to* $V_{A^{\dagger -1}p}$. Indeed, if $v \in V_p$, by (4.22) — recalling that $A\gamma$ is really $\kappa(A)\gamma$ — we have

$$(A^{\dagger -1}p)_\mu \gamma^\mu \Phi(A)v = p_\mu \kappa(A)^\mu_\nu \gamma^\nu \Phi(A)v = p_\mu \Phi(A)\gamma^\mu v = m\Phi(A)v,$$

so $\Phi(A)v \in V_{A^{\dagger -1}p}$. Moreover, since γ^j is skew-adjoint for $j > 0$, for $v \in V_p$ we have $m|v|^2 = p_\mu \langle \gamma^\mu v | v\rangle = p_0 \langle \gamma^0 v | v \rangle$, so $\|v\|_p^2 = \langle \gamma^0 v | v\rangle$. (Here $\langle \cdot | \cdot \rangle$ is the Euclidean inner product on $\mathbb{C}^4$. The condition $v \in V_p$ is essential for the positivity of $\langle \gamma^0 v | v\rangle$!) The fact that $\Phi(A)$ is an isometry now follows since

$$\Phi(A)^\dagger \gamma^0 \Phi(A) = \begin{pmatrix} A^{-1} & 0 \\ 0 & A^\dagger \end{pmatrix} \begin{pmatrix} 0 & I \\ I & 0 \end{pmatrix} \begin{pmatrix} A^{\dagger -1} & 0 \\ 0 & A \end{pmatrix} = \begin{pmatrix} 0 & I \\ I & 0 \end{pmatrix} = \gamma^0.$$

[9]Photons, gravitons, and some gluons are their own antiparticles; other gluons are distinguished from their antiparticles by their color charge.

We now transfer the representation π_D to a space of sections of V. Let $\mathcal{H}_\Pi$ be the Hilbert space of sections f of V that are square-integrable with respect to the pointwise norm $\|\cdot\|_p$ and the invariant measure $d^3\mathbf{p}/(2\pi)^3\omega_\mathbf{p}$ on X_m^+, that is,

$$\int \|f(\omega_\mathbf{p},\mathbf{p})\|_p^2 \frac{d^3\mathbf{p}}{(2\pi)^3\omega_\mathbf{p}} = \int \frac{m|f(\omega_\mathbf{p},\mathbf{p})|^2}{\omega_\mathbf{p}^2}\frac{d^3\mathbf{p}}{(2\pi)^3} < \infty.$$

In view of the results of the preceding paragraph, the following formula defines a unitary representation Π of $\mathbb{R}^4 \ltimes SL(2,\mathbb{C})$ on $\mathcal{H}_\Pi$:

$$\Pi(b,B)f(p) = e^{-i\langle p,b\rangle}\Phi(B)f(B^\dagger p).$$

We claim that *the map T defined by*

$$Tf(a,A) = \exp[i\langle p_m^+, A^{-1}a\rangle]\Phi(A^{-1})f(A^{\dagger-1}p_m^+)$$

gives a unitary equivalence of Π *and* π_D.

To prove this, first observe that Tf has the right $SU(2)$-covariance property: if $B \in SU(2)$,

$$\begin{aligned} Tf((a,A)(b,B)) &= Tf(a+Ab, AB) \\ &= \exp[i\langle p_m^+, (AB)^{-1}(a+Ab)\rangle]\Phi((AB)^{-1})f((AB)^{\dagger-1}p_m^+) \\ &= \exp[i\langle B^{-1\dagger}p_m^+, A^{-1}a+b\rangle]\Phi(B^{-1})\Phi(A^{-1})f(A^{-1\dagger}B^{-1\dagger}p_m^+) \\ &= \exp[i\langle p_m^+, b\rangle]\Phi(B^{-1})Tf(a,A), \end{aligned}$$

since $B^{-1\dagger}p_m^+ = p_m^+$. Next, observe that since $p_m^+ = (m,0,0,0)$ and $\gamma^0 = \left(\begin{smallmatrix} 0 & I \\ I & 0 \end{smallmatrix}\right)$, the fiber $V_{p_m^+}$ is just $\{v : \gamma^0 v = v\} = \{v : (v_1,v_2) = (v_3,v_4)\}$, and $\|v\|_{p_m^+} = |v|$. Since $\Phi(A^{-1})$ is an isometry from $V_{A^{\dagger-1}p_m^+}$ to $V_{p_m^+}$, it follows that T is an isometry from $\mathcal{H}_\Pi$ into $\mathcal{H}_{\pi_D}$. Since π_D is irreducible, it remains only to check that T is an intertwining operator from Π to π_D:

$$\begin{aligned} T[\Pi(b,B)f](a,A) &= \exp[i\langle p_m^+, A^{-1}a\rangle S(A^{-1})[\Pi(b,B)f](A^{\dagger-1}p_m^+) \\ &= \exp[i\langle p_m^+, A^{-1}a\rangle - i\langle A^{\dagger-1}p_m^+, b\rangle]\Phi(A^{-1})\Phi(B)f(B^\dagger A^{\dagger-1}p_m^+) \\ &= \exp[i\langle p_m^+, A^{-1}(a-b)\rangle]\Phi((B^{-1}A)^{-1})f((B^{-1}A)^{\dagger-1}p_m^+) \\ &= Tf(B^{-1}(a-b), B^{-1}A) = Tf((b,B)^{-1}(a,A)) \\ &= [\pi_D(b,B)Tf](a,A). \end{aligned}$$

The Hilbert space $\mathcal{H}_\Pi$ consists of momentum-space wave functions, and one can return to the position-space wave functions via the Fourier transform. That is, if $f \in \mathcal{H}_\Pi$, let

$$\psi(t,\mathbf{x}) = \int \exp(-i\omega_\mathbf{p}t + \mathbf{p}\cdot\mathbf{x})\frac{\sqrt{m}\,f(\omega_\mathbf{p},\mathbf{p})}{\omega_\mathbf{p}}\frac{d^3\mathbf{p}}{(2\pi)^3}. \tag{4.44}$$

The map $f \mapsto \psi(t,\cdot)$ is an isometry from $\mathcal{H}_\Pi$ into $L^2(\mathbb{R}^3,\mathbb{C}^4)$ for each t, and the fact that $p_\mu\gamma^\mu f(p) = mf(p)$ implies that ψ satisfies the free Dirac equation $i\gamma^\mu\partial_\mu\psi = m\psi$. The space of ψ's thus obtained is the space of positive-energy solutions to the Dirac equation; that is, for each t, $\psi(t,\cdot)$ belongs to the subspace of $L^2(\mathbb{R}^3,\mathbb{C}^4)$ on which the Dirac Hamiltonian is a positive operator (called $\mathcal{H}_+$ in §4.3).

One obtains the negative-energy solutions by playing the same game with the representation $\pi_{m,1/2}^- = \pi_{p_m^-,\sigma_{1/2}}$ and the orbit X_m^-. Since the equation $p_\mu\gamma^\mu v = mv$ becomes $-\gamma^0 v = v$ when $p = p_m^-$, the analogue of π_D here is the subrepresentation of $\pi^-_{m,1/2} \oplus \pi^-_{m,1/2}$ on the space of functions f such that $(f_3, f_4) = -(f_1, f_2)$. The resolution of the negative-energy paradox by a reinterpretation of the Dirac equation in terms of fields will be explained in the next chapter.

4.5. Multiparticle state spaces

Having discussed state spaces for single particles, we now turn to the question of constructing state spaces for systems of particles. As we mentioned in §3.1, if we have k particles with state spaces $\mathcal{H}_1, \ldots, \mathcal{H}_k$, we can describe states of the k-particle system using the tensor product $\mathcal{H}_1 \otimes \cdots \otimes \mathcal{H}_k$. This is the completion of the algebraic tensor product of the $\mathcal{H}_j$'s with respect to the inner product

$$\langle u_1 \otimes u_2 \cdots \otimes u_k,\, v_1 \otimes v_2 \cdots \otimes v_k \rangle = \langle u_1 | v_1 \rangle \langle u_2 | v_2 \rangle \cdots \langle u_k | v_k \rangle.$$

To put it another way, if $\{e^j_n\}_{n=1}^\infty$ is an orthonormal basis for $\mathcal{H}_j$, then

$$\{e^1_{n_1} \otimes \cdots \otimes e^k_{n_k} : 1 \le n_1, \ldots, n_k < \infty\}$$

is an orthonormal basis for $\mathcal{H}_1 \otimes \cdots \otimes \mathcal{H}_k$. In this setting, $u_1 \otimes \cdots \otimes u_k$ describes the compound state in which the jth particle is in state u_j for each j. Superpositions of states that are not reducible to single products also occur; for example, if $u_\pm$ are states in which particle A has spin "up" or "down" along some axis, and $v_\pm$ similarly for particle B, then $(u_+ \otimes v_+ + u_- \otimes v_-)/\sqrt{2}$ is a state in which the spins of the particles are aligned but the direction is unspecified. On the macroscopic level, the famous Schrödinger cat paradox [**67**] involves states of the form

$$(\text{undecayed atom}) \otimes (\text{live cat}) + (\text{decayed atom}) \otimes (\text{dead cat}).$$

If we are describing a system of k particles of the same species (k electrons, for example), we can take the Hilbert spaces $\mathcal{H}_j$ all to be the same space $\mathcal{H}$. However, in this case the putative k-particle state space $\bigotimes^k \mathcal{H}$ is too big. It is a fundamental fact that particles of the same species are truly indistinguishable; there is no way to attach labels to two electrons so as to tell which one is which. Mathematically, the permutation group on k letters, S_k, acts on $\bigotimes^k \mathcal{H}$ in a canonical way,

$$\sigma(u_1 \otimes \cdots \otimes u_k) = u_{\sigma(1)} \otimes \cdots \otimes u_{\sigma(k)},$$

and the only unit vectors that can represent sets of k identical particles are those that are mapped into scalar multiples of themselves by the permutation group: $\sigma(\mathbf{v}) = R(\sigma)\mathbf{v}$, where $R(\sigma)$ is a scalar of absolute value 1. Clearly R is a one-dimensional representation of S_k, of which there are only two, the trivial representation and the representation $R(\sigma) = \operatorname{sgn}\sigma$, which correspond to symmetric and antisymmetric tensors respectively. In short, we are led to consider the subspaces

$$\text{Ⓢ}^k\mathcal{H} = \left\{\mathbf{v} \in \bigotimes^k \mathcal{H} : \sigma(\mathbf{v}) = \mathbf{v},\ \forall \sigma \in S_k\right\},$$
$$\bigwedge^k \mathcal{H} = \left\{\mathbf{v} \in \bigotimes^k \mathcal{H} : \sigma(\mathbf{v}) = (\operatorname{sgn}\sigma)\mathbf{v},\ \forall \sigma \in S_k\right\}.$$

Which of these is appropriate depends on the species of particle in question. Particles whose multiparticle states are symmetric are called *Bosons*; those whose multiparticle states are antisymmetric are called *Fermions*. It turns out that Bosons are precisely those particles whose spin is an integer; this is the *spin-statistics theorem*, and we shall say more about it in §5.3 and §6.5.

Let us take a closer look at the Boson space $\circledS^k\mathcal{H}$. There is a canonical projection P_s (s for symmetric) from $\bigotimes^k\mathcal{H}$ onto $\circledS^k\mathcal{H}$:

$$P_s(u_1 \otimes \cdots \otimes u_k) = \frac{1}{k!}\sum_{\sigma\in S_k} u_{\sigma(1)} \otimes \cdots \otimes u_{\sigma(k)}.$$

To analyze P_s it is convenient to consider an orthonormal basis $\{e_j\}$ for $\mathcal{H}$. Suppose $u_1, \ldots, u_k$ are all elements of this basis, say $u_j = e_{i_1}$ for n_1 values of j, ..., $u_j = e_{i_m}$ for n_m values of j, where $i_1, \ldots, i_m$ are distinct and $n_1 + \cdots + n_m = k$, and let $\mathbf{u} = u_1 \otimes \cdots \otimes u_k$. The sum defining $P_s(\mathbf{u})$ breaks up into groups of identical terms, corresponding to permutations that only permute the factors of e_i among themselves. Each group has $n_1! \cdots n_m!$ terms, there are $k!/n_1! \cdots n_m!$ groups, and terms in different groups are orthogonal to each other. Hence the sum is a sum of $k!/n_1! \cdots n_m!$ distinct terms, all orthogonal to each other, each of norm $n_1! \cdots n_m!$, and so

$$\|P_s(\mathbf{u})\| = \frac{n_1! \cdots n_m!}{k!}\sqrt{\frac{k!}{n_1! \cdots n_m!}} = \sqrt{\frac{n_1! \cdots n_m!}{k!}}.$$

Moreover, if $v_1, \ldots, v_k$ are also elements of $\{e_j\}$ and $\mathbf{v} = v_1 \otimes \cdots \otimes v_k$, we have

$$\langle P_s(\mathbf{u})|\mathbf{v}\rangle = \left\{\begin{matrix} n_1! \cdots n_m!/k! & \text{if } \mathbf{v} = \sigma(\mathbf{u}) \text{ for some } \sigma \in S_k \\ 0 & \text{otherwise} \end{matrix}\right\} = \langle \mathbf{u}|P_s(\mathbf{v})\rangle,$$

Thus P_s is an *orthogonal* projection.

For any sequence $\{n_j\}$ of nonnegative integers with $\sum_1^\infty n_j = k$, let

$$|n_1, n_2, \ldots\rangle = \sqrt{\frac{n_1! n_2! \cdots}{k!}} P_s\big((\otimes^{n_1} e_1) \otimes (\otimes^{n_2} e_2) \otimes \cdots\big), \tag{4.45}$$

where $\otimes^n e_j = e_j \otimes \cdots \otimes e_j$ (n factors). (Here we are using the convenient Dirac convention of labeling a vector simply by the indices that specify it.) Then

$$\{|n_1, n_2, \ldots\rangle : \sum n_j = k\}$$

is an orthonormal basis for $\circledS^k\mathcal{H}$.

The Boson Fock space. In order to form a state space that can accomodate any number of identical Bosons, we put all the spaces $\circledS^k\mathcal{H}$ together to form the complete symmetric tensor algebra over $\mathcal{H}$, also known as the *Boson Fock space* over $\mathcal{H}$:

$$\mathcal{F}_s(\mathcal{H}) = \bigoplus_{k=0}^{\infty} \circledS^k\mathcal{H}.$$

Here $\circledS^0\mathcal{H} = \mathbb{C}$, and the object on the right is the orthogonal direct sum of Hilbert spaces. We combine the orthonormal bases of the preceding paragraph for the spaces $\circledS^k\mathcal{H}$, together with

$$|0, 0, \ldots\rangle = 1 \in \circledS^0\mathcal{H},$$

to obtain the orthonormal basis

$$\Big\{|n_1, n_2, \ldots\rangle : n_j \geq 0, \sum n_j < \infty\Big\}$$

for $\mathcal{F}_s(\mathcal{H})$. $\mathcal{F}_s(\mathcal{H})$ is a state space that describes arbitrary numbers of identical Bosons, including none at all. The basis vector $|n_1, n_2, \ldots\rangle$ specifies the state in which there are n_j particles in state e_j for each j.

There are no states in $\mathcal{F}_s(\mathcal{H})$ that contain infinitely many particles, but there are states in which there is a positive probability of finding arbitrarily large numbers of particles, namely, infinite linear combinations of basis vectors $|n_1, n_2, \ldots\rangle$ with $\sum n_j$ arbitrarily large. It is convenient also to consider the *finite-particle subspace* of states in which the total number of particles is bounded above, that is,

$$\mathcal{F}_s^0(\mathcal{H}) = \text{ the algebraic direct sum of the spaces } \textcircled{S}^k\mathcal{H}.$$

On this space we define the *number operator* N by

$$N\left(\sum a_{n_1n_2\ldots}|n_1, n_2, \ldots\rangle\right) = \sum (n_1 + n_2 + \cdots)a_{n_1n_2\ldots}|n_1, n_2, \ldots\rangle.$$

If $\mathbf{u} \in \textcircled{S}^k\mathcal{H}$ we have $N\mathbf{u} = k\mathbf{u}$, so N is the observable that tells how many particles are in a given state. (N has a unique extension to an unbounded self-adjoint operator on $\mathcal{F}_s(\mathcal{H})$.)

We now introduce some operators that raise or lower the number of particles. For this purpose we retreat temporarily to the full tensor algebra $\mathcal{F}(\mathcal{H}) = \bigoplus_0^\infty \bigotimes^k\mathcal{H}$ and its dense subspace $\mathcal{F}^0(\mathcal{H})$, the algebraic direct sum of the tensor spaces $\bigotimes^k\mathcal{H}$. The number operator N is defined on $\mathcal{F}^0(\mathcal{H})$ just as above:

$$N = kI \text{ on } \bigotimes^k\mathcal{H}.$$

For $v \in \mathcal{H}$, the operators $B(v)$ and $B(v)^\dagger$ on $\mathcal{F}^0(\mathcal{H})$ are defined by

$$B(v)(u_1 \otimes \cdots \otimes u_k) = \langle v|u_1\rangle u_2 \otimes \cdots \otimes u_k,$$
$$B(v)^\dagger(u_1 \otimes \cdots \otimes u_k) = v \otimes u_1 \otimes \cdots \otimes u_k.$$

(When $k = 0$, $B(v)1 = 0$ and $B(v)^\dagger 1 = v$, and when $k = 1$, $B(v)u = \langle v, u\rangle 1$, where 1 is the unit element in $\mathbb{C} = \bigotimes^0\mathcal{H}$.) It is easily verified that $\langle B(v)\mathbf{u}|\mathbf{w}\rangle = \langle \mathbf{u}|B(v)^\dagger\mathbf{w}\rangle$ for $\mathbf{u}, \mathbf{w} \in \mathcal{F}^0(\mathcal{H})$, so the notation is consistent. Note that $B(v)$ depends anti-linearly on v, whereas $B(v)^\dagger$ depends linearly on v.

The operator $B(v)^\dagger$ does not preserve the symmetric subspace $\mathcal{F}_s^0(\mathcal{H})$, as the factor v is inserted only on the left, but $B(v)$ does preserve this subspace:

$$\begin{aligned} B(v)P_s(u_1 \otimes \cdots \otimes u_k) &= \frac{1}{k!}\sum_{j=1}^k \langle v, u_j\rangle \sum_{\sigma(1)=j} u_{\sigma(2)} \otimes \cdots \otimes u_{\sigma(k)} \\ &= \frac{1}{k}\sum_{j=1}^k \langle v, u_j\rangle P_s(u_1 \otimes \cdots \widehat{u}_j \cdots \otimes u_k), \end{aligned}$$

where the hat indicates that the term is omitted. If we take the u_j and v to be elements of an orthonormal basis $\{e_l\}$ for $\mathcal{H}$, say $v = e_l$ and $u_j = e_l$ for n values of j, the sum on the right consists of n identical nonzero terms and $k - n$ zero terms. Taking into account the normalizations in (4.45), we see that the action of $B(e_j)$ on the basis $\{|n_1, n_2, \ldots\rangle\}$ for $\mathcal{F}_s^o(\mathcal{H})$ is

$$B(e_j)|n_1, \ldots, n_j, \ldots\rangle = \sqrt{\frac{n_j}{k}}|n_1, \ldots, n_j - 1, \ldots\rangle \quad (k = n_1 + n_2 + \cdots).$$

For reasons that will become clearer shortly, it is convenient to get rid of the factor of $1/\sqrt{k}$ in this formula by defining the operator $A(v)$ on $\mathcal{F}_s^0(\mathcal{H})$ as

$$A(v) = B(v)\sqrt{N} = \sqrt{N+1}B(v) \quad \text{on } \mathcal{F}_s^0(\mathcal{H}), \tag{4.46}$$

that is,

$$A(v)\mathbf{w} = \sqrt{k}\,B(v)\mathbf{w} \text{ for } \mathbf{w} \in ⓢ^k\mathcal{H},$$

so that

$$A(e_j)|n_1,\ldots,n_j,\ldots\rangle = \sqrt{n_j}|n_1,\ldots,n_j-1,\ldots\rangle.$$

The adjoint of $A(v)$ as an operator on $\mathcal{F}_s^0(\mathcal{H})$ (rather than $\mathcal{F}_0(\mathcal{H})$) is given by

$$A(v)^\dagger = P_s B(v)^\dagger \sqrt{N+I} = P_s\sqrt{N}\,B(v)^\dagger \quad \text{on } \mathcal{F}_s^0(\mathcal{H}). \tag{4.47}$$

(An extra factor P_s could be inserted on the right of the formula defining $A(v)$; it has no effect since the domain of $A(v)$ is contained in the range of P_s.) Informally speaking, $A(v)^\dagger$ creates a particle in the state v, whereas $A(v)$ destroys a particle in the state v and annihilates any multiparticle state in which no particle has any probability of being in the state v. For this reason, $A(v)^\dagger$ and $A(v)$ are called *creation* and *annihilation* operators.

The most important consequence of introducing the factor of $\sqrt{N}$ into $A(v)$ is that the operators $A(v)$ satisfy the following variant of the canonical commutation relations:

$$[A(v), A(w)^\dagger] = \langle v|w\rangle I, \qquad [A(v), A(w)] = [A(v)^\dagger, A(w)^\dagger] = 0 \tag{4.48}$$

for any $v, w \in \mathcal{H}$. To see this, we compute the action of $A(v)A(w)^\dagger$ and $A(w)^\dagger A(v)$ on $\mathbf{u} = P_s(u_1 \otimes \cdots \otimes u_k) \in ⓢ^k\mathcal{H}$:

$$\begin{aligned} A(v)A(w)^\dagger\mathbf{u} &= (k+1)B(v)P_sB(w)^\dagger P_s(u_1\otimes\cdots\otimes u_k) \\ &= (k+1)B(v)P_s\left[\frac{1}{k!}\sum_{\sigma\in S_k} w\otimes u_{\sigma(1)}\otimes\cdots\otimes u_{\sigma(k)}\right] \\ &= \frac{1}{k!}B(v)\sum_{j=0}^{k}\sum_{\sigma\in S_k} u_{\sigma(1)}\otimes\cdots\otimes w\otimes\cdots\otimes u_{\sigma(k)} \quad (w \text{ in } j\text{th place}) \\ &= \frac{1}{k!}\sum_{\sigma\in S_k}\langle v|w\rangle u_{\sigma(1)}\otimes\cdots\otimes u_{\sigma(k)} \\ &\quad + \frac{1}{k!}\sum_{j=1}^{k}\sum_{\sigma\in S_k}\langle v|u_{\sigma(1)}\rangle u_{\sigma(2)}\otimes\cdots\otimes w\otimes\cdots\otimes u_{\sigma(k)} \\ &= \langle v, w\rangle P_s(u_1\otimes\cdots\otimes u_k) + \sum_{j=1}^{k}\langle v|u_j\rangle P_s(w\otimes u_1\otimes\cdots\widehat{u}_j\cdots\otimes u_k), \end{aligned}$$

where $\widehat{u}_j$ means that u_j is omitted. On the other hand,

$$\begin{aligned} A(w)^\dagger A(v)\mathbf{u} &= kP_sB(w)^\dagger B(v)P_s(u_1\otimes\cdots\otimes u_k) \\ &= \frac{k}{k!}P_sB(w)^\dagger\sum_{\sigma\in S_k}\langle v, u_{\sigma(1)}\rangle u_{\sigma(2)}\otimes\cdots\otimes u_{\sigma(k)} \\ &= P_sB(w)^\dagger\sum_{j=1}^{k}\langle v|u_j\rangle P_s(u_1\otimes\cdots\widehat{u}_j\cdots\otimes u_k) \\ &= \sum_{j=1}^{k}\langle v|u_j\rangle P_s(w\otimes u_1\otimes\cdots\widehat{u}_j\cdots\otimes u_k). \end{aligned}$$

Thus $A(v)A(w)^\dagger\mathbf{u} - A(w)^\dagger A(v)\mathbf{u} = \langle v|w\rangle\mathbf{u}$.

This proves the first equality in (4.48). Moreover, an easy calculation similar to the preceding ones shows that the restrictions of $B(v)$ and $B(w)$ to $\mathcal{F}_s^0$ commute, and hence

$$A(v)A(w) = \sqrt{N+I}\,B(v)B(w)\sqrt{N} = \sqrt{N+I}\,B(w)B(v)\sqrt{N} = A(w)A(v).$$

By taking adjoints it follows that $A(v)^\dagger A(w)^\dagger = A(w)^\dagger A(v)^\dagger$, so (4.48) is established.

Rather than considering $A(v)$ for arbitrary $v \in \mathcal{H}$, it is often convenient to take v to belong to some useful orthonormal basis. Thus, given an orthonormal basis $\{e_j\}$ for $\mathcal{H}$, we set

$$A_j = A(e_j), \qquad A_j^\dagger = A(e_j)^\dagger. \tag{4.49}$$

The action of these operators on the basis (4.45) is simple:

$$\begin{aligned} A_j|n_1, \ldots, n_j, \ldots\rangle &= \sqrt{n_j}\,|n_1, \ldots, n_j - 1, \ldots\rangle, \\ A_j^\dagger|n_1, \ldots, n_j, \ldots\rangle &= \sqrt{n_j + 1}\,|n_1, \ldots, n_j + 1, \ldots\rangle. \end{aligned} \tag{4.50}$$

It follows that

$$A_j^\dagger A_j|n_1, \ldots, n_j, \ldots\rangle = n_j|n_1, \ldots, n_j, \ldots\rangle,$$

and hence

$$\sum A_j^\dagger A_j = N. \tag{4.51}$$

Moreover, by (4.48),

$$[A_j, A_k^\dagger] = \delta_{jk} I, \qquad [A_j, A_k] = [A_j^\dagger, A_k^\dagger] = 0. \tag{4.52}$$

There is one other important construction on Fock space that needs to be discussed here. Suppose U is a unitary operator on $\mathcal{H}$. Then U induces a unitary operator on all the tensor powers of $\mathcal{H}$ and hence a unitary operator $\mathcal{F}(U)$ on the full tensor algebra $\mathcal{F}(\mathcal{H}) = \bigoplus_0^\infty \bigotimes^k \mathcal{H}$, by the formula

$$\mathcal{F}(U)(u_1 \otimes \cdots \otimes u_k) = Uu_1 \otimes \cdots \otimes Uu_k. \tag{4.53}$$

$\mathcal{F}(U)$ clearly commutes with the number operator N and with the projections onto the symmetric and antisymmetric subspaces. Moreover, for any $v \in \mathcal{H}$,

$$\begin{aligned} \mathcal{F}(U)B(v)\mathcal{F}(U)^{-1}(u_1 \otimes \cdots \otimes u_k) &= \mathcal{F}(U)\langle v|U^{-1}u_1\rangle U^{-1}u_2 \otimes \cdots \otimes U^{-1}u_k \\ &= \langle Uv|u_1\rangle u_2 \otimes \cdots \otimes u_k = B(Uv)(u_1 \otimes \cdots \otimes u_k), \end{aligned}$$

so that $\mathcal{F}(U)B(v)\mathcal{F}(U)^{-1} = B(Uv)$. It follows that $\mathcal{F}(U)B(v)^\dagger\mathcal{F}(U)^{-1} = B(Uv)^\dagger$ also. Combining these facts, we obtain

$$\mathcal{F}(U)A(v)\mathcal{F}(U)^{-1} = A(Uv), \qquad \mathcal{F}(U)A(v)^\dagger\mathcal{F}(U)^{-1} = A(Uv)^\dagger. \tag{4.54}$$

The Fermion Fock space. Just as for Bosons, we can define a state space for arbitrary numbers of identical Fermions: the *Fermion Fock space*

$$\mathcal{F}_a(\mathcal{H}) = \bigwedge\mathcal{H} = \bigoplus_0^\infty \bigwedge^k\mathcal{H} \qquad \left(\bigwedge^0\mathcal{H} = \mathbb{C}\right).$$

As before, the direct sum on the right is an orthogonal sum of Hilbert spaces, and the corresponding algebraic direct sum is denoted by $\mathcal{F}_a^0$.

All of the preceding notions have analogues here. First, we have the natural orthogonal projection $P_a : \bigotimes^k \mathcal{H} \to \bigwedge^k \mathcal{H}$:

$$P_a(u_1 \otimes \cdots \otimes u_k) = u_1 \wedge \cdots \wedge u_k = \frac{1}{k!} \sum_{\sigma \in S_k} (\operatorname{sgn} \sigma) u_{\sigma(1)} \otimes \cdots \otimes u_{\sigma(k)}.$$

The number operator N is again defined by $N = kI$ on $\bigwedge^k \mathcal{H}$, and for $v \in \mathcal{H}$, the annihilation and creation operators $A(v)$ and $A(v)^*$ are defined on $\mathcal{F}_a^0(\mathcal{H})$ by

$$A(v) = B(v)\sqrt{N}, \qquad A(v)^\dagger = P_a \sqrt{N}\, B(v)^\dagger. \tag{4.55}$$

The exterior product notation leads to simple formulas for $A(v)$ and $A(v)^*$:

$$A(v)(u_1 \wedge \cdots \wedge u_k) = \frac{1}{\sqrt{k}} \sum_{j=1}^{k} (-1)^j \langle v | u_j \rangle u_1 \wedge \cdots \widehat{u}_j \cdots \wedge u_k,$$
$$A(v)^\dagger (u_1 \wedge \cdots \wedge u_k) = \sqrt{k+1}\, v \wedge u_1 \wedge \cdots \wedge u_k.$$

$A(v)$ destroys a particle in the state v and gives 0 if there is no such particle; $A(v)^\dagger$ creates a particle in the state v but gives 0 if there is already a particle in that state.

The big difference here is that the analogue of (4.48) involves *anticommutation* relations. To wit,

$$\{A(v), A(w)^\dagger\} = \langle v, w \rangle I, \qquad \{A(v), A(w)\} = \{A(v)^\dagger, A(w)^\dagger\} = 0 \tag{4.56}$$

for any $v, w \in \mathcal{H}$, where

$$\{A, B\} = AB + BA.$$

To see this, observe that since $v \wedge w + w \wedge v = 0$, we have $\{A(v)^\dagger, A(w)^\dagger\} = 0$ and hence also $\{A(v), A(w)\} = 0$. Next,

$$\begin{aligned} A(v)A(w)^\dagger (u_1 \wedge \cdots \wedge u_k) &= \sqrt{k+1}\, A(v) w \wedge u_1 \wedge \cdots \wedge u_k \\ &= \langle v, w \rangle u_1 \wedge \cdots \wedge u_k + \sum_1^k (-1)^j \langle v, u_j \rangle w \wedge u_1 \wedge \cdots \widehat{u}_j \cdots \wedge u_k, \end{aligned}$$

whereas

$$A(w)^\dagger A(v)(u_1 \wedge \cdots \wedge u_k) = \sum_1^k (-1)^{j-1} \langle v | u_j \rangle w \wedge u_1 \wedge \cdots \widehat{u}_j \cdots \wedge u_k.$$

(The $\sqrt{k}$'s or $\sqrt{k+1}$'s cancel out in both cases.) Hence $\{A(v), A(w)^\dagger\} = \langle v, w \rangle I$.

By the way, in marked contrast to the Boson case, the operators $A(v)$ and $A(v)^\dagger$ extend to bounded operators on the whole Fermion Fock space. Indeed, for any $\mathbf{w} \in \mathcal{F}_a^0(\mathcal{H})$, (4.56) implies that

$$\|A(v)\mathbf{w}\|^2 + \|A(v)^\dagger \mathbf{w}\|^2 = \langle \mathbf{w} | A(v)^\dagger A(v) \mathbf{w} \rangle + \langle \mathbf{w} | A(v) A(v)^\dagger \mathbf{w} \rangle = \|v\|^2 \|\mathbf{w}\|^2,$$

so that

$$\|A(v)\| \le \|v\|, \qquad \|A(v)^\dagger\| \le \|v\|. \tag{4.57}$$

The operators $\mathcal{F}(U)$ defined by (4.53) can be considered as unitary operators on $\mathcal{F}_a(\mathcal{H})$, and the formulas (4.54) remain valid in this setting (with the same proof).

As in the Boson case, it is often convenient to express things in terms of an orthonormal basis for $\mathcal{H}$. First, if $\{e_j\}$ is such a basis,

$$\{e_{i_1} \wedge \cdots \wedge e_{i_k} : i_1 < \cdots < i_k\}$$

is an orthogonal basis for $\bigwedge^k \mathcal{H}$. Moreover, since $e_{i_1} \wedge \cdots \wedge e_{i_k}$ is $1/k!$ times a sum of $k!$ orthogonal terms $(\operatorname{sgn}\sigma)e_{i_{\sigma(1)}} \otimes \cdots \otimes e_{i_{\sigma(k)}}$, each of which has norm 1, we have $\|e_{i_1} \wedge \cdots \wedge e_{i_k}\| = 1/\sqrt{k!}$. The corresponding orthonormal basis for $\mathcal{F}_a(\mathcal{H})$ is therefore

$$\Big\{|n_1, n_2, \ldots\rangle : n_j = 0 \text{ or } 1, \sum n_j < \infty\Big\}$$

given by

$$|n_1, n_2, \ldots\rangle = \sqrt{k!}\, e_{i_1} \wedge \cdots \wedge e_{i_k},$$

where $i_1 < \cdots < i_k$ are the indices i for which $n_i = 1$. The fact that each n_j must be 0 or 1 is a precise statement of the *Pauli exclusion principle*, which says that "no two identical Fermions can occupy the same state."

As before, we set

$$A_j = A(e_j), \qquad A_j^\dagger = A(e_j)^\dagger.$$

By (4.56), we then have

$$\{A_j, A_k^\dagger\} = \delta_{jk} I, \qquad \{A_j, A_k\} = \{A_j^\dagger, A_k^\dagger\} = 0. \tag{4.58}$$

It is also easily verified that

$$A_j|n_1, \ldots, n_j, \ldots\rangle = \begin{cases} (-1)^m |n_1, \ldots, n_j - 1, \ldots\rangle & \text{if } n_j = 1, \\ 0 & \text{if } n_j = 0, \end{cases}$$

$$A_j^\dagger|n_1, \ldots, n_j, \ldots\rangle = \begin{cases} (-1)^m |n_1, \ldots, n_j + 1, \ldots\rangle & \text{if } n_j = 0, \\ 0 & \text{if } n_j = 1, \end{cases}$$

where m is the integer such that $n_i = 1$ for m values of i with $i < j$. It follows that

$$A_j^\dagger A_j |n_1, n_2, \ldots\rangle = n_j |n_1, n_2, \ldots\rangle,$$

and hence, as in (4.51),

$$\sum A_j^\dagger A_j = N.$$

Finally, we consider the state spaces for arbitrary numbers of particles of several different species. More precisely, we consider systems of *free* particles. The state space for a rigorous model of an *interacting* system might turn out to be rather different. However, our approach to interactions will be to apply perturbation theory to noninteracting systems, so we still need the free-particle state space to get started.

There is not much new to be said here. If one has K species of particles, one takes an appropriate single-particle state space $\mathcal{H}_k$ for each species, $k = 1, \ldots, K$, and forms the appropriate (Boson or Fermion) Fock space $\mathcal{F}_k$ over $\mathcal{H}_k$ for each k. Then the state space for the whole system is

$$\mathcal{F} = \mathcal{F}_1 \otimes \cdots \otimes \mathcal{F}_K.$$

For each k one has annihilation and creation operators $A_k(v)$, $A_k^\dagger(v)$ $(v \in \mathcal{H}_k)$ on $\mathcal{F}_k$, and these are taken to act on $\mathcal{F}$ by the identification

$$A_k(v) \leftrightarrow I \otimes \cdots \otimes I \otimes A_k(v) \otimes I \otimes \cdots \otimes I.$$

With these definitions, creation and/or annihilation operators pertaining to two different species of particle always commute with each other. However, it simplifies some things to postulate that such operators pertaining to *Fermions* should *anti-commute*, and this convention is commonly adopted in the physics literature. This can be achieved by a small modification in the definitions of the Fermionic annihilation and creation operators. To wit, it suffices to list the Fermion Fock spaces in some definite order, say $\mathcal{F}_1, \ldots, \mathcal{F}_J$, and then to replace the operators $A_j(v)$ on $\mathcal{F}_j$ by $A_j(v)(-1)^{N_1+\cdots+N_{j-1}}$ where N_i is the number operator on $\mathcal{F}_i$, and likewise for $A_j(v)^\dagger$. That is, the operators $A_j(v)$ and $A_j^\dagger(v)$ acquire an extra minus sign when applied to states where there is an odd number of particles of the previously listed Fermion species. This does not affect the states defined by these operators in a significant way, as $\mathbf{v}$ and $-\mathbf{v}$ define the same physical state; for the same reason, the choice of ordering is immaterial. Only the (anti)commutation relations among the operators are important.

A footnote for fans of category theory. There are some abstract structural aspects of the constructions in this section that may be worth pointing out. To begin with, let $\mathbf{H}$ be the category whose objects are Hilbert spaces and whose morphisms are (linear) contractions, i.e., linear maps of norm ≤ 1. (It might seem more natural to take the morphisms to be the linear isometries, and one could do so. But if one wants the set of morphisms to be closed under adjoints, and orthogonal direct sums to be products in the sense of category theory, one needs to include partial isometries; and every contraction can be approximated in norm by compositions of partial isometries. Either way, the isomorphisms are the unitary maps.) To each $\mathcal{H} \in \mathbf{H}$ one can associate the full Fock space $\mathcal{F}(\mathcal{H}) = \bigoplus_0^\infty \bigotimes^k \mathcal{H}$ as well as the Boson and Fermion Fock spaces $\mathcal{F}_s(\mathcal{H})$ and $\mathcal{F}_a(\mathcal{H})$, and to each morphism $A : \mathcal{H}_1 \to \mathcal{H}_2$ one can associate the morphism $\mathcal{F}(A) : \mathcal{F}(\mathcal{H}_1) \to \mathcal{F}(\mathcal{H}_2)$, defined just as in (4.53), as well as its restrictions to $\mathcal{F}_s(\mathcal{H}_1)$ and $\mathcal{F}_a(\mathcal{H}_1)$. (It is obvious that $\mathcal{F}(A)$ is a contraction whenever A is. Note, however, that if one tries to define $\mathcal{F}(A)$ on $\mathcal{F}(\mathcal{H}_1)$ by (4.53) when $\|A\| > 1$, the result is generally an unbounded operator on $\mathcal{F}(\mathcal{H}_1)$.) In short, $\mathcal{F}$, $\mathcal{F}_s$, and $\mathcal{F}_a$ are *functors* from $\mathbf{H}$ to itself. They are, in fact, "exponential functors" in the sense that they convert direct sums into tensor products: $\mathcal{F}(\mathcal{H}_1 \oplus \mathcal{H}_2) \cong \mathcal{F}(\mathcal{H}_1) \otimes \mathcal{F}(\mathcal{H}_2)$, and likewise for $\mathcal{F}_s$ and $\mathcal{F}_a$. (We leave the verification of this as an exercise for the interested reader.) See Nelson [**86**] for a related functorial construction that is relevant to quantum field theory.

CHAPTER 5

Free Quantum Fields

He had bought a large map representing the sea,
  Without the least vestige of land:
And the crew were much pleased when they found it to be
  A map they could all understand.
—Lewis Carroll, *The Hunting of the Snark* (Fit the Second)

In this chapter we take our first step into quantum field theory by discussing free fields, i.e., fields without interactions. The good thing about free fields is that they can be constructed with complete mathematical rigor, although that task requires more sophistication than one might imagine at the outset; the bad thing is that they don't exhibit any interesting physics. But the time spent on them is not wasted, because they are an essential ingredient in the interacting field theories that we shall consider in the next chapter.

5.1. Scalar fields

We begin by constructing the free quantum scalar fields of mass $m > 0$. What this means is the following: we think of solutions ϕ of the Klein-Gordon equation $(\partial^2 + m^2)\phi = 0$ as representing a classical field, and we are going to construct the corresponding quantum field. The quanta of these fields, as we shall see, are particles of spin zero and mass m.

There is a little conceptual barrier here in that there is no actual *classical physical* field that is described by the Klein-Gordon equation, and spin-zero elementary particles are also rather scarce.[1] (There are some; the pions are the least exotic examples, but since the charged pions have a mean lifetime of about 10^{-8} seconds and the neutral ones have a mean lifetime of about 10^{-16} seconds, they aren't often seen by anyone but particle physicists.) The reader is best advised to consider what we are doing as constructing a toy model for the electromagnetic field. The free electromagnetic potential (not interacting with any charged matter) in Lorentz gauge satisfies the wave equation $\partial^2 A_\mu = 0$, which is the Klein-Gordon equation but (i) with $m = 0$ and (ii) for a vector-valued rather than scalar-valued function. These conditions correspond to the fact that photons have mass 0 and spin 1. The vector nature of the field necessitates some additional algebra that does not seriously affect the underlying ideas but makes their execution a little messier, and the condition $m = 0$ complicates the situation considerably. The scalar field

[1]The ideas of quantum field theory also have interesting applications to condensed matter physics, however, and certain excitations of crystal lattices ("phonons") can be modeled by spin-zero particles. We refer the reader to Mattuck [**81**] and Zee [**136**] for more information about this subject.

with positive mass exhibits all the essential ideas and difficulties of quantum fields in general, but in the simplest possible context.

We start by deriving the quantum field as physicists (going back to Dirac) usually do, paying more attention to intuitive clarity than mathematical rigor; then we shall show how to perform the construction rigorously. The point of following the physicists' path is twofold: first, it shows how to arrive at the quantum field starting from more familiar objects and makes clear what are the technical difficulties along the way; second, it leads to the standard notations that are used throughout the physics literature and in this book. In any case, the rigorous construction is easier to understand if one already knows what one is trying to construct.

We shall first construct a *neutral* scalar field, which corresponds to a *real-valued* classical Klein-Gordon field, and whose quanta are their own antiparticles (such as neutral pions). We shall then modify the construction to obtain a "charged" scalar field, which corresponds to a *complex-valued* classical Klein-Gordon field, and whose quanta have distinct antiparticles (although they need not possess electric charge).

To make the technical details a bit easier to begin with, we consider the Klein-Gordon equation not in all of $\mathbb{R}^4$ but in a box. That is, we fix a bounded region $\mathcal{B} \subset \mathbb{R}^3$ and consider the differential equation $(\partial_t^2 - \nabla^2 + m^2)\phi(t, \mathbf{x}) = 0$ for $t \in \mathbb{R}$ and $\mathbf{x} \in \mathcal{B}$, subject to real boundary conditions that make $-\nabla^2$ positive and self-adjoint. (We shall make more specific choices of $\mathcal{B}$ and the boundary conditions later.) The boundedness of $\mathcal{B}$ guarantees that $-\nabla^2$ has only a discrete spectrum, that is, there is an orthonormal eigenbasis $\{f_j\}$ for $-\nabla^2$ with eigenvalues $\{\lambda_j^2\}$. Since the differential equation and boundary conditions are real, the eigenfunctions f_j may be chosen to be real, and for the time being we assume that they are.

Now, a classical real field $\phi_{\mathrm{clas}}(t, \mathbf{x})$ can be expanded in terms of the eigenfunctions f_j:

$$\phi_{\mathrm{clas}}(t, \mathbf{x}) = \sum q_j(t) f_j(\mathbf{x}),$$

where the coefficients $q_j(t)$ are also real. ϕ_{clas} satisfies the Klein-Gordon equation if and only if these coefficients satisfy

$$q_j''(t) + \omega_j^2 q_j = 0, \qquad \omega_j^2 = \lambda_j^2 + m^2. \tag{5.1}$$

But this is the equation for a classical harmonic oscillator. *To turn ϕ_{clas} into a quantum field, we replace these classical oscillators with quantum oscillators.*

This necessitates a digression to discuss systems of quantum harmonic oscillators. First let us consider a system of K one-dimensional oscillators, where K is a positive integer. The state space is $L^2(\mathbb{R}^K)$ and the Hamiltonian is

$$H = -\tfrac{1}{2}\nabla^2 + \tfrac{1}{2}\sum \omega_j x_j^2 = \sum \frac{\omega_j}{2}\left(-\frac{1}{\omega_j}\frac{\partial^2}{\partial x_j^2} + \omega_j x_j^2\right),$$

where $\omega_1, \ldots, \omega_K$ are positive numbers. The substitution $y_j = \sqrt{\omega_j}\, x_j$ makes

$$H = \sum \frac{\omega_j}{2}\left(-\frac{\partial^2}{\partial y_j^2} + y_j^2\right),$$

a sum of standard one-dimensional oscillator Hamiltonians, to which we can apply the analysis of §3.4. Namely, let X_j and P_j be the position and momentum operators for the jth oscillator,

$$X_j = \text{mult. by}\, x_j, \qquad P_j = \frac{1}{i}\frac{\partial}{\partial x_j},$$

and let Y_j and Q_j be the corresponding operators for the y-coordinates,

$$Y_j = \sqrt{\omega_j}\, X_j = \text{mult. by}\sqrt{\omega_j}\, x_j, \qquad Q_j = \frac{1}{\sqrt{\omega_j}} P_j = \frac{1}{i\sqrt{\omega_j}}\frac{\partial}{\partial x_j}.$$

We then set

$$A_j = \frac{1}{\sqrt{2}}(Y_j + iQ_j), \qquad A_j^\dagger = \frac{1}{\sqrt{2}}(Y_j - iQ_j).$$

For future reference we note that

$$X_j = \frac{1}{\sqrt{2\omega_j}}(A_j + A_j^\dagger), \qquad P_j = \frac{1}{i}\sqrt{\frac{\omega_j}{2}}(A_j - A_j^\dagger). \tag{5.2}$$

These operators satisfy the canonical commutation relations

$$[X_j, P_k] = i\delta_{jk} I, \qquad [A_j, A_k^\dagger] = \delta_{jk} I,$$
$$[X_j, X_k] = [P_j, P_k] = [A_j, A_k] = [A_j^\dagger, A_k^\dagger] = 0,$$

and the Hamiltonian is

$$H = \sum \omega_j (A_j^\dagger A_j + \tfrac{1}{2}).$$

The eigenfunctions of H are simply the products of the one-dimensional eigenfunctions of the one-dimensional oscillator given by (3.27):

$$\phi_{n_1 n_2 \cdots n_K}(x) = \frac{(A_1^\dagger)^{n_1} \cdots (A_K^\dagger)^{n_K}}{\sqrt{n_1! \cdots n_K!}} \phi_{00\cdots 0}(x),$$

where $\phi_{00\cdots 0}(x) = \pi^{-n/4} \exp(-\frac{1}{2}\sum \omega_j^2 x_j^2)$. By (3.28) and (3.29) we have

$$A_j \phi_{n_1 \cdots n_K} = \sqrt{n_j}\, \phi_{n_1 \cdots n_j - 1 \cdots n_K}, \qquad A_j^\dagger \phi_{n_1 \cdots n_K} = \sqrt{n_j + 1}\, \phi_{n_1 \cdots n_j + 1 \cdots n_K},$$
$$H\phi_{n_1 \cdots n_K} = (n_1 + \cdots + n_K + \tfrac{1}{2}K)\phi_{n_1 \cdots n_K}.$$

If the reader has studied §4.5, this should look very familiar. Indeed, the correspondence

$$\phi_{n_1 \ldots n_K} \mapsto |n_1, \ldots, n_K\rangle$$

defines a unitary map from $L^2(\mathbb{R}^K)$ to the Boson Fock space $\mathcal{F}_s(\mathbb{C}^K)$ that intertwines the operators A_j and $A_j^\dagger$ with the creation and annihilation operators on $\mathcal{F}_s(\mathbb{C}^K)$! But what we really need for field theory is the case $K = \infty$, an *infinite* collection of harmonic oscillators. The material of §4.5 gives us exactly the mathematical machinery that we need: the Boson Fock space $\mathcal{F}_s(\mathcal{H})$, where now $\mathcal{H}$ is a separable infinite-dimensional Hilbert space, equipped with its creation and annihilation operators A_j and $A_j^\dagger$ relative to a fixed orthonormal basis with the same index set as the eigenfunctions f_j.

There is just one problem that has to be addressed. Formally, the Hamiltonian ought to be $H = \sum_1^\infty \omega_j (A_j^\dagger A_j + \frac{1}{2})$, but this does not make sense when the series $\sum \omega_j$ diverges, as it will in the situations we need since $\omega_j^2 = \lambda_j^2 + m^2 \geq m^2$ for all j. This is our first brush with the dreaded divergences of quantum field theory, but unlike most of them, this one is easy to fix. Adding a constant to the Hamiltonian does not affect the dynamics, so we simply throw away the infinite $\frac{1}{2}\sum\omega_j$ and consider the "renormalized" Hamiltonian

$$H = \sum \omega_j A_j^\dagger A_j \tag{5.3}$$

instead. This is a well defined positive operator on the finite-particle space $\mathcal{F}_s^0(\mathcal{H})$, and it has a unique extension to a positive self-adjoint operator on $\mathcal{F}_s(\mathcal{H})$; the vectors $|n_1, n_2, \cdots\rangle$ are an eigenbasis for it with eigenvalues $\sum \omega_j n_j$ (a finite sum since only finitely many n_j are nonzero).

We are now in a position to quantize the classical field (5.1). By (5.2), the position operator for the jth oscillator is $X_j = (A_j + A_j^\dagger)/\sqrt{2\omega_j}$. Substituting this into (5.1) in place of the classical position variables q_j, we obtain the quantum field, denoted by ϕ:

$$\phi(\mathbf{x}) = \sum \frac{1}{\sqrt{2\omega_j}} f_j(\mathbf{x})(A_j + A_j^\dagger). \tag{5.4}$$

(We shall not worry about the convergence of this series until a little later.)

Before proceeding further, we need to free ourselves from the assumption that the eigenfunctions f_j are real, that is, to obtain a formula for $\phi(\mathbf{x})$ in terms of an arbitrary complex-valued orthonormal eigenbasis $\{g_j\}$ for $-\nabla^2$ on the box $\mathcal{B}$ subject to the given boundary conditions. It cannot be right simply to substitute g_j for f_j in (5.4) because the resulting operator will not be Hermitian, and "Hermitian" is the quantum analogue of the classical "real-valued." Rather, let us observe that $g_j = \sum_k u_{jk} f_k$ where $U = (u_{jk})$ is a unitary matrix such that $u_{jk} = 0$ unless the eigenvalue for g_j is the same as the eigenvalue for f_k, and hence the corresponding ω_j's are also equal. Since f_k is real and $U^{-1} = U^\dagger$,

$$\sum_j u_{jk}^* g_j = f_k = f_k^* = \sum_j u_{jk} g_j^*.$$

Substituting these equations into (5.4), we obtain

$$\phi = \sum_k \frac{1}{\sqrt{2\omega_k}} f_k(A_k + A_k^\dagger) = \sum_{j,k} \frac{1}{\sqrt{2\omega_j}} (g_j u_{jk}^* A_k + g_j^* u_{kj} A_j^\dagger).$$

Thus if we set

$$A_j' = \sum_k u_{jk}^* A_k,$$

we have

$$\phi = \sum_j \frac{1}{\sqrt{2\omega_j}} (g_j A_j' + g_j^* (A_j')^\dagger).$$

Moreover, we recall from §4.5 that $A_j = A(e_j)$ where $\{e_j\}$ is some orthonormal basis of $\mathcal{H}$. Since the map $v \mapsto A(v)$ is antilinear, we then have $A_j' = A(e_j')$ where $e_j' = \sum_k u_{jk} e_k$, and $\{e_j'\}$ is again an orthonormal basis of $\mathcal{H}$.

In short, we see that no matter what basis $\{f_j\}$ we pick, the correct formula for the field is

$$\phi(\mathbf{x}) = \sum_j \frac{1}{\sqrt{2\omega_j}} (f_j(\mathbf{x}) A_j + f_j(\mathbf{x})^* A_j^\dagger), \tag{5.5}$$

where the A_j's and $A_j^\dagger$'s are the annihilation and creation operators for an orthonormal basis of $\mathcal{H}$. The choice of such a basis is irrelevant; only the formal structure matters.

The t-dependence of the classical field seems to have disappeared, but it is restored by passing from the Schrödinger-picture operators $A_j, A_j^\dagger$ to the time-dependent Heisenberg-picture operators

$$A_j(t) = e^{itH}A_je^{-itH}, \qquad A_j^\dagger(t) = e^{itH}A_j^\dagger e^{-itH}. \tag{5.6}$$

This can easily be made more concrete. Indeed, from (4.52) and the fact that $H = \sum \omega_j A_j^\dagger A_j$, we see that

$$[A_j, H] = \omega_j A_j, \qquad [A_j^\dagger, H] = -\omega_j A_j^\dagger.$$

In view of this, (5.6) implies that

$$\frac{dA_j(t)}{dt} = \frac{1}{i}[A_j, H] = -i\omega_j A_j, \qquad \frac{dA_j^\dagger(t)}{dt} = \frac{1}{i}[A_j^\dagger, H] = i\omega_j A_j^\dagger,$$

and hence

$$A_j(t) = e^{-i\omega_j t}A_j, \qquad A_j^\dagger(t) = e^{i\omega_j t}A_j^\dagger.$$

The Heisenberg-picture field $\phi(t, \mathbf{x})$ is therefore

$$\phi(t, \mathbf{x}) = \sum \frac{1}{\sqrt{2\omega_j}}(f_j(\mathbf{x})e^{-i\omega_j t}A_j + f_j(\mathbf{x})^* e^{i\omega_j t}A_j^\dagger). \tag{5.7}$$

Now we make some specific choices. We take the box $\mathcal{B}$ to be the cube $[-\frac{1}{2}L, \frac{1}{2}L]^3$ in $\mathbb{R}^3$, and we impose periodic boundary conditions. (Don't worry about the nature of the boundary conditions; we're going to let $L \to \infty$ presently.) Then we can take the normalized eigenfunctions to be $f_{\mathbf{p}}(\mathbf{x}) = L^{-3/2}e^{i\mathbf{p}\cdot\mathbf{x}}$, with eigenvalues $-|\mathbf{p}|^2$, where $\mathbf{p}$ lies in the lattice

$$\Lambda = \left[\frac{2\pi}{L}\mathbb{Z}\right]^3.$$

The corresponding numbers $\omega_{\mathbf{p}}$ are

$$\omega_{\mathbf{p}} = \sqrt{|\mathbf{p}|^2 + m^2}, \tag{5.8}$$

and the field takes the form

$$\phi(t, \mathbf{x}) = \sum_\Lambda \frac{1}{\sqrt{2\omega_{\mathbf{p}}L^3}}(e^{i\mathbf{p}\cdot\mathbf{x} - i\omega_{\mathbf{p}}t}A_{\mathbf{p}} + e^{-i\mathbf{p}\cdot\mathbf{x} + i\omega_{\mathbf{p}}t}A_{\mathbf{p}}^\dagger). \tag{5.9}$$

This is starting to look promising. Although our calculations have been completely non-Lorentz-invariant, we can see Lorentz form re-emerging in the exponents. Moreover, it is clear that (5.9) formally satisfies the Klein-Gordon equation. However, it is time we paid a little attention to questions of convergence, and here the situation is not so happy. Suppose, for instance, we apply the operator $\phi(t, \mathbf{x})$ to the vacuum state:

$$\phi(t, \mathbf{x})|0, 0, \ldots\rangle = \sum_\Lambda \frac{1}{\sqrt{2\omega_{\mathbf{p}}L^3}}e^{-i\mathbf{p}\cdot\mathbf{x} + i\omega_{\mathbf{p}}t}|0, \ldots, 0, 1, 0, \ldots\rangle \quad \text{(1 in the } \mathbf{p}\text{th slot)}.$$

The square of the norm of this alleged vector is $\sum_\Lambda 1/2(|\mathbf{p}|^2 + m^2)$, which is infinite (by comparison to $\int_{\mathbb{R}^3} d\mathbf{p}/(|\mathbf{p}|^2 + m^2) = 4\pi \int_0^\infty r^2\, dr/(r^2 + m^2)$). If we apply $\phi(t, \mathbf{x})$ to a multiparticle state, the result is even worse, because the operators A_j and A_j^* introduce factors of $\sqrt{n_j}$ that are generally bigger than 1.

The way out of this is to interpret ϕ as an operator-valued distribution rather than an operator-valued function. That is, ϕ is the linear map that assigns to each compactly supported C^∞ function χ_1 on $\mathbb{R}$ and each C^∞ Λ-periodic function χ_2 on $\mathbb{R}^3$ the operator

$$\int_B \int_{\mathbb{R}} \phi(t,\mathbf{x})\chi_1(t)\chi_2(\mathbf{x})\,dt\,d\mathbf{x} \\ = \sum_\Lambda \frac{1}{\sqrt{2\omega_{\mathbf{p}} L^3}} \big[\widehat{\chi}_1(\omega_{\mathbf{p}})\widehat{\chi}_2(-\mathbf{p})A_{\mathbf{p}} + \widehat{\chi}_1(-\omega_{\mathbf{p}})\widehat{\chi}_2(\mathbf{p})A^\dagger_{\mathbf{p}}\big],$$

with the obvious interpretation of the Fourier coefficients $\widehat{\chi}_1$ and $\widehat{\chi}_2$. The rapid decay of these coefficients as $|\mathbf{p}| \to \infty$ guarantees that this series converges nicely as an operator on the finite-particle space $\mathcal{F}^0_s(\mathcal{H})$.

The final step in the construction of the quantum field is the removal of the box $\mathcal{B}$. That is, we think of (5.9) as a Riemann sum for an integral over $\mathbf{p}$-space and pass to the limit as $L \to \infty$. The discrete volume element implicit in (5.9) is $\Delta V = (2\pi/L)^3$, the volume of a fundamental cube of the lattice Λ, so we can rewrite (5.9) as

$$\phi(t,\mathbf{x}) = \sum_\Lambda \frac{1}{\sqrt{2\omega_{\mathbf{p}}}} L^{3/2}\left(e^{i\mathbf{p}\cdot\mathbf{x}-i\omega_{\mathbf{p}}t}A_{\mathbf{p}} + e^{-i\mathbf{p}\cdot\mathbf{x}+i\omega_{\mathbf{p}}t}A^*_{\mathbf{p}}\right)\frac{\Delta V}{(2\pi)^3}.$$

As $L \to \infty$, the sum becomes an integral over $\mathbb{R}^3$, and the rescaled annihilation and creation operators $L^{3/2}A_{\mathbf{p}}$ and $L^{3/2}A^\dagger_{\mathbf{p}}$ turn into *operator-valued distributions* on $\mathbb{R}^3$, denoted by a and $a^\dagger$ (actually, the action of a on test functions is antilinear rather than linear). Indeed, if we take the underlying Hilbert space $\mathcal{H}$ to be $L^2(\mathbb{R}^3, d^3\mathbf{p}/(2\pi)^3)$, the evaluation of the distribution a on the test function $\chi \in C^\infty_c(\mathbb{R}^3)$ is the operator denoted by $A(\chi)$ in §4.5. They satisfy the distributional commutation relations

$$(5.10) \qquad [a(\mathbf{p}), a^\dagger(\mathbf{p}')] = (2\pi)^3\delta(\mathbf{p}-\mathbf{p}')I, \qquad [a(\mathbf{p}), a(\mathbf{p}')] = [a^\dagger(\mathbf{p}), a^\dagger(\mathbf{p}')] = 0.$$

which are a restatement of (4.48).

Maintaining the notational fiction that distributions are functions, we can therefore write the field with the box removed as

$$(5.11) \qquad \phi(t,\mathbf{x}) = \int_{\mathbb{R}^3} \frac{1}{\sqrt{2\omega_{\mathbf{p}}}}\left(e^{i\mathbf{p}\cdot\mathbf{x}-i\omega_{\mathbf{p}}t}a(\mathbf{p}) + e^{-i\mathbf{p}\cdot\mathbf{x}+i\omega_{\mathbf{p}}t}a^\dagger(\mathbf{p})\right)\frac{d^3\mathbf{p}}{(2\pi)^3}$$

or, with $x = (t,\mathbf{x})$ and $p = (\omega_{\mathbf{p}}, \mathbf{p})$,

$$(5.12) \qquad \phi(x) = \int \frac{1}{\sqrt{2\omega_{\mathbf{p}}}}\left(e^{-ip_\mu x^\mu}a(\mathbf{p}) + e^{ip_\mu x^\mu}a^\dagger(\mathbf{p})\right)\frac{d^3\mathbf{p}}{(2\pi)^3}.$$

Warning: It is a sad fact of life that if one examines n books on quantum field theory, one will probably find n variants of (5.12) that differ in the factors of 2π and $\omega_{\mathbf{p}}$. The point is that one can redefine $a(\mathbf{p})$ by incorporating such factors into it. If one wants to think of (5.12) as an integral over the mass shell X^+_m with its Lorentz-invariant measure $d^3\mathbf{p}/(2\pi)^3\omega_{\mathbf{p}}$, for example, one can replace $a(\mathbf{p})$ with $\widetilde{a}(\mathbf{p}) = \omega_{\mathbf{p}}^{1/2}a(\mathbf{p})$; or if one wants to eliminate the 2π's from (5.10), one can replace $a(\mathbf{p})$ with $a(\mathbf{p})/(2\pi)^{3/2}$. In this book we shall keep the canonical relations (5.10) for the annihilation and creation operators and let the consequences be what they may.

The good news about the formula (5.12) is its obviously Lorentz-friendly form. The bad news is its analytic prickliness: the operator-valued functions in the integrand are actually distributions, and the integral itself must be interpreted in a distributional sense. We shall reassure the reader by showing how to construct ϕ by rigorous mathematics in the next section. However, it must be understood that formulas such as (5.12) are quite convenient for the sort of calculations one actually has to perform to extract the physics, and they are ubiquitous in the physics literature. They will appear frequently in this book, and hopefully the reader will acquire greater comfort with them by seeing how they are used in practice.

Now, the Hilbert space on which $\phi(x)$ (or rather the regularized version obtained by integrating against a test function) acts is $\mathcal{F}_s(\mathcal{H})$, the Boson Fock space over a Hilbert space $\mathcal{H}$. We introduced this Fock space simply because it is the place where the required operators arising from the harmonic oscillators live. But the physics of the quantum field comes from taking the Fock space seriously as a multiparticle state space built from a single-particle state space $\mathcal{H}$; the particles in question are the quanta of the field. The operators $a(\mathbf{p})$ and $a^\dagger(\mathbf{p})$ annihilate and create particles with momentum $\mathbf{p}$ (where again this is an idealization; a and $a^\dagger$ are distributions, and real particles don't have completely definite momenta), and $\phi(x)$ creates and annihilates particles at the space-time point x. The physical meaning of these assertions may seem rather obscure, but we must ask the reader to accept them for the time being on the level of hand-waving. We are building a house, and at present we are merely laying the foundation.

Let us now turn to the quantization of a *complex* Klein-Gordon field. We employ the informal language of distributions as in (5.12).

If ϕ_{clas} is a complex-valued solution of the Klein-Gordon equation, its real and imaginary parts $\phi_{1,\text{clas}}$ and $\phi_{2,\text{clas}}$ are also solutions. To quantize them, we introduce annihilation and creation operators $a_1(\mathbf{p})$, $a_1^\dagger(\mathbf{p})$ and $a_2(\mathbf{p})$, $a_2^\dagger(\mathbf{p})$, which are taken to satisfy the canonical commutation relations (5.10) among themselves and to commute with each other:

$$[a_1(\mathbf{p}), a_1^\dagger(\mathbf{p}')] = [a_2(\mathbf{p}), a_2^\dagger(\mathbf{p}')] = (2\pi)^3\delta(\mathbf{p}-\mathbf{p}'), \quad \text{all other commutators } = 0.$$

(The Hilbert space $\mathcal{F}$ on which these operators act is obtained by starting with two Fock spaces $\mathcal{F}_1$ and $\mathcal{F}_2$ on which the a_1's and a_2's act, respectively, and taking $\mathcal{F} = \mathcal{F}_1 \otimes \mathcal{F}_2$.) We can then form the quantum fields

$$\phi_j(x) = \int \frac{1}{\sqrt{2\omega_{\mathbf{p}}}}\left(e^{-ip_\mu x^\mu} a_j(\mathbf{p}) + e^{ip_\mu x^\mu} a_j^\dagger(\mathbf{p})\right) \frac{d^3\mathbf{p}}{(2\pi)^3} \qquad (j=1,2).$$

We put these together into a "complex" quantum field by setting

(5.13)
$$\phi(x) = \frac{\phi_1(x) + i\phi_2(x)}{\sqrt{2}}, \quad a(\mathbf{p}) = \frac{a_1(\mathbf{p}) + ia_2(\mathbf{p})}{\sqrt{2}}, \quad b(\mathbf{p}) = \frac{a_1(\mathbf{p}) - ia_2(\mathbf{p})}{\sqrt{2}},$$

so that

$$\phi^\dagger(x) = \frac{\phi_1(x) - i\phi_2(x)}{\sqrt{2}}, \quad a^\dagger(\mathbf{p}) = \frac{a_1^\dagger(\mathbf{p}) - ia_2^\dagger(\mathbf{p})}{\sqrt{2}}, \quad b^\dagger(\mathbf{p}) = \frac{a_1^\dagger(\mathbf{p}) + ia_2^\dagger(\mathbf{p})}{\sqrt{2}},$$

The factors of $\sqrt{2}$ are there to normalize the commutation relations so that

(5.14)
$$[a(\mathbf{p}), a^\dagger(\mathbf{p}')] = [b(\mathbf{p}), b^\dagger(\mathbf{p}')] = (2\pi)^3\delta(\mathbf{p}-\mathbf{p}'), \quad \text{all other commutators } = 0.$$

The fields ϕ and $\phi^\dagger$ are expressed in terms of the a's and b's as follows:

$$\begin{aligned}\phi(x) &= \int \frac{1}{\sqrt{2\omega_{\mathbf{p}}}}\big(e^{-ip_\mu x^\mu}a(\mathbf{p}) + e^{ip_\mu x^\mu}b^\dagger(\mathbf{p})\big)\frac{d^3\mathbf{p}}{(2\pi)^3},\\ \phi^\dagger(x) &= \int \frac{1}{\sqrt{2\omega_{\mathbf{p}}}}\big(e^{-ip_\mu x^\mu}b(\mathbf{p}) + e^{ip_\mu x^\mu}a^\dagger(\mathbf{p})\big)\frac{d^3\mathbf{p}}{(2\pi)^3}.\end{aligned} \tag{5.15}$$

Let us look at this from a slightly different angle. Suppose ϕ is a tempered distribution on $\mathbb{R}^4$ that satisfies the Klein-Gordon equation $(\partial^2 + m^2)\phi = 0$. Then the Fourier transform of ϕ is a tempered distribution that satisfies $(-p^2+m^2)\widehat{\phi}(p) = 0$ and hence is supported on the two-sheeted hyperboloid

$$\begin{gathered}\{p : p^2 = m^2\} = X_m^+ \cup X_m^-,\\ X_m^+ = \{p : p^2 = m^2,\ p^0 > 0\}, \qquad X_m^- = \{p : p^2 = m^2,\ p^0 < 0\}.\end{gathered}$$

In particular, we may consider the space of solutions of the Klein-Gordon equation whose Fourier transforms are of the form $u\lambda$ where λ is a Lorentz-invariant measure on $X_m^+ \cup X_m^-$, considered as a distribution on $\mathbb{R}^4$, and u is a (reasonable) function. If ϕ is such a solution, we have

$$\begin{aligned}\phi(x) &= \int_{X_m^+\cup X_m^-} e^{-ip_\mu x^\mu}u(p)\,d\lambda(p)\\ &= \int_{X_m^+} \big(e^{-ip_\mu x^\mu}u(p) + e^{ip_\mu x^\mu}u(-p)\big)\,d\lambda(p).\end{aligned}$$

Now, if ϕ is real-valued, we have $\widehat{\phi}(-p) = \widehat{\phi}(p)^*$, so ϕ is completely determined by the restriction of $\widehat{\phi}$ to X_m^+. We obtain the corresponding quantum field by replacing the Fourier coefficients $u(p)$, $p \in X_m^+$, by suitably normalized annihilation operators and their complex conjugates $u(-p)$ by their adjoint creation operators. However, if ϕ is allowed to be complex valued, the restrictions of $\widehat{\phi}$ to X_m^+ and X_m^- are independent, and both of them are needed to recover ϕ. Hence the Fourier coefficients $u(-p)$ must be replaced by a *different* set of creation operators.

The physical interpretation is as follows: $a, a^\dagger$ and $b, b^\dagger$ are the annihilation and creation operators for two *different* species of particle with mass m and spin 0, which are *antiparticles* of each other. We make the conventional choice that $a, a^\dagger$ are associated with "particles" and $b, b^\dagger$ are associated with "antiparticles." Thus the field ϕ destroys particles and creates antiparticles; vice versa for $\phi^\dagger$.

This interpretation gives the solution of the problem of "negative energy states," which plagued the early attempts to use the Klein-Gordon equation just as it did the Dirac equation. When the Klein-Gordon equation is used as a single-particle wave equation, the negative energy states present a real difficulty. But when we think of the Klein-Gordon equation as describing a "classical" field and pass to the corresponding quantum field, there is a way out: the part of the solution whose Fourier transform lives on the negative energy shell contributes not particle states with negative energy but antiparticle states with positive energy. We shall see how this works in more detail in the next section.

5.2. The rigorous construction

We now present the rigorous construction of a neutral (real) quantum scalar field with mass parameter m. Afterwards we shall modify the construction to obtain a charged (complex) field.

We have already assembled all the ingredients we need. The first is the state space for a single spin-zero particle of mass m (see §4.4):

$$\mathcal{H} = L^2(X_m^+, \lambda),$$

where X_m^+ is the mass shell and λ is the normalized Lorentz-invariant measure on it:

$$d\lambda(p) = d\lambda(\omega_{\mathbf{p}}, \mathbf{p}) = \frac{d^3\mathbf{p}}{(2\pi)^3\omega_{\mathbf{p}}} \qquad (\omega_{\mathbf{p}} = p^0 = \sqrt{|\mathbf{p}|^2 + m^2}).$$

Based on this, we construct the Boson Fock space $\mathcal{F}_s(\mathcal{H})$, on which we have annihilation and creation operators $A(v)$ and $A(v)^\dagger$ for $v \in \mathcal{H}$ as defined in §4.5. Finally, we define $R : \mathcal{S}(\mathbb{R}^4) \to \mathcal{H}$ by

$$Rf = \widehat{f}|X_m^+,$$

where $\widehat{f}(p) = \int e^{ip_\mu x^\mu} f(x)\, d^4x$ is the Lorentz-covariant Fourier transform of f.

The quantum field Φ (which differs from the ϕ of the preceding section by a change of variable, as we explain below) is defined as a real tempered distribution on $\mathbb{R}^4$ with values in the space of operators on the finite-particle Fock space $\mathcal{F}_s^0(\mathcal{H})$ — that is, an $\mathbb{R}$-linear map that takes a *real-valued* Schwartz-class function f on $\mathbb{R}^4$ to an operator $\Phi(f)$ on $\mathcal{F}_s^0(\mathcal{H})$ — by

$$\Phi(f) = \frac{1}{\sqrt{2}}[A(Rf) + A(Rf)^\dagger]. \tag{5.16}$$

(One can extend Φ to a $\mathbb{C}$-linear map on complex-valued Schwartz functions by setting $\Phi(f + ig) = \Phi(f) + i\Phi(g)$, of course, but (5.16) is only $\mathbb{R}$-linear as it stands because A is antilinear.)

Observe that Φ is a distribution solution of the Klein-Gordon equation, that is, $\Phi((\partial^2 + m^2)f) = 0$ for any $f \in \mathcal{S}$, because $[(\partial^2 + m^2)f]\widehat{\ } = (-p^2 + m^2)\widehat{f} = 0$ on X_m^+.

Let us see how to go from (5.16) to (5.12). First, we map X_m^+ to $\mathbb{R}^3$ in the obvious way, $(\omega_{\mathbf{p}}, \mathbf{p}) \mapsto \mathbf{p}$, and correspondingly identify $L^2(X_m^+, \lambda)$ with $L^2(\mathbb{R}^3)$, where the measure on $\mathbb{R}^3$ is taken to be $d^3\mathbf{p}/(2\pi)^3$. Explicitly, this correspondence is the unitary map $J : L^2(X_m^+, \lambda) \to L^2(\mathbb{R}^3)$ defined by

$$Ju(\mathbf{p}) = \frac{1}{\sqrt{\omega_{\mathbf{p}}}} u(\omega_{\mathbf{p}}, \mathbf{p}).$$

We then have the induced unitary map $\mathcal{F}(J) : \mathcal{F}_s(L^2(X_m^+, \lambda)) \to \mathcal{F}_s(L^2(\mathbb{R}^3))$, as explained in §4.5. We use it to transfer the annihilation, creation, and field operators to $L^2(\mathbb{R}^3)$; that is, we define

$$a(v) = \mathcal{F}(J)A(J^{-1}v)\mathcal{F}(J)^{-1}, \quad a^\dagger(v) = \mathcal{F}(J)A(J^{-1}v)^\dagger\mathcal{F}(J)^{-1} \qquad (g \in L^2(\mathbb{R}^3)),$$

(it will be more convenient to write $a^\dagger(v)$ rather than $a(v)^\dagger$) and

$$\phi(f) = \mathcal{F}(J)\Phi(f)\mathcal{F}(J)^{-1} \qquad (f \in \mathcal{S}_{\mathbb{R}}(\mathbb{R}^4)).$$

Now, $a^\dagger$ can be interpreted as an operator-valued distribution on $\mathbb{R}^3$ by restricting its argument to $\mathcal{S}(\mathbb{R}^3)$. Adopting the notational fiction that it is a function, we write

$$a^\dagger(v) = \int a^\dagger(\mathbf{p})v(\mathbf{p})\,\frac{d^3\mathbf{p}}{(2\pi)^3}.$$

Likewise, a is an anti-linear distribution, and we write

$$a(v) = \int a(\mathbf{p})v(\mathbf{p})^*\,\frac{d^3\mathbf{p}}{(2\pi)^3}.$$

By (4.48) we have

$$\begin{aligned}\iint [a(\mathbf{p}), a^\dagger(\mathbf{q})]u(\mathbf{p})^* v(\mathbf{q})\,\frac{d^3\mathbf{p}\,d^3\mathbf{q}}{(2\pi)^6} &= [a(u), a^\dagger(v)] = \langle u|v\rangle = \int u(\mathbf{p})^* v(\mathbf{p})\,\frac{d^3\mathbf{p}}{(2\pi)^3}\\ &= \iint (2\pi)^3\delta(\mathbf{p}-\mathbf{q})v(\mathbf{p})^* v(\mathbf{q})\,\frac{d^3\mathbf{p}\,d^3\mathbf{q}}{(2\pi)^6}\end{aligned}$$

for all u and v, and similarly for $[a(\mathbf{p}), a(\mathbf{q})]$ and $[a^\dagger(\mathbf{p}), a^\dagger(\mathbf{q})]$. (The last three expressions are written as scalars but, as usual, are to be interpreted as scalar multiples of the identity.) Hence the commutation relations (4.48) expressed in colloquial distribution language are

$$[a(\mathbf{p}), a^\dagger(\mathbf{q})] = (2\pi)^3\delta(\mathbf{p}-\mathbf{q}), \qquad [a(\mathbf{p}), a(\mathbf{q})] = [a^\dagger(\mathbf{p}), a^\dagger(\mathbf{q})] = 0. \tag{5.17}$$

Next, observe that for $f \in \mathcal{S}_{\mathbb{R}}(\mathbb{R}^4)$ we have

$$\phi(f) = \frac{1}{\sqrt{2}}\mathcal{F}(J)\big[A(Rf) + A(Rf)^\dagger\big]\mathcal{F}(J)^{-1} = \frac{1}{\sqrt{2}}\big[a(JRf) + a^\dagger(JRf)\big],$$

and

$$JRf(\mathbf{p}) = \frac{1}{\sqrt{\omega_\mathbf{p}}}\widehat{f}(\omega_\mathbf{p}, \mathbf{p}),$$

so

$$\begin{aligned}\phi(f) &= \int \frac{1}{\sqrt{2\omega_\mathbf{p}}}\big(\widehat{f}(\omega_\mathbf{p}, \mathbf{p})^* a(\mathbf{p}) + \widehat{f}(\omega_\mathbf{p}, \mathbf{p})a^\dagger(\mathbf{p})\big)\,\frac{d^3\mathbf{p}}{(2\pi)^3}\\ &= \iint \frac{1}{\sqrt{2\omega_\mathbf{p}}}\big(e^{-ip_\mu x^\mu}a(\mathbf{p}) + e^{ip_\mu x^\mu}a^\dagger(\mathbf{p})\big)f(x)\,d^4x\,\frac{d^3\mathbf{p}}{(2\pi)^3}.\end{aligned}$$

In other words,

$$\phi(x) = \int \frac{1}{\sqrt{2\omega_\mathbf{p}}}\big(e^{-ip_\mu x^\mu}a(\mathbf{p}) + e^{ip_\mu x^\mu}a^\dagger(\mathbf{p})\big)\,\frac{d^3\mathbf{p}}{(2\pi)^3},$$

which is (5.12).

We now modify the construction to obtain the charged (complex) field. For this purpose we need both the positive and negative energy mass shells X_m^+ and X_m^-, each equipped with the invariant measure $d\lambda(p) = d^3\mathbf{p}/(2\pi)^3\omega_\mathbf{p}$. We set

$$\mathcal{H}_+ = L^2(X_m^+, \lambda), \qquad \mathcal{H}_- = L^2(X_m^-, \lambda), \qquad \mathcal{H} = L^2(X_m^+ \cup X_m^-) = \mathcal{H}_+ \oplus \mathcal{H}_-,$$

and define the *anti-unitary* operator $C : \mathcal{H}_\pm \to \mathcal{H}_\mp$ by

$$Cu(p) = u(-p)^*.$$

(On the inverse Fourier transform side, i.e., in position space, C is just complex conjugation, $f \mapsto f^*$.) $\mathcal{H}_+$ is the state space for a single particle of mass m and spin 0; we build the Fock space $\mathcal{F}_+ = \mathcal{F}_s(\mathcal{H}_+)$ and the annihilation and creation operators $A(u)$, $A(u)^\dagger$ for $u \in \mathcal{H}_+$ on it as before. Now, here comes the twist. Let $\mathcal{F}_-$ be *another copy of* $\mathcal{F}_s(\mathcal{H}_+)$ (*not* $\mathcal{F}_s(\mathcal{H}_-)$) and define annihilation and creation operators $B(v)$, $B(v)^\dagger$ for $v \in \mathcal{H}_-$ (*not* $\mathcal{H}_+$) on it by

$$B(v) = A(Cv), \qquad B(v)^\dagger = A(Cv)^\dagger.$$

We think of $\mathcal{F}_+$ and $A, A^\dagger$ as the Fock space and annihilation-creation operators for particles, and $\mathcal{F}_-$ and $B, B^\dagger$ as the Fock space and annihilation-creation operators for antiparticles. C is the "charge-conjugation" operator (although there may be no electric charge involved) that turns a particle into an antiparticle and vice versa. Observe that since $A(v)$ and $A(v)^\dagger$ depend antilinearly and linearly on v, respectively, and C is antilinear, $B(v)$ and $B(v)^\dagger$ depend linearly and antilinearly on v, respectively.

We combine the two Fock spaces into a single Fock space $\mathcal{F} = \mathcal{F}_+ \otimes \mathcal{F}_-$. Observe that $\mathcal{F}_\pm = \bigoplus_0^\infty \mathcal{F}_\pm^k$ where $\mathcal{F}_\pm^k = \circledS^k \mathcal{H}_\pm$ is the space of k-particle (for $+$) or k-antiparticle (for $-$) states, and hence

$$\mathcal{F} = \bigoplus_{j,k=0}^{\infty} \mathcal{F}_+^j \otimes \mathcal{F}_-^k,$$

where the summands on the right are the spaces of states with j particles and k antiparticles. We consider $A(u)$ and $B(v)$ and their adjoints as operators on $\mathcal{F}$ by letting $A(u)$ act on the first factor and $B(v)$ on the second one. Finally, we define $R_\pm : \mathcal{S}(\mathbb{R}^4) \to \mathcal{H}_\pm$ by

$$R_\pm f = \widehat{f}\,\big|X_m^\pm,$$

and for $f \in \mathcal{S}(\mathbb{R}^4)$ (now complex-valued) we define the field operators $\Phi(f)$ and $\Phi^\dagger(f)$ on the finite-particle subspace of $\mathcal{F}$ by

$$\begin{aligned} \Phi(f) &= \frac{1}{\sqrt{2}}\big[A(R_+f) + B(R_-f)^\dagger\big], \\ \Phi^\dagger(f) &= \frac{1}{\sqrt{2}}\big[A(R_+f)^\dagger + B(R_-f)\big]. \end{aligned} \tag{5.18}$$

Thus $\Phi^\dagger$ is a distribution on $\mathbb{R}^4$ with values in the space of operators on the finite-particle subspace of $\mathcal{F}$; Φ is too, except that it depends *antilinearly* on its argument f. The reader may verify by a calculation similar to the one we performed for the neutral field that (5.18) is equivalent to (5.15).

Further discussion of the mathematics of free scalar fields can be found in Reed and Simon [**94**], §X.7.

5.3. Lagrangians and Hamiltonians

In nonrelativistic quantum mechanics the dynamics of a system is described by the Schrödinger equation, which we derived by starting with the Hamiltonian formulation of the classical mechanical system and quantizing the Hamiltonian. Quantum field theory, on the other hand, is relativistic from the outset, so the Lagrangian formulation provides a better starting point. That is, one begins with a Lagrangian L that is the integral of a Lorentz-invariant Lagrangian density $\mathcal{L}$, derives a Hamiltonian from it, and then quantizes the latter. The requirement of Lorentz invariance places severe constraints on the possible form of the Lagrangian and hence provides much guidance on the possible forms of quantum field theories.

Let us see how this works for free scalar fields. There is no interesting dynamics here, but it is a useful exercise to see how the process works in this simple case. It will lead to some useful formulas and provide some practice with the sort of freewheeling calculations with operator-valued distributions that physicists like to perform. We emphasize that all such calculations *in this chapter* can be made perfectly rigorous. The reader may wish to do so as an exercise, but here the idea is to see how the physicists operate.

We begin with the real scalar field. The Lagrangian that yields the Klein-Gordon equation is

$$L = \int \mathcal{L}\, d^3\mathbf{x}, \qquad \mathcal{L} = \tfrac{1}{2}[(\partial_\mu\phi)(\partial^\mu\phi) - m^2\phi^2] = \tfrac{1}{2}[(\partial_t\phi)^2 - |\nabla_{\mathbf{x}}\phi|^2 - m^2\phi^2].$$

Here, by analogy with Lagrangian mechanics with finitely many degrees of freedom, $\phi(\cdot, \mathbf{x})$ plays the role of a "position" variable labeled by $\mathbf{x}$; the corresponding canonically conjugate "momentum" is

$$\pi = \frac{\partial\mathcal{L}}{\partial(\partial_t\phi)} = \partial_t\phi,$$

and the Hamiltonian is

$$H = \int \mathcal{H}\, d^3\mathbf{x}, \qquad \mathcal{H} = \pi(\partial_t\phi) - \mathcal{L} = \tfrac{1}{2}[(\partial_t\phi)^2 + |\nabla_{\mathbf{x}}\phi|^2 + m^2\phi^2].$$

We can quantize these things. Let ϕ now denote the quantum field (5.12). The canonically conjugate field (*not* the physical momentum associated with the field) is

$$\pi(x) = \partial_t\phi(x) = \frac{1}{i}\int\sqrt{\frac{\omega_{\mathbf{p}}}{2}}\,(e^{ip_\mu x^\mu}a(\mathbf{p}) - e^{-ip_\mu x^\mu}a^\dagger(\mathbf{p}))\,\frac{d^3\mathbf{p}}{(2\pi)^3}. \tag{5.19}$$

We could also arrive at this formula in another way. At the beginning of our derivation of the quantum field ϕ, we replaced Fourier coefficients $q_j(t)$ for the classical field with position operators $(A_j + A_j^\dagger)/\sqrt{2\omega_j}$ for harmonic oscillators. The conjugate momentum operators are $\sqrt{\omega_j/2}(A_j - A_j^\dagger)$ (see (5.2)); if we insert them in the Fourier series in place of the position operators and proceed as before, we end up with (5.19). Similarly, one can modify the rigorous construction of $\phi(f)$ for $f \in \mathcal{S}(\mathbb{R}^4)$ to obtain $\pi(f)$; we leave the details to the reader. (They involve replacing Rf by $\widetilde{R}f(p) = \omega_{\mathbf{p}}Rf(p)$.)

The fields ϕ and π at a given time t satisfy "canonical commutation relations." The formal calculation is as follows: $\phi(t, \mathbf{x})$ and $\pi(t, \mathbf{y})$ are each the sum of two integrals, the first involving annihilation operators and the second involving creation operators. If one writes out the commutator $[\phi(t, \mathbf{x}), \pi(t, \mathbf{y})]$ as a double integral and uses the commutation relations (5.10) to simplify it, one finds that it is equal to

$$
\begin{aligned}
&\frac{1}{2i}\iint \sqrt{\frac{\omega_{\mathbf{q}}}{\omega_{\mathbf{p}}}}(e^{-ip_\mu x^\mu + iq_\mu y^\mu}[a^\dagger(\mathbf{p}), a(\mathbf{q})] - e^{ip_\mu x^\mu - iq_\mu y^\mu}[a(\mathbf{p}), a^\dagger(\mathbf{q})])\,\frac{d^3\mathbf{p}\,d^3\mathbf{q}}{(2\pi)^6}\\
=&\frac{i}{2}\iint \sqrt{\frac{\omega_{\mathbf{q}}}{\omega_{\mathbf{p}}}}(e^{-ip_\mu x^\mu + iq_\mu y^\mu} + e^{ip_\mu x^\mu - iq_\mu y^\mu})\delta(\mathbf{p}-\mathbf{q})\,\frac{d^3\mathbf{p}\,d^3\mathbf{q}}{(2\pi)^3}\\
=&\frac{i}{2}\iint (e^{ip_\mu(y^\mu - x^\mu)} + e^{ip_\mu(x^\mu - y^\mu)})\,\frac{d^3\mathbf{p}}{(2\pi)^3}\\
=&i\delta(\mathbf{x}-\mathbf{y}).
\end{aligned}
$$

The last equality is (1.5), taking into account that $x^0 - y^0 = t - t = 0$. Similarly one sees that $\phi(t,\mathbf{x})$ commutes with $\phi(t,\mathbf{y})$, and $\pi(t,\mathbf{x})$ with $\pi(t,\mathbf{y})$, for all $\mathbf{x}$ and $\mathbf{y}$. In short:

$$
(5.20)\qquad [\phi(t,\mathbf{x}),\pi(t,\mathbf{y})] = i\delta(\mathbf{x}-\mathbf{y}),\qquad [\phi(t,\mathbf{x}),\phi(t,\mathbf{y})] = [\pi(t,\mathbf{x}),\pi(t,\mathbf{y})] = 0.
$$

(One may worry about the fact that (5.20) is not Lorentz-invariant, as the notion of simultaneity at different points of space does not make sense relativistically. In fact, one may replace $(t,\mathbf{y})$ by $(s,\mathbf{y})$ in (5.20) as long as $|t-s| < |\mathbf{x}-\mathbf{y}|$, i.e., as long as $(t,\mathbf{x})$ and $(s,\mathbf{y})$ are space-like separated. We shall return to this point later.)

The formulation of these results in mathematicians' language is a rather easy consequence of (5.16). There is one little technicality to be overcome: the relations (5.20) hold only for a fixed time t, whereas smearing out $\phi(t,\mathbf{x})$ into $\phi(f) = \int \phi(t,\mathbf{x})f(t,\mathbf{x})\,dt\,d^3\mathbf{x}$ involves the values of ϕ at different times. But in fact, there is no need to smear ϕ out in t. A glance at the definition of $\phi(f)$ shows that it makes sense not just for $f \in \mathcal{S}(\mathbb{R}^4)$ but for distributions of the form $f(t,\mathbf{x}) = \delta(t-t_0)F(\mathbf{x})$ with $F \in \mathcal{S}(\mathbb{R}^3)$. Indeed, for such f we have $\widehat{f}(\omega_p,\mathbf{p}) = e^{-it_0\omega_{\mathbf{p}}}\widehat{F}(-\mathbf{p})$ where $\widehat{F}$ denotes the Euclidean Fourier transform on $\mathbb{R}^3$, and this is still a nice function on the mass shell X_m^+.

Now let us calculate the Hamiltonian. There are three possible starting points. First, the Hamiltonian is the infinitesimal generator of the time-translation group. Thus, on the one-particle space $L^2(X_m^+,\lambda)$ it is multiplication by $\omega_{\mathbf{p}}$, and on the k-particle space it is the sum of multiplications by $\omega_{\mathbf{p}}$ on all the factors:

$$
(5.21)\ H(P_s(u_1\otimes\cdots\otimes u_k)) = P_s((\omega_{\mathbf{p}}u_1)\otimes\cdots\otimes u_k) + \cdots + P_s(u_1\otimes\cdots\otimes(\omega_{\mathbf{p}}u_k)).
$$

Second, we can start from the renormalized Hamiltonian (5.3) for a discrete but infinite system of harmonic oscillators and pass to the continuum limit, obtaining

$$
(5.22)\qquad H = \int \omega_{\mathbf{p}} a^*(\mathbf{p})a(\mathbf{p})\,\frac{d^3\mathbf{p}}{(2\pi)^3}.
$$

It is not hard to see that (5.21) and (5.22) are equivalent. Third, we can quantize the formula for the classical Hamiltonian derived from the Lagrangian:

$$
(5.23)\qquad H = \frac{1}{2}\int \left[(\partial_t\phi(t,\mathbf{x}))^2 + |\nabla_{\mathbf{x}}\phi(t,\mathbf{x}))|^2 + m^2\phi(t,\mathbf{x})^2\right] d^3\mathbf{x},
$$

by substituting the expression (5.12) for $\phi(t,\mathbf{x})$ in (5.23).

Let us see how this works. We have $\phi = \phi^+ + \phi^-$, where

$$
\phi^+(x) = \int \frac{1}{\sqrt{2\omega_{\mathbf{p}}}} e^{-ip_\mu x^\mu} a(\mathbf{p})\,\frac{d^3\mathbf{p}}{(2\pi)^3},\qquad \phi^-(x) = \phi^+(x)^\dagger.
$$

Now,

$$\partial_t \phi^+(x) = \int \frac{1}{\sqrt{2\omega_{\mathbf{p}}}}(-i\omega_{\mathbf{p}})e^{-ip_\mu x^\mu} a(\mathbf{p}) \frac{d^3\mathbf{p}}{(2\pi)^3},$$

so

$$\int (\partial_t \phi^+(x))^2 \, d^3\mathbf{x} = -\iiint \frac{\sqrt{\omega_{\mathbf{p}}\omega'_{\mathbf{p}}}}{2} e^{-ip_\mu x^\mu} e^{-ip'_\mu x^\mu} a(\mathbf{p})a(\mathbf{p}') \frac{d^3\mathbf{p}\, d^3\mathbf{p}'}{(2\pi)^6} \, d^3\mathbf{x}.$$

By the Fourier inversion formula

$$\int e^{i(\mathbf{p}+\mathbf{p}')\cdot\mathbf{x}} \, d^3\mathbf{x} = (2\pi)^3 \delta(\mathbf{p}+\mathbf{p}'),$$

this reduces to

$$\int (\partial_t \phi^+(x))^2 \, d^3\mathbf{x} = -\int \frac{\omega_{\mathbf{p}}}{2} e^{-2i\omega_{\mathbf{p}} t} a(\mathbf{p})a(-\mathbf{p}) \frac{d^3\mathbf{p}}{(2\pi)^3}.$$

After similar calculations to evaluate the integrals of $(\partial^t \phi^-)^2$, $(\partial_t \phi^+)(\partial_t \phi^-)$, and $(\partial_t \phi^-)(\partial_t \phi^+)$, we obtain

$$\begin{aligned}&\int (\partial_t \phi(x))^2 \, d^3\mathbf{x} \\ &= \int \frac{\omega_{\mathbf{p}}^2}{2\omega_{\mathbf{p}}} \left[-e^{-2i\omega_{\mathbf{p}} t} a(\mathbf{p})a(-\mathbf{p}) + a(\mathbf{p})a^\dagger(\mathbf{p}) + a^\dagger(\mathbf{p})a(\mathbf{p}) - e^{2i\omega_{\mathbf{p}} t} a^\dagger(\mathbf{p})a^\dagger(-\mathbf{p})\right] \frac{d^3\mathbf{p}}{(2\pi)^3}.\end{aligned}$$

Likewise,

$$\nabla_{\mathbf{x}} \phi^+(x) = \int \frac{1}{\sqrt{2\omega_{\mathbf{p}}}}(i\mathbf{p})e^{-ip_\mu x^\mu} a(\mathbf{p}) \frac{d^3\mathbf{p}}{(2\pi)^3},$$

so an analogous calculation gives

$$\begin{aligned}&\int |\nabla_{\mathbf{x}} \phi(x)|^2 \, d^3\mathbf{x} \\ &= \int \frac{|\mathbf{p}|^2}{2\omega_{\mathbf{p}}} \left[e^{-2i\omega_{\mathbf{p}} t} a(\mathbf{p})a(-\mathbf{p}) + a(\mathbf{p})a^\dagger(\mathbf{p}) + a^\dagger(\mathbf{p})a(\mathbf{p}) + e^{2i\omega_{\mathbf{p}} t} a^\dagger(\mathbf{p})a^\dagger(-\mathbf{p})\right] \frac{d^3\mathbf{p}}{(2\pi)^3}.\end{aligned}$$

The signs of the coefficients of $a(\mathbf{p})a(-\mathbf{p})$ and $a^\dagger(\mathbf{p})a^\dagger(-\mathbf{p})$ are different here because the $\delta(\mathbf{p}+\mathbf{p}')$ makes $\mathbf{p}\cdot\mathbf{p}'$ into $-|\mathbf{p}|^2$ rather than $|\mathbf{p}|^2$. Yet another calculation of the same sort gives

$$\begin{aligned}&\int m^2 \phi(x)^2 \, d^3\mathbf{x} \\ &= \int \frac{m^2}{2\omega_{\mathbf{p}}} \left[e^{-2i\omega_{\mathbf{p}} t} a(\mathbf{p})a(-\mathbf{p}) + a(\mathbf{p})a^\dagger(\mathbf{p}) + a^\dagger(\mathbf{p})a(\mathbf{p}) + e^{2i\omega_{\mathbf{p}} t} a^\dagger(\mathbf{p})a^\dagger(-\mathbf{p})\right] \frac{d^3\mathbf{p}}{(2\pi)^3}.\end{aligned}$$

Adding these results and recalling that $\omega_{\mathbf{p}}^2 = |\mathbf{p}|^2 + m^2$, we see that the expression on the right of (5.23) is equal to

$$\frac{1}{2} \int \omega_{\mathbf{p}} \left[a(\mathbf{p})a^\dagger(\mathbf{p}) + a^\dagger(\mathbf{p})a(\mathbf{p})\right] \frac{d^3\mathbf{p}}{(2\pi)^3}.$$

Finally, since $[a(\mathbf{p}), a^\dagger(\mathbf{p}')] = (2\pi)^3 \delta(\mathbf{p}-\mathbf{p}')$, $a(\mathbf{p})a^\dagger(\mathbf{p})$ is equal to $a^\dagger(\mathbf{p})a(\mathbf{p})$ plus an infinite constant times the identity (an "infinite c-number" in physicists' parlance). After discarding the constant, we obtain (5.22). This infinite constant is exactly the same one we discarded in renormalizing the Hamiltonian for the infinite family of oscillators in §5.1.

The correction that needs to be applied to (5.23) to remove the infinite constant can be simply described as follows: replace all $a(\mathbf{p})a^\dagger(\mathbf{p}')$ by $a^\dagger(\mathbf{p}')a(\mathbf{p})$. This procedure is sufficiently commonly encountered that it has a name. In general, a sum of products of creation and annihilation operators is said to be *Wick ordered* or *normally ordered* if all creation operators occur to the left of all annihilation operators in each product, and to *Wick order* such a sum of products is to replace it with the corresponding Wick ordered product. This is well defined since creation operators commute with each other, as do annihilation operators. Wick ordering is indicated by putting a colon on either side of the expression in question. Thus, for example,

$$:a(\mathbf{p})a^\dagger(\mathbf{p}): = :a^\dagger(\mathbf{p})a(\mathbf{p}): = a^\dagger(\mathbf{p})a(\mathbf{p}),$$

and the corrected version of (5.23) is

$$H = \frac{1}{2}\int :(\partial_t\phi(t,\mathbf{x}))^2 + |\nabla_{\mathbf{x}}\phi(t,\mathbf{x}))|^2 + m^2\phi(t,\mathbf{x})^2: d^3\mathbf{x}. \tag{5.24}$$

We now turn to the complex Klein-Gordon field. The Lagrangian for a classical complex field is the same as for a real field except that squares of real numbers are replaced by absolute squares of complex numbers. Taking into account the factor of $\sqrt{2}$ that we introduced in (5.13), we find that the classical Lagrangian is

$$L = \int \mathcal{L}\, d^3\mathbf{x}, \qquad \mathcal{L} = (\partial_\mu\phi)\partial^\mu\phi^* - m^2|\phi|^2,$$

so the canonically conjugate field is

$$\pi = \frac{\partial\mathcal{L}}{\partial(\partial_t\phi)} = \partial_t\phi^*,$$

and the Hamiltonian is

$$H = \int \mathcal{H}\, d^3\mathbf{x}, \qquad \mathcal{H} = |\partial_t\phi|^2 + |\nabla_{\mathbf{x}}\phi|^2 + m^2|\phi|^2.$$

When we quantize, the complex conjugates turn into adjoints: The canonical conjugate of the quantum field ϕ of (5.15) is

$$\pi(x) = \partial_t\phi^\dagger(x) = \int \frac{i\omega_{\mathbf{p}}}{\sqrt{2\omega_{\mathbf{p}}}}\left[e^{ip_\mu x^\mu}a^\dagger(\mathbf{p}) - e^{-ip_\mu x^\mu}b(\mathbf{p})\right]\frac{d^3\mathbf{p}}{(2\pi)^3}.$$

The Hamiltonian can be expressed by formulas analogous to (5.22) and (5.24). The analogue of (5.22) is

$$H = \int \omega_{\mathbf{p}}\left[a^\dagger(\mathbf{p})a(\mathbf{p}) + b^\dagger(\mathbf{p})b(\mathbf{p})\right]\frac{d^3\mathbf{p}}{(2\pi)^3}, \tag{5.25}$$

and the analogue of (5.24) is

$$H = \int :\partial_t\phi^\dagger\partial_t\phi + \nabla_{\mathbf{x}}\phi^\dagger\cdot\nabla_{\mathbf{x}}\phi + m^2\phi^\dagger\phi: d^3\mathbf{x}, \tag{5.26}$$

Let us examine this a little more closely. If one starts from the classical Hamiltonian and quantizes the fields as in (5.23), without Wick ordering, one finds two possibilities for H depending on whether one replaces $|\phi|^2$ by $\phi^\dagger\phi$ or $\phi\phi^\dagger$:

$$\int(\partial_t\phi^\dagger\partial_t\phi + \nabla_{\mathbf{x}}\phi^\dagger\cdot\nabla_{\mathbf{x}}\phi + m^2\phi^\dagger\phi)\,d^3\mathbf{x}$$

$$\text{and} \int(\partial_t\phi\partial_t\phi^\dagger + \nabla_{\mathbf{x}}\phi\cdot\nabla_{\mathbf{x}}\phi^\dagger + m^2\phi\phi^\dagger)\,d^3\mathbf{x}.$$

Calculations like the ones above that led from (5.23) to (5.22) show that these two expressions are respectively equal to

$$\int \omega_{\mathbf{p}} \left[a^\dagger(\mathbf{p})a(\mathbf{p}) + b(\mathbf{p})b^\dagger(\mathbf{p})\right] \frac{d^3\mathbf{p}}{(2\pi)^3}$$

(5.27)

$$\text{and} \int \omega_{\mathbf{p}} \left[a(\mathbf{p})a^\dagger(\mathbf{p}) + b^\dagger(\mathbf{p})b(\mathbf{p})\right] \frac{d^3\mathbf{p}}{(2\pi)^3}.$$

Wick ordering turns both of these into (5.25); either way, there is an infinite constant to be discarded.

There is another quantity of interest here. Recall that when we introduced the Klein-Gordon equation in §4.1 we found the conserved quantity $\int \operatorname{Im} \phi^* \partial_t \phi \, d^3\mathbf{x}$. The quantum analogue of this is

$$Q = i \int :\phi^\dagger \partial_t \phi - \partial_t \phi^\dagger \phi: d^3\mathbf{x},$$

which, by a calculation like the one leading from (5.23) to (5.22), is equal to

$$\int \left[a^\dagger(\mathbf{p})a(\mathbf{p}) - b^\dagger(\mathbf{p})b(\mathbf{p})\right] \frac{d^3\mathbf{p}}{(2\pi)^3} = N_a - N_b.$$

Here N_a and N_b are simply the number operators for particles and antiparticles (the operators with eigenvalue k on the states with k particles and k antiparticles, respectively). It is easily verified that $N_a - N_b$ commutes with the Hamiltonian, so it represents a conserved quantity. (In fact, N_a and N_b commute with H separately, but that is because we are dealing with a free field where nothing is happening. The net particle number $N_a - N_b$ is conserved in many situations where N_a and N_b are not.) If the particles and antiparticles carry unit electric charges of opposite signs, then $N_a - N_b$ is the total net charge (if the charge of the "particles" is taken as positive). In this situation, it is clear that the *classical* quantity $\operatorname{Im}[\phi^*(\partial_t \phi)]$ should be interpreted as the charge density of the field, and the corresponding expression with space derivatives, $\operatorname{Im}[\phi^* \nabla_{\mathbf{x}} \phi]$, as the current density.

One final remark. In classical mechanics, the Lagrangian and Hamiltonian are close cousins. They are used for different purposes, but they are similar in form, and it is usually easy to pass from one to the other. In quantum field theory, on the other hand, the Hamiltonian turns into an operator on some Hilbert space, while the Lagrangian remains a "classical," unquantized object. The Lagrangian is, so to speak, the classical root from which the quantum theory grows. This idea will be illustrated in the following chapters, and it will attain a deeper and more compelling significance in Chapter 8.

5.4. Spinor and vector fields

Spinor fields. The construction of the Dirac spinor field whose quanta are spin-$\frac{1}{2}$ particles is similar to the construction of the complex scalar field, with only minor algebraic complications, so we shall be brief. As before, the idea is to think of a solution of the Dirac equation as a *classical* field, write it as a Fourier integral, and quantize it by replacing the coefficients by creation and annihilation operators.

We begin with the Fourier representation of a (reasonably) general solution of of the Dirac equation $i\gamma^\mu \partial_\mu \psi = m\psi$. That is, we take ψ to be the inverse Fourier transform of a $\mathbb{C}^4$-valued function f on the extended mass shell $X_m^+ \cup X_m^-$ times the invariant measure $d\lambda(p) = d^3\mathbf{p}/(2\pi)^3\omega_\mathbf{p}$, satisfying $p_\mu\gamma^\mu f = mf$:

(5.28)

$$\psi(x) = \int_{X_m^+\cup X_m^-} e^{-ip_\mu x^\mu} f(p)\,\frac{\sqrt{m}\,d^3\mathbf{p}}{(2\pi)^3\omega_\mathbf{p}} = \int_{X_m^+} [e^{-ip_\mu x^\mu} f(p) + e^{ip_\mu x^\mu} f(-p)]\,\frac{\sqrt{m}\,d^3\mathbf{p}}{(2\pi)^3\omega_\mathbf{p}}.$$

(The $\sqrt{m}$ is as in (4.44); we shall comment on its significance later.) Our first job is to rewrite this in a way that displays the spin states explicitly and encodes the equation $p_\mu\gamma^\mu f = mf$.

Recall the discussion of the Dirac equation at the end of §4.4 — in particular, the action of $SL(2,\mathbb{C})$ on $X_m^\pm$, which we denoted by $(A,p) \mapsto A^{\dagger-1}p$; the base points $p_m^\pm = (\pm m, 0, 0, 0)$; the spin representation Φ of $SL(2,\mathbb{C})$ defined by (4.21); and the fact that if $v \in \mathbb{C}^4$ satisfies $p_\mu\gamma^\mu v = mv$, then $w = \Phi(A)v$ satisfies $p'_\mu\gamma^\mu w = mw$ where $p' = A^{\dagger-1}p$. Let

(5.29)

$$u(\mathbf{0},+) = \begin{pmatrix}1\\0\\1\\0\end{pmatrix},\quad u(\mathbf{0},-) = \begin{pmatrix}0\\1\\0\\1\end{pmatrix},\quad v(\mathbf{0},+) = \begin{pmatrix}1\\0\\-1\\0\end{pmatrix},\quad v(\mathbf{0},-) = \begin{pmatrix}0\\1\\0\\-1\end{pmatrix}.$$

Then $u(\mathbf{0},\pm)$ and $v(\mathbf{0},\pm)$, respectively, are bases for the subspaces of $\mathbb{C}^4$ defined by the equations $\gamma^0 v = v$ and $-\gamma^0 v = v$, i.e., $(p_m^+)_\mu\gamma^\mu v = v$ and $(p_m^-)_\mu\gamma^\mu v = v$, with γ^μ in the Weyl representation. Here the $\mathbf{0}$ is the space component of $p_m^\pm$ and the $\pm$ is a label for the spin state. (More precisely, $\pm\frac{1}{2}$ is the eigenvalue of the x^3-component of the spin, since the u's and v's are eigenvectors of $\left(\begin{smallmatrix}\sigma & 0\\ 0 & \sigma\end{smallmatrix}\right)$ where $\sigma = \sigma_3$ is the 3rd Pauli matrix.) Now, for each $p \in X_m^\pm$ there is a unique positive Hermitian $B_p \in SL(2,\mathbb{C})$ such that $B_p^{\dagger-1}p_m^\pm = p$ — or rather $\kappa(B_p^{\dagger-1})p_m^\pm = p$. (Indeed, let A be any element of $SL(2,\mathbb{C})$ such that $A^{\dagger-1}p_m^\pm = p$, and let $A = BU$ be its polar decomposition. Since $SU(2)$ is the subgroup that fixes $p_m^\pm$, we can take $B_p = B$, and any other A' with $(A')^{\dagger-1}p_m^\pm = p$ gives the same B because $A' = AU'$ for some $U \in SU(2)$. The Lorentz transformation defined by B_p is a pure boost with no rotation.) We set

$$u(\mathbf{p},\pm) = \Phi(B_{(\omega_\mathbf{p},\mathbf{p})})u(\mathbf{0},\pm),\quad v(\mathbf{p},\pm) = \Phi(B_{(-\omega_\mathbf{p},\mathbf{p})})v(\mathbf{0},\pm). \tag{5.30}$$

Then $u(\mathbf{p},\pm)$ and $v(\mathbf{p},\pm)$ are bases for the subspaces of $\mathbb{C}^4$ defined by the equation $p_\mu\gamma^\mu v = mv$ for $p \in X_m^+$ and $p \in X_m^-$, respectively. Thus, any $\mathbb{C}^4$-valued function $f(p)$ on X_m^+ or X_m^- that satisfies $p_\mu\gamma^\mu f(p) = mf(p)$ can be written as a linear combination of $u(\mathbf{p},\pm)$ or $v(\mathbf{p},\pm)$ with scalar coefficients that depend on $\mathbf{p}$.

We can therefore rewrite (5.28) as

$$\psi(x) = \int \sum_{s=\pm} \sqrt{\frac{m}{2\omega_\mathbf{p}}}[e^{-ip_\mu x^\mu} f(\mathbf{p},s)u(\mathbf{p},s) + e^{ip_\mu x^\mu} g(\mathbf{p},s)v(\mathbf{p},s)]\,\frac{d^3\mathbf{p}}{(2\pi)^3},$$

where $p = (\omega_\mathbf{p},\mathbf{p})$ and the coefficients $f(\mathbf{p},s)$ and $g(\mathbf{p},s)$ are scalar-valued functions in which a factor of $\sqrt{2/\omega_\mathbf{p}}$ has been incorporated. Just as in the case of the complex scalar field, the quantum Dirac field is now obtained by replacing the coefficient $f(\mathbf{p},s)$ by the annihilation operator $a(\mathbf{p},s)$ for a particle with momentum $\mathbf{p}$ and spin state s, and the coefficient $g(\mathbf{p},s)$ by the creation operator $b^\dagger(\mathbf{p},s)$ for an *antiparticle* with momentum $\mathbf{p}$ and spin state s. The quanta of the Dirac field are going to be Fermions (a point to which we shall return shortly), so these operators need to satisfy the *anticommutation* relations

$$\begin{aligned}\{a(\mathbf{p},s),a^\dagger(\mathbf{p}',s')\} = \{b(\mathbf{p},s),b^\dagger(\mathbf{p}',s')\} = (2\pi)^3\delta(\mathbf{p}-\mathbf{p}')\delta_{ss'},\\ \{a(\mathbf{p},s),a(\mathbf{p}',s')\} = \{b(\mathbf{p},s),b(\mathbf{p}',s')\} = \{a^\sharp(\mathbf{p},s),b^\flat(\mathbf{p}',s')\} = 0,\end{aligned} \tag{5.31}$$

where $a^\sharp$ is either a or $a^\dagger$ and $b^\flat$ is either b or $b^\dagger$. Thus, the quantized Dirac field is

$$\psi(x) = \int \sum_{s=\pm} \sqrt{\frac{m}{2\omega_{\mathbf{p}}}}[e^{-ip_\mu x^\mu}u(\mathbf{p},s)a(\mathbf{p},s) + e^{ip_\mu x^\mu}v(\mathbf{p},s)b^\dagger(\mathbf{p},s)]\,\frac{d^3\mathbf{p}}{(2\pi)^3}, \tag{5.32}$$

and its Dirac adjoint (recall (4.14)) is

$$\overline{\psi}(x) = \int \sum_{s=\pm} \sqrt{\frac{m}{2\omega_{\mathbf{p}}}}[e^{ip_\mu x^\mu}\overline{u}(\mathbf{p},s)a^\dagger(\mathbf{p},s) + e^{-ip_\mu x^\mu}\overline{v}(\mathbf{p},s)b(\mathbf{p},s)]\,\frac{d^3\mathbf{p}}{(2\pi)^3}. \tag{5.33}$$

(5.31)–(5.33) can be expressed in rigorous mathematical language just as we did for the scalar field: the operators $a(\mathbf{p},s)$, $b(\mathbf{p},s)$, and $\psi(x)$ are actually distributions with values in the operators on the finite-particle subspace of $\mathcal{F} = \mathcal{F}_1 \otimes \mathcal{F}_2$, where $\mathcal{F}_1$ and $\mathcal{F}_2$ are copies of the Fermion Fock space over the state space for a single Dirac particle as discussed in §4.4, §4.5, and §5.2. These distributions are actually better behaved than their Bosonic analogues in the following respect: The smeared-out creation and annihilation operators, $a(f,s) = \int a(\mathbf{p},s)f(\mathbf{p})\,d^3\mathbf{p}$ and so forth, are bounded operators on $\mathcal{F}$ because of (4.57), and hence so are the spatially smeared-out field operators $\psi(t,f) = \int \psi(t,\mathbf{x})f(\mathbf{x})\,d^3\mathbf{x}$.

At this point we insert a calculation of "spin sums" that will be used here and later. Recall that $u(\mathbf{p},\pm)$ and $v(\mathbf{p},\pm)$ are column vectors; their Dirac adjoints $\overline{u}(\mathbf{p},\pm) = u(\mathbf{p},\pm)^\dagger\gamma_0$ and $\overline{v}(\mathbf{p},\pm) = v(\mathbf{p},\pm)^\dagger\gamma^0$ are row vectors, and the products $u(\mathbf{p},\pm)\overline{u}(\mathbf{p},\pm)$ and $v(\mathbf{p},\pm)\overline{v}(\mathbf{p},\pm)$ are therefore 4×4 matrices. An easy calculation from the definitions (5.29) shows that

$$\sum_{s=\pm} u(\mathbf{0},s)\overline{u}(\mathbf{0},s) = \gamma^0 + I, \qquad \sum_{s=\pm} v(\mathbf{0},s)\overline{v}(\mathbf{0},s) = \gamma^0 - I.$$

From this we can obtain the corresponding sums for arbitrary $\mathbf{p}$. By (5.30), with $B = B_{(\omega_{\mathbf{p}},\mathbf{p})}$ we have

$$\sum_{s=\pm} u(\mathbf{p},s)\overline{u}(\mathbf{p},s) = \sum_{s=\pm} \Phi(B)u(\mathbf{0},s)u(\mathbf{0},s)^\dagger\Phi(B)^\dagger\gamma^0.$$

But by (4.12) and (4.15) we have $\Phi(B)^\dagger = \Phi(B^\dagger) = \gamma^0\Phi(B)^{-1}\gamma^0$, so

$$\sum_{s=\pm} u(\mathbf{p},s)\overline{u}(\mathbf{p},s) = \sum_{s=\pm} \Phi(B)u(\mathbf{0},s)\overline{u}(\mathbf{0},s)\Phi(B)^{-1} = \Phi(B)\gamma^0\Phi(B)^{-1} + I.$$

Moreover, if $L = \kappa(B)^{-1} \in SO(1,3)$, by (4.22) we have $\Phi(B)\gamma^0\Phi(B)^{-1} = L^0_\mu\gamma^\mu$, and in view of the fact that B is Hermitian, the defining condition $B^{\dagger-1}p^+_m = p$ just means that $mL^0_\mu = p_\mu$. Therefore, after an isomorphic calculation with the v's, we conclude that

$$\sum_{s=\pm} u(\mathbf{p},s)\overline{u}(\mathbf{p},s) = m^{-1}p_\mu\gamma^\mu + I, \qquad \sum_{s=\pm} v(\mathbf{p},s)\overline{v}(\mathbf{p},s) = m^{-1}p_\mu\gamma^\mu - I, \tag{5.34}$$

with $p_0 = \omega_{\mathbf{p}}$.

We recall from §4.1 that the Lagrangian density associated to the free Dirac equation is

$$\mathcal{L} = i\overline{\psi}\gamma^\mu\partial_\mu\psi - m\overline{\psi}\psi,$$

so the field π canonically conjugate to ψ is

$$\pi = \frac{\partial \mathcal{L}}{\partial(\partial_t \psi)} = i\psi^\dagger.$$

The fields ψ and π satisfy *canonical anticommutation relations* analogous to (5.20):

$$\begin{aligned} &\{\psi_j(t,\mathbf{x}), \pi_k(t,\mathbf{y})\} = i\delta(\mathbf{x}-\mathbf{y})\delta_{jk}, \\ &\{\psi_j(t,\mathbf{x}), \psi_k(t,\mathbf{y})\} = \{\pi_j(t,\mathbf{x}), \pi_k(t,\mathbf{y})\} = 0. \end{aligned} \qquad (j,k=1,\cdots,4) \tag{5.35}$$

To prove this, observe that by (5.32), $\{\psi_j(t,\mathbf{x}), i\psi_k^\dagger(t,\mathbf{y})\}$ is equal to

$$i \iint \sum_{s,s'=\pm} \frac{me^{-i\omega_{\mathbf{p}}t + i\mathbf{p}\cdot\mathbf{x}} e^{i\omega_{\mathbf{p}'}t - i\mathbf{p}'\cdot\mathbf{y}}}{2\sqrt{\omega_{\mathbf{p}}\omega_{\mathbf{p}'}}} u(\mathbf{p},s)_j u(\mathbf{p}',s')_k^\dagger \{a(\mathbf{p},s), a^\dagger(\mathbf{p}',s')\} \frac{d^2\mathbf{p}\, d^3\mathbf{p}'}{(2\pi)^6}$$

plus a similar term with u's and a's replaced by v's and b's and the signs in the exponents reversed. By (5.31), (5.34), and the fact that

$$u(\mathbf{p},s)_j u(\mathbf{p},s)_k^\dagger = u(\mathbf{p},s)_j [\overline{u}(\mathbf{p},s)\gamma^0]_k = [u(\mathbf{p},s)\overline{u}(\mathbf{p},s)\gamma^0]_{jk},$$

integration over $\mathbf{p}'$ and summation over s yield

$$\begin{aligned} \{\psi_j(t,\mathbf{x}), i\psi_k^\dagger(t,\mathbf{y})\} =& i\int \frac{m}{2\omega_{\mathbf{p}}} e^{i\mathbf{p}\cdot(\mathbf{x}-\mathbf{y})} [(m^{-1}p_\mu\gamma^\mu + mI)\gamma^0]_{jk} \frac{d^3\mathbf{p}}{(2\pi)^3} \\ &+ i\int \frac{m}{2\omega_{\mathbf{p}}} e^{-i\mathbf{p}\cdot(\mathbf{x}-\mathbf{y})} [(m^{-1}p_\mu\gamma^\mu - mI)\gamma^0]_{jk} \frac{d^3\mathbf{p}}{(2\pi)^3}. \end{aligned}$$

The substitution $\mathbf{p} \to -\mathbf{p}$ in the second integral shows that the terms involving mI and $p_\mu\gamma^\mu$ with $\mu > 0$ in the two integrals cancel out. The terms involving $p_0\gamma^0$ add up, and since $(\gamma^0)^2 = I$ and $p_0 = \omega_{\mathbf{p}}$, the integrals reduce to

$$\{\psi_j(t,\mathbf{x}), i\psi_k^\dagger(t,\mathbf{y})\} = i\int e^{i\mathbf{p}\cdot(\mathbf{x}-\mathbf{y})}\delta_{jk} \frac{d^3\mathbf{p}}{(2\pi)^3} = i\delta(\mathbf{x}-\mathbf{y})\delta_{jk}.$$

The other equations in (5.35) are obtained similarly.

The factor $\sqrt{m}$ in (5.32) that we carried over from (4.44) is what gives the proper normalization in (5.35). One can also see that the Dirac field (5.32) should contain a factor with dimensions $[m^{1/2}]$ in comparison to the scalar field (5.15) from the fact that the mass terms in their respective Lagrangians are $m\overline{\psi}\psi$ and $m^2|\phi|^2$, both of which are supposed to yield an energy density. However, there are even more variations on the formula (5.32) in the literature than there are for the scalar field, because one can use different normalizations not only for a and b but also for the spinors u and v. In particular, some people incorporate the factor of $\sqrt{m}$ into u and v.

The Hamiltonian density for the Dirac field ψ is

$$\mathcal{H} = \pi\partial_t\psi - \mathcal{L} = \pi\partial_t\psi = i\psi^\dagger\partial_t\psi,$$

since $\mathcal{L} = \mathcal{L}(\psi, \partial_\mu\psi) = 0$ when ψ is a solution of the Dirac equation. After substituting (5.32) for ψ in this expression, a short calculation involving the Fourier inversion formula $\int e^{i\mathbf{p}\cdot\mathbf{x}}\, d^3\mathbf{x} = (2\pi)^3\delta(\mathbf{p})$ shows that the Hamiltonian is

$$H = \int \mathcal{H}\, d^3\mathbf{x} = \int \sum_\sigma \omega_{\mathbf{p}} [a^\dagger(\mathbf{p},s)a(\mathbf{p},s) - b(\mathbf{p},s)b^\dagger(\mathbf{p},s)] \frac{d^3\mathbf{p}}{(2\pi)^3}. \tag{5.36}$$

At first sight this looks like it might not be bounded below, which would be most unfortunate. But by (5.31) we have

$$b(\mathbf{p},s)b^\dagger(\mathbf{p}',s') = -b^\dagger(\mathbf{p}',s')b(\mathbf{p},s) + \delta(\mathbf{p}-\mathbf{p}')\delta_{ss'},$$

so the cost of throwing away an infinite constant, as we did for the scalar field, we can rewrite (5.36) as

$$H = \int \sum_\sigma \omega_{\mathbf{p}}[a^\dagger(\mathbf{p},s)a(\mathbf{p},s) + b^\dagger(\mathbf{p},s)b(\mathbf{p},s)]\, \frac{d^3\mathbf{p}}{(2\pi)^3}, \tag{5.37}$$

which is manifestly positive. (5.37) is, of course, the Wick-ordered form of (5.36) — with the understanding that where Fermionic operators are concerned, an interchange of two operators in the Wick-ordering process introduces a minus sign.

This brings up an important point. If we had tried to quantize the Dirac field as a *Boson* field, so that the anticommutation relations (5.31) were replaced by commutation relations, this trick would not work and we would end up with energies that are not bounded below. For exactly the same reason, if we had tried to quantize the scalar Klein-Gordon field as a *Fermion* field, we would be in trouble after arriving at the expressions (5.27). This is one aspect of the spin-statistics theorem, which we shall discuss further in §5.5 and §6.5.

Vector fields. We would now like to quantize the electromagnetic field. The first point that has to be appreciated is that the field to be quantized is not the electromagnetic field $F_{\mu\nu}$ itself but its potential A_μ. For one thing, the vector field A_μ rather than the tensor field $F_{\mu\nu}$ is the appropriate thing to yield spin-one quanta, which photons are known by experiment to be. Moreover, the Aharonov-Bohm effect (see Sakurai [**102**]) shows that the potential A_μ has a real effect on quantum phenomena even in regions where $F_{\mu\nu}$ vanishes. As Weinberg [**129**], p. 340, observes, one could construct a consistent theory "by demanding that all interactions involve only $F_{\mu\nu}(x) = \partial_\mu A_\nu(x) - \partial_\nu A_\mu(x)$ and its derivatives, not $A_\mu(x)$, but this is not the most general possibility, and not the one realized in nature."

But various difficulties arise from the fact that A_μ is defined only up to gauge transformations and from the closely related fact that the photon is massless. To get a handle on the situation, let us first consider an analogous field whose quanta have nonzero mass. (Such fields describe the vector bosons that appear in the Weinberg-Salam theory of weak interactions.) On the classical level, such a field is a an $\mathbb{C}^4$-valued function A on space-time that transforms under Lorentz transformations by the identity representation of the Lorentz group: $A(x) \mapsto LA(L^{-1}x)$. Its Lagrangian is

$$\mathcal{L} = -\tfrac{1}{4}F_{\mu\nu}F^{\mu\nu} + \tfrac{1}{2}m^2 A_\mu A^\mu \qquad (F^{\mu\nu} = \partial^\mu A^\nu - \partial^\nu A^\mu), \tag{5.38}$$

and the resulting field equations are a massive version of Maxwell's equations known as the *Proca equations*:

$$\partial_\mu F^{\mu\nu} + m^2 A^\nu = 0. \tag{5.39}$$

Since $\partial_\mu\partial_\nu F^{\mu\nu} = 0$, (5.39) implies that $\partial_\nu A^\nu = 0$, and since $\partial_\mu F^{\mu\nu} = \partial^2 A^\nu - \partial^\nu(\partial_\mu A^\mu)$, we see that (5.39) is equivalent to the Klein-Gordon equation plus the condition of zero 4-divergence:

$$\partial^2 A^\nu + m^2 A^\nu = 0, \qquad \partial_\nu A^\nu = 0. \tag{5.40}$$

In the limit $m \to 0$, this yields Maxwell's equations with the Lorentz gauge condition. On the Fourier transform side, the Klein-Gordon equation implies that $\widehat{A}$ lives on the mass shells $X_m^\pm$, and the equation $\partial_\nu A^\nu = 0$ becomes the linear constraint $p_\nu \widehat{A}^\nu(p) = 0$, which reduces the number of independent components of A from 4 to 3. With this in mind, one sees that the action of the rotation group on A is given by its (complexified) identity representation on $\mathbb{C}^3$, i.e., the representation associated to spin one.

The quantization of the massive vector field proceeds in much the same way as that of the Dirac field. The result is

(5.41)
$$A^\nu(x) = \int \sum_{j=1}^{3} \frac{1}{\sqrt{2\omega_{\mathbf{p}}}}[e^{-ip_\mu x^\mu}\boldsymbol{\epsilon}^\nu(\mathbf{p},j)a(\mathbf{p},j) + e^{ip_\mu x^\mu}\boldsymbol{\epsilon}^\nu(\mathbf{p},j)^* b^\dagger(\mathbf{p},j)]\frac{d^3\mathbf{p}}{(2\pi)^3},$$

where a and b represent annihilation operators for particles and antiparticles — the possibility $b = a$ for neutral particles is allowed — and the $\boldsymbol{\epsilon}^\nu(\mathbf{p},j)$ are *polarization vectors*. The nontrivial commutation relations for a and b are the usual ones,

$$[a(\mathbf{p},j), a^\dagger(\mathbf{p}',j')] = [b(\mathbf{p},j), b^\dagger(\mathbf{p}',j')] = (2\pi)^3\delta(\mathbf{p}-\mathbf{p}')\delta_{jj'},$$

and the polarization vectors are specified as follows. For $\mathbf{p} = \mathbf{0}$, $\{\boldsymbol{\epsilon}^\nu(\mathbf{0},j) : j = 1,2,3\}$ is a given orthonormal basis for $\{0\}\times\mathbb{C}^3$ (the same one for each ν), and then $\boldsymbol{\epsilon}^\nu(\mathbf{p},j) = B^\nu_\mu\boldsymbol{\epsilon}^\mu(\mathbf{0},j)$ where $B \in SO^\uparrow(1,3)$ is the pure boost that takes $p_m^+ = (m,\mathbf{0})$ to $(\omega_{\mathbf{p}},\mathbf{p})$ (similarly to the spinor coefficients of the Dirac field). (This automatically guarantees that $p_\nu\boldsymbol{\epsilon}^\nu(\mathbf{p},j) = 0$.)

Now, what happens if we let $m \to 0$? The Proca equations (5.40) turn into Maxwell's equations with the Lorentz gauge condition, which is fine. But the single-particle states for massive vector particles have three independent components, corresponding to the three polarization vectors in (5.41), whereas the single-particle states for massless ones have only two, specified by the positive or negative helicity along the direction of motion. (On the classical level, this corresponds to the fact that the electromagnetic field admits only transverse oscillations, not longitudinal ones.) Thus if we try to quantize the electromagnetic field in the form (5.41) (with $b = a$), we have too many components, and some of them will represent unphysical "ghost fields" that create "ghost states" whose contributions to later computations must ultimately cancel out. It is indeed possible to proceed in this way, picking one's way among the technical obstacles carefully; the result is known as *Gupta-Bleuler quantization.*

Another possibility is to discard Lorentz invariance in favor of a quantization that will yield the physical fields more directly. On the classical level, the Lorentz gauge condition $\partial_\mu A^\mu = 0$ does not specify A^μ completely; one can still add a term $\partial^\mu\chi$ where $\partial^2\chi = 0$. For a free field, in the absence of charges, one can thereby arrange to make $A^0 = 0$, that is, to put A into the *Coulomb gauge* or *radiation gauge*:

$$A = (0,\mathbf{A}), \qquad \nabla\cdot\mathbf{A} = 0.$$

Quantization of *this* field yields

$$\mathbf{A}(x) = \int \sum_{j=1}^{2} \frac{1}{\sqrt{2|\mathbf{p}|}}[e^{-ip_\mu x^\mu}\boldsymbol{\epsilon}(\mathbf{p},j)a(\mathbf{p},j) + e^{ip_\mu x^\mu}\boldsymbol{\epsilon}(\mathbf{p},j)^* a^\dagger(\mathbf{p},j)]\frac{d^3\mathbf{p}}{(2\pi)^3}, \tag{5.42}$$

where $\boldsymbol{\epsilon}(\mathbf{p},1)$ and $\boldsymbol{\epsilon}(\mathbf{p},2)$ are an orthonormal basis for $\{\boldsymbol{\epsilon} : \boldsymbol{\epsilon}\cdot\mathbf{p} = 0\}$. (One can choose these vectors to depend smoothly on $\mathbf{p} \in \mathbb{R}^3 \setminus \{0\}$ by allowing them to be complex — that is, by allowing general elliptical polarizations rather than just linear ones. Note also that $\omega_{\mathbf{p}} = |\mathbf{p}|$ since $m = 0$ here.) In effect, the condition $\boldsymbol{\epsilon}\cdot\mathbf{p} = 0$ means that one is keeping only that part of the field (5.41) whose polarization is transverse to the direction of motion.

Using the Coulomb gauge, of course, destroys the manifest Lorentz invariance of the theory. Moreover, both this procedure and the Lorentz-invariant quantization with ghost fields present some technical problems in connection with constructing canonically conjugate fields for use in constructing the Hamiltonian. There is no general agreement among physics texts about the best way of handling these problems, and as mathematical tourists we are better off not getting into these muddy waters. The most essential information that must be extracted from the fields is the associated particle propagators; we shall address that problem in §6.5 and §8.3.

Concluding remarks. One can construct a free quantum field whose quanta have any mass $m > 0$ and spin $s \in \frac{1}{2}\mathbb{Z}_+$. It acts on the Fock space over the one-particle state space associated to the representation $\pi_{m,s}$ of §4.4 (or the tensor product of two copies of the Fock space for particle-antiparticle pairs). The general idea is the same as in the cases $s = 0, \frac{1}{2}, 1$ that we have described; only the algebra is different. See Streater and Wightman [**114**] and Weinberg [**129**].

Let us review what we have accomplished. We started from the idea of a classical field, that is, a function ϕ on (some region in) space-time whose value at a point x is an observable quantity — say, a force, a velocity, a temperature — that can be determined by measurements performed at x. One would expect quantization to yield an function Φ on space-time whose value at x is the quantum observable — i.e., self-adjoint operator — corresponding to the classical observable $\phi(x)$. Thus, when the system is in a state v the expectation value $\langle v|\Phi(x)|v\rangle$ should be somehow closely related to the classical field $\phi(x)$ in the corresponding classical state.

What we seem to have ended up with, however, is rather different. The fact that we need operator-valued distributions rather than functions is a technical obstacle rather than a conceptual one, for even classically the notion of making measurements at a single point specified to infinite precision is an idealization. What is more serious is that the values of the quantum fields we have constructed are self-adjoint (or, more loosely, Hermitian) only when the quanta of the field are their own antiparticles; in other cases the values of the fields cannot directly represent observable quantities. This is perhaps just as well, for what would we be observing? The "classical fields" we have quantized to produce the scalar and Dirac quantum fields have no meaning in classical physics; even for the electromagnetic field, what we have quantized is not the directly observable field $F_{\mu\nu}$ but the more abstract potential A_μ. Rather, these "classical fields" arise as frameworks for describing single quantum particles, and the quantum fields we have constructed turn out to be machines for creating and destroying such particles.

At this point readers may well be wondering what these quantum fields have to do with physical reality. So far, the answer is: not much. We have constructed only *free* fields, whereas all the real physics comes from *interactions*. Getting to the interesting stuff will require a little more patience and a lot more willingness to accept mathematical incompleteness. We shall take the plunge in the next chapter.

5.5. The Wightman axioms

The success of quantum electrodynamics naturally led to attempts to develop the theory in a mathematically rigorous way and to generalize it to encompass other kinds of quantum fields. In the late 1950s Wightman and Gårding formulated a list of basic properties that any physically reasonable and mathematically well-defined quantum field theory should have. This list has come to be known as the *Gårding-Wightman axioms* or the *Wightman axioms.*

The ingredients for the Wightman axioms are a Hilbert space $\mathcal{F}$ equipped with

i. a dense subspace $\mathcal{D}$,
ii. a set of operator-valued distributions $\phi_1, \dots, \phi_N$ on $\mathbb{R}^4$ (the meaning of this is explained more precisely in Axiom 1 below),
iii. a unitary representation U of the double cover $\mathbb{R}^4 \ltimes SL(2,\mathbb{C})$ of the restricted Poincaré group on $\mathcal{F}$, and
iv. a (nonunitary) representation S of $SL(2,\mathbb{C})$ on $\mathbb{C}^N$.

Since the unitary operators $U(a,I)$ $(a \in \mathbb{R}^4)$ all commute, they have a common spectral resolution; that is, there is a projection-valued measure E on $\mathbb{R}^4$ such that $U(a,I) = \int e^{-ip_\mu a^\mu}\, dE(p)$.

The axioms are as follows. All of them should look rather familiar except perhaps the last one, which we shall discuss later.

1. Each ϕ_n is a map from the Schwartz space $\mathcal{S}(\mathbb{R}^4)$ to the set of (perhaps unbounded) linear operators on $\mathcal{F}$ such that
   i. for all $f \in \mathcal{S}(\mathbb{R}^4)$ we have
   $$\mathcal{D} \subset \mathrm{Dom}(\phi_n(f)) \cap \mathrm{Dom}(\phi_n(f)^\dagger), \qquad \phi_n(f)\mathcal{D} \cup \phi_n(f)^\dagger\mathcal{D} \subset \mathcal{D};$$
   ii. for all $\xi, \eta \in \mathcal{D}$, the map $f \mapsto \langle \xi | \phi_n(f) | \eta \rangle$ is a tempered distribution on $\mathbb{R}^4$.
2. There is a unit vector $\Omega \in \mathcal{D}$ (the "vacuum state"), unique up to a phase factor, such that $U(a,A)\Omega = \Omega$ for all $(a,A) \in \mathbb{R}^4 \ltimes SL(2,\mathbb{C})$.
3. The set of vectors of the form $\phi_{n_1}(f_1)\cdots\phi_{n_k}(f_k)\Omega$ with $k \geq 0$, $n_j \in \{1,\dots,N\}$, and $f_j \in \mathcal{S}(\mathbb{R}^4)$ is dense in $\mathcal{H}$.
4. For any $f \in \mathcal{S}(\mathbb{R}^4)$, $(a,A) \in \mathbb{R}^4 \ltimes SL(2,\mathbb{C})$, and $n \in \{1,\dots,N\}$,
   $$U(a,A)\phi_n(f)U(a,A)^{-1} = \sum_{m=1}^{N} S(A^{-1})^m_n \phi_m((a,A)\cdot f)$$
   as operators on $\mathcal{D}$, where $[(a,A)\cdot f](x) = f(\kappa(A)^{-1}(x-a))$ and $\kappa : SL(2,\mathbb{C}) \to SO^\uparrow(1,3)$ is the covering map. If we adopt the notational pretense that ϕ_n is a function rather than a distribution, this means that
   $$U(a,A)\phi_n(x)U(a,A)^{-1} = \sum_{m=1}^{N} S(A^{-1})^m_n \phi_m(Ax+a).$$
5. The support of the projection-valued measure E is contained in the region $\{(p^0,\mathbf{p}) : p^0 \geq |\mathbf{p}|\}$ inside the forward light cone. (Equivalently: if P_μ is the infinitesimal generator of the one-parameter group $t \mapsto U(te_\mu, I)$, where $e_0,\dots,e_3$ are the standard basis for $\mathbb{R}^4$, then P_0 and $P_0^2 - \sum_1^3 P_j^2$ are both positive operators.)

6. If f and g are in $\mathcal{S}(\mathbb{R}^4)$ and the supports of f and g are spacelike separated (i.e., $x-y$ is spacelike whenever $f(x)g(y)\neq 0$), and $m,n\in\{1,\dots,N\}$, then as operators on $\mathcal{D}$,

$$\text{either}\quad [\phi_m(f),\phi_n(g)] = [\phi_m(f),\phi_n(g)^\dagger] = 0$$
$$\text{or}\quad \{\phi_m(f),\phi_n(g)\} = \{\phi_m(f),\phi_n(g)^\dagger\} = 0. \tag{5.43}$$

(Which of these two possibilities occurs may depend on m and n.) Again, if we pretend that the ϕ_n are functions, this means that $\phi_m(x)$ (and $\phi_m(x)^\dagger$) commutes or anticommutes with $\phi_n(y)$ and $\phi_n(y)^\dagger$ whenever $x-y$ is spacelike.

This framework can accomodate any number of fields representing any number of particle types, because there is no assumption that the representation S should be irreducible. If S is a direct sum of irreducible representations S_k on $\mathbb{C}^{n_k}$ with $k=1,\dots,K$ and $n_1+\cdots n_K=N$, the N-tuple of fields ϕ_n separates out into n_k-tuples, each of which may be the components of the field representing one of the particles of the theory.

The free massive scalar, Dirac, and vector fields that we have discussed earlier in this chapter are all examples of systems that satify the Wightman axioms, as are the free fields of arbitrary mass $m>0$ and spin $s\in\frac{1}{2}\mathbb{Z}_+$. The Hilbert space $\mathcal{F}$ is the (Boson or Fermion) Fock space over the appropriate single-particle state space $\mathcal{H}$, and the dense subspace $\mathcal{D}$ is the finite-particle subspace. The vacuum Ω is the no-particle state, i.e., a unit vector in $\bigotimes^0\mathcal{H}=\mathbb{C}$, and the closed linear span of the vectors $\phi_{n_1}(f_1)\cdots\phi_{n_k}(f_k)\Omega$ with $k\le K$ is the sum of the k-particle spaces for $k\le K$. The representation U is given by $U(a,A)=\mathcal{F}(\pi(a,A))$, where π is the appropriate irreducible representation of $\mathbb{R}^4\ltimes SL(2,\mathbb{C})$ on $\mathcal{H}$ as described in §4.4 and $\mathcal{F}(\pi(\cdot))$ is the corresponding representation on $\mathcal{F}$ given by (4.53). The representation S is the trivial representation on $\mathbb{C}$ for the scalar field, the representation Φ defined by (4.21) for the Dirac field, and the covering map $\kappa : SL(2,\mathbb{C})\to SO^\uparrow(1,3)$ for the vector field. The joint spectrum of the position-momentum operators on the single-particle space $\mathcal{H}$ is the mass shell X_m^+, and hence the joint spectrum on the k-particle space is $\{p_1+\cdots+p_k : p_1,\dots,p_k\in X_m^+\}$. All of these sets lie inside the forward light cone, and the support of the measure E in Axiom 5 is the closure of their union. The verification of Axiom 6 is a calculation that we shall perform in §6.5. (See Streater and Wightman [**114**] or Bogolubov et al. [**11**], [**12**];[2] also Reed and Simon [**94**] for a thorough treatment of the free scalar field.)

Let us examine the meaning of Axiom 6, known as the *microscopic causality* condition, in more detail. The idea is that for $x\in\mathbb{R}^4$ the fields $\phi_n(x)$ are supposed to represent phenomena that take place at x; more precisely, if $f\in\mathcal{S}(\mathbb{R}^4)$, the $\phi_n(f)$ are supposed to represent phenomena that take place in the space-time support of f. (For example, if one unravels all the Fourier analysis in the definition of the free scalar field ϕ, one finds that $\phi(t,\mathbf{x})\Omega$ is the state containing one particle located at position $\mathbf{x}$ at time t.) If x and y are space-like separated, nothing that happens at x can influence what happens at y and vice versa, so (according to the discussion of uncertainty in §3.3) observables connected with phenomena at x and y must commute. This argument does not immediately imply (5.43), for as we

[2]As with many Russian names, there are several possible spellings of "Bogolubov" in the Roman alphabet; the most common variations involve the replacement of the "u" by "iu," "ju," or "yu." We use the spelling in the works just cited.

have observed in the preceding section, the field operators usually do not represent observable quantities. However, as we shall see in the next chapter, quantities of direct physical significance are constructed out of field operators, and for them these (anti)commutation relations are important.[3] In particular, they are needed to guarantee the Lorentz invariance of the scattering matrix; according to Weinberg [**129**], p. 198, this is the real reason to insist on microscopic causality.

In this setting one can prove a rigorous form of the *spin-statistics theorem* that particles of integer spin must be Bosons and particles of half-integer spin must be Fermions.[4] The precise statement is as follows. Suppose $\phi_1, \ldots, \phi_N$ are the fields of a theory satisfying the Wightman axioms that describes one or several particle types; thus each ϕ_n is a component of an irreducible subset of fields that describes one of the particle types. If ϕ_n is a component of a field describing a particle of half-integer spin and $[\phi_n(f), \phi_n(g)^\dagger] = 0$ whenever f and g have spacelike-separated supports, then $\phi_n(f)\Omega = 0$ for all f; likewise if ϕ_n is a component of a field describing a particle of integer spin and $\{\phi_n(f), \phi_n(g)^\dagger\} = 0$. In either case, if the different fields of the theory all either commute or anticommute (normally the case), it follows that $\phi_n = 0$.

The Wightman axioms have served as the foundation for a large body of research in the mathematically rigorous theory of quantum fields. We shall not pursue this subject here, but we would be remiss not to mention an especially fundamental result, the so-called *PCT theorem.* This states that any quantum field theory satisfying the axioms is invariant under the combined operation PCT (the factors may be permuted in any order) where P and T are the space and time inversion operators coming from the action of the Lorentz group and C is the charge-conjugation operator that interchanges particles and antiparticles. (We discussed the latter for Dirac wave functions in §4.2 and for scalar fields in §5.2; the definition in general is similar.) A half-century ago it was believed that these three operators individually should be symmetries of any reasonable physical theory, so that the PCT theorem might have seemed like much ado about the obvious. But then it was discovered that the weak interaction has a definite "handedness" and so is not P-invariant (see §9.4), and more recently, experimentalists found that certain meson decays are not CP-invariant — at which point the significance of the PCT theorem could no longer be doubted.

A different but related mathematical framework in which rigorous quantum field theory can be studied is provided by the notion of algebras of *local observables.* The fundamental data here are:

i. a C* algebra $\mathcal{A}$ of operators on a Hilbert space $\mathcal{H}$,
ii. a C* subalgebra $\mathcal{A}(O) \subset \mathcal{A}$ for each bounded open set $O \subset \mathbb{R}^4$, and
iii. a representation α of the restricted Poincaré group $\mathcal{P}_0$ as automorphisms of $\mathcal{A}$,

subject to the following axioms:

1. $\mathcal{A}(O_1) \subset \mathcal{A}(O_2)$ whenever $O_1 \subset O_2$.
2. $\bigcup_O \mathcal{A}(O)$ is dense in $\mathcal{A}$.
3. $\alpha(g)(\mathcal{A}(O)) = \mathcal{A}(g(O))$ for all $g \in \mathcal{P}_0$.

---

[3]These quantities almost always involve products of even numbers of Fermion field operators, and anticommutation relations for the latter yield commutation relations for the products.

[4]This result is due to Fierz and Pauli; the rigorous proof in the framework of the Wightman axioms is due to Burgoyne and to Lüders and Zumino. See Streater and Wightman [**114**] for references.

4. $[A_1, A_2] = 0$ for all $A_1 \in \mathcal{A}(O_1)$ and $A_2 \in \mathcal{A}(O_2)$ whenever O_1 and O_2 are spacelike separated.

For example, if $\{\phi_1, \ldots, \phi_N\}$ is a set of fields satisfying the Wightman axioms for Bosons (i.e., with commutators in (5.43)), one can take $\mathcal{A}(O)$ to be the C* algebra generated by the operators $\phi_j(f)$ with $1 \leq j \leq N$ and $f \in C_c^\infty(O)$, with $\alpha(g)A = U(g)AU(g)^{-1}$. The study of quantum fields in this framework is known as *local* or *algebraic quantum field theory.* (In spite of initial appearances, it can also accomodate Fermion fields.)

There is an extensive mathematical theory based on the Wightman or local-observable axioms. The reader who wishes to learn more about it may consult Streater and Wightman [**114**], Jost [**69**], Bogolubov et al. [**11**], [**12**], and Araki [**3**]. All of these contain proofs of the spin-statistics and PCT theorems. In addition, Haag [**59**] gives a comprehensive survey of the subject, with many arguments sketched or omitted.

The bad news is that it has turned out to be a remarkably difficult task to construct examples of field theories that satisfy the Wightman axioms and have nontrivial interactions. Attempts to produce such examples in physical space-time $\mathbb{R}^4$ have yet to succeed, so most of the work has dealt with field theories in $\mathbb{R}^d$ (equipped with the Lorentz group $O(1, d-1)$) for $d = 2$ or 3, where the singularities that plague quantum fields tend to be somewhat tamer. (It is easy to adapt the Wightman axioms to any number of space dimensions.) Even there, things are not easy, and for several years after the formulation of the Wightman axioms it was an open question whether they were satisfied by any systems other than the free fields and minor modifications thereof. The first example, a self-interacting scalar field in dimension $d = 2$, was constructed by Glimm and Jaffe. Since then a variety of other examples have been explored: other self-interacting fields, a Dirac field and a scalar field with a Yukawa interaction, and various gauge fields — but all in dimensions 2 and 3. Much of this work has made use of the functional integral approach, and we shall say more about it in Chapter 8. There is a lot of interesting mathematics in it as well as interesting physics, including connections to other areas of physics such as statistical mechanics, but none of it does much to help with the down-to-earth calculations of quantum electrodynamics that we are aiming towards. We therefore leave it aside and refer the reader to the appendix of Streater and Wightman [**114**] and Rosen [**100**] for concise surveys and to Glimm and Jaffe [**54**] and Simon [**111**] for more extensive accounts.

CHAPTER 6

Quantum Fields with Interactions

To have to stop to formulate rigorous demonstrations would put a stop to most physico-mathematical inquiries. ... The physics will guide the physicist along somehow to useful and important results, by the constant union of physical and geometrical or analytical ideas.
—Oliver Heaviside, *Electromagnetic Theory* (§224)

MEPHISTOPHELES: Allwissend bin ich nicht; doch viel ist mir bewusst.
[I am not omniscient, but much is known to me.]
—J. W. Goethe, *Faust* (Part I, 4th scene)

The last temptation is the greatest treason:
To do the right deed for the wrong reason.
—T. S. Eliot, *Murder in the Cathedral* (Part I)

Everything we have done so far is mathematically respectable, although some of the results have been phrased in informal language. To make further progress, however, it is necessary to make a bargain with the devil. The devil offers us effective and conceptually meaningful techniques for calculating physically interesting quantities. In return, however, he requires us to compromise our mathematical souls by accepting the validity of certain approximation procedures and certain formal calculations without proof and — what is a good deal more disconcerting — by working with some putative mathematical objects that lack a rigorous definition. The situation is in some ways similar to the mathematical analysis of the eighteenth century, which developed without the support of a rigorous theory of limits and with the use of poorly defined infinitesimals.

The first and most essential tool the devil offers us is perturbation theory.

6.1. Perturbation theory

Suppose we have a quantum system whose Hamiltonian H is the sum of two terms:

$$H = H_0 + H_I.$$

We take H_0 to be known and understood, and we wish to study the effect of adding in the extra term H_I. In most of our applications, H_0 will be the Hamiltonian for a free field and H_I will be the interaction term, but the discussion below applies also to other situations where H is taken to be a perturbation of H_0. There is a considerable mathematical literature dealing with this kind of situation, in which precise hypotheses are placed on H, H_0, and H_I. Unfortunately, it is largely irrelevant for our purposes, because the H_I's that we shall need are too singular. We simply have to proceed step by step, taking some care to recognize when we are taking something on faith.

We begin with some general considerations. For psychological comfort, the reader may wish to think of the case where H_0 is a self-adjoint operator and H_I is a *bounded* self-adjoint operator, in which case H is a self-adjoint operator with the same domain as H_0. However, this is far from the case we shall need in the sequel.

What we are really interested in is not H and H_0 but the one-parameter unitary groups they generate, i.e., the time evolution operators for the unperturbed system and the perturbed system:

$$U_0(t) = e^{-itH_0}, \qquad U(t) = e^{-itH}.$$

We assume that the group $U_0(t)$ is "well known," and our problem is to compute $U(t)$. For this purpose it is convenient to adopt a point of view that is intermediate between the Schrödinger picture (in which the states evolve in time, $\psi(t) = U(t)\psi$, while the observables remain fixed) and the Heisenberg picture (in which the states remain fixed while the observables evolve in time, $A(t) = U(-t)AU(t)$), called the *interaction picture*. Namely, we let the observables evolve in time according to the *unperturbed* Hamiltonian,

$$A(t) = U_0(-t)AU_0(t), \tag{6.1}$$

and the states evolve in such a way as to correct $U_0(t)$ to $U(t)$:

$$\psi(t) = V(t)\psi, \quad \text{where} \quad V(t) = U_0(-t)U(t). \tag{6.2}$$

The matrix element of the observable A between the states ψ_1 and ψ_2 at time t is thus

$$\langle\psi_2(t)|A(t)|\psi_1(t)\rangle = \langle\psi_2|V(t)^\dagger U_0(-t)AU_0(t)V(t)|\psi\rangle = \langle\psi_2|U(-t)AU(t)|\psi_1\rangle,$$

which is the matrix element of A between $U(t)\psi_1$ and $U(t)\psi_2$ (Schrödinger picture) or the matrix element of $U(-t)AU(t)$ between ψ_1 and ψ_2 (Heisenberg picture), as it should be.

In the interaction picture, the problem is to compute $V(t)$ (from which $U(t) = U_0(t)V(t)$ is easily derived since $U_0(t)$ is assumed known). Now, $V(t)$ satisfies the differential equation

$$\frac{dV(t)}{dt} = iU_0(-t)H_0U(t) - iU_0(-t)HU(t) = \frac{1}{i}U_0(-t)H_IU(t) = \frac{1}{i}H_I(t)V(t), \tag{6.3}$$

where the t-dependence of $H_I(t)$ is defined by (6.1). This is the real basis for our calculations. That is, instead of starting with the Hamiltonians H and H_0, we start with a one-parameter unitary group $U_0(t)$ and a Hermitian operator H_I, and we look for a one-parameter family $V(t)$ of unitary operators such that $iV'(t) = H_I(t)V(t)$. (H_I may be unbounded, in which case there are issues about domains that we ignore here.) Then $U(t) = U_0(t)V(t)$ will be a one-parameter unitary group, and H_0 and H will merely be the infinitesimal generators of $U_0(t)$ and $U(t)$; we need not worry about them further.

The differential equation $iV'(t) = H_I(t)V(t)$ is equivalent to the integral equation

$$V(t) = V(0) + \int_0^t \frac{d}{d\tau}V(\tau)\,d\tau = I + \frac{1}{i}\int_0^t H_I(\tau)V(\tau)\,d\tau,$$

which may be iterated in the usual way:

$$V(t) = I + \frac{1}{i}\int_0^t \left[H_I(\tau)\,d\tau + \frac{1}{i}\int_0^\tau H_I(\tau)H_I(\tau')V(\tau')\,d\tau'\right]d\tau,$$

and so forth, leading to the *formal* expansion

$$V(t) \sim I + \sum_{1}^{\infty} V_n(t),$$
(6.4)
$$V_n(t) = \frac{1}{i^n} \int_0^t \int_0^{\tau_n} \cdots \int_0^{\tau_2} H_I(\tau_n) H_I(\tau_{n-1}) \cdots H_I(\tau_1)\, d\tau_1 \cdots d\tau_{n-1}\, d\tau_n.$$

The series $I + \sum_1^\infty V_n(t)$ is called the *Dyson series*[1] for $V(t)$.

What does this really mean? If H_I is actually a bounded operator, the series $I + \sum V_n(t)$ converges in the norm topology since $\|V_n(t)\| \le \|H_I\|^n/n!$, and its sum is indeed the operator $V(t)$. If H_I is unbounded, but there is a dense domain $\mathcal{D}$ mapped into itself by H_I and the operators $U_0(t)$, then $V_n(t)$ at least makes sense as an operator on $\mathcal{D}$. If H_I also contains a coefficient λ (i.e., $H_I = \lambda \widetilde{H}_I$), it may be that the formula $V(t) \sim I + \sum V_n(t)$ is an asymptotic expansion in powers of λ, valid on $\mathcal{D}$. Of course, making this into a rigorous theorem requires some estimates for the error term. All this is mathematically interesting, but it does not really serve our purposes.

The gist of perturbation theory as a practical art is to take the first few partial sums of the series (6.4) as effective approximations to the operator $V(t)$. More precisely, the theory should generate numbers representing physically significant quantities that can be compared with experiment. Typically these numbers come from matrix elements of $V(t)$; in any case, the series (6.4) is used to generate numerical series $\sum_0^\infty c_n$ that formally represent a physical quantity C. The first part of our bargain with the devil is this:

(6.5) *If the first few terms c_n decrease rapidly in magnitude as n increases, the corresponding partial sums of the series $\sum c_n$ are to be accepted as effective approximations to the quantity C.*

The hypothesis of this *Ansatz* is a serious business. The progress of quantum field theory remained almost at a standstill for twenty years because the terms c_n, at first sight, have an unfortunate predilection for being infinite, and it takes some hard work to prune away the divergences and generate c_n's that are finite and meaningful. Even then, there is no guarantee that they will be suitably small. In quantum electrodynamics they are, and the theory is very successful; but for the interaction that holds atomic nuclei together they are usually not, and our understanding of nuclear phenomena at a fundamental level remains incomplete for this reason.

Let us be perfectly clear about one thing here: the validity of (6.5) has *absolutely nothing* to do with the convergence of the series $\sum c_n$ or the series $\sum V_n$ from which it is derived. To bring this point home, here is a simple parable. Consider the two series

$$\sum_0^\infty \frac{(-100)^n}{n!}, \qquad \sum_0^\infty \frac{n!}{(-100)^n}.$$

The first series converges to e^{-100}, but its first few partial sums are 1, -99, 4901, $-161765\frac{2}{3}$, 4004901, etc. To call these "approximations" to e^{-100} ($\approx 10^{-44}$) is nothing but a bad joke. On the other hand, the second series diverges, but its

[1] This notion antedates Dyson, of course; the name is more properly applied to the perturbation series for the S-matrix that we shall study later in this chapter.

first few partial sums — 1, 0.99, 0.9902, 0.990194, 0.9902065 — provide excellent approximations to the integral

$$\int_0^\infty \frac{100e^{-t}\,dt}{100+t}.$$

Indeed, the error is less in magnitude than the first neglected term of the series (to see this, write $1/(1+(t/100))$ as a finite geometric series plus a remainder), and these magnitudes decrease up to the 100th term $100!/(100)^{100} \approx 10^{-42}$.

We return to the Dyson series. The formulas (6.4) can be rewritten in a convenient way by using the notion of time-ordered product. Suppose $A_1(t), \ldots, A_n(t)$ are operators depending on $t \in \mathbb{R}$. If $t_1, \ldots, t_n$ are distinct points in $\mathbb{R}$, the *time-ordered product* $\mathcal{T}[A_1(t_1)\cdots A_n(t_n)]$ is defined by

$$\mathcal{T}[A_1(t_1)\cdots A_n(t_n)] = A_{i_1}(t_{i_1})\cdots A_{i_n}(t_{i_n}), \text{ where } t_{i_1} > t_{i_2} \cdots > t_{i_n}.$$

That is, the factors are ordered so that the operators are applied in the order of increasing time parameter. We also declare the time-ordering operation to be linear, so that the time-ordering of a sum or integral of such products is the sum or integral of the time-ordered products. If $V_n(t)$ is defined as in (6.4), we then have

$$\begin{aligned} V_n(t) &= \frac{1}{i^n n!}\int_0^t\int_0^t\cdots\int_0^t \mathcal{T}[H_I(\tau_1)\cdots H_I(\tau_n)]d\tau_1\cdots d\tau_n \\ &= \mathcal{T}\left[\frac{1}{i^n n!}\int_0^t\int_0^t\cdots\int_0^t H_I(\tau_1)\cdots H_I(\tau_n)d\tau_1\cdots d\tau_n\right]. \end{aligned} \tag{6.6}$$

Indeed, the integral over the cube $(0,t)^n$ is the sum of the integrals over the $n!$ simplices where the coordinates have a particular ordering such as $\tau_1 > \tau_2 > \cdots > \tau_n$, and each of the latter integrals is equal to $V_n(t)$. The expansion of $V(t)$ then takes the form

$$\begin{aligned} V(t) &\sim \mathcal{T}\left[\sum_0^\infty \frac{1}{i^n n!}\int_0^t\int_0^t\cdots\int_0^t H_I(\tau_1)\cdots H_I(\tau_n)d\tau_1\cdots d\tau_n\right] \\ &\equiv \mathcal{T}\exp\left[\frac{1}{i}\int_0^t H_I(\tau)\,d\tau\right]. \end{aligned} \tag{6.7}$$

This last expression is called the *time-ordered exponential* of $i^{-1}\int_0^t H_I(\tau)\,d\tau$, and it is, by definition, the sum of the time-orderings of the terms in the Taylor series of the exponential. *This is merely a convenient and suggestive notation, not a new formula for V.* A time-ordered exponential has just as much, or as little, meaning as the series that defines it.

Recall that what we really want to calculate is the group $U(t)$. Since $U(t) = U_0(t)V(t)$, the Dyson series (6.4) yields an expansion of $U(t)$ in terms of the free evolution group $U_0(t)$ and the interaction Hamiltonian H_I. Indeed, we have

$$U(t) \sim U_0(t) + \sum_1^\infty U_n(t), \text{ where } U_n(t) = U_0(t)V_n(t) \text{ for } n \geq 1.$$

Recalling that $H_I(t) = U_0(-t)H_IU_0(t)$, for $t > 0$ we have

$$U_1(t) = \frac{1}{i}\int_0^t U_0(t-\tau)H_IU_0(\tau)\,d\tau,$$

$$U_2(t) = \frac{1}{i^2}\int_{t>\tau_2>\tau_1>0} U_0(t-\tau_2)H_IU(\tau_2-\tau_1)H_IU_0(\tau_1)\,d\tau_1\,d\tau_2,$$

and so forth. These formulas admit a very suggestive graphical interpretation, shown in Figure 6.1.

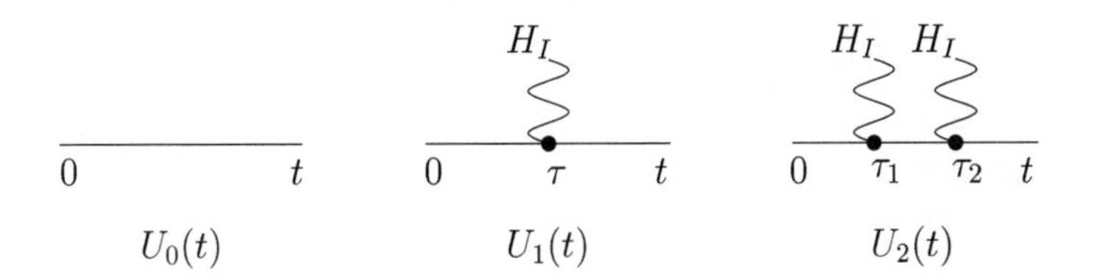

FIGURE 6.1. Graphical interpretation of the Dyson series.

In the formula for $U_1(t)$, the integrand $U_0(t-\tau)H_IU_0(\tau)$ represents the system propagating freely until time τ, being hit with the interaction H_I, then propagating freely again until time t. Likewise, in $U_2(t)$, the system propagates freely until time τ_1, is hit with H_I, propagates freely again until time τ_2, is hit with H_I again, and then propagates freely until time t. Thus, at least to the extent to which the Dyson series can be trusted, the picture that emerges is as follows. The system can propagate freely ($U_0(t)$)over the time interval $(0,t)$, interact once at some intermediate time ($U_1(t)$), interact twice ($U_2(t)$), and so forth. We add up all the ways these things can happen over all possible intermediate times, and the result is the interacting propagator $U(t)$. This is the simplest version of the philosophy for computing time evolutions that was pioneered by Feynman, and Figure 6.1 consists of embryonic Feynman diagrams.

We conclude this section with an elementary but technical calculation that will be needed in the next section. Let us rewrite the interaction Hamiltonian as gH_I, where g is a small parameter. Suppose that the unperturbed Hamiltonian H_0 has an eigenvalue λ_0 with unit eigenvector v_0, and suppose that $H = H_0 + gH_I$ has an eigenvalue λ_g with eigenvector v_g, normalized so that $\langle v_0|v_g\rangle = 1$, such that λ_g and v_g depend smoothly on g. (Of course, there are theorems that guarantee that this will happen under suitable hypotheses — for example, if H_I is bounded, or just bounded relative to H_0, and the eigenvalue λ_0 is simple; see Reed and Simon [**96**], §XII.2.) Thus we have

$$\lambda_g = \lambda_0 + \epsilon_1 g + \epsilon_2 g^2 + O(g^3), \qquad v_g = v_0 + gw_1 + g^2w_2 + O(g^3),$$

for some $\epsilon_j \in \mathbb{C}$ and $w_j \in \mathcal{H}$. We wish to calculate ϵ_1 and ϵ_2. (This calculation can be extended recursively to find the higher-order coefficients in the Taylor expansion of λ_g too.)

We first observe that since

$$1 = \langle v_0|v_g\rangle = 1 + g\langle v_0|w_1\rangle + g^2\langle v_0|w_2\rangle + \cdots$$

for all small g, we have

$$\langle v_0|w_1\rangle = \langle v_0|w_2\rangle = 0. \tag{6.8}$$

Now,

$$\begin{aligned} Hv_g &= (H_0 + gH_I)(v_0 + gw_1 + g^2 w_2 + \cdots) \\ &= \lambda_0 v_0 + g(H_I v_0 + H_0 w_1) + g^2(H_I w_1 + H_0 w_2) + \cdots, \end{aligned}$$

and on the other hand,

$$Hv_g = \lambda_g v_g = \lambda_0 v_0 + g(\epsilon_1 v_0 + \lambda_0 w_1) + g^2(\epsilon_2 v_0 + \epsilon_1 w_1 + \lambda_0 w_2) + \cdots .$$

Hence

$$H_0 w_1 + H_I v_0 = \epsilon_1 v_0 + \lambda_0 w_1, \tag{6.9}$$

$$H_0 w_2 + H_I w_1 = \epsilon_2 v_0 + \epsilon_1 w_1 + \lambda_0 w_2. \tag{6.10}$$

Since $\langle v_0|H_0|w_1\rangle = \lambda_0\langle v_0|w_1\rangle = 0$, by taking the inner product of (6.9) with v_0 and using (6.8) we obtain

$$\epsilon_1 = \langle v_0|H_I|v_0\rangle, \tag{6.11}$$

and hence

$$(H_0 - \lambda_0)w_1 = -(H_I - \epsilon_1)v_0 = -(H_I v_0 - \langle v_0|H_I|v_0\rangle v_0) = -(I - P)H_I v_0,$$

where P is the orthogonal projection onto $\mathbb{C}v_0$. The restriction of $H_0 - \lambda_0$ to the orthogonal complement of v_0, i.e., the range of $I - P$, is invertible on that space, so with a small abuse of notation,

$$w_1 = -(H_0 - \lambda_0)^{-1}(I - P)H_I v_0 = -(I - P)(H_0 - \lambda_0)^{-1}(I - P)H_I v_0.$$

Finally, we take the inner product of (6.10) with v_0 and use (6.8) as above to obtain

$$\epsilon_2 = \langle v_0|H_I|w_1\rangle = -\langle v_0|H_I(I - P)(H_0 - \lambda_0)^{-1}(I - P)H_I|v_0\rangle. \tag{6.12}$$

6.2. A toy model for electrons in an atom

In this section we analyze a nonrelativistic particle of mass M moving in a potential V and interacting with a neutral scalar field ϕ of mass μ, as a toy model for an electron in an atom interacting with the electromagnetic radiation field. In more detail, in the latter situation we take the atomic nucleus to be infinitely heavy and therefore stationary. The potential V is the Coulomb potential generated by the nucleus (perhaps modified by the presence of other electrons), the particle is an electron moving nonrelativistically (according to the Schrödinger equation) in the potential V, and the field ϕ is the rest of the electromagnetic field — which classically is the radiation field produced by the motion of the electron, but here is treated as a *quantum* field. However, in our toy model we neglect the complications due to the spins of electrons and photons, and hence take the wave function for the particle to be scalar-valued and the field to be a scalar field. We also assign a positive mass μ to the field quanta, but this should be envisioned as being very small in comparision with all other masses and energies under consideration. (An analogous discussion for genuine electromagnetism, without the simplifications, can be found in Sakurai [**102**], Chapter 2.) In what follows, it will be tempting to refer to the nonrelativistic particle as an electron and the particles associated to the field as photons, but in order not to create false impressions, we shall refer to them as "the particle" and "the (field) quanta," respectively.

We are going to perform two calculations with this model: the transition rate for the particle to emit or absorb field quanta, and the shift in the energy levels E_n due to the presence of the field. This is not fundamental physics, but it will provide a useful illustration of the workings of a quantum field and the art of approximation as well as some insight into real phenomena of atomic physics.

Here is the setup. To minimize technicalities, we take the particle and the field to live in a large cubical box $\mathcal{B}$ of side length L, with periodic boundary conditions, as in §5.1 — that is, on a 3-torus rather than $\mathbb{R}^3$. The state space is $\mathcal{H} = \mathcal{H}_{\text{part}} \otimes \mathcal{H}_{\text{field}}$, where $\mathcal{H}_{\text{part}} = L^2(\mathcal{B})$ and $\mathcal{H}_{\text{field}}$ is the Boson Fock space over $L^2(\mathcal{B})$. The Hamiltonian is $H = H_{\text{part}} + H_{\text{field}} + H_I$, where $H_{\text{part}} = -\nabla^2/2M + V$ acting on $\mathcal{H}_{\text{part}}$, $\mathcal{H}_{\text{field}} = \sum_{\mathbf{p}\in\Lambda} \omega_{\mathbf{p}} a^*(\mathbf{p})a(\mathbf{p})$ as in (5.3) with $\Lambda = [(2\pi/L)\mathbb{Z}]^3$ being the lattice of allowable momenta, and H_I is the interaction term. Our first choice for H_I is the simplest possible thing: $H_I = g\phi$, where g is a scalar called the *coupling constant.* It corresponds to the charge of the electron, and the form $H_I = g\phi$ of the interaction is motivated by classical electrodynamics, where the interaction of a charged body with the electromagnetic field is proportional to the charge and to the field strength.

More precisely, if $\phi(x)$ is the Schrödinger-picture field,

$$\phi(\mathbf{x}) = \sum_{\mathbf{p}\in\Lambda} \frac{1}{\sqrt{2\omega_{\mathbf{p}}}}(e^{i\mathbf{p}\cdot\mathbf{x}}a(\mathbf{p}) + e^{-i\mathbf{p}\cdot\mathbf{x}}a^\dagger(\mathbf{p})), \tag{6.13}$$

we would like to take

$$H_I(\psi\otimes\mathbf{v})(x) = g\psi(x)\otimes\phi(x)\mathbf{v} \qquad (\psi\in L^2(\mathcal{B}) = \mathcal{H}_{\text{part}},\ \mathbf{v}\in\mathcal{H}_{\text{field}}).$$

To put it another way, we can think of $\mathcal{H}$ as $L^2(\mathcal{B}, \mathcal{H}_{\text{field}})$, and then H_I should be simply pointwise multiplication by the operator-valued function $\phi(\cdot)$.

But already there is a difficulty: $\phi(\cdot)$ is not a genuine function, as the series (6.13) converges only in the sense of distributions, as we pointed out in §5.1. (If we replaced the box by $\mathbb{R}^3$, the situation would be even worse.) When we deal with interacting fields in a relativistically correct manner, we shall have to bite the bullet and proceed somehow in the face of such singularities. But since we are presently treating the particle as nonrelativistic, there cannot be much harm in discarding the high-frequency (i.e., high-energy) components of the field and replacing (6.17) by a finite sum over the $\mathbf{p}$ such that $|\mathbf{p}| \le K$, for some large K. Another way to look at this is as follows: since we are taking the the particle's velocity to be much less than 1 (the speed of light) in magnitude, its momentum must be much less than its mass M in magnitude; in particular, the *uncertainty* in momentum is much less than M. But then by (3.19), the uncertainty in *position* must be larger than (roughly) $\hbar/M = 1/M$. It is therefore reasonable to smooth the field ϕ out by convolving it with a smooth approximation χ to the delta-function on $\mathbb{R}^3$ that is negligibly small outside the set $|\mathbf{x}| < 1/M$:

$$\phi * \chi(\mathbf{x}) = \frac{1}{L^{3/2}} \sum_{\mathbf{p}} \frac{\widehat{\chi}(\mathbf{p})}{\sqrt{2\omega_{\mathbf{p}}}}(e^{i\mathbf{p}\cdot\mathbf{x}}a_{\mathbf{p}} + e^{-i\mathbf{p}\cdot\mathbf{x}}a_{\mathbf{p}}^\dagger),$$

We can specify such a χ by requiring that its Fourier transform $\widehat{\chi}(\mathbf{p})$ be equal to 1 on a ball $|\mathbf{p}| \le K$ and supported in a slightly larger ball, where $K \ge M$, and then $\phi * \chi$ is essentially the finite sum suggested above. The details of the smoothing will be of no importance to us, so we shall simply replace ϕ by this finite sum and take

$$H_I = \frac{g}{L^{3/2}} \sum_{\mathbf{p}\in\Lambda,\ |\mathbf{p}|\le K} \frac{1}{\sqrt{2\omega_{\mathbf{p}}}}(e^{i\mathbf{p}\cdot\mathbf{x}}a(\mathbf{p}) + e^{-i\mathbf{p}\cdot\mathbf{x}}a^\dagger(\mathbf{p})). \tag{6.14}$$

Since the system is in a box, the spectra of $H_{\rm part}$ and $H_{\rm field}$ are discrete. Let $E_0 \le E_1 \le E_2 \le \cdots$ be the eigenvalues of $H_{\rm part}$, with eigenvectors $\psi_0, \psi_1, \psi_2, \ldots$. (The eigenvalues of interest are the negative ones, corresponding to the bound states, and the nonrelativistic condition entails their being much less than the mass M in absolute value.) Also, let Ω be the vacuum state in the Fock space $\mathcal{H}_{\rm field}$. Then a basis for the state space $\mathcal{H}$ is obtained by taking the tensor products of the ψ_n's with the k-quantum states $a^\dagger(\mathbf{p}_1)\cdots a^\dagger(\mathbf{p}_k)\Omega$ ($k = 0, 1, 2, \ldots$) as the $\mathbf{p}_j$'s range over the lattice Λ of allowable momenta. We employ Dirac-style shorthand notation for these basis vectors:

$$|n;\mathbf{p}_1,\ldots,\mathbf{p}_k\rangle = c_{\mathbf{p}_1,\ldots,\mathbf{p}_k}\psi_n \otimes a^*(\mathbf{p}_1)\cdots a^*(\mathbf{p}_k)\Omega, \qquad |n\rangle = \psi_n \otimes \Omega.$$

($c_{\mathbf{p}_1,\ldots,\mathbf{p}_k}$ is a normalization constant, equal to $[\prod_{\mathbf{p}\in\Lambda} n_{\mathbf{p}}!]^{-1/2}$ where $n_{\mathbf{p}}$ is the number of $\mathbf{p}_j$ equal to $\mathbf{p}$.) These vectors are eigenvectors for the free Hamiltonian $H_0 = H_{\rm part} + H_{\rm field}$:

$$(6.15) \qquad (H_{\rm part} + H_{\rm field})|n;\mathbf{p}_1,\ldots,\mathbf{p}_k\rangle = (E_n + \omega_{\mathbf{p}_1} + \cdots + \omega_{\mathbf{p}_k})|n;\mathbf{p}_1,\ldots,\mathbf{p}_k\rangle,$$

where $\omega_{\mathbf{p}} = \sqrt{|\mathbf{p}|^2+\mu^2}$. On the other hand, the interaction Hamiltonian is given by

$$(6.16) \quad H_I|n;\mathbf{p}_1,\ldots,\mathbf{p}_k\rangle(\mathbf{x}) = \frac{g}{L^{3/2}}\sum_{|\mathbf{p}|\le K}\frac{1}{\sqrt{2\omega_{\mathbf{p}}}}\big[e^{i\mathbf{p}\cdot\mathbf{x}}\sqrt{m_{\mathbf{p}}}\,|n;\mathbf{p}'_1,\ldots,\mathbf{p}'_{k-1}\rangle + e^{-i\mathbf{p}\cdot\mathbf{x}}\sqrt{m_{\mathbf{p}}+1}|n;\mathbf{p},\mathbf{p}_1,\ldots,\mathbf{p}_k\rangle\big],$$

where $m_{\mathbf{p}}$ is the number of $\mathbf{p}_j$ that are equal to $\mathbf{p}$ and, when $m_{\mathbf{p}} > 0$, $(\mathbf{p}'_1,\ldots,\mathbf{p}'_{k-1})$ is obtained from $(\mathbf{p}_1,\ldots,\mathbf{p}_k)$ by omitting one of the $\mathbf{p}_j$'s that is equal to $\mathbf{p}$.

Emission and absorption of quanta. Let us find the transition probability for emission of a field quantum when there are no quanta present initially. That is, we assume that at time 0 the state is $|n\rangle$ (the particle has energy E_n, and there are no quanta), and ask for the probability that at time $t > 0$ the state is $|m,\mathbf{p}\rangle$ (the particle has energy E_m and there is one quantum with momentum $\mathbf{p}$), namely, $|\langle m,\mathbf{p}|U(t)|n\rangle|^2$. (We assume that $|E_n|$, $|E_m|$, and $|\mathbf{p}|$ are less than the cutoff energy K.) According to (6.4), the first-order approximation to $\langle m,\mathbf{p}|U(t)|n\rangle$ is

$$\langle m,\mathbf{p}|U(t)|n\rangle \approx \langle m,\mathbf{p}|n\rangle + \frac{1}{i}\int_0^t \langle m,\mathbf{p}|U_0(t-\tau)H_IU_0(\tau)|n\rangle\,d\tau.$$

The term $\langle m,\mathbf{p}|n\rangle$ vanishes because the vacuum state of the field is orthogonal to the 1-quantum states. The action of $U_0(t) = e^{-itH_0}$ and H_I in the other term are given by (6.15) and (6.16):

$$\begin{aligned}\frac{1}{i}\int_0^t & \langle m,\mathbf{p}|U_0(t-\tau)H_IU_0(\tau)|n\rangle\,d\tau \\ &= e^{-i(E_m+\omega_{\mathbf{p}})t}\int_0^t e^{i(E_m+\omega_{\mathbf{p}}-E_n)\tau}\langle m,\mathbf{p}|H_I|n\rangle\,d\tau \\ &= \frac{g}{L^{3/2}i}e^{-i(E_m+\omega_{\mathbf{p}})t}\frac{e^{i(E_m-E_n+\omega_{\mathbf{p}})t}-1}{i(E_m-E_n+\omega_{\mathbf{p}})}\int_B \psi_m(\mathbf{x})^*\frac{e^{-i\mathbf{p}\cdot\mathbf{x}}}{\sqrt{2\omega_{\mathbf{p}}}}\psi_n(\mathbf{x})\,d^3\mathbf{x}.\end{aligned}$$

Thus, if we set

$$\alpha = E_m - E_n + \omega_{\mathbf{p}}, \qquad C(m,n,\mathbf{p}) = \int_B \psi_m(\mathbf{x})^* e^{-i\mathbf{p}\cdot\mathbf{x}} \psi_n(\mathbf{x})\, d^3\mathbf{x},$$

the transition probability to first order is
(6.17)

$$|\langle m, \mathbf{p}|U(t)|n\rangle|^2 \approx \frac{g^2}{L^3} \frac{|C(m,n,\mathbf{p})|^2}{2\omega_{\mathbf{p}}} \left| \frac{e^{i\alpha t} - 1}{i\alpha} \right|^2 = \frac{g^2}{L^3} \frac{|C(m,n,\mathbf{p})|^2}{2\omega_{\mathbf{p}}} \frac{\sin^2(\alpha t/2)}{(\alpha/2)^2}.$$

Let us examine the meaning of this. First, when t is reasonably large the function

$$f_t(\alpha) = \frac{\sin^2(\alpha t/2)}{(\alpha/2)^2}$$

has a spike at $\alpha = 0$ of height t^2 and width $\approx 1/t$, and is small elsewhere; moreover, by a standard contour integral, $\int_{-\infty}^{\infty} f_t(\alpha)\, d\alpha = 2\pi t$. Therefore, $f_t(\alpha) \approx 2\pi t \delta(\alpha)$. This expresses the fact that for the transition to take place we must have (approximately) $\alpha = 0$, that is, the energy $\omega_{\mathbf{p}}$ of the quantum must equal the difference $E_n - E_m$ between the initial and final energies of the particle. The fact that this relation becomes blurred for t small is a reflection of the time-energy uncertainty principle: for very short times there is an unremovable uncertainty in the energy. In real life, this phenomenon manifests itself in the fact that the photons emitted when an atomic electron makes a transition from one energy level to another one are not all of exactly the same frequency; the spectroscopists speak of the "line width" in the resulting spectral plot.

The approximation $f_t(\alpha) \approx 2\pi t \delta(\alpha)$ is useful provided t is large enough so that almost all the mass of $f_t(\alpha)$ is concentrated in an interval where α (the deviation from perfect energy conservation) is small in comparison with the energy difference $E_n - E_m$ of the of the initial and final states of the particle, that is, when $t \gg 1/(E_n - E_m)$. In this situation, (6.17) says that the transition probability *per unit time* is approximately

$$\frac{\pi g^2}{L^3} \frac{|C(m,n,\mathbf{p})|^2}{\omega_{\mathbf{p}}} \delta(E_m - E_n + \omega_{\mathbf{p}}). \tag{6.18}$$

Of course this makes sense only when t is small enough so that the probability at time t is substantially less than one. Thus we are in a slightly uncomfortable situation where t must be neither too small nor too large; without some more specific numbers it is impossible to tell whether we have accomplished anything at all. We shall consider this question in more detail after carrying out the calculation one step further for the limiting situation in which we dispense with the box, i.e., let $L \to \infty$.

We now have the emission rate for the particular quantum whose momentum is $\mathbf{p}$. However, if we are just interested in the decay of an excited state, a quantum with momentum $\mathbf{p}'$ is just as good as a quantum with momentum $\mathbf{p}$ provided that $\omega_{\mathbf{p}} = \omega_{\mathbf{p}'} = E_n - E_m$. The number of quantum states per unit volume in momentum space is $(L/2\pi)^3$ (for L very large), so the number of quantum states whose momentum points into a small solid angle $d\Omega$ and whose energy is in a small band $[\omega, \omega + d\omega]$ (that is, $\mathbf{p}/|\mathbf{p}| \in d\Omega$, $\omega_{\mathbf{p}} \in [\omega, \omega + d\omega]$) is

$$\left(\frac{L}{2\pi}\right)^3 |\mathbf{p}|^2 \frac{d|\mathbf{p}|}{d\omega}\, d\omega\, d\Omega = \left(\frac{L}{2\pi}\right)^3 \omega |\mathbf{p}|\, d\omega\, d\Omega,$$

since $|\mathbf{p}| = \sqrt{\omega_{\mathbf{p}}^2 - \mu^2}$, μ being the rest mass of a field quantum. The rate of emission of *any* quantum in this region of momentum space is the quantity (6.18) multiplied by the number of states in this region, that is,

$$\frac{g^2}{8\pi^2}|C(m,n,p)|^2|\mathbf{p}|\delta(E_n - E_m + \omega)\,d\omega\,d\Omega.$$

Observe that at this point the box has disappeared, and the total emission rate for a quantum in any direction, written out explicitly, is

$$\Gamma = \frac{g^2}{8\pi^2}\int_{|\mathbf{p}|^2=(E_n-E_m)^2-\mu^2} |\mathbf{p}|\left|\int \psi_m(\mathbf{x})^* e^{-i\mathbf{p}\cdot\mathbf{x}}\psi_n(\mathbf{x})\,d^3\mathbf{x}\right|^2 d\Omega(\mathbf{p}). \tag{6.19}$$

We emphasize that $d\Omega(\mathbf{p})$ is solid angle measure, not surface area measure; i.e., $\int_{|\mathbf{p}|=r} d\Omega(\mathbf{p}) = 4\pi$ no matter what r is.

As we said above, Γ is a "transition probability per unit time," so that for reasonably short times Δt the probability of a transition in time Δt is approximately $\Gamma\Delta t$ — assuming that the restriction $\Delta t \gg 1/(E_n - E_m)$ doesn't interfere with the notion of "reasonably short." Under this assumption, the interpretation of Γ for longer time intervals is easily obtained. Suppose we start with a large number N identical systems (N atoms, if you will) with particles in the state ψ_n. After a short time interval Δt, $N\Gamma\Delta t$ of the particles will have undergone the transition to state ψ_m. This leads to the differential equation $dN/dt = -\Gamma N$, so that $N(t) = N_0 e^{-\Gamma t}$. That is, Γ is the *transition rate* or *decay rate* of the excited state.

Now let us plug in numbers for the parameters in this simplified model that correspond to the quantities in the real physical situation that inspired it: an electron in the Coulomb potential of an atomic nucleus interacting with the electromagnetic field to make a transition from an excited state with energy E_n to the ground state with energy E_0. That is, we take $g^2/4\pi$ to be the fine structure constant 1/137 (so $g^2 \approx 10^{-1}$), M to be the mass of the electron ($M = 5 \times 10^5$ eV), and μ to be the mass of the photon ($\mu = 0$). (Yes, our model requires $\mu \neq 0$, but we can assume $\mu \ll E_n - E_0$ and replace it by 0 as an approximation here.) The difference in energy levels $E_n - E_0$ is on the order of a few eV, or $10^{-5}M$. The magnitude of the integral $\int \psi_0(\mathbf{x})^* e^{i\mathbf{p}\cdot\mathbf{x}}\psi_n(\mathbf{x})\,d^3\mathbf{x}$ depends strongly on the shape of the excited state wave function ψ_n. The integral always vanishes at $\mathbf{p} = \mathbf{0}$, so when ψ_n is concentrated near the origin (i.e., when n is not too large), it can be estimated by replacing $e^{i\mathbf{p}\cdot\mathbf{x}}$ by $i\mathbf{p}\cdot\mathbf{x}$:

$$\left|\int \psi_0(\mathbf{x})^*(\mathbf{p}\cdot\mathbf{x})\psi_n(\mathbf{x})\,d^3\mathbf{x}\right| \le |\mathbf{p}|\langle r\rangle$$

where $\langle r\rangle$ is the root-mean-square distance of the particle from the origin in the state ψ_n. (If the integral vanishes to higher order in $\mathbf{p}$, it is even smaller.) We may take $\langle r\rangle$ to be at most a few times the Bohr radius of the hydrogen atom, $r_0 = \hbar^2/Mg^2 \approx 10/M$ (with M and g as above), so with $|\mathbf{p}| = E_n - E_0 \approx 10^{-5}M$, we have $|\mathbf{p}|\langle r\rangle \approx 10^{-3}$ or less. Thus the integral (6.19) is estimated by

$$\Gamma \lesssim \frac{g^2}{2\pi}(10^{-5}M)(10^{-3})^2 \sim 10^{-11}M.$$

This is some 10^6 times smaller than $E_n - E_0$. In other words, the decay time is some 10^6 times longer than the time $1/(E_n - E_0)$ needed to dispose of the energy uncertainty, so the assumption that there are times that are neither too small nor too large is tenable. The mass-time conversion (1.6) shows that the decay time $(10^{-11}M)^{-1}$ is on the order of 10^{-10} second, which is in the right ball park for emission of photons in real atoms.

To do a thorough job of analyzing this situation, one should go further. By a more careful analysis of the operator $Pe^{-itH}P$, where P is the orthogonal projection onto the state $|n\rangle$, one can show that $\langle n|e^{-itH}|n\rangle \approx e^{-(\Gamma+i\beta)t}$ where Γ is the decay rate (6.19) found above and $\beta = E_n + \Delta E_n$, ΔE_n being a correction to the energy level E_n that we shall discuss below. See Messiah [**82**], §XXI.13.

We took the initial state to be one with no field quanta present. If we take the initial state to contain field quanta, the calculations are very similar, but there are some points to be noted. First, if the initial state has k quanta, there is a nonzero transition probability to states with either $k+1$ or $k-1$ quanta, that is, a new quantum can be emitted or an existing quantum can be absorbed. If all the existing quanta are in different states initially, the calculations go through with no essential change. But if some of them — say, m of them — are in the *same* state, there is an extra factor of $\sqrt{m+1}$ or $\sqrt{m}$ in (6.16) that shows up as an extra factor of $m+1$ or m in the transition rate. It is no surprise that a quantum in a particular state is more likely to be absorbed when there are lots of such quanta around; the interesting thing is that the presence of quanta in a particular state speeds up the rate at which such quanta are emitted too. This is the principle behind the laser.

What happens if we go to higher orders? The second-order term in the Dyson series (6.4) contains the interaction Hamiltonian — that is, the field operators — twice, so it has nonzero matrix elements between states with k quanta and states with $k+2$, k, or $k-2$ quanta. It therefore has no effect on the transition rate for emission or absorption of a single quantum, but it introduces new processes: emission of two quanta, absorption of two quanta, and emission and reabsorption (or absorption and reemission) of a single quantum. If one works out the calculations, in the first two cases one finds formulas for the emission or absorption rate of the same nature as (6.18); they involve delta-functions (or approximate delta-functions) that express conservation of energy. But in the third case, the uncertainty principle allows the particle and/or quantum to be temporarily "off mass shell" between the times of emission and absorption or vice versa. In particular, when a quantum is emitted and reabsorbed, it can have *any* energy E as long as it is reabsorbed in a time $\approx 1/E$. We say that it is a *virtual quantum*. This is the paradigm for all "virtual particles," which we shall discuss in more detail in §6.6.

Rather than considering absorption and emission processes further, we shall do another calculation to show how the presence of the field affects the energy levels of the particle. This will provide an introduction to the idea of renormalization.

Energy level shifts. We consider an eigenvalue E_n of H_{part}, which is also an eigenvalue of $H_0 = H_{\text{part}} + H_{\text{field}}$ with eigenvector $|n\rangle$, the state where the particle has energy E_n and there are no field quanta present. We assume as before that $|E_n| \ll M$ so that the nonrelativistic approximation is valid, and we wish to determine how the presence of the field affects E_n, to second order in the coupling constant g. We indicated at the end of §6.1 how to perform this calculation, provided that the perturbed eigenvalue and eigenvector depend smoothly on g. Here we proceed on the assumption that this condition is valid.[2]

---

[2]No physicist would waste a moment worrying about this, but here are a few remarks for the mathematically fastidious. The calculations that follow involve only states with at most two

The results at the end of §6.1, with $\lambda_0 = E_n$ and $v_0 = |n\rangle$, show that the perturbed eigenvalue is of the form $E_n + \epsilon_1 g + \epsilon_2 g^2$ to second order in g, where ϵ_1 and ϵ_2 involve $H_I|n\rangle$. We have

$$H_I|n\rangle = L^{-3/2} \sum_{|\mathbf{p}|\le K} (2\omega_{\mathbf{p}})^{-1/2} e^{-i\mathbf{p}\cdot\mathbf{x}} |n;\mathbf{p}\rangle,$$

where $e^{-i\mathbf{p}\cdot\mathbf{x}}$ denotes the operation of pointwise multiplication by the function $e^{-i\mathbf{p}\cdot(\cdot)}$ on $L^2(B, \mathcal{H}_{\text{field}})$. The 1-quantum state $H_I|n\rangle$ is orthogonal to the 0-quantum state $|n\rangle$, so $\epsilon_1 = 0$ by (6.11). Hence, by (6.12), second-order correction is

$$\Delta E_n = \epsilon_2 g^2 = -\frac{g^2}{2L^3} \sum_{\mathbf{p},\mathbf{p}'} \frac{1}{\sqrt{\omega_{\mathbf{p}'}\omega_{\mathbf{p}}}} \langle n;\mathbf{p}'| e^{i\mathbf{p}'\cdot\mathbf{x}} (I-P)(H_0 - E_n)^{-1}(I-P) e^{-i\mathbf{p}\cdot\mathbf{x}} |n;\mathbf{p}\rangle,$$

where P is the orthogonal projection onto $|n\rangle$. The operator $e^{-i\mathbf{p}\cdot\mathbf{x}}$ takes $|n;\mathbf{p}\rangle$ to a linear combination of the vectors $|m;\mathbf{p}\rangle$ with different m's but the same $\mathbf{p}$. These are all orthogonal to $|n\rangle$, so the factors $I - P$ can be omitted; they are also preserved up to scalar multiples by H_0, so the relation $\langle \mathbf{p}', \mathbf{p}\rangle = \delta_{\mathbf{p}'\mathbf{p}}$ implies that only the terms with $\mathbf{p}' = \mathbf{p}$ survive, and we have

$$\Delta E_n = -\frac{g^2}{2L^3} \sum_{|\mathbf{p}|\le K} \frac{1}{\omega_{\mathbf{p}}} \langle n;\mathbf{p}| e^{i\mathbf{p}\cdot\mathbf{x}} (H_0 - E_n)^{-1} e^{-i\mathbf{p}\cdot\mathbf{x}} |n;\mathbf{p}\rangle. \tag{6.20}$$

Moreover, the operator $e^{i\mathbf{p}\cdot\mathbf{x}}$ commutes with H_{field} and with multiplication by the potential $V(\mathbf{x})$, and $e^{i\mathbf{p}\cdot\mathbf{x}} \circ \nabla \circ e^{-i\mathbf{p}\cdot\mathbf{x}} = \nabla - \mathbf{p}$, so

$$\begin{aligned} e^{i\mathbf{p}\cdot\mathbf{x}} (H_0 - E_n)^{-1} e^{-i\mathbf{p}\cdot\mathbf{x}} &= \left(-\frac{(\nabla - \mathbf{p})^2}{2M} + V(\mathbf{x}) + H_{\text{field}} - E_n \right)^{-1} \\ &= \left(H_0 - E_n - \frac{\mathbf{p}\cdot\nabla}{M} + \frac{|\mathbf{p}|^2}{2M} \right)^{-1}. \end{aligned}$$

Therefore,

$$\Delta E_n = -\frac{g^2}{2L^3} \sum_{|\mathbf{p}|\le K} \frac{1}{\omega_{\mathbf{p}}} \left\langle n;\mathbf{p} \middle| \left(\omega_{\mathbf{p}} + \frac{|\mathbf{p}|^2}{2M} - \frac{\mathbf{p}\cdot\nabla}{M} \right)^{-1} \middle| n;\mathbf{p} \right\rangle. \tag{6.21}$$

Now, ∇ / iM is the velocity operator for the particle, and the nonrelativistic approximation entails the velocity being small. Thus, as a crude estimate of ΔE_n, we drop the term $\mathbf{p}\cdot\nabla/M$:

$$\Delta E_n \sim -\frac{g^2}{2L^3} \sum_{|\mathbf{p}|\le K} \frac{1}{\omega_{\mathbf{p}}(\omega_{\mathbf{p}} + (|\mathbf{p}|^2/2M))}.$$

field quanta, so they are unaffected if we multiply the interaction Hamiltonian (6.14) fore and aft by the orthogonal projection onto the space of these states. Since we have already cut off the high frequencies, the result is a bounded operator. Hence standard results of perturbation theory quoted in §6.1 apply provided that E_n is a simple eigenvalue of H_0, and this can be artificially arranged by adding a small generic perturbation to the potential V. Then the perturbed eigenvalue depends smoothly on g for sufficiently small g, but whether the physical value $g = \sqrt{4\pi/137}$ is "sufficiently small" is another matter.

At this point we might as well pass to the limit $L \to \infty$, so that the sum becomes an integral and $(2\pi/L)^3$ becomes the volume element $d^3\mathbf{p}$:

$$\Delta E_n \sim -\frac{g^2}{16\pi^3}\int_{|\mathbf{p}|\le K}\frac{d^3\mathbf{p}}{\omega_{\mathbf{p}}(\omega_{\mathbf{p}}+(|\mathbf{p}|^2/2M)}. \tag{6.22}$$

We can calculate this exactly in the limit of massless field quanta, for which $\omega_{\mathbf{p}} = |\mathbf{p}|$:

$$\begin{aligned} -\frac{g^2}{16\pi^3}\int_{|\mathbf{p}|\le K}\frac{d^3\mathbf{p}}{|\mathbf{p}|^2(1+(|\mathbf{p}|/2M)} &= -\frac{g^2}{4\pi^2}\int_0^K \frac{d\rho}{1+(\rho/2M)} \\ &= -\frac{g^2}{2\pi^2}M\log\left(1+\frac{K}{2M}\right). \end{aligned}$$

There is clearly something fishy about this result, because it depends on the artifical cutoff parameter K (although it is quite insensitive to the precise value of K). In fact, if we remove the cutoff by letting $K \to \infty$, it diverges. Moreover, if we take $K = M$ and use values for the other parameters appropriate to atomic electrons ($g^2/4\pi = 1/137$, $M = 5\times 10^5$ eV, $E_n \sim -10$ eV) we get $\Delta E_n \sim -2\times 10^{-3}M = -10^3$ eV, which is around 100 times as big as E_n itself. Thus the validity of (6.22) as a "correction" to E_n seems highly dubious.

On the other hand, the integral (6.22) doesn't depend on the particular energy eigenstate $|n\rangle$, or even on the potential V. In fact, it is *exactly* the second-order correction (in the limit $L \to \infty$) to the ground-state energy level of a *free* particle. Indeed, in this case the eigenfunctions for $H_{\text{part}} = -\nabla^2/2M$ are indexed by the same lattice in momentum space as the one-quantum states of the field; the normalized eigenfunction with momentum $\mathbf{p}$ is $e_{\mathbf{p}} = L^{-3/2}e^{-i\mathbf{p}\cdot\mathbf{x}}$, with eigenvalue $E_{\mathbf{p}} = |\mathbf{p}^2|/2M$. Noting that $e_{\mathbf{p}}$ is the result of applying the *operator* $e^{-i\mathbf{p}\cdot\mathbf{x}}$ to the ground state $|0\rangle = e_{\mathbf{0}}$, we see that (6.20) for the ground state of a free particle (i.e., $V = 0$, $n = 0$, and $E_n = 0$) becomes

$$\begin{aligned} \Delta E_0 &= \frac{g^2}{2L^3}\sum_{\mathbf{p}}\frac{1}{\omega_{\mathbf{p}}}\langle e_{\mathbf{p}};\mathbf{p}|(H_{\text{part}}+H_{\text{field}})^{-1}|e_{\mathbf{p}};\mathbf{p}\rangle \\ &= -\frac{g^2}{2L^3}\sum_{|\mathbf{p}|\le K}\frac{1}{\omega_{\mathbf{p}}(\omega_{\mathbf{p}}+(|\mathbf{p}|^2/2M))}, \end{aligned}$$

which is (6.22) in the limit $L \to \infty$.

What can this shift in the ground state energy of a free particle possibly mean? The only energy such a particle has is its mass, so ΔE_0 must be interpreted as a change in the *rest mass* of the particle due to its interaction with the field:

$$\Delta E_0 = \Delta M.$$

Now, thinking in terms of the real world of atomic electrons of which this is a simplified model, we can move an electron into or out of an atomic potential well and measure what happens when we do so, but *we cannot decouple it from the electromagnetic field*, as its electric charge is always present. Hence the physical mass that is measured in the laboratory is not the M in our Hamiltonian but the corrected — or *renormalized* — mass $M + \Delta M$. (The same issue arose in classical electrodynamics, where it was an unresolved puzzle to figure out how much of an electron's mass should be attributed to the energy of the electrostatic field that it generates. We shall return to this point in §7.10.)

With this in mind, it is clear that the mass shift ΔM accounts for a large fraction of *all* the energy level shifts for a particle in an arbitrary potential. Indeed, the approximation we made by neglecting the term $\mathbf{p}\cdot\nabla/M$ in (6.21) amounts precisely to replacing ΔE_n by ΔM, and the physically interesting part of the shift ΔE_n is the remainder $\Delta E_n - \Delta M$. Actually this is not quite right either. The numbers E_n are the eigenvalues of the Hamiltonian $H_{\text{part}} = -(\nabla^2/2M) + V$ containing the original "bare" mass M, but the physically meaningful energy levels are the eigenvalues E_n' of the Hamiltonian obtained by replacing M by the physical mass $M + \Delta M$. The "true" energy level shifts are the differences

$$\Delta' E_n = (\Delta E_n - \Delta M) + (E_n' - E_n),$$

and it is *their* differences $\Delta' E_n - \Delta' E_m$, as corrections to the differences $E_n' - E_m'$, that are observed spectroscopically when particles make a transition from one state to another.

We shall not carry out the calculation of these corrections here, as the differences between our model and real atomic electrons are too great to make it worthwhile. But nonrelativistic calculations of this sort by Bethe gave the first reasonably accurate theoretical calculation of the Lamb shift, i.e., the difference in the energy levels between the $2S_{1/2}$ and the $2P_{1/2}$ states of the hydrogen atom. The interested reader can consult Bethe's original account [**9**] or the retelling of it in Sakurai [**102**], §2.8. Of course, to obtain really accurate and theoretically sound results along these lines, one must proceed from a relativistically correct theory and take account of the contributions of high-energy virtual quanta, so that the results do not depend on the choice of cutoff. We shall say a little more about this in §7.11.

One tends to think of renormalization as a way of "subtracting off infinities," and indeed it plays that role here if we define H_I in terms of the original quantum field without cutting off the high frequencies. But the fact that it still has a substantial role to play when the cutoff is employed highlights the fact that it has a deeper significance. The presence of interactions really does change the effective parameters of a system, and those changes must be taken into account whether they be finite or not.

6.3. The scattering matrix

After this warmup, we now turn to the real subject at hand: quantum fields with interactions. The path we follow here is the one we followed in discussing free fields, "canonical quantization." That is, we start out with classical field equations and a relativistically invariant Lagrangian from which they are derived, then replace the classical field variables by quantum fields, which are Fourier integrals of creation and annihilation operators that satisfy suitable commutation or anticommutation relations. The Lagrangian will be a sum of free-field terms and interaction terms, and the same will be true of the corresponding Hamiltonian. We then make the assumption that perturbation theory will yield some meaningful results, so that we can calculate the time evolution by using the Dyson series.

What are we trying to calculate, anyhow? In the typical particle-physics experiment some incoming particles with certain momenta, initially far apart, come together and interact with each other, producing some outgoing particles with certain momenta that again become far apart and eventually arrive at detectors. The phrase "far apart" is meant to indicate that the particles no longer interact and can be treated as free particles, and we are emphasizing that the momenta of the particles are usually a lot more important than their positions. Speaking loosely, there is an "incoming" state $|\text{in}\rangle$ consisting of free particles and an "outgoing" state $|\text{out}\rangle$ consisting of free particles, and we wish to find the transformation $|\text{in}\rangle \mapsto |\text{out}\rangle$.

Setting up a precise description of this situation is a rather subtle business. We shall do so quite informally and refer the reader to Weinberg [**129**], §3.1, for a more careful analysis. Let $U(t) = e^{-itH}$ be the time evolution operator; we assume that H is the sum of a free-field Hamiltonian H_0 and an interaction Hamiltonian H_I. If $|v\rangle$ is the state vector of the system at the present time, one might at first think that $|\text{in}\rangle$ and $|\text{out}\rangle$ would be $\lim_{t\to-\infty} U(t)|v\rangle$ and $\lim_{t\to+\infty} U(t)|v\rangle$, but these limits are unlikely to exist, as the particles move off to infinity as $t \to \pm\infty$. But if $U(t)|v\rangle$ is essentially a collection of free particles for $\pm t$ large, we can keep that collection in our field of view by moving it back to the present using the free-field evolution $U_0(-t) = e^{iH_0t}$. Thus we are led to the *interaction-picture* time dependence $|v(t)\rangle = U_0(-t)U(t)|v\rangle$ as discussed in §6.1, and $|\text{in}\rangle$ and $|\text{out}\rangle$ should be the limits of $|v(t)\rangle$ as $t \to -\infty$ or $t \to +\infty$. In short, the transformation $|\text{in}\rangle \to |\text{out}\rangle$ that we are interested in is the *scattering operator*

$$S = \lim_{t_0\to-\infty,\ t_1\to+\infty} U_0(-t_1)U(t_1 - t_0)U_0(t_0). \tag{6.23}$$

More practically, we are interested in the matrix elements of S, $\langle\text{out}|S|\text{in}\rangle$, which are collectively known as the *scattering matrix* or *S-matrix.*

From the point of view of rigorous mathematics, the existence and unitarity of the scattering operator is an interesting and highly nontrivial question. (There is a rigorous scattering theory in the context of the Wightman axioms due to Haag and Ruelle; see Jost [**69**], Bogolubov et al. [**11**], [**12**], and Araki [**3**]. The scattering theory of nonrelativistic quantum mechanics is developed in Reed and Simon [**95**].) But we have already made our bargain with the devil, so we shall not stop to worry about this, but rather proceed directly to try to calculate the matrix elements by perturbation theory. By (6.3), the operator $V(t) = U_0(-t)U(t)$ is the solution of the initial value problem $iV'(t) = H_I(t)V(t)$, $V(0) = I$, where $H_I(t) = U_0(-t)H_IU_0(t)$ is the interaction-picture form of H_I. Therefore, the operator

$$V(t, t_0) = U_0(-t)U(t - t_0)U_0(t_0) = V(t)V(t_0)^{-1}$$

is the solution of the initial value problem

$$i\frac{d}{dt}V(t, t_0) = H_I(t)V(t)V(t_0)^{-1} = H_I(t)V(t, t_0), \qquad V(t_0, t_0) = I,$$

so by the same calculation that leads to (6.4), it is formally given by the Dyson series

$$\begin{aligned} V(t, t_0) &\sim \mathcal{T}\exp\frac{1}{i}\int_{t_0}^{t} H_I(t)\,dt \\ &= I + \sum_1^\infty \frac{1}{i^n n!}\int_{t_0}^{t}\cdots\int_{t_0}^{t} \mathcal{T}[H_I(\tau_1)\cdots H_I(t_n)]\,d\tau_1\cdots d\tau_n. \end{aligned}$$

The corresponding formal expansion for the scattering operator (6.23) is then

$$S = V(+\infty, -\infty) \sim I + \sum_1^\infty \frac{1}{i^n n!}\int_{-\infty}^{\infty}\cdots\int_{-\infty}^{\infty} \mathcal{T}[H_I(\tau_1)\cdots H_I(t_n)]\,d\tau_1\cdots d\tau_n. \tag{6.24}$$

Proceeding on the faith that formula (6.24) has some useful content, we proceed to calculate its matrix elements $\langle \text{out}|S|\text{in}\rangle$ by calculating the corresponding elements for the operators on the right. In line with the ideas sketched above, we take the in and out states to be idealized states describing particles with definite momenta, namely, states obtained by applying a finite sequence of creation operators $a_j(\mathbf{p}_j)$ to the vacuum state. More precisely, the "vacuum state" here is the no-particle state in the tensor product of the Fock spaces for the free fields in question, which we denote by $|0\rangle$,[3] and the in and out states are taken to be of the form

$$a_1^\dagger(\mathbf{p}_1)\cdots a_k^\dagger(\mathbf{p}_k)|0\rangle,$$

where the subscripts $1, \dots, k$ on the $a^\dagger$'s specify the particle species, spin state, and any other relevant parameters.

Now, there are both mathematical and physical drawbacks to this. Mathematically, one must recall that $a_j^\dagger(\cdot)$ is a distribution rather than a function; applying it to $|0\rangle$ yields what physicists call a "non-normalizable state," that is, a generalized state whose Fourier-transformed wave function is a delta-function in momentum space. Physicists have no *a priori* objection to such states, but they bear only a vague resemblance to a collection of distinct particles that are supposed to be "far apart" from each other, as they are not localized in position at all.[4] Nonetheless, we brush all such objections aside for the time being and proceed; once we get some results it will be time to ask what significance they have.

Before proceeding with the quantum field theory, let us briefly examine how these ideas work in a simpler situation: the scattering of a particle by a fixed potential. Let $V(\mathbf{x})$ be a potential function on $\mathbb{R}^3$, which we assume to be negligibly small outside some bounded set Σ; the minimal condition we need is that $V \in L^1(\mathbb{R}^3)$. We send in a particle with momentum $\mathbf{p}$ from far outside Σ; the potential deflects it, and it emerges with some momentum $\mathbf{q}$. What is the amplitude for this process?

Here the Hilbert space is $L^2(\mathbb{R}^3)$, the (nonrelativistic) state space for a single particle; the free Hamiltonian H_0 is $-\nabla^2/2m$, and the "interaction" Hamiltonian H_I is multiplication by the function V. The state $|\mathbf{p}\rangle$ of a free particle with momentum $\mathbf{p}$ is given by the wave function $e^{i\mathbf{p}\cdot\mathbf{x}}$, and we wish to calculate $\langle \mathbf{q}|S|\mathbf{p}\rangle$. Let us consider just the first-order approximation to S arising from the series (6.24):

$$S \approx I + \frac{1}{i}\int_{-\infty}^{\infty} H_I(t)\,dt = I + \int_{-\infty}^{\infty} e^{iH_0t}Ve^{-iH_0t}\,dt.$$

Since $H_0|\mathbf{p}\rangle = (|\mathbf{p}|^2|/2m)|\mathbf{p}\rangle$, we have

$$\begin{aligned}\langle \mathbf{q}|S|\mathbf{p}\rangle &\approx \langle \mathbf{q}|\mathbf{p}\rangle + \frac{1}{i}\int_{\infty}^{\infty} \langle \mathbf{q}|e^{i|\mathbf{q}|^2t/2m}Ve^{-i|\mathbf{p}|^2t/2m}|\mathbf{p}\rangle\,dt \\ &= \int e^{i(\mathbf{p}-\mathbf{q})\cdot\mathbf{x}}\,d^3\mathbf{x} + \frac{1}{i}\int_{-\infty}^{\infty} e^{i(|\mathbf{q}|^2-|\mathbf{p}|^2)t/2m}\,dt\int_{\mathbb{R}^3} e^{-i(\mathbf{q}-\mathbf{p})\cdot\mathbf{x}}V(\mathbf{x})\,d^3\mathbf{x},\end{aligned}$$

[3]The other common symbol for the vacuum, Ω, will be used for a different vacuum state in §6.11.

[4]Still, given that space is infinite in extent, particles whose position is uniformly distributed over space should be far apart on the average.

so by the Fourier inversion formula (1.5),

$$\langle \mathbf{q}|S|\mathbf{p}\rangle \approx (2\pi)^3\delta(\mathbf{q}-\mathbf{p}) - 2\pi i\delta\left(\frac{|\mathbf{q}|^2-|\mathbf{p}|^2}{2m}\right)\widehat{V}(\mathbf{q}-\mathbf{p}). \tag{6.25}$$

This is known as the *Born approximation* to the scattering amplitude. We shall find it useful later on in comparing the results of quantum field theory with more classical descriptions of various processes.

The formula (6.25) requires some commentary. First, the delta-functions are the price we pay for using the idealized states $\langle\mathbf{q}|$ and $|\mathbf{p}\rangle$. In this situation, of course, it is easy to restate the result in terms of honest L^2 states with wave functions g and f: just multiply both sides by $\widehat{g}(\mathbf{q})\widehat{f}(\mathbf{p})$ and integrate over $\mathbf{p}$ and $\mathbf{q}$ to obtain

$$\langle g|S|f\rangle \approx \langle g|f\rangle - 2\pi i\int_{\mathbb{R}^3}\delta\left(\frac{|\mathbf{q}|^2-|\mathbf{p}|^2}{2m}\right)\widehat{g}(\mathbf{q})\widehat{V}(\mathbf{q}-\mathbf{p})\widehat{f}(\mathbf{p})\,\frac{d^3\mathbf{p}\,d^3\mathbf{q}}{(2\pi)^6}.$$

(The integral on the right makes sense without further interpretation if f and g have a little regularity.) The delta-function $\delta(\mathbf{q}-\mathbf{p})$ is always there simply because the perturbation series starts with the identity, but it disappears as soon as $\mathbf{q}$ differs from $\mathbf{p}$ even slightly, and it is the second term in (6.25), the matrix element for $S-I$, that contains the interesting information. The delta-function in it expresses conservation of energy; we shall refer to the rest — that is, $-2\pi i\widehat{V}(\mathbf{q}-\mathbf{p})$ — as the *Born amplitude* for scattering by the potential V. We refer to Reed and Simon [**95**], §XI.6, for a detailed mathematical treatment of the Born approximation and the full perturbative series of which it is the first-order part.

We shall find that the S-matrix in quantum field processes has a form similar to (6.25). More specifically, for incoming particles with 4-momenta $p_1,\dots,p_m$ and outgoing particles with 4-momenta $q_1,\dots,q_n$, we will have

$$\langle q_1,\dots,q_n|S-I|p_1,\dots,p_m\rangle = i(2\pi)^4\delta(\textstyle\sum q_k - \sum p_j)M(q_1,\dots,q_n;p_1,\dots,p_m), \tag{6.26}$$

where the quantity of real interest is M. Note that here the delta-function expresses conservation of total energy *and* momentum. In (6.25) there was conservation of energy only, because the fixed potential does not respect conservation of momentum.

Let us now return to quantum fields. We have in mind a field theory with k interacting fields, corresponding to k species of particles and antiparticles. Each field will contribute a free-field term to the classical Lagrangian density. For a real scalar field ϕ of mass m, this term is $\frac{1}{2}[(\partial\phi)^2 - m^2\phi^2]$; for a Dirac field ψ of mass m it is $\overline{\psi}(i\gamma^\mu\partial_\mu - m)\psi$; for the electromagnetic field it is $-\frac{1}{4}F_{\mu\nu}F^{\mu\nu}$, and so forth. In addition, there will be interaction terms, which we assume to be polynomials in the fields and their conjugates. (Here the "conjugate" of a field is the field that becomes the operator adjoint on the quantum level.) Since the Lagrangian must be real, nonreal fields (i.e., non-Hermitian ones — ones whose particles have distinct antiparticles) and their conjugates must appear symmetrically in the Lagrangian. This is true also for Fermion fields. (Fermions that are their own antiparticles are theoretically possible; none are definitely known in nature so far, although they may have something to do with neutrinos.) These interaction terms will contain numerical coefficients, the "coupling constants."

The two examples that we shall keep returning to for most of the rest of this book are as follows:

- *Quantum electrodynamics* (*QED*). Here we have one Dirac (spin-$\frac{1}{2}$) field ψ of mass $m > 0$ (representing charged particles of some species and their antiparticles) and one neutral vector field A_μ of mass 0 (representing photons), and the Lagrangian density is

$$\mathcal{L} = \overline{\psi}(i\gamma^\mu \partial_\mu - m)\psi - \tfrac{1}{4}F_{\mu\nu}F^{\mu\nu} - eA_\mu \overline{\psi}\gamma^\mu \psi, \tag{6.27}$$

where $F_{\mu\nu} = \partial_\mu A_\nu - \partial_\nu A_\mu$ and e (the coupling constant) is the charge of the particles in question. More generally, we can allow several Dirac fields ψ_j here, each one representing a different type of charged particle; we simply include one term $\overline{\psi}_j(i\gamma^\mu\partial_\mu - m_j)\psi_j$ and one term $-e_j A_\mu \overline{\psi}_j \gamma^\mu \psi_j$ in the Lagrangian for each such field. The resulting theory describes *only* the electromagnetic interactions of these various charged particles. (If the particles interact with each other in other ways, one must include additional interaction terms in the Lagrangian, resulting in a more complicated theory.) Most of the time it is enough to consider just one species of charged particle, and the most important case is that of electrons, for which e is conventionally taken to be negative (about $-\sqrt{4\pi/137}$). When we speak of QED in the sequel, we shall always mean the one-Fermion form arising from (6.27) unless we say otherwise, and we shall take the Fermion in question to be the electron.

- *The ϕ^4 scalar field theory.* Here we have just one self-interacting neutral scalar field ϕ of mass $m > 0$. The Lagrangian density is

$$\mathcal{L} = \frac{1}{2}(\partial_\mu \phi)(\partial^\mu \phi) - \frac{1}{2}m^2\phi^2 - \frac{\lambda}{4!}\phi^4, \tag{6.28}$$

where the coupling constant λ is assumed positive (and small). (The factor of 4! is just for convenience.) This corresponds to the nonlinear Klein-Gordon equation $\partial^2\phi = m^2\phi + \lambda\phi^3/3!$. Fields of this sort do occur in real-world physical theories; in particular, the Higgs Boson that plays a crucial role in gauge field theories is a neutral scalar particle whose self-interaction is of a similar type. But the main reason for the ubiquitous appearance of the ϕ^4 scalar field theory in quantum field theory texts is that it is the simplest example of an interacting field theory, which offers a setting in which to learn how quantum fields work without the various complicating factors of QED. In space-time dimensions 2 and 3 it is also one of the few quantum field theories that admits a mathematically rigorous model.

We shall meet other physically important fields in Chapter 9. In addition, we shall occasionally make a few remarks about two other field theories:

- *Yukawa field theory.* In the simplest version of this, we have one Dirac field ψ of mass $m_\psi > 0$ and one neutral scalar field ϕ of mass $m_\phi > 0$, and the Lagrangian density is

$$\mathcal{L} = \overline{\psi}(i\gamma^\mu\partial_\mu - m_\psi)\psi - \tfrac{1}{2}(\partial\phi)^2 - \tfrac{1}{2}m_\phi^2\phi^2 - g\phi\overline{\psi}\psi. \tag{6.29}$$

The coupling constant is g. With a little elaboration, this is the model needed for the first reasonably successful theory of the strong interaction, which accounts for the attraction between nucleons in an atomic nucleus in terms of exchange of (virtual) pions. In the simple Lagrangian (6.29), ψ can represent either protons or neutrons and ϕ represents neutral pions.

For a more realistic theory, one needs two Dirac fields ψ_p and ψ_n for protons and neutrons, as well as two scalar fields ϕ_0 (real) and ϕ_1 (complex) for neutral and charged pions. Moreover, pions have negative parity (they are "pseudoscalars" rather than "scalars"), which entails an extra factor of γ^5 in the interactions. Besides the free-field terms for each particle type, the Lagrangian includes the interactions

$$-g\big[\phi_0(\overline{\psi}_p\gamma^5\psi_p + \overline{\psi}_n\gamma^5\psi_n) + \phi_1\overline{\psi}_p\gamma^5\psi_n + \phi_1^*\overline{\psi}_n\gamma^5\psi_p\big]. \tag{6.30}$$

The first two terms represent a proton or neutron emitting or absorbing a neutral pion; the next one represents a neutron absorbing a positive pion or emitting a negative pion and turning into a proton, and the last one a proton absorbing a negative pion or emitting a positive pion and turning into a neutron.[5]

- *The Fermi model for beta decay.* Here we have four Dirac fields ψ_e, ψ_p, ψ_n, and ψ_ν representing electrons, protons, neutrons, and neutrinos. The free-field Lagrangian is the sum of the four $\overline{\psi}(i\gamma^\mu\partial_\mu - m_\psi)\psi$'s, and the interaction terms are

$$-G[\overline{\psi}_p\Gamma\psi_n\overline{\psi}_e\Gamma\psi_\nu + (\overline{\psi}_p\Gamma\psi_n\overline{\psi}_e\Gamma\psi_\nu)^*], \tag{6.31}$$

where G is the coupling constant and Γ is a suitable combination of Dirac matrices. (There are several possibilities here; we shall be more specific when we discuss this model in §9.4.) This interaction is meant to model the decay of the neutron, $n \to e + p + \overline{\nu}$, and related processes.

In all of these cases (as well as others of importance), the passage from the Lagrangian to the Hamiltonian on the classical level is quite simple, following the paradigm

$$L = T - V \quad \longrightarrow \quad H = T + V$$

with T and V the kinetic and potential energies. The kinetic terms are those involving the time derivatives of the fields in question, and the potential terms are everything else. In particular, the kinetic terms are always part of the free-field terms, so the interaction Hamiltonian — the crucial ingredient for the perturbation-theoretic treatment — is the spatial integral of a Hamiltonian density,

$$H_I(t) = \int_{\mathbb{R}^3} \mathcal{H}_I(t,\mathbf{x})\,d^3\mathbf{x}, \tag{6.32}$$

and $\mathcal{H}_I(t,\mathbf{x})$ in turn is simply the negative of the sum of the interaction terms in the Lagrangian density. For the three examples above the interaction Hamiltonian densities are, respectively,

$$\mathcal{H}_I = e\overline{\psi}\gamma^\mu\psi A_\mu, \qquad \mathcal{H}_I = \frac{\lambda}{4!}\phi^4, \qquad \mathcal{H}_I = g\overline{\psi}\psi\phi.$$

Thus in all these cases, as well as more complicated ones that arise in the standard model, *the classical interaction Hamiltonian density is a sum of products of field variables.*

When we pass to the quantum picture, the field variables become operator-valued functions (or rather distributions). Suppose the theory involves k fields $\phi_1, \ldots, \phi_k$, each of which acts on its own Fock space $\mathcal{F}_j$. The Hilbert space for the

---

[5]The fact that pions are unstable particles that readily decay into leptons is largely irrelevant here; the lifetime of the pion is large in comparison with the characteristic time scale of the strong interaction.

interacting theory is then the tensor product $\mathcal{F} = \mathcal{F}_1 \otimes \cdots \otimes \mathcal{F}_k$, and the fields act on this space in the obvious way, with the jth field acting "in the jth variable." Each Fock space $\mathcal{F}_j$ comes with its own creation and annihilation operators $a_j^\dagger$ and a_j for particles and $b_j^\dagger$ and b_j for antiparticles (if particles and antiparticles are distinct), and the field ϕ_j is a Fourier integral of such operators. (As noted in §4.5, this picture needs to be modified slightly so that the creation/annihilation operators for different Fermion fields anticommute rather than commute.)

Our notation for creation and annihilation operators will be as follows. For each particle species, its creation and annihilation operators are functions of momentum $\mathbf{p} \in \mathbb{R}^3$ and a discrete variable σ that specifies the spin state of the particle. (Actually they are distributions rather than functions in $\mathbf{p}$, but we ignore this distinction here. And of course σ is absent for spin-zero fields.) It will be convenient to introduce a label for the particle species as a third variable π. (If a particle has a distinct antiparticle, these two are considered different species for this purpose. Any other discrete parameters relevant to the problem should also be incorporated into π.) Thus, we write $a(\mathbf{p}, \sigma, \pi)$ or $a^\dagger(\mathbf{p}, \sigma, \pi)$ for the operator that annihilates or creates a particle of species π with momentum $\mathbf{p}$ and spin state σ. The canonical (anti)commutation relations are then

$$
\begin{aligned}
a(\mathbf{p},\sigma,\pi)a^\dagger(\mathbf{p}',\sigma',\pi') &= \pm a^\dagger(\mathbf{p}'\,\sigma',\pi')a(\mathbf{p},\sigma,\pi) + (2\pi)^3\delta(\mathbf{p}-\mathbf{p}')\delta_{\sigma\sigma'}\delta_{\pi\pi'}, \\
a(\mathbf{p},\sigma,\pi)a(\mathbf{p}',\sigma',\pi') &= \pm a(\mathbf{p}',\sigma',\pi')a(\mathbf{p},\sigma,\pi), \\
a^\dagger(\mathbf{p},\sigma,\pi)a^\dagger(\mathbf{p}',\sigma',\pi') &= \pm a^\dagger(\mathbf{p}',\sigma',\pi')a^\dagger(\mathbf{p},\sigma,\pi).
\end{aligned}
\tag{6.33}
$$

Here the $\pm$ sign is $+$ if at least one of the species π, π' is Bosonic and $-$ if both species are Fermionic.

What is the quantum interaction Hamiltonian density? The classical $\mathcal{H}_I(t, \mathbf{x})$ is a product of field variables, but the corresponding quantum objects are operator-valued distributions, so it is not clear how to multiply them, or how to integrate the resulting product over $\mathbb{R}^3$ to obtain the actual interaction Hamiltonian $H_I(t)$ as a well-defined operator on the state space. In fact, a rigorous mathematical construction of $H_I(t)$ is generally not available, but fortunately it is not really needed to extract meaningful results from the theory. Rather, one has to regard the *formal* expression for $\mathcal{H}_I(t, \mathbf{x})$ simply as a way of encoding certain information to be used as input for calculations that do have a well-defined meaning. One might draw a parallel with a formal power series, which encodes certain useful algebraic information about its sequence of coefficients that does not depend on the convergence of the series.

Of course, if H_I cannot be taken seriously as an operator, then neither can the full Hamiltonian H, the unitary operators e^{-itH} it generates, the scattering operator S, or the individual terms in its Dyson series $\sum_0^\infty S_n$. What *can* be taken seriously, as it turns out, are the S-matrix elements in perturbation theory, that is, the quantities $\sum_0^N \langle \text{out}|S_n|\text{in}\rangle$ for N finite. They are given by sums of integrals of well-defined functions over $\mathbb{R}^{4k}$ for suitable k. Many of these integrals diverge (a reflection of the ill-definedness of the operators just mentioned), but in favorable cases the divergences can be removed by renormalization, and the resulting quantities can then be taken to the lab and compared with experiment. Our goal, therefore, is to compute these matrix elements.

6.4. Evaluation of the S-matrix: first steps

By (6.24) and (6.32), the scattering operator S is formally given by

$$S \sim I + \sum_1^\infty \frac{1}{i^n n!} \int_{\mathbb{R}^4} \cdots \int_{\mathbb{R}^4} \mathcal{T}[H_I(x_1) \cdots H_I(x_n)]\, d^4x_1 \cdots d^4x_n. \tag{6.34}$$

It is not clear whether this makes any sense at all, but at least on a formal level it is rather nice. In particular, the fact that it is given by integrals over space-time suggests at least the possibility of relativistic invariance, except for the time-ordering inside the integrals. We shall have to examine this point in more detail later; for now, we just forge resolutely ahead. Recall that $\mathcal{H}_I$ is a sum of products of field operators. If

$$|\text{in}\rangle = \prod_1^{k_{\text{in}}} a^\dagger(\mathbf{p}_j^{\text{in}}, \sigma_j^{\text{in}}, \pi_j^{\text{in}})|0\rangle, \qquad |\text{out}\rangle = \prod_1^{k_{\text{out}}} a^\dagger(\mathbf{p}_j^{\text{out}}, \sigma_j^{\text{out}}, \pi_j^{\text{out}})|0\rangle,$$

the S-matrix element $\langle\text{out}|S|\text{in}\rangle$ is given by a sum of integrals of terms of the form

$$\left\langle 0 \middle| \prod_1^{k_{\text{out}}} a(\mathbf{p}_j^{\text{out}}, \sigma_j^{\text{out}}, \pi_j^{\text{out}}) \mathcal{T}\left[\prod_1^{k_{\text{field}}} \phi_j(x_j)\right] \prod_1^{k_{\text{in}}} a^\dagger(\mathbf{p}_j^{\text{in}}, \sigma_j^{\text{in}}, \pi_j^{\text{in}}) \middle| 0 \right\rangle. \tag{6.35}$$

One small technical point arises here. In passing from the classical to the quantum interaction Hamiltonian, one is replacing classical functions, which commute, with operators, which may not, so how is one to order the factors? Actually, this is rarely an issue. Since operators pertaining to particles of different species always commute or anticommute, the only possible problem arises when there is a product of a non-Hermitian field with its own adjoint. In the specific theories we are concerned with, the only such terms are of the form $\overline{\psi}\gamma^\mu\psi$ where ψ is a Dirac field, and here we take the order dictated by the matrix algebra, with the $\overline{\psi}$ on the left. Another prescription that has some theoretical justification is to declare that all products of fields with themselves or their adjoints should be Wick ordered; see Weinberg [**129**], p. 200. We shall say more about this later, but it will not be a serious concern for us.

Our first job is to evaluate expressions of the form (6.35). The field operators $\phi_j(x_j)$ are themselves Fourier integrals of the operators $a(\mathbf{p}, \sigma, \pi)$ and $a^\dagger(\mathbf{p}, \sigma, \pi)$, so we have to evaluate the vacuum expectation values of products of creation and annihilation operators. The method for doing this is simple: use the commutation relations (6.33) to move each creation operator to the left and each annihilation operator to the right. When any annihilation operator reaches the right end and acts on $|0\rangle$ it yields 0, and when any creation operator reaches the left end and acts on $\langle 0|$ (via its adjoint) it yields 0. The only nonzero terms will arise when each creation operator (in the initial ket $|\text{in}\rangle$ or in one of the fields) is paired with an annihilation operator (in the final bra $\langle\text{out}|$ or in one of the fields); their commutator will yield a numerical factor (containing delta-functions), and the resulting product of these factors is what must be integrated to get the final result.

Let us work out a very simple example (too simple to be useful all by itself, although it occurs as a part of larger calculations). Consider a single neutral scalar field ϕ and its associated creation operator $a^\dagger(\mathbf{p})$, and let $|\text{in}\rangle = a^\dagger(\mathbf{p})|0\rangle$ and

$|\text{out}\rangle = |0\rangle$. We have

$$\langle 0|\phi(t,\mathbf{x})a^\dagger(\mathbf{p})|0\rangle = \int \frac{1}{\sqrt{2\omega_{\mathbf{q}}}}\langle 0|[e^{iq_\mu x^\mu}a(\mathbf{q}) + e^{-iq_\mu x^\mu}a^\dagger(\mathbf{q})]a^\dagger(\mathbf{p})|0\rangle \frac{d^3\mathbf{q}}{(2\pi)^3}.$$

Now $\langle 0|a^\dagger(\mathbf{q}) = \langle 0|a^\dagger(\mathbf{p}) = 0$, and $a(\mathbf{q})a^\dagger(\mathbf{p}) = a^\dagger(\mathbf{p})a(\mathbf{q}) + (2\pi)^3\delta(\mathbf{q}-\mathbf{p})$, so

$$\langle 0|\phi(t,\mathbf{x})a^\dagger(\mathbf{p})|0\rangle = \int \frac{1}{\sqrt{2\omega_{\mathbf{q}}}}e^{iq_\mu x^\mu}\delta(\mathbf{q}-\mathbf{p})\,d^3\mathbf{q} = \frac{e^{ip_\mu x^\mu}}{\sqrt{2\omega_{\mathbf{p}}}}. \tag{6.36}$$

The reader may be reassured to see that the final answer is a perfectly well-defined function!

Returning to the general case, we need a generalization of the notion of Wick ordering that was introduced in §5.3. First, suppose $A_1,\dots,A_n$ are operators that are sums or integrals of *Bosonic* creation and annihilation operators; that is, $A_j = A_j^a + A_j^c$ where A_j^a (resp. A_j^c) is a sum or integral of annihilation (resp. creation) operators satisfying the commutation relations (6.33) with the $\pm$ signs taken to be $+$. For example, A_j could be a field operator $\phi(t,\mathbf{x})$ or a single creation operator $a^\dagger(\mathbf{p})$. The *Wick ordered product* or *normally ordered product* $:A_1\cdots A_n:$ is the product $A_1\cdots A_n$ with the terms rearranged so that all creation operators are to the left of all annihilation operators. For example, when $n=2$,

$$\begin{aligned} :A_1A_2: \;&=\; :(A_1^a + A_1^c)(A_2^a + A_2^c): \;=\; A_1^aA_2^a + A_1^cA_2^a + A_2^cA_1^a + A_1^cA_2^c \\ &= A_1A_2 - [A_1^a, A_2^c], \end{aligned} \tag{6.37}$$

and for $n=3$,

$$\begin{aligned} :A_1A_2A_3: \;=\; &A_1^aA_2^aA_3^a + A_3^cA_1^aA_2^a + A_2^cA_1^aA_3^a + A_1^cA_2^aA_3^a \\ &+ A_1^cA_2^cA_3^a + A_1^cA_3^cA_2^a + A_2^cA_3^cA_1^a + A_1^cA_2^cA_3^c. \end{aligned} \tag{6.38}$$

(The order of the A^a's and A^c's is immaterial, as creation operators commute with each other.)

If some or all of the A_n involve Fermionic creation and annihilation operators, the operation of Wick ordering is defined as above except that *a factor of -1 is attached to any term when two Fermionic operators are interchanged.* Thus, the analogues of (6.37) and (6.38) when all the A_j are Fermionic are

$$\begin{aligned} :A_1A_2: \;&=\; :(A_1^a + A_1^c)(A_2^a + A_2^c): \;=\; A_1^aA_2^a + A_1^cA_2^a - A_2^cA_1^a + A_1^cA_2^c \\ &= A_1A_2 - \{A_1^a, A_2^c\} \end{aligned}$$

and

$$\begin{aligned} :A_1A_2A_3: \;=\; &A_1^aA_2^aA_3^a + A_3^cA_1^aA_2^a - A_2^cA_1^aA_3^a + A_1^cA_2^aA_3^a \\ &+ A_1^cA_2^cA_3^a - A_1^cA_3^cA_2^a + A_2^cA_3^cA_1^a + A_1^cA_2^cA_3^c. \end{aligned}$$

In all cases, the essential feature of Wick ordered products is that their vacuum expectation values vanish:

$$\langle 0|{:}A_1\cdots A_n{:}|0\rangle = 0. \tag{6.39}$$

We shall also modify our definition of time-ordered products where Fermion fields are concerned, again by introducing a minus sign when two Fermionic operators are interchanged. Namely, if $A_1(t),\dots,A_n(t)$ are Fermionic operators depending on a time parameter t, we set

$$\mathcal{T}[A_1(t_1)\cdots A_n(t_n)] = (\operatorname{sgn}\sigma)A_{\sigma(1)}(t_{\sigma(1)})\cdots A_{\sigma(n)}(t_{\sigma(n)}), \tag{6.40}$$

where σ is the permutation of $1, \ldots, n$ such that $t_{\sigma(1)} > \cdots > t_{\sigma(n)}$. This convention has no effect on the time-ordered products of interaction Hamiltonians that occur in the Dyson series, because Fermion fields always occur in pairs in such Hamiltonians, so interchanging $\mathcal{H}_I(t_1)$ with $\mathcal{H}_I(t_2)$ always involves an even number of Fermion interchanges.

Next, suppose A_1 and A_2 are sums or integrals of Bosonic or Fermionic creation and annihilation operators as above, and suppose that one or both of them depend on time parameters. If both do, and their time parameters are different, $\mathcal{T}(A_1A_2)$ denotes their time-ordered product; otherwise, $\mathcal{T}(A_1A_2) = A_1A_2$. The *contraction* of A_1 with A_2 is the scalar $\overbrace{A_1A_2}$ defined by

$$\overbrace{A_1A_2}\, I = \mathcal{T}(A_1A_2) - {:}A_1A_2{:}\,. \tag{6.41}$$

To see that the difference on the right is indeed a scalar multiple of I, suppose that at most one of the A_j's is Fermionic. By (6.37), if at most one of the A's is time-dependent, then

$$\mathcal{T}(A_1A_2) - {:}A_1A_2{:} = [A_1^a, A_2^c],$$

whereas if $A_1 = A_1(t_1)$ and $A_2 = A_2(t_2)$, then

$$\mathcal{T}(A_1A_2) - {:}A_1A_2{:} = \begin{cases} [A_1^a, A_2^c] & \text{if } t_1 > t_2, \\ [A_2^a, A_1^c] & \text{if } t_2 > t_1. \end{cases}$$

If both A_j's are Fermionic, the result is the same except that commutators are replaced by anticommutators, and there is an extra minus sign from interchanging A_1 and A_2 in the case $t_2 > t_1$. In all cases, these commutators or anticommutators are scalars by (6.33). (As usual, we shall be cavalier about identifying the scalar $\overbrace{A_1A_2}$ with the operator $\overbrace{A_1A_2}\, I$.)

Recall that we are trying to evaluate vacuum expectation values of integrals of products of operators A_j of the types we have been discussing. More precisely, the products involve annihilation operators on the left, time-ordered field operators in the middle, and creation operators on the right: symbolically, $A^a\mathcal{T}[A^f]A^c$. As a matter of notational convenience, we agree that the time-ordering on A^f may encompass the other operators without changing anything:

$$\mathcal{T}[A^aA^fA^c] = A^a\mathcal{T}[A^f]A^c.$$

Moreover, the individual fields that make up the Hamiltonian $H_I(t)$, which all involve the same time t, are taken in the order in which they appear in $H_I(t)$. With this understanding, the result of moving all the annihilation operators to the right as described above is known as *Wick's theorem*:

$\mathcal{T}(A_1 \cdots A_n)$ *is equal to the sum of all operators obtained by contracting* k *pairs of* A_j*'s,* $0 \leq k \leq [n/2]$*, and Wick-ordering the product of the remaining* A_j*'s.*

The proof is a straightforward but tedious induction on n, and we omit it. But to make the meaning clear, we shall write out the result for small n. For $n = 2$, the theorem states that $\mathcal{T}(A_1A_2) = {:}A_1A_2{:} + \overbrace{A_1A_2}$, which is just the definition of $\overbrace{A_1A_2}$. For $n = 3$, we have

$$\mathcal{T}(A_1A_2A_3) = {:}A_1A_2A_3{:} + \overbrace{A_1A_2}\, A_3 + \overbrace{A_1A_3}\, A_2 + \overbrace{A_2A_3}\, A_1,$$

and for $n = 4$ we have

$$\begin{aligned}\mathcal{T}(A_1A_2A_3A_4) = {:}A_1A_2A_3A_4{:}\\ &+ \overbrace{A_1A_2}\,{:}A_3A_4{:} + \overbrace{A_1A_3}\,{:}A_2A_4{:} + \overbrace{A_1A_4}\,{:}A_2A_3{:}\\ &+ \overbrace{A_2A_3}\,{:}A_1A_4{:} + \overbrace{A_2A_4}\,{:}A_1A_3{:} + \overbrace{A_3A_4}\,{:}A_1A_2{:}\\ &+ \overbrace{A_1A_2}\,\overbrace{A_3A_4} + \overbrace{A_1A_3}\,\overbrace{A_2A_4} + \overbrace{A_1A_4}\,\overbrace{A_2A_3}\,.\end{aligned}$$

As an immediate corollary of Wick's theorem and (6.39), we achieve our goal of evaluating the vacuum expectation value of a time-ordered product $\mathcal{T}(A_1, \ldots, A_n)$:

The vacuum expectation value of $\mathcal{T}(A_1 \cdots A_n)$ *is*

$$\langle 0|\mathcal{T}(A_1 \cdots A_n)|0\rangle = \begin{cases} 0 & \text{if } n \text{ is odd,} \\ \sum \overbrace{A_{j_1}A_{j_2}}\,\overbrace{A_{j_3}A_{j_4}} \cdots \overbrace{A_{j_{n-1}}A_{j_n}} & \text{if } n \text{ is even,} \end{cases} \tag{6.42}$$

where the sum in the even case is over all $1 \cdot 3 \cdots (n-1)$ *ways of grouping the* A_j*'s into* $n/2$ *ordered pairs with the ordering in each pair being the same as in the original product* ($j_1 < j_2$, $j_3 < j_4$, *etc.).*

Our interest with these vacuum expectation values is as ingredients in the Dyson series for the scattering matrix. In that situation the operators A_j are either field operators (from the interaction Hamiltonians) or individual creation or annihilation operators (from the in and out states). The former are time-dependent; the latter are not. So we need to compute $\overbrace{A_1A_2}$ where each A_j is either a creation or annihilation operator or a field operator. If both A_j's are individual creation or annihilation operators, the result is simply given by the canonical commutation relations (6.33). If one A_j is a field operator, the calculation is still simple; we essentially did it in (6.36). To wit, if the field operator is

$$\phi_\pi(x) = \sum_\tau \int f(\mathbf{q})\big[u(\mathbf{q},\tau,\pi)a(\mathbf{q},\tau,\pi)e^{-iq_\mu x^\mu} + v(\mathbf{q},\tau,\pi)a^\dagger(\mathbf{q},\tau,\overline{\pi})e^{iq_\mu x^\mu}\big]\,\frac{d^3\mathbf{q}}{(2\pi)^3}$$

($\overline{\pi}$ being the antiparticle of π, and $f(\mathbf{q})$ incorporating the remaining scalar factors such as $\sqrt{2\omega_{\mathbf{q}}}$), then

$$\begin{aligned}\overbrace{a(\mathbf{p},\sigma,\pi')\phi_\pi(x)} &= f(\mathbf{p})v(\mathbf{p},\sigma,\pi)e^{ip_\mu x^\mu}\delta_{\overline{\pi}\pi'},\\ \overbrace{\phi_\pi(x)a^\dagger(\mathbf{p},\sigma,\pi')} &= f(\mathbf{p})u(\mathbf{p},\sigma,\pi)e^{-ip_\mu x^\mu}\delta_{\pi\pi'},\end{aligned} \tag{6.43}$$

(The contractions $\overbrace{a^\dagger(\mathbf{p},\sigma,\pi)\phi(x_\mu)}$ and $\overbrace{\phi(x_\mu)a(\mathbf{p},\sigma,\pi)}$ don't arise in practice, and in any case they both vanish.)

The interesting case is where both A_j's are field operators. Here each A_j involves annihilation operators for some species of particle and creation operators for its antiparticle. Unless the particles of one A_j are the antiparticles for the other one, these operators all commute or anticommute, with the result that the contraction vanishes. Thus *the only nontrivial contractions are those involving a field and its adjoint.* These contractions (up to a conventional factor of i) are known as *propagators* for the field in question.

The reason for the name is that by (6.39) and (6.41),

$$\overbrace{\phi(t,\mathbf{x})\phi^\dagger(s,\mathbf{y})} = \langle 0|\mathcal{T}[\phi(t,\mathbf{x})\phi^\dagger(s,\mathbf{y})]|0\rangle.$$

The expression on the right is the amplitude for the field to create a particle out of the vacuum at $\mathbf{y}$ at time s and annihilate it at $\mathbf{x}$ at time t if $t > s$, and the amplitude for the field to create an antiparticle out of the vacuum at $\mathbf{x}$ at time t and annihilate it at $\mathbf{y}$ at time s if $s > t$; in either case the particle or antiparticle "propagates" from $\mathbf{y}$ to $\mathbf{x}$ or from $\mathbf{x}$ to $\mathbf{y}$. We shall see that this interpretation remains valid when propagators are incorporated into more realistic processes.

We must now pause to study propagators in some detail.

6.5. Propagators

We begin by considering a neutral scalar field ϕ of mass m:

$$\phi(t,\mathbf{x}) = \int \frac{1}{\sqrt{2\omega_\mathbf{p}}}[e^{-i\omega_\mathbf{p}t+i\mathbf{p}\cdot\mathbf{x}}a(\mathbf{p}) + e^{i\omega_\mathbf{p}t-i\mathbf{p}\cdot\mathbf{x}}a^\dagger(\mathbf{p})]\,\frac{d^3\mathbf{p}}{(2\pi)^3}.$$

If $t_1 > t_2$, $\overbrace{\phi(t_1,\mathbf{x}_1)\phi(t_2,\mathbf{x}_2)}$ is equal to

$$\begin{aligned}
&\iint \frac{1}{2\sqrt{\omega_{\mathbf{p}_1}\omega_{\mathbf{p}_2}}} e^{-i\omega_{\mathbf{p}_1}t_1+i\mathbf{p}_1\cdot\mathbf{x}_1+i\omega_{\mathbf{p}_2}t_2-i\mathbf{p}_2\cdot\mathbf{x}_2}[a(\mathbf{p}_1),a^\dagger(\mathbf{p}_2)]\,\frac{d^3\mathbf{p}_1\,d^3\mathbf{p}_2}{(2\pi)^6}\\
&= \iint \frac{1}{2\sqrt{\omega_{\mathbf{p}_1}\omega_{\mathbf{p}_2}}} e^{-i\omega_{\mathbf{p}_1}t_1+i\mathbf{p}_1\cdot\mathbf{x}_1+i\omega_{\mathbf{p}_2}t_2-i\mathbf{p}_2\cdot\mathbf{x}_2}\delta(\mathbf{p}_1-\mathbf{p}_2)\,\frac{d^3\mathbf{p}_1\,d^3\mathbf{p}_2}{(2\pi)^3}\\
&= \int \frac{1}{2\omega_\mathbf{p}} e^{-i\omega_\mathbf{p}(t_1-t_2)+i\mathbf{p}\cdot(\mathbf{x}_1-\mathbf{x}_2)}\,\frac{d^3\mathbf{p}}{(2\pi)^3}.
\end{aligned}$$

If $t_1 < t_2$ we must replace $t_1 - t_2$ in this last integral by $t_2 - t_1$, but the $\mathbf{x}_1 - \mathbf{x}_2$ can remain as it is because the substitution $\mathbf{p} \to -\mathbf{p}$ shows that the integral is even in $\mathbf{x}_1 - \mathbf{x}_2$. In short,

$$\overbrace{\phi(t_1,\mathbf{x}_1)\phi(t_2,\mathbf{x}_2)} = -i\Delta_F(t_1-t_2,\,\mathbf{x}_1-\mathbf{x}_2), \tag{6.44}$$

where

$$\Delta_F(t,\mathbf{x}) = i\int \frac{e^{-i\omega_\mathbf{p}|t|+i\mathbf{p}\cdot\mathbf{x}}}{2\omega_\mathbf{p}}\,\frac{d^3\mathbf{p}}{(2\pi)^3}. \tag{6.45}$$

This is to be interpreted as the tempered distribution on $\mathbb{R}^4$ whose Fourier transform in the space variables $\mathbf{x}$ is

$$(\mathcal{F}_\mathbf{x}\Delta_F)(t,\mathbf{p}) = \frac{ie^{-i\omega_\mathbf{p}|t|}}{2\omega_\mathbf{p}}. \tag{6.46}$$

(The foregoing derivation of (6.44) is easily made rigorous by integrating both sides against a test function $f(\mathbf{x}_1)g(\mathbf{x}_2)$.)

Δ_F is called the *Feynman propagator*. It is both physically and mathematically interesting. Here is a list of its most important properties:

i. *Δ_F is a fundamental solution of the Klein-Gordon operator:*

$$(\partial_t^2 - \nabla_\mathbf{x}^2 + m^2)\Delta_F = \delta.$$

ii. *The full space-time Fourier transform of* Δ_F *is*

$$\widehat{\Delta}_F(\omega,\mathbf{p}) = \lim_{\epsilon\to 0+} \frac{1}{-\omega^2+|\mathbf{p}|^2+m^2-i\epsilon}, \tag{6.47}$$

that is,

$$\widehat{\Delta}_F(p) = \lim_{\epsilon\to 0+} \frac{1}{-p^2+m^2-i\epsilon}, \tag{6.48}$$

where the limit is taken in the weak topology of tempered distributions.

iii. Δ_F *is invariant under the full Lorentz group.*

iv. Δ_F *is a* C^∞ *function away from the light cone, and its support is* $\mathbb{R}^4$.

Let us prove these assertions. (Mathematical readers should find this a refreshing change of pace!) For (i), by taking the Fourier transform in $\mathbf{x}$ it is equivalent to show that $g(t) = ie^{-i\omega_{\mathbf{p}}|t|}$ satisfies $(\partial_t^2 + |\mathbf{p}|^2 + m^2)g(t) = 2\omega_{\mathbf{p}}\delta(t)$. This is elementary: $\partial_t g = \omega_{\mathbf{p}}(\operatorname{sgn} t)e^{-i\omega_{\mathbf{p}}|t|}$, so

$$\partial_t^2 g(t) = 2\omega_{\mathbf{p}}\delta(t)e^{-i\omega_{\mathbf{p}}|t|} - i\omega_{\mathbf{p}}^2 e^{-i\omega_{\mathbf{p}}|t|} = 2\omega_{\mathbf{p}}\delta(t) - (|\mathbf{p}|^2+m^2)g(t).$$

Next, given $\epsilon > 0$ and $\mathbf{p}\in\mathbb{R}^3$, let $a = a(\epsilon,\mathbf{p})$ be the square root of $i\epsilon - |\mathbf{p}|^2 - m^2$ with positive real part. Then, for $\omega\in\mathbb{R}$,

$$\begin{aligned}\frac{1}{-\omega^2+|\mathbf{p}|^2+m^2-i\epsilon} &= -\frac{1}{2a}\left[\frac{1}{a+i\omega}+\frac{1}{a-i\omega}\right]\\ &= -\frac{1}{2a}\left[\int_0^\infty e^{-(a+i\omega)t}\,dt + \int_{-\infty}^0 e^{(a-i\omega)t}\,dt\right].\end{aligned}$$

That is, as a function of ω, $(-\omega^2+|\mathbf{p}|^2+m^2-i\epsilon)^{-1}$ is the Fourier transform in t of $-(1/2a)e^{-a|t|}$. Now let $\epsilon\to 0+$: we have $a\to i\sqrt{|\mathbf{p}|^2+m^2} = i\omega_{\mathbf{p}}$, so $-(1/2a)e^{-a|t|}\to(i/2\omega_{\mathbf{p}})e^{-i\omega_{\mathbf{p}}|t|}$. This convergence takes place in the topology of tempered distributions on $\mathbb{R}^4$, by the dominated convergence theorem, so (ii) follows from (6.46). Also, the function $h(\omega,\mathbf{p}) = (-\omega^2+|\mathbf{p}|^2+m^2-i\epsilon)^{-1}$ is manifestly Lorentz-invariant on $\mathbb{R}^4$; hence so is its limit as $\epsilon\to 0+$, and so is the latter's Fourier transform Δ_F, which proves (iii).

One can compute Δ_F explicitly by replacing $|t|$ by $|t| - i\epsilon$ in (6.45), evaluating the resulting absolutely convergent integral, and letting $\epsilon\to 0+$ at the end of the calculation. Since $e^{-\omega_{\mathbf{p}}(\epsilon+i|t|)}/\omega_{\mathbf{p}}$ is a radial function of $\mathbf{p}$, the integration reduces to a one-dimensional one that is part of the classic lore of special functions, and the result is that

$$\Delta_F(t,\mathbf{x}) = \frac{im}{4\pi^2\sqrt{|\mathbf{x}|^2-t^2}}K_1(m\sqrt{|\mathbf{x}|^2-t^2}) \text{ on } \mathbb{R}^4\setminus\{(t,\mathbf{x}) : |t| = |\mathbf{x}|\},$$

where K_1 is the modified Bessel function of order 1. We omit the details since we shall have no use for this explicit formula, but it implies (iv).

The fact that Δ_F is C^∞ on the complement of the light cone can also be deduced from general considerations about differential equations (see Folland [**45**], §8G). Since Δ_F is a fundamental solution for the Klein-Gordon operator, its wave front set is contained in the characteristic variety of that operator (i.e., the union of the light cones in the cotangent spaces at all points of $\mathbb{R}^4$) together with the wave front set of δ (i.e., the cotangent space at the origin). On the other hand, since Δ_F is Lorentz-invariant, its wave front set is contained in the union of the conormal bundles of the orbits of the Lorentz group. Since the normal to the orbit through

$(t, \mathbf{x})$ does not lie on the light cone unless $(t, \mathbf{x})$ itself does, Δ_F has no wave front set away from the light cone and is therefore C^∞ there.

Important remark: It is common in the physics literature to suppress the "$\lim_{\epsilon\to 0+}$" in (6.47) and (6.48). In general, in formulas of this sort that contain an unspecified ϵ, a "$\lim_{\epsilon\to 0+}$" should be understood, with the limit taken in a suitable sense. We shall often avail ourselves of this bit of shorthand.

Property (i) is the reason for factoring out the $-i$ in the definition of Δ_F.

By the way, the preceding results apply equally to the case $m = 0$, where Δ_F is a fundamental solution of the wave operator $\partial_t^2 - \nabla^2$. It is quite different from the fundamental solution that usually arises in classical wave mechanics, namely, the one supported on the forward light cone. Let us denote that solution by G_r (for "retarded Green's function"), and let G_a ("advanced Green's function") be the corresponding solution supported on the backward light cone, $G_a(t, \mathbf{x}) = G_r(-t, \mathbf{x})$. Then $\Delta_F - \frac{1}{2}(G_r + G_a)$ is a solution of the homogeneous wave equation; in fact,

$$\Delta_F(t, \mathbf{x})\big|_{m=0} = \tfrac{1}{2}(G_r + G_a)(t, \mathbf{x}) + \frac{i}{4\pi^2}(\partial_t^2 - \nabla_{\mathbf{x}}^2)\log\big|\,|\mathbf{x}|^2 - t^2\big|,$$

which agrees off the light cone with the function $i/4\pi^2(|\mathbf{x}|^2 - t^2)$. It may be instructive to compare the Fourier transforms of G_r and G_a with that of Δ_F:

$$\widehat{G}_r(\omega, \mathbf{p}) = \lim_{\epsilon\to 0+} \frac{1}{-(\omega - i\epsilon)^2 + |\mathbf{p}|^2}, \qquad \widehat{G}_a(\omega, \mathbf{p}) = \lim_{\epsilon\to 0+} \frac{1}{-(\omega + i\epsilon)^2 + |\mathbf{p}|^2},$$

$$\widehat{\Delta}_F(\omega, \mathbf{p}) = \lim_{\epsilon\to 0+} \frac{1}{-\omega^2 + |\mathbf{p}|^2 - i\epsilon}.$$

See Folland **[46]** for a complete derivation of these facts.

We return to the calculation of contractions of field operators. For a charged scalar field

$$\phi(x) = \int \frac{1}{\sqrt{2\omega_{\mathbf{p}}}}[e^{-ip_\mu x^\mu} a(\mathbf{p}) + e^{itp_\mu x^\mu} b^\dagger(\mathbf{p})] \frac{d^3\mathbf{p}}{(2\pi)^3},$$

where a and b are the annihilation operators for particles and antiparticles, the calculation is similar to the neutral case. Since $[a(\mathbf{p}), b^\dagger(\mathbf{p}')] = 0$ for any $\mathbf{p}, \mathbf{p}'$, whereas $[a(\mathbf{p}), a^\dagger(\mathbf{p}')] = [b(\mathbf{p}), b^\dagger(\mathbf{p}')] = (2\pi)^3\delta(\mathbf{p} - \mathbf{p}')$, it follows easily that

$$\begin{aligned} &\overbrace{\phi(t_1, \mathbf{x}_1)\phi(t_2, \mathbf{x}_2)} = \overbrace{\phi^\dagger(t_1, \mathbf{x}_1)\phi^\dagger(t_2, \mathbf{x}_2)} = 0, \\ &\overbrace{\phi(t_1, \mathbf{x}_1)\phi^\dagger(t_2, \mathbf{x}_2)} = \overbrace{\phi^\dagger(t_1, \mathbf{x}_1)\phi(t_2, \mathbf{x}_2)} = -i\Delta_F(t_1 - t_2, \mathbf{x}_1 - \mathbf{x}_2). \end{aligned} \tag{6.49}$$

Thus the propagator $-i\Delta_F$ is the same as for the neutral field.

For Dirac fields, the important quantity is the contraction of a component ψ_j of a Dirac field ψ with a component $\overline{\psi}_k$ of its Dirac adjoint $\overline{\psi} = \psi^\dagger\gamma_0$ $(j, k = 1, \ldots, 4)$ (in either order). We have

$$\begin{aligned} \overbrace{\psi_j(t_1, \mathbf{x}_1)\overline{\psi}_k(t_2, \mathbf{x}_2)} &= \mathcal{T}[\psi_j(t_1, \mathbf{x}_1)\overline{\psi}_k(t_2, \mathbf{x}_2)] - {:}\psi_j(t_1, \mathbf{x}_1)\overline{\psi}_k(t_2, \mathbf{x}_2){:} \\ &= \begin{cases} \{\psi_j^a(t_1, \mathbf{x}_1), \overline{\psi}_k^c(t_2, \mathbf{x}_2)\} & \text{if } t_1 > t_2, \\ -\{\psi_k^a(t_2, \mathbf{x}_2), \overline{\psi}_j^c(t_1, \mathbf{x}_1)\} & \text{if } t_1 < t_2, \end{cases} \end{aligned}$$

where the superscripts a and c indicate the parts of the fields involving annihilation and creation operators, respectively. These anticommutators are evaluated by using the Fourier expansions (5.32)–(5.33) of the fields and the spin sums (5.34) in much the same way as in the calculation of (5.35). In fact, setting $t = t_1 - t_2$ and $\mathbf{x} = \mathbf{x}_1 - \mathbf{x}_2$, for $t_1 > t_2$ we have

$$\begin{aligned}
\overbrace{\psi_j(t_1,\mathbf{x}_1)\overline{\psi}_k(t_2,\mathbf{x}_2)} &= \int \sum_{s=\pm} \frac{m}{2\omega_{\mathbf{p}}} e^{-i\omega_{\mathbf{p}}t+i\mathbf{p}\cdot\mathbf{x}} u(\mathbf{p},s)_j \overline{u}(\mathbf{p},s)_k \frac{d^3\mathbf{p}}{(2\pi)^3} \\
&= \int \frac{m}{2\omega_{\mathbf{p}}} e^{-i\omega_{\mathbf{p}}t+i\mathbf{p}\cdot\mathbf{x}} (m^{-1}p_\mu\gamma^\mu + I)_{jk} \frac{d^3\mathbf{p}}{(2\pi)^3} \\
&= (i\gamma^\mu\partial_\mu + mI)_{jk} \int \frac{e^{-i\omega_{\mathbf{p}}t+i\mathbf{p}\cdot\mathbf{x}}}{2\omega_{\mathbf{p}}} \frac{d^3\mathbf{p}}{(2\pi)^3},
\end{aligned}$$

and by (6.45), the last integral is equal to $-i\Delta_F(t,\mathbf{x})$. On the other hand, for $t_1 < t_2$ we have

$$\begin{aligned}
\overbrace{\psi_j(t_1,\mathbf{x}_1)\overline{\psi}_k(t_2,\mathbf{x}_2)} &= -\int \sum_{s=\pm} \frac{m}{2\omega_{\mathbf{p}}} e^{i\omega_{\mathbf{p}}t-i\mathbf{p}\cdot\mathbf{x}} \overline{v}(\mathbf{p},s)_k v(\mathbf{p},s)_j \frac{d^3\mathbf{p}}{(2\pi)^3} \\
&= -\int \frac{m}{2\omega_{\mathbf{p}}} e^{i\omega_{\mathbf{p}}t-i\mathbf{p}\cdot\mathbf{x}} (m^{-1}p_\mu\gamma^\mu - mI)_{jk} \frac{d^3\mathbf{p}}{(2\pi)^3} \\
&= (i\gamma^\mu\partial_\mu + mI)_{jk} \int \frac{e^{i\omega_{\mathbf{p}}t-i\mathbf{p}\cdot\mathbf{x}}}{2\omega_{\mathbf{p}}} \frac{d^3\mathbf{p}}{(2\pi)^3}.
\end{aligned}$$

The substitution $\mathbf{p} \to -\mathbf{p}$ shows that the $e^{-i\mathbf{p}\cdot\mathbf{x}}$ can be replaced by $e^{i\mathbf{p}\cdot\mathbf{x}}$ in this last integral, so it is again equal to $-i\Delta_F(t,\mathbf{x})$. The calculation with ψ and $\overline{\psi}$ switched is similar, and the result is the same except for a minus sign.

We have shown that

(6.50)

$$\overbrace{\psi_j(t_1,\mathbf{x}_1)\overline{\psi}_k(t_2,\mathbf{x}_2)} = -\overbrace{\overline{\psi}_j(t_1,\mathbf{x}_1)\psi_k(t_2,\mathbf{x}_2)} = -i\big[\Delta_{\text{Dirac}}(t_1-t_2,\, \mathbf{x}_1-\mathbf{x}_2)\big]_{jk},$$

where the *Dirac propagator* Δ_{Dirac}, a 4×4 matrix-valued distribution, is

$$\Delta_{\text{Dirac}} = (i\gamma^\mu\partial_\mu + m)\Delta_F,$$

that is,

$$\widehat{\Delta}_{\text{Dirac}}(p) = \frac{(p_\mu\gamma^\mu + m)}{-p^2 + m^2 - i\epsilon}. \tag{6.51}$$

In this last formula there is an implicit "$\lim_{\epsilon\to0+}$," as we described in connection with Δ_F.

The Fourier-transformed propagator (6.51) is a matrix-valued distribution that agrees with the function $(p_\mu\gamma^\mu + m)/(-p^2+m^2)$ off the mass shell $p^2 = m^2$. Since

$$(p_\mu\gamma^\mu + m)(p_\mu\gamma^\mu - m) = p^2 - m^2,$$

one often finds (6.51) rewritten as

$$-i\widehat{\Delta}_{\text{Dirac}}(p) = \frac{i}{p_\mu\gamma^\mu - m} \qquad (p^2 \neq m^2). \tag{6.52}$$

(Of course, the expression on the right denotes the matrix $i(p_\mu\gamma^\mu - mI)^{-1}$.) For the same reason, Δ_{Dirac} is a fundamental solution for the Dirac operator $-i\gamma^\mu\partial_\mu + m$.

Similar calculations yield the propagators for field operators for fields of arbitrary spin; see Weinberg [**129**], §5.7. Their Fourier transforms all turn out to be of the form $F(p)/(-p^2+m^2-i\epsilon)$ where F is a polynomial with values in the appropriate spinor space. The one we will need for quantum electrodynamics is the photon field, which is a spin-1 field of mass 0. Unfortunately, the absence of mass complicates the picture, and a detailed analysis of the situation is rather difficult. Readers may consult physics texts for more information, but they will not find any unanimity about the best way to explain this matter. For the present we content ourselves with a brief heuristic discussion leading to the final result. We shall give another derivation of it — still nonrigorous but perhaps more convincing — in §8.3.

Let us begin with the easier case of a massive vector field instead. Just as the propagators for spin-0 and spin-$\frac{1}{2}$ fields are given by fundamental solutions for the corresponding field operators — the Klein-Gordon and Dirac operators — so we expect the propagator for the massive vector field to be given by a fundamental solution of the Proca operator. The Proca equation (5.39) equation can be written

$$(-g^{\mu\nu}(\partial^2 + m^2) + \partial^\mu\partial^\nu)A_\mu = 0,$$

so we are looking for a fundamental solution of $-g^{\mu\nu}(\partial^2 + m^2) + \partial^\mu\partial^\nu$, that is, a matrix-valued distribution whose Fourier transform evaluated at p is the inverse of the matrix

$$D^{\mu\nu} = g^{\mu\nu}(p^2 - m^2) - p^\mu p^\nu.$$

This inverse is readily computed:

$$(D^{-1})_{\mu\nu} = \frac{-g_{\mu\nu} + p_\mu p_\nu/m^2}{-p^2 + m^2},$$

and indeed, the propagator for the massive vector field is given by

$$\Delta^{\text{Proca}}_{\mu\nu} = -(g_{\mu\nu} + \partial_\mu\partial_\nu/m^2)\Delta_F, \text{ i.e., } \widehat{\Delta}^{\text{Proca}}_{\mu\nu}(p) = \frac{-g_{\mu\nu} + p_\nu p_\mu/m^2}{-p^2 + m^2 - i\epsilon}. \tag{6.53}$$

Of course, to verify this, and to see that the overall sign and normalization are correct, one must actually calculate the relevant contractions; the reader may try this as an exercise or consult Weinberg [**129**], §5.3.

Now, (6.53) blows up as $m \to 0$; the trouble is that when $m = 0$ the matrix $D_{\mu\nu}$ is not invertible, for $D_{\mu\nu}q^\nu = 0$. This has to do, once again, with the fact that the elecromagnetic field cannot propagate longitudinal waves, and with the gauge-invariance of the theory. If one imposes the Landau gauge condition $\partial_\mu A^\mu = 0$, however, the term $\partial_\mu\partial_\nu A^\mu$ in Maxwell's equation disappears. In fact, under this condition one can replace the Maxwell equation $(-g^{\mu\nu}\partial^2 + \partial^\mu\partial^\nu)A_\mu = 0$ by

$$(-g^{\mu\nu}\partial^2 + (1-b)\partial^\mu\partial^\nu)A_\mu = 0$$

for any $b \in \mathbb{R}$, and as long as $b \neq 0$ the corresponding matrix

$$D^{\mu\nu}_{(b)} = g^{\mu\nu}p^2 + (b-1)p^\mu p^\nu$$

is invertible: its inverse is

$$\frac{g_{\mu\nu} + (a-1)p_\mu p_\nu/p^2}{p^2}, \qquad a = b^{-1}.$$

And, in fact, this device gives the right answer: in computing S-matrix elements one can take the contractions of photon fields to be given by

$$\overbrace{A_\mu(t_1, \mathbf{x}_1)A_\nu(t_2, \mathbf{x}_2)} \longrightarrow -i\Delta^{(a)}_{\mu\nu}(t_1 - t_2, \mathbf{x}_1 - \mathbf{x}_2), \tag{6.54}$$

where

$$\widehat{\Delta}^{(a)}_{\mu\nu}(p) = \frac{g_{\mu\nu} + (a-1)p_\mu p_\nu/p^2}{p^2 + i\epsilon}. \tag{6.55}$$

We write an arrow rather than an equal sign in (6.54) because a can be taken to be *any* real number. This indeterminacy may seem unsettling, but as it turns out, the gauge-invariance of the theory guarantees that the contributions of the $q_\mu q_\nu$ term in the photon propagators ultimately cancel out when one computes a physically meaningful quantity. (We shall explain this more fully in §7.7.) As a result, one can use the freedom in choosing a to simplify calculations as needed. In particular, the default choice is $a = 1$ (sometimes called *Feynman gauge*), which gives the photon propagator

$$-i\Delta_{\mu\nu} = ig_{\mu\nu}\Delta_F\big|_{m=0}, \quad \text{i.e.,} \quad \widehat{\Delta}_{\mu\nu}(p) = \frac{g_{\mu\nu}}{p^2 + i\epsilon}. \tag{6.56}$$

Propagators and the spin-statistics theorem. The terms of the Dyson series (6.24) are integrals over space-time of time-ordered products of interaction Hamiltonian densities $\mathcal{H}_I(x_j)$ ($x_j \in \mathbb{R}^4$). If these integrals are to have a relativistically invariant meaning, it is necessary to examine the time-ordering in more detail. For $x, y \in \mathbb{R}^4$ it makes Lorentz-invariant sense to say that $x^0 > y^0$ if $x - y$ is timelike ($(x-y)^2 > 0$), but it $x - y$ is spacelike there are Lorentz transformations that reverse the sign of $x^0 - y^0$ or make it vanish. Since the time-ordered products are to be integrated over *all* space-time, the result will have a relativistically invariant meaning only if the time ordering is irrelevant when the space-time points in question are spacelike separated. In other words, we must require of the Hamiltonian density $\mathcal{H}_I$ that

$$[\mathcal{H}_I(x), \mathcal{H}_I(y)] = 0 \text{ when } (x-y)^2 \le 0.$$

Because of the way $\mathcal{H}_I$ is formed from products of field operators, the way to guarantee this is to require that the field operators $\phi(x)$ and $\phi(y)$ commute or anticommute when $x - y$ is spacelike. More precisely, if ϕ is a field that occurs an odd number of times in some term in $\mathcal{H}_I$ (which can only happen if $\phi = \phi^\dagger$), we must have $[\phi(x), \phi(y)] = 0$ when $x - y$ is spacelike. On the other hand, if ϕ and $\phi^\dagger$ together occur an even number of times in each term, it is enough to have either $[\phi(x), \phi(y)] = 0$ or $\{\phi(x), \phi(y)\} = 0$ for $x - y$ spacelike, as two minus signs will cancel in the latter case. (For example, in the interaction for QED, the electromagnetic field occurs once and the Dirac field occurs twice; the field ϕ occurs four times in the ϕ^4 interaction.)

Calculating these commutators and anticommutators is easy enough; it is essentially the same as the calculation of contractions that we did earlier, except that the time-ordering is absent. For a neutral scalar field ϕ of mass m given by (5.11), with $p = (\omega_{\mathbf{p}}, \mathbf{p})$ we have

$$\begin{aligned}[\phi(x), \phi(y)] &= \int \frac{e^{-ip_\mu x^\mu + iq_\mu y^\mu}}{2\sqrt{\omega_{\mathbf{p}}\omega_{\mathbf{q}}}} [a(\mathbf{p}), a^\dagger(\mathbf{q})] \frac{d^3\mathbf{p}\, d^3\mathbf{q}}{(2\pi)^6} \\ &\quad + \int \frac{e^{ip_\mu x^\mu - iq_\mu y^\mu}}{2\sqrt{\omega_{\mathbf{p}}\omega_{\mathbf{q}}}} [a^\dagger(\mathbf{p}), a(\mathbf{q})] \frac{d^3\mathbf{p}\, d^3\mathbf{q}}{(2\pi)^6},\end{aligned}$$

and since $[a(\mathbf{p}), a^\dagger(\mathbf{q})] = -[a^\dagger(\mathbf{p}), a(\mathbf{q})] = (2\pi)^3\delta(\mathbf{p} - \mathbf{q})$, this boils down to

$$[\phi(x), \phi(y)] = -i\Delta_+(x-y) + i\Delta_+(y-x),$$

where

$$\Delta_+(x) = i\int \frac{e^{-ip_\mu x^\mu}}{2\omega_{\mathbf{p}}} \frac{d^3\mathbf{p}}{(2\pi)^3}.$$

Δ_+ is just like the Feynman propagator Δ_F except that one has $x^0 = t$ in the exponent instead of $|t|$; in particular, the Fourier integral is to be interpreted in the sense of distributions as before. (Rather than being a fundamental solution for the Klein-Gordon operator like Δ_F, though, Δ_+ is a solution of the *homogeneous* Klein-Gordon equation.)

Δ_+ is just i times the Fourier transform of the invariant measure $d^3\mathbf{p}/2\omega_\mathbf{p}$ on the positive mass shell $p_0 = \omega_\mathbf{p}$, and hence it is invariant under the orthochronous Lorentz group. (It is not invariant under time inversion, which interchanges the mass shell with $p_0 = -\omega_\mathbf{p}$.) Moreover, it agrees with Δ_F in the region $t > 0$ (where $|t| = t$), and hence — since Δ_+ and Δ_F are both invariant under the orthochronous Lorentz group — it agrees with Δ_F in the whole region outside the backward light cone, and in particular in the spacelike region. Therefore, if $x - y$ is spacelike,

$$[\phi(x), \phi(y)] = -i\Delta_+(x-y) + i\Delta_+(y-x) = -i\Delta_F(x-y) + i\Delta_F(y-x) = 0,$$

because Δ_F is even. This is exactly what we were looking for.

By the same calculation, for a charged scalar field we have $[\phi(x), \phi^\dagger(y)] = 0$ when $x - y$ is spacelike. (That $[\phi(x), \phi(y)] = 0$ is automatic since ϕ contains annihilation operators for a particle and creation operators for its antiparticle; these operators commute.)

However, suppose we tried to construct a quantum theory involving *Fermions* with spin zero. The field describing these Fermions would be constructed out of creation and annihilation operators that satisfy *anticommutation* relations, so we would have to use *anticommutators* instead of commutators in this calculation. With this change, the calculation proceeds much as before, except that one crucial minus sign changes to a plus sign, giving

$$\{\phi(x), \phi(y)\} = -i\Delta_+(x-y) - i\Delta_+(y-x) = -2i\Delta_F(x-y) \text{ for } x-y \text{ spacelike.}$$

As we have observed earlier, Δ_F is not zero in the spacelike region, so relativistic invariance is destroyed. *Particles of spin zero must be Bosons.*

For a Dirac field ψ of mass m, calculations similar to the ones above show that

$$\{\psi_a(x), \overline{\psi}_b(y)\} = -i\big[(i\gamma^\mu\partial_\mu + m)\Delta_+(x-y) + (i\gamma^\mu\partial_\mu - m)\Delta_+(y-x)\big]_{ab}. \tag{6.57}$$

Since Δ_+ is even in the spacelike region, the terms involving m cancel out there, and since $\partial_\mu\Delta_+$ is odd, so do the terms involving $\gamma^\mu\partial_\mu$. However, if we tried to construct a Dirac field out of creation and annihilation operators that satisfy *commutation* relations, we would have to consider $[\psi_a(x), \overline{\psi}_b(y)]$ instead, and it would be given by the right side of (6.57) with the plus sign between the two distributions replaced by a minus sign. Those two terms would add up instead of canceling, and relativistic invariance would be violated again. *Particles of spin $\frac{1}{2}$ must be Fermions.*

Similar considerations apply to particles of higher spin: for fields of spin s, the commutators or anticommutators will involve derivatives of order $2s$ of Δ_+, which will be even or odd according as $2s$ is even or odd. Hence, in order to obtain relativistic invariance, one must use commutators when $2s$ is even and anticommutators when s is odd. This gives another way of arriving at the conclusion that *particles of integer spin must be Bosons; particles of half-integer spin must be Fermions.* See Weinberg [**129**], §5.7, for a more detailed treatment.

6.6. Feynman diagrams

We now return to the Dyson series for a scattering matrix element between initial and final states consisting of free particles with definite momenta. The Nth term in the series is

$$\frac{(-i)^N}{N!}\int\cdots\int \langle 0|a_1(\mathbf{p}_1^{\text{out}})\cdots a_J(\mathbf{p}_J^{\text{out}})\mathcal{T}[\mathcal{H}_I(x_1)\cdots\mathcal{H}_I(x_N)] \cdot a_1^\dagger(\mathbf{p}_1^{\text{in}})\cdots a_K^\dagger(\mathbf{p}_K^{\text{in}})|0\rangle\, d^4x_1\cdots d^4x_N, \tag{6.58}$$

Each $\mathcal{H}_I(x_n)$ is a sum of products of field operators, and (6.58) is the sum of the integrals obtained from it by replacing each $\mathcal{H}_I(x_n)$ by a single one of these products (including the relevant coupling constant). There can be any number of different fields here, and the initial and final creation and annihilation operators can be for any of the corresponding particle or antiparticle types. By the corollary (6.42) of Wick's theorem, when each $\mathcal{H}_I(x_n)$ is taken to consist of a single product, (6.58) is equal to $(-i)^N/N!$ times the sum of all the terms obtained by pairing up all the creation operators $a_k^\dagger(\mathbf{p}_k)$, the annihilation operators $a_j(\mathbf{p}_j')$, and the field operators in the $\mathcal{H}_I(x_m)$ in all possible ways and integrating the resulting product of contractions. Each such term is the integral, with respect to $x_1,\ldots,x_N$, of a product of contractions of the form

$$\overbrace{a_j(\mathbf{p}_j^{\text{out}})a_k^\dagger(\mathbf{p}_k^{\text{in}})},\qquad \overbrace{a_j(\mathbf{p}_j^{\text{out}})\phi(x_n)},\qquad \overbrace{\phi(x_n)a_k^\dagger(\mathbf{p}_k^{\text{in}})},\qquad \overbrace{\phi(x_n)\phi^\dagger(x_{n'})},$$

where ϕ is one of the fields contributing to $\mathcal{H}_I$. (The notation here is very stripped-down. If ϕ is a field involved in $\mathcal{H}_I$ then so is $\phi^\dagger$, so contractions of the form $\overbrace{a_j(\mathbf{p}_j^{\text{out}})\phi^\dagger(x_n)}$ and $\overbrace{\phi^\dagger(x_n)a_k^\dagger(\mathbf{p}_k^{\text{in}})}$ are also included in this scheme. Moreover, there may be other parameters indicating components of vectors, spin states, particle species, and other entities such as Dirac matrices, that are being suppressed. Finally, the a's and $a^\dagger$'s are completely unrelated; for instance, $a_1(\mathbf{p}_1^{\text{out}})$ and $a_1^\dagger(\mathbf{p}_1^{\text{in}})$ may pertain to different spin states or different particles.)

We now know how to evaluate all of these:

- $\overbrace{a_j(\mathbf{p}_j^{\text{out}})a_k^\dagger(\mathbf{p}_k^{\text{in}})}$ is zero if a_j and $a_k^\dagger$ pertain to different spin states or different particles and is $(2\pi)^3\delta(\mathbf{p}_j^{\text{out}}-\mathbf{p}_k^{\text{in}})$ otherwise.
- $\overbrace{a_j(\mathbf{p}_j^{\text{out}})\phi(x_n)}$ is zero unless ϕ creates particles of the type annihilated by a_j. In that case

  $$\phi(x) = \int e^{-ip_\mu x^\mu}u(\mathbf{p}_j)a_j^\dagger(\mathbf{p})\,d^3\mathbf{p}/(2\pi)^3 + \cdots \qquad (p=(\omega_{\mathbf{p}},\mathbf{p})),$$

  where any further ingredients such as spin components and $\sqrt{2\omega_{\mathbf{p}}}$ are incorporated in $u(\mathbf{p})$, and we have

  $$\overbrace{a_j(\mathbf{p}_j^{\text{out}})\phi(x_n)} = e^{-ip_\mu^{\text{out}}x^\mu}u(\mathbf{p}_j^{\text{out}}).$$

- Similarly, $\overbrace{\phi(x_n)a_k^\dagger(\mathbf{p}_k^{\text{in}})}$ is zero unless ϕ annihilates particles of the type created by $a_k^\dagger$. In that case

  $$\phi(x) = \int e^{-ip_\mu x^\mu}u(\mathbf{p}_j)a_j(\mathbf{p})\,d^3\mathbf{p}/(2\pi)^3 + \cdots \qquad (p=(\omega_{\mathbf{p}},\mathbf{p})),$$

and we have

$$\overbrace{\phi(x_n)a_k^\dagger(\mathbf{p}_k^{\mathrm{in}})} = e^{ip^{\mathrm{in}}_\mu x^\mu}u(\mathbf{p}_k^{\mathrm{in}}).$$

- Finally, $\overbrace{\phi(x_n)\phi^\dagger(x_{n'})} = -i\Delta(x_n - x_{n'})$ where Δ is the propagator for the field ϕ.

So now it remains to multiply these together and integrate, but we postpone this task for a bit. First, we present the marvelous graphical interpretation of these integrals due to Richard Feynman, which makes it much easier to see what is going on and keep track of all the different terms.

Each product of *nonzero* contractions of the above types corresponds uniquely to a graph with some additional structure called a *Feynman diagram*, according to the following rules.

- For each initial creation operator $a_j^\dagger(\mathbf{p}_k^{\mathrm{in}})$, each final annihilation operator $a^j(\mathbf{p}_j^{\mathrm{out}})$, and each point x_n where field operators act there is a vertex. The vertices corresponding to the initial (resp. final) operators are called *initial* (resp. *final*); initial and final vertices are both called *external*; and the vertices corresponding to the field operators are called *internal.* The external vertices are labeled with the on-mass-shell 4-momenta $p = (\omega_{\mathbf{p}}, \mathbf{p})$ corresponding to their 3-momenta $\mathbf{p}_j^{\mathrm{out}}$ and $\mathbf{p}_k^{\mathrm{in}}$, and the internal vertices are labeled with their points x_n; other parameters such as spin states and Lorentz indices may be included too. When drawing a Feynman diagram, it is customary to put the vertices for creation operators at the left or bottom of the picture, the vertices for annihilation operators at the right or top of the picture, and the internal vertices in between. In this book we almost always use the left-right convention.
- For each contraction there is an edge of the graph between the appropriate vertices: between the external vertices with labels p_j^{out} and p_k^{in} for $\overbrace{a_j(\mathbf{p}_j^{\mathrm{out}})a_k^\dagger(\mathbf{p}_k^{\mathrm{in}})}$, between the internal vertices with labels x_n and $x_{n'}$ for $\overbrace{\phi(x_n)\phi^\dagger(x_{n'})}$, etc. Edges with one end at an external vertex are called *external lines*; edges with both ends at internal vertices are called *internal lines.*
- Each line is associated to a particular particle species, namely, the one to which the operators in its contraction pertain. It should be labeled in some way to indicate this species. In practice this is done by using solid lines, dotted lines, wavy lines, etc., for the various species. (In particular, in QED electrons are denoted by solid lines and photons by wavy lines.) However, *for this purpose, particles and antiparticles are considered as belonging to the same species*; the distinction between then comes in the next item.
- At this point, for each particle-antiparticle pair that occurs in the calculation, one must decide once and for all which one is the "particle" and which one is the "antiparticle." (In practice this is a matter of iron-clad convention. In particular, electrons and protons are "particles," and the determination for other leptons and baryons is determined by laws to the effect that in any interaction the number of leptons [or baryons] minus the number of antileptons [or antibaryons] is conserved.) The lines of the

graph associated to particles with distinct antipaticles are equipped with arrows as follows. (No arrows are drawn on lines corresponding to a particle without a distinct antiparticle, such as a photon; however, arrows may be drawn adjacent to them to indicate the direction of momentum flow, as we shall explain later.)

i. The arrow of a line with one end at an initial vertex p^{in} points away from that vertex if the entity created there is a particle and toward that vertex if it is an antiparticle.
ii. The arrow of a line with one end at a final vertex p^{out} points toward from that vertex if the entity annihilated there is a particle and away from that vertex if it is an antiparticle. (Note that this is consistent with (1) for lines that join initial and final vertices.)
iii. The arrow of an internal line joining the vertices x and x' points from x' to x if the field ϕ in the associated contraction $\overbrace{\phi(x)\phi^\dagger(x')}$ annihilates particles and creates antiparticles, and from x to x' if it creates particles and annihilates antiparticles.

We observe that the following conditions always hold:

- Every external vertex is contained in exactly one line, and the other end of that line is either an internal vertex or an external vertex of the opposite sort (initial or final).
- Recall that the integral giving rise to a Feynman diagram is obtained from (6.58) by replacing each $\mathcal{H}_I(x_n)$ by one of its constituent products of field operators. The number of lines that meet the vertex x_n is the number of terms in that product. More precisely, if the contribution of a particular Hermitian field ϕ to the product is ϕ^m, there will be m lines of type ϕ at x_n; if the contribution of non-Hermitian field ϕ is $(\phi\phi^\dagger)^m$, there will be m lines of type ϕ with arrows pointing toward the vertex x_n and m with arrows pointing away.

Part of what makes this graphical representation useful is that one can go the other way, from Feynman diagrams to terms in the Dyson series. To wit, suppose one is given an interaction Hamiltonian $\mathcal{H}_I$ of the sort described above and some initial and final states of free particles with definite momenta. One builds a Feynman diagram by starting with K initial vertices labeled with momenta p_k^{in}, J final vertices labeled with momenta p_j^{out}, and N internal vertices labeled with spacetime points x_n, and choosing one product in $\mathcal{H}_I(x_n)$ for each n to be the interaction at the vertex x_n. One then connects the vertices subject to the conditions detailed above. Each way of doing this yields a product of contractions whose values can be read off from the list compiled above; the integral of this product, multiplied by $(-i)^N \prod_1^N \lambda_n$, where λ_n is the coupling constant for the vertex x_n, yields one of the terms whose sum is (6.58).

This is a very neat way to catalogue all the integrals that need to be calculated! However, there are some combinatorial issues that must be addressed. First, we do not distinguish between two Feynman diagrams that differ only in the relabeling of their internal vertices. With that understanding, the factor of $1/N!$ in (6.58) disappears. (If one compares the present discussion with the derivation of the Dyson series in §6.1, one sees that this $1/N!$ is there precisely to cancel the $N!$ equal contributions of these permuted diagrams.) Second, if the interaction involves powers higher than one of a single field, there are additional counting problems. For example, in the ϕ^4 scalar field theory, each of the four factors of $\phi(x)$ in an interaction term $\mathcal{H}_I(x)$ must be contracted with some other operator, and there are generally 4! ways to arrange these contractions; it is precisely to compensate for this fact that the factor of 1/4! is inserted into (6.28). However, if two or three of these factors are contracted with $\phi(y)$'s that are all in the same $\mathcal{H}_I(y)$, this compensation is off by a factor of 2! or 3!, which must therefore be adjusted accordingly. (The corresponding diagrams have two or three lines connecting the vertices x and y, and the point is that the diagram is unchanged by the permutation of these lines.) These *symmetry factors*, which can also arise in other ways, can be annoying to deal with, but they will cause no difficulty in the calculations we shall perform here. (See Weinberg [**129**], §6.1, for a more extended discussion of these issues.)

One also has to be careful about keeping track of minus signs when the interactions involve Fermions. For one thing, changing the order of the creation operators that define the initial and final states can introduce factors of -1, so one has to order the operators in the initial and final states consistently. Moreover, when one expresses an S-matrix element as a product of contractions $\overbrace{AB}$, one must permute the operators involved so that each A is adjacent to its partner B, and permuting Fermionic operators introduces a factor of the sign of the permutation. In the next section we shall see an example of the importance of these minus signs.

At this point the reader is undoubtedly hungry for some specific examples, so let us look at the lowest-order terms for an S-matrix element in ϕ^4 scalar field theory in which the initial and final states have two particles each: $|\text{in}\rangle = a^\dagger(\mathbf{p}_1)a^\dagger(\mathbf{p}_2)|0\rangle$ and $|\text{out}\rangle = a^\dagger(\mathbf{p}_3)a^\dagger(\mathbf{p}_4)|0\rangle$. The zeroth term in the perturbation series is

$$\begin{aligned}
&\langle 0|a(\mathbf{p}_3)a(\mathbf{p}_4)a^\dagger(\mathbf{p}_1)a^\dagger(\mathbf{p}_2)|0\rangle \\
&\qquad = \overbrace{a(\mathbf{p}_3)a^\dagger(\mathbf{p}_1)}\,\overbrace{a(\mathbf{p}_4)a^\dagger(\mathbf{p}_2)} + \overbrace{a(\mathbf{p}_3)a^\dagger(\mathbf{p}_2)}\,\overbrace{a(\mathbf{p}_4)a^\dagger(\mathbf{p}_1)} \\
&\qquad = (2\pi)^6\big[\delta(\mathbf{p}_3-\mathbf{p}_1)\delta(\mathbf{p}_4-\mathbf{p}_2) + \delta(\mathbf{p}_3-\mathbf{p}_2)\delta(\mathbf{p}_4-\mathbf{p}_1)\big],
\end{aligned}$$

which corresponds to the two Feynman diagrams in Figure 6.2. No surprises here: nothing happens, and the two outgoing particles are the two incoming particles.

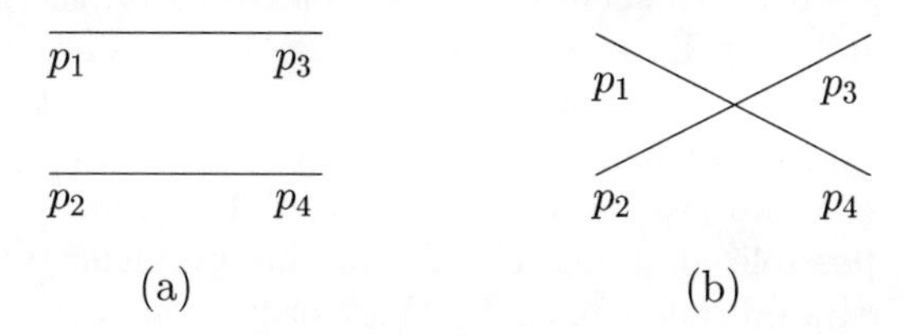

FIGURE 6.2. The trivial interaction in ϕ^4 theory.

Note: We shall always indicate internal vertices of Feynman diagrams by black dots. If two lines cross but there is no black dot at the intersection, there is no vertex there.

The first-order term is

$$\frac{\lambda}{4!}\int \langle 0|a(\mathbf{p}_3)a(\mathbf{p}_4)\phi(x)\phi(x)\phi(x)\phi(x)a^\dagger(\mathbf{p}_1)a^\dagger(\mathbf{p}_2)|0\rangle\, d^4x.$$

The various ways of contracting the eight terms in the integrand yield diagrams of the types shown in Figure 6.3. The first one, representing the most basic scattering process of the ϕ^4 interaction, comes from contracting each of the $\phi(x)$'s with one of the external creation or annihilation operators. There are 4! ways to do this, which all give the same result; this cancels the 4! in the coupling constant. By (5.12) we have

$$\overbrace{a(\mathbf{p}_j)\phi(x)} = \int \frac{e^{ip_\mu x^\mu}}{\sqrt{2\omega_{\mathbf{p}}}}\delta(\mathbf{p}-\mathbf{p}_j)\,d^3\mathbf{p} = \frac{e^{ip_{j\mu}x^\mu}}{\sqrt{2\omega_{\mathbf{p}_j}}}$$

and likewise $\overbrace{\phi(x)a^\dagger(\mathbf{p}_j)} = e^{-ip_{j\mu}x^\mu}/\sqrt{2\omega_{\mathbf{p}_j}}$, so the value of Figure 6.3a is

$$\lambda\int\frac{e^{i(p_{3\mu}+p_{4\mu}-p_{1\mu}-p_{2\mu})x^\mu}}{4\sqrt{\omega_{\mathbf{p}_1}\omega_{\mathbf{p}_2}\omega_{\mathbf{p}_3}\omega_{\mathbf{p}_4}}}\,d^4x = \lambda\frac{(2\pi)^4}{4\sqrt{\omega_{\mathbf{p}_1}\omega_{\mathbf{p}_2}\omega_{\mathbf{p}_3}\omega_{\mathbf{p}_4}}}\delta(p_3+p_4-p_1-p_2).$$

The delta-function embodies the overall conservation of energy and momentum in the process.

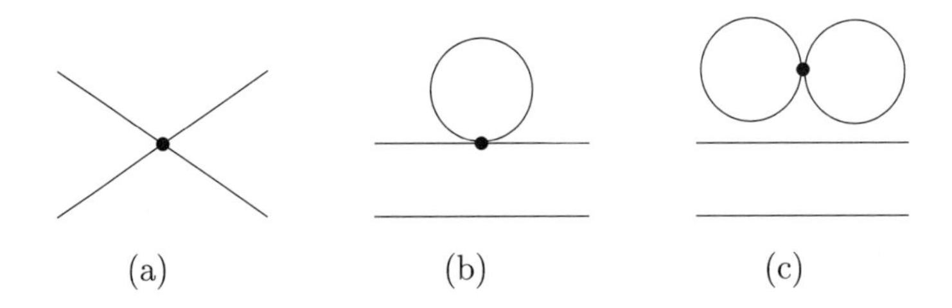

FIGURE 6.3. First-order interactions in ϕ^4 theory.

Figure 6.3b corresponds to contracting two of the field operators with external creation or annihilation operators and the other two with each other. There are actually four diagrams of this type, depending on which initial vertices are connected to which final ones and which line gets the extra loop. Each of them has a symmetry factor of 2, because there is only one way to pair two factors of $\phi(x)$ rather than two. Figure 6.3c corresponds to contracting all of the field operators with each other; there are two such diagrams, depending on the pairing of initial and final vertices, and they have a symmetry factor of 8 since there are only 3 ways of grouping the four ϕ's into two pairs. But there is something highly peculiar about these diagrams, because the contraction that corresponds to the loop is $-i\Delta_F(x-x) = \Delta_F(0)$, and Δ_F has a singularity at the origin.

There are two possible attitudes to take to this problem. One is to declare that the interaction Hamiltonian must be Wick ordered at the outset — that is, to take $\mathcal{H}_I(x)$ to be $(\lambda/4!){:}\phi(x)^4{:}$ rather than $(\lambda/4!)\phi(x)^4$. This has the effect of removing all contractions of operators within the same $\mathcal{H}_I(x)$ from consideration; diagrammatically, it amounts to adding a rule that no line can begin and end at the same vertex. This prescription is always followed in the study of rigorous models in space-time dimensions 2 and 3, and there is a theoretical justification for it noted by Weinberg [**129**], p. 200. On the other hand, the Fourier representation (6.47) of Δ_F gives a formula for $\Delta_F(0)$ as a divergent integral,

$$\Delta_F(0) = \int\frac{1}{-p^2+m^2}\frac{d^4p}{(2\pi)^4}.$$

As such it is only one of many divergent integrals that arise from Feynman diagrams containing loops, and one can include it in the general renormalization scheme to be discussed in Chapter 7; it then affects the counterterms that must be added to cancel the divergences. In the end, these procedures lead to physically equivalent results as long as the field ϕ is considered in isolation, although one-line loops can have a nontrivial effect when there are external fields present. (See, e.g., Weinberg [**129**], p. 576.) We shall take the easy way out by henceforth forbidding lines to start and end at the same vertex.

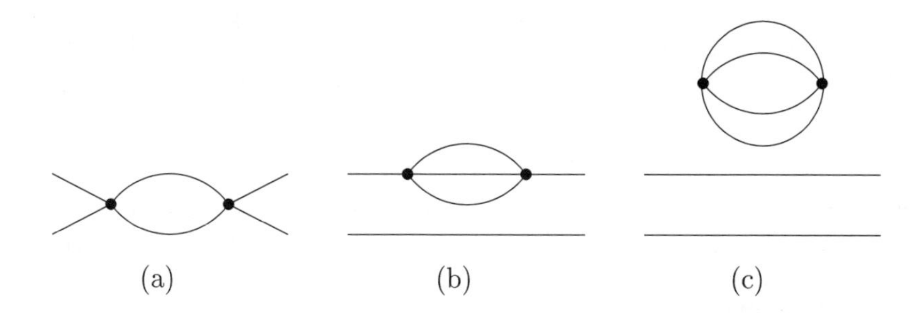

FIGURE 6.4. Second-order interactions in ϕ^4 theory

With this simplification, the reader can easily verify that the second-order term

$$\iint \langle 0|a(\mathbf{p}_3)a(\mathbf{p}_4)\mathcal{T}[{:}\phi(x)^4{:}\,{:}\phi(y)^4{:}]a^\dagger(\mathbf{p}_1)a^\dagger(\mathbf{p}_2)|0\rangle\, d^4x\, d^4y$$

gives rise to the three kinds of diagrams in Figure 6.4. They have symmetry factors 2!, 3!, and 4!, respectively, from the permutation of the internal lines. We shall analyze the first of these in detail in §7.4 and the second in a more qualitative way in §7.6. The last one contains a "vacuum bubble," one of the quantum fluctuations of the vacuum; we shall say more about such diagrams in §6.11.

For QED the basic ideas are the same, but the algebra is more complicated. There are two fields here, the photon field A and the electron field ψ (although one can substitute another charged spin-$\frac{1}{2}$ particle for the electron); the interaction Hamiltonian is $\mathcal{H}_I = eA_\mu\overline{\psi}\gamma^\mu\psi$ where e is the charge of the electron. These fields are both vector-valued, and the vector spaces for both of them are 4-dimensional, but they are not the same: for the photon field the space is physical space-time or its dual momentum space, whereas for the electron field it is the space of Dirac spinors. Likewise, the photon propagator and the electron propagator are both 4×4 matrices, but acting on these two different spaces. The standard way to avoid massive confusion is to display *all* Lorentz indices but *no* spinor indices explicitly. Here is a list of the main features to keep in mind:

i. Each internal vertex, labeled by a space-time point x, has its own Lorentz index μ, which is the Lorentz index of the Dirac matrix γ^μ in $\mathcal{H}_I(x)$; it forms part of the label of the vertex. If two internal vertices lableled by (x,μ) and (y,ν) are joined by a photon line, the corresponding photon propagator is $-i\Delta_{\mu\nu}(x-y)$ (which is symmetric in μ and ν and even in $x-y$, so the order doesn't matter).
ii. The contraction $\overbrace{A_\mu(x)a^\dagger(\mathbf{p})}$ or $\overbrace{a(\mathbf{p})A_\mu(x)}$ corresponding to an external photon line has the form $\boldsymbol{\epsilon}_\mu(\mathbf{p})/\sqrt{2|\mathbf{p}|}$ or $\boldsymbol{\epsilon}^*_\mu(\mathbf{p})/\sqrt{2|\mathbf{p}|}$ where $\boldsymbol{\epsilon}_\mu$ is the polarization

vector for the incoming or outgoing photon.[6] This will be contracted into the Dirac matrix γ^μ for the internal vertex x.

iii. If an electron line joins the vertices labeled by x and y and its arrow points from x to y,, the corresponding propagator is $-i\Delta_{\mathrm{Dirac}}(y-x)$. (Here the order does matter!)

iv. The contraction $\overbrace{\psi(x)a^\dagger(\mathbf{p})}$ or $\overbrace{a(\mathbf{p})\overline{\psi}(x)}$ has the form $\sqrt{m/2\omega_{\mathbf{p}}}\, u(\mathbf{p})$ (a column vector) or $\sqrt{m/2\omega_{\mathbf{p}}}\overline{u}(\mathbf{p})$ (a row vector), where $u(\mathbf{p})$ is the appropriate Dirac spinor defined by (5.29) and (5.30).

v. The electron lines in any diagram fall into two kinds of groups: chains of lines connecting an incoming electron vertex to an outgoing one, and internal loops. The total contribution of a chain connecting external vertices through n internal vertices with labels (x_j, μ_j) has the form

$$(-ie)^n \overline{u}(\mathbf{p}^{\mathrm{out}})\gamma^{\mu_n}\Delta_{\mathrm{Dirac}}(x_n - x_{n-1})\gamma^{\mu_{n-1}}\cdots\gamma^{\mu_2}\Delta_{\mathrm{Dirac}}(x_2 - x_1)\gamma^{\mu_1}u(\mathbf{p}^{\mathrm{in}}).$$

Note that this is a product of a row vector, some matrices, and a column vector, and hence is a scalar as far as spinor space is concerned. The total contribution of a loop with vertices labeled $(x_1, \mu_1), \ldots, (x_n, \mu_n)$ (and $(x_{n+1}, \mu_{n+1}) = (x_1, \mu_1)$) has the form

$$-(-ie)^n \operatorname{tr}\left[\gamma^{\mu_n}\Delta_{\mathrm{Dirac}}(x_n - x_{n-1})\gamma^{\mu_{n-1}}\cdots\gamma^{\mu_2}\Delta_{\mathrm{Dirac}}(x_2 - x_1)\gamma^{\mu_1}\Delta_{\mathrm{Dirac}}(x_1 - x_n)\right], \tag{6.59}$$

which is also a scalar vis-a-vis spinor space. The extra minus sign comes from the fact that all the propagators from contractions $\overbrace{\psi(x_{j+1})\overline{\psi}(x_j)}$ except the last one, which comes from $\overbrace{\overline{\psi}(x_n)\psi(x_1)} = -\overbrace{\psi(x_1)\overline{\psi}(x_n)}$. The fact that the formula involves a trace comes from the workings of the linear algebra; the point is that the product of two matrices A and B is given by $(AB)_{ik} = \sum_j A_{ij}B_{jk}$, but when one completes the loop at the end, one sets $k = i$ and sums to obtain $\sum_{i,j} A_{ij}B_{ji} = \operatorname{tr}(AB)$.

vi. The Lorentz indices in the expressions in (5) are all summed out against photon propagators or external polarization vectors. Hence the final result is a *scalar.*

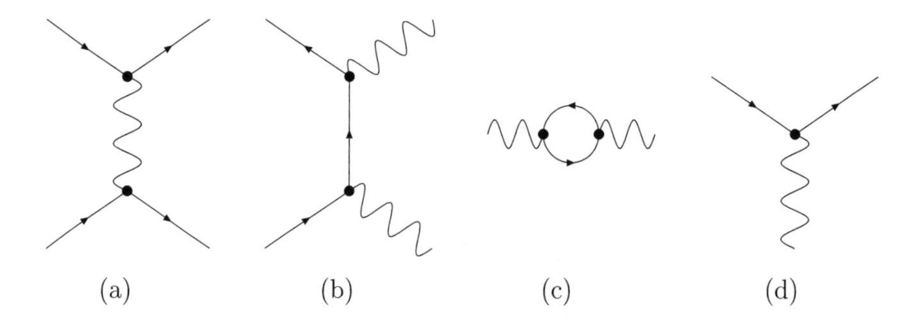

FIGURE 6.5. Some Feynman diagrams for QED.

[6]This is exactly what one would expect, and exactly what one gets from the quantization of the photon field in Coulomb gauge (5.42). Even if one quantizes in another way, the external photons must be in physically correct states with polarization vectors that have no time component and are orthogonal to the momentum. See Weinberg [**129**], §5.9, for further discussion.

The algebra is sufficiently tedious that we shall not write out a list of examples here, but rather consider individual examples as the occasion arises. However, a few simple Feynman diagrams for QED are shown in Figure 6.5. The first one is the main contribution to electron-electron scattering; the second is the main contribution to electron-positron annihilation; the third represents a photon turning into an electron-positron pair and back into a photon. The last one represents *the* fundamental interaction of QED, but unlike the analogous diagram (Figure 6.3a) for the ϕ^4 interaction, it cannot stand on its own. One doesn't need quantum field theory, merely conservation of energy, to see that the amplitude for the depicted process, a (real) electron emitting or absorbing a (real) photon with nothing else happening, is zero. (Hint: Consider the reference frame in which the electron is initially at rest.)

By the way, if one rotates the diagram in Figure 6.5b through a right angle clockwise, it depicts the scattering of a photon off an electron, a process that we shall study in detail in §6.10. It follows that the corresponding amplitudes for electron-positron annihilation and electron-photon scattering are closely related. As the reader may verify, the precise relation is as follows. The values of these diagrams are functions of the external particle momenta; the value of the original diagram with the incoming positron assigned (4-)momentum p and the bottom photon assigned momentum k is equal to the value of the rotated diagram with the outgoing electron assigned momentum $-p$ and the incoming photon assigned momentum $-k$. Note that p and $-p$, or k and $-k$, cannot both be physically allowed, as $p_0 > 0$ for physical particles, so the relation is a mathematical one rather than a directly physical one, but it is useful nonetheless. An analogous result holds for any pair of diagrams that are obtained from one another by replacing some incoming particles by outgoing antiparticles or vice versa, a phenomenon known as *crossing symmetry*.

The correspondence between the integrals that make up the Dyson series and Feynman diagrams is perfectly precise and well-defined. However, it is customary to go further and think of the Feynman diagrams as schematic pictures of physical processes, and here the interpretation acquires a more imaginative character. The lines starting at the initial or final vertices can clearly be thought of as the trajectories of the initial and final particles entering or leaving the interaction. Once one has made this connection, the temptation is irresistible to interpret the lines joining internal vertices also as trajectories of particles involved in intermediate steps of the interaction. Some of them form chains pertaining to the same particle species that connect an incoming line to an outgoing one by following the arrows; they can be interpreted as the trajectories of the real particles that enter, interact, and leave. The particles that are created and annihilated within the diagram, however, have a much more tenuous claim to reality. They are not observed directly; they sometimes travel faster than the speed of light; they may violate conservation of energy quite flagrantly. They are, in short, the infamous *virtual particles* that are so ubiquitous in physicists' discourse. In the final analysis, the only existence they possess for certain is as picturesque ways of thinking about the ingredients of the integrals in the Dyson series.

However, it must be said that the line between virtual particles and real ones is not completely sharp. A real particle may be observed to be a little off mass shell because of the uncertainty principle, and a photon emitted by an electron in the sun that manages to reach the earth before being absorbed by another electron has every right to claim reality even though the photon in Figure 6.5a is presumably virtual.

The arrows in a Feynman diagram indicate the direction of travel of the (virtual) particles in question: they point in the direction of travel for particles and in the opposite direction for antiparticles. The convention for the direction of the arrow on an edge joining two internal vertices is explained as follows. If ϕ is a field that destroys particles and creates antiparticles, the contraction $\overbrace{\phi(x)\phi^\dagger(y)}$ involves either the commutator of the annihilation operators in $\phi(x)$ with the creation operators in $\phi^\dagger(y)$ or the commutator of the creation operators in $\phi(x)$ with the annihilation operators in $\phi^\dagger(y)$, depending on the time-ordering of x and y. That is, if $x_0 > y_0$, $\overbrace{\phi(x)\phi^\dagger(y)}$ represents the creation of a (virtual) particle at y followed by its destruction at x, whereas if $y_0 > x_0$, it represents the creation of a (virtual) antiparticle at x followed by its destruction at y. Both processes are represented by the same directed line in the Feynman diagram. In fact, Feynman argued for their physical indistinguishability; he was fond of asserting that antiparticles are just particles traveling backward in time.

6.7. Feynman diagrams in momentum space

What we have described so far is the "position space" correspondence between Feynman diagrams and multiple integrals that contribute to an S-matrix element; that is, the vertices in the diagram are labeled by position-space variables, and these are the variables of integration. However, when one actually performs the calculation, it is easier to use a "momentum space" representation instead, because the most convenient way of expressing the propagators is by their Fourier expansions. The way this works is as follows.

Each integral represented by a Feynman diagram is of the form

$$\int \cdots \int F_1(x, \mathbf{p}^{\text{out}}) F_2(x) F_3(x, \mathbf{p}^{\text{in}})\, d^4x_1 \cdots d^4x_N, \tag{6.60}$$

where $x = (x_1, \ldots, x_n) \in \mathbb{R}^{4n}$ and $\mathbf{p}^{\text{out}}$ and $\mathbf{p}^{\text{in}}$ are likewise concatenations of the outgoing and incoming momentum vectors. More precisely,

i. $F_1(x, \mathbf{p}^{\text{out}})$ is a product of factors $\exp(ip^{\text{out}}_{j\mu} x^\mu_n)$, multiplied by coefficients that depend on p^{out}_j and the field types but not on x_n, coming from the final lines (i.e., the contractions $\overbrace{a(\mathbf{p}^{\text{out}}_j)\phi(x_n)}$);

ii. $F_2(x)$ is a product of propagators coming from the internal lines (i.e., the contractions $\overbrace{\phi(x_n)\phi^\dagger(x_{n'})}$);

iii. $F_3(x, \mathbf{p}^{\text{in}})$ is a product of factors $\exp(-ip^{\text{in}}_{k\mu} x^\mu_n)$, again multiplied by certain coefficients, coming from the initial lines (i.e., the contractions $\overbrace{\phi(x_n)a^\dagger(\mathbf{p}^{\text{in}}_k)}$).

(There may also be some combinatorial factors.) We express each of the propagators in $F_2(x)$ as a Fourier integral:

$$-i\Delta(x_n - x_{n'}) = -i \int \exp[-iq_\mu (x_n - x_{n'})^\mu] \widehat{\Delta}(q) \frac{d^4q}{(2\pi)^4}, \tag{6.61}$$

in which $\widehat{\Delta}(q)$ is a rational function of q with denominator $-q^2+m^2-i\epsilon$ (and there is an implicit $\lim_{\epsilon\to 0}$). Substituting these integrals for the propagators turns (6.60) into

(6.62)

$$\int\cdots\iint\cdots\int F_1(x,\mathbf{p}^{\text{out}})F_2'(x,q)F_2''(q)F_3(x,\mathbf{p}^{\text{in}})\,d^4x_1\cdots d^4x_N\,\frac{d^4q_1\cdots d^4q_M}{(2\pi)^{4M}},$$

where there is one 4-momentum variable q_m for each internal line, and

i. $F_2'(x,q)$ is the product of factors $\exp[-iq_{m\mu}(x_n-x_{n'})^\mu]$;

ii. $F_2''(q)$ is the product of the Fourier transforms of the propagators.

At this point the x-dependence in the integrand is simply a product of imaginary exponentials, so the x-integrations can be performed by the Fourier inversion formula

$$\int e^{-ix_\mu a^\mu}\,d^4x=(2\pi)^4\delta(a).$$

The result is that the integral (6.62) becomes

(6.63)

$$G(\mathbf{p}^{\text{out}},\mathbf{p}^{\text{in}})\int\cdots\int(2\pi)^{4N}\delta(\Sigma_1)\cdots\delta(\Sigma_N)R_1(q_1)\cdots R_m(q_M)\,\frac{d^4q_1\cdots d^4q_M}{(2\pi)^{4M}},$$

where

i. $G(\mathbf{p}^{\text{out}},\mathbf{p}^{\text{in}})$ is the product of field coefficients coming from the external lines (and any combinatorial factors),

ii. the R_m are the Fourier-transformed propagators ("R" stands for "rational function"), and

iii. Σ_n is the algebraic sum (see below) of the 4-momenta (p_j^{out}, p_k^{in}, and/or q_m) associated to the lines that meet at the vertex x_n.

In addition, when any initial vertex $\mathbf{p}_k^{\text{in}}$ is connected directly to a final vertex $\mathbf{p}_j^{\text{out}}$, there will be an extra factor of $\delta(\mathbf{p}_k^{\text{in}}-\mathbf{p}_j^{\text{out}})$.

The term "algebraic sum" requires some explanation. Some lines that meet at the vertex x_n have arrows attached to them; a line is considered "incoming" if its arrow points toward x_n and "outgoing" if it points away from x_n.[7] Lines without arrows (corresponding to neutral particles) are arbitrarily assigned arrows for this purpose; one merely has to be consistent in using the same arrow for the calculations at both ends of the line. (These arrows are usually drawn adjacent and parallel to the lines rather than on them.) The *algebraic sum* Σ_n *of the momenta at* x_n is then the sum of the outgoing momenta *minus* the sum of the incoming momenta. Thus, the delta-function $\delta(\Sigma_n)$ expresses "conservation of energy-momentum at the vertex x_n."

When one thinks of a Feynman diagram as representing the momentum-space integral (6.63), instead of labeling the internal vertices with the position variables, one labels the lines with 4-momentum variables. The conversion from position to momentum labels is as follows. The line from the initial vertex with momentum $\mathbf{p}_k^{\text{in}}$ is labeled by the corresponding 4-momentum $p_k^{\text{in}}=(\omega_{\mathbf{p}_k^{\text{in}}},\mathbf{p}_k^{\text{in}})$, and likewise for lines to final vertices. The internal line from the vertex $x_{n'}$ to the vertex x_n (representing the propagator $-i\Delta(x_n-x_{n'})$) is labeled by the momentum variable q in the Fourier expansion (6.61). We think of q as flowing in the direction of the arrow associated to

[7]One can reverse this convention for lines that are considered to represent antiparticles; it doesn't matter as long as the convention is consistent from vertex to vertex.

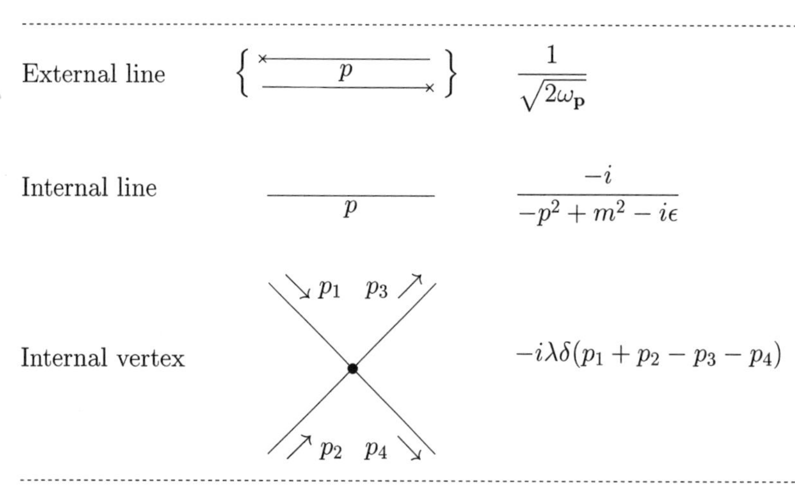

TABLE 6.1. Momentum-space Feynman rules for ϕ^4 scalar field theory. (External vertices are denoted by crosses.)

the line (that is, "from $x_{n'}$ to x_n") if the line is considered as representing a particle (including neutral particles) and in the opposite direction if the line is considered as representing an antiparticle, as in the preceding paragraph. This distinction is unimportant for internal Boson lines, as their propagators are even functions and $-q_m$ is just as good a variable as q_m in (6.63); but it is significant for internal Fermion lines (whose propagators are not even) and external lines (whose momenta p must be on mass shell so that $p^0 > 0$).

The integral (6.63) corresponding to a diagram labeled in this way is built up as follows:

i. For each external line labeled by the momentum p there is a factor $u(p)$ consisting of the appropriate field coefficient arising from the contraction associated to the line.
ii. For the internal line labeled with q_m $(m = 1, \ldots, M)$ there is a factor $-i\widehat{\Delta}(q_m)$, the Fourier-transformed propagator for the field associated to the line.
iii. For each vertex there is a factor $-i(2\pi)^4\lambda\delta(\Sigma)$, where Σ is the algebraic sum of the 4-momenta at that vertex and λ incorporates the coupling constant and any other algebraic factors such as Dirac matrices for the interaction at that vertex.

One now multiplies all these factors together, integrates over all the internal momenta, and also sums over whatever discrete indices (spin, etc.) are present. One must also include combinatorial factors and Fermionic minus signs as needed, just as for the position-space integral.

In this setting the "virtualness" of the particles associated to the internal lines manifests itself in the fact that their 4-momenta q_m are completely arbitrary, not on mass shell.

For future reference, we give a summary of the momentum-space Feynman rules for the ϕ^4 scalar field theory and for QED in Tables 6.1 and 6.2.

We return to the general discussion of the evaluation of integrals associated to Feynman diagrams. Because of the delta-functions in the integrand, some, and

Incoming electron line	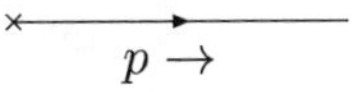 ×——▸—— $p \to$	$\sqrt{m/2\omega_{\mathbf{p}}}\, u(\mathbf{p}, s)$
Incoming positron line	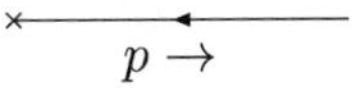 ×——◂—— $p \to$	$\sqrt{m/2\omega_{\mathbf{p}}}\, v(\mathbf{p}, s)$
Incoming photon line	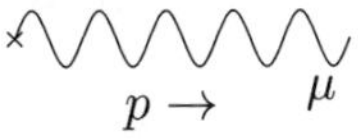 ×∿∿∿ μ, $p \to$	$\epsilon_\mu(\mathbf{p})$
Outgoing electron line	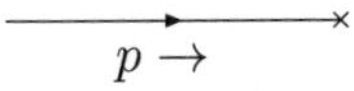 ——▸——× $p \to$	$\sqrt{m/2\omega_{\mathbf{p}}}\, \overline{u}(\mathbf{p}, s)$
Outgoing positron line	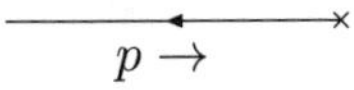 ——◂——× $p \to$	$\sqrt{m/2\omega_{\mathbf{p}}}\, \overline{v}(\mathbf{p}, s)$
Outgoing photon line	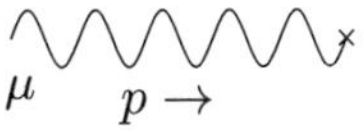 μ ∿∿∿× $p \to$	$\epsilon^*_\mu(\mathbf{p})$
Internal electron or positron line	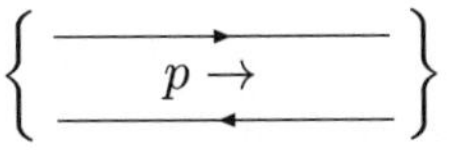 {——▸—— / ——◂——} $p \to$	$\dfrac{-i(p_\mu \gamma^\mu + m)}{-p^2 + m^2 - i\epsilon}$
Internal photon line	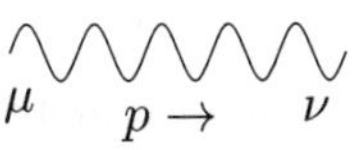 μ ∿∿∿ ν, $p \to$	$\dfrac{-ig_{\mu\nu}}{p^2 + i\epsilon}$
Internal vertex	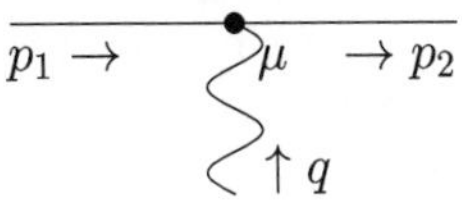 $p_1 \to$ •μ $\to p_2$, $\uparrow q$	$-ie\gamma^\mu \delta(p_1 + q - p_2)$

TABLE 6.2. Momentum-space Feynman rules for QED. (External vertices are denoted by crosses. The Lorentz indices at internal ends of photon lines contract with the Lorentz indices at internal vertices.)

perhaps all, of the integrations in (6.63) collapse immediately. Each internal momentum variable q_m occurs in precisely two of the sums Σ_n (the ones coming from the vertices at the ends of the line associated to q), so repeated application of the formula

$$\int (2\pi)^4 \delta(q - q')(2\pi)^4 \delta(q' - q'')\, \frac{d^4 q'}{(2\pi)^4} = (2\pi)^4 \delta(q - q'')$$

(this is honest mathematics: the convolution of two measures) serves to eliminate some or all of the internal momenta, which are then expressed in terms of the external momenta and remaining internal momenta (if any) in the remaining part of the integrand.

The form of the result of performing these trivial integrations depends on the topology of the Feynman diagram. First, if the diagram is disconnected, the integral it represents is just the product of the integrals represented by its connected subdiagrams, so it suffices to consider connected diagrams. For these we quote a basic topological result for one-dimensional cell complexes, the analogue of the Euler formula $V - E + F = 2 - 2g$ for polyhedra. (A proof can be found in any introductory book on algebraic topology; see, e.g., Croom [**19**], Theorem 2.5.)

Let V and E be the number of vertices and edges, respectively, in a connected graph G, and let L be the number of independent loops in G (i.e., the rank of the homology group $H_1(G)$).[8] *Then $V - E = 1 - L$.*

In a Feynman diagram, each external vertex lies on only one line, so deletion of external vertices and lines does not affect the connectedness or the numbers $V - E$ and L. By applying the preceding theorem to the subgraph consisting of the internal vertices and lines, we obtain:

Let V and I be the number of internal vertices and internal lines, respectively, in a connected Feynman diagram, and let L be the number of independent loops. Then

$$V - I = 1 - L. \tag{6.64}$$

In particular, we always have $I \geq V - 1$, so there are always enough q-integrals to eat up all but one of the delta-functions $\delta(\Sigma_n)$. As the reader may verify by keeping more careful track of the different internal and external momenta in the sums Σ_n, once these $V - 1$ integrations have been performed, the argument of the remaining delta-function is precisely $\sum_j p_j^{\text{out}} - \sum_k p_k^{\text{in}}$, the sum of the final 4-momenta minus the sum of the initial 4-momenta. Hence this delta-function simply expresses the fact that overall momentum and energy, for the real (nonvirtual) initial and final particles, must be conserved by the interaction. It contains no internal momenta and so plays no role in any further integrations.

If $L = 0$, that is, if the diagram is a *tree*, then $I = V - 1$, there are no more integrations after the delta-functions have been reduced, and the calculation is complete. The result is a product of (1) field coefficients evaluated at the external momenta, (2) propagators whose internal momentum variable has been expressed in terms of the external momenta via the conditions $\Sigma_n = 0$; (3) the overall momentum conservation delta-function $(2\pi)^4\delta(\sum p_j^{\text{out}} - \sum p_k^{\text{in}})$.

If $L > 0$, there are L additional integrations to be performed, and the integrand is a rational function of the remaining internal momenta (a product of propagators). These integrals have an unfortunate propensity for diverging, so their analysis must be postponed until the next chapter.

There are a couple of modifications to the rules for evaluating external lines in a Feynman diagram that are frequently useful. First, one may wish to consider a part of a Feynman diagram in isolation, so that its external lines are actually internal lines of a larger diagram. In that case the external particles are not on mass shell and the external lines correspond to propagators rather than field coefficients. (The same idea is relevant to evaluation of vacuum expectation values of products of fields, as we shall explain in §6.11.) On the other hand, sometimes one wants to concentrate on the integrations associated to the internal lines without worrying about the quantities associated to one or more external lines at all, so one omits the

[8] For example, the diagram in Figure 6.4b has three loops, one formed by the two curved arcs and the other two formed by one of the arcs and the line segment in between, of which any two are independent; the third is the sum or the difference of the other two in $H_1(G)$.

corresponding field coefficients or propagators entirely. This is known as *amputating* the external line.

The integral associated to any Feynman diagram will be called the *value* of the Feynman diagram. However, it is often convenient to take a short-cut and identify a diagram with its value. Thus, we may speak of a "sum of Feynman diagrams," a "divergent Feynman diagram," etc.

6.8. Cross sections and decay rates

At this point, let us see what we have accomplished. We have shown how to compute the scattering matrix elements between initial and final states consisting of collections of particles with definite momenta in terms of quantities represented by Feynman diagrams; more precisely, our expectation is that the sum of finitely many of these quantities will yield a good approximation to the matrix element. Some of the quantities are divergent, but there are methods to be discussed in the next chapter for removing the divergence. Let us assume either that this has been done or that we are content to use only diagrams without loops where the divergence problem does not occur (the "tree level approximation"), so that we have obtained a formula for the scattering matrix element. Now what? That is, *what does this matrix element have to do with quantities that the experimentalists can measure?*

The question of interpretation is clearly nontrivial. A scattering matrix element is a transition *amplitude* between the initial and final states, and the actual transition *probability* is obtained by taking the absolute square of this amplitude. But the matrix element always contains a momentum-conservation delta-function, so how is one supposed to interpret its square? Of course, some such problem is not unexpected, since the initial and final states with definite momenta are not actual elements of the appropriate Hilbert space. To get more rigorous results, one should replace them with honest wave packets. However, if the final answer is to have any experimental relevance, it had better not depend in any serious way on the precise structure of the wave packets. In the typical situation where an interaction results from the collision of two initial particles prepared in a particle accelerator, one usually has very little precise information about the wave functions of these particles except that their momenta are well localized about certain values determined by the experiment — so that, on some level, the assumption of definite momenta has to be a good approximation after all.

In short, the bad news is that mathematical precision is difficult to achieve, but the good news is that it is also somewhat irrelevant: the end result needs to be insensitive to a certain amount of sloppiness. As Weinberg [**129**], p. 134, says, "(as far as I know) no interesting open problems in physics hinge on getting the fine points right regarding these matters." We therefore follow him in taking a quick and dirty route to the main results, using the same device we employed in §5.1 and §6.2: putting the particles in a box. This also has the advantage of clarifying the normalization constants. (See Peskin and Schroeder [**88**], §4.5, or Itzykson and Zuber [**65**], §5-1-1, for alternate derivations using wave packets. However, one should be aware that different authors use different normalization conventions; in particular, factors of 2π are likely not to be invariant under change of reference text.)

Before getting started on this, we need to make a general observation about scattering amplitudes. If one of the initial particles is in exactly the same state as one of the final particles (same species, same momentum, same spin, etc.), there is a possibility — in fact, an overwhelming likelihood — that one incoming particle simply goes in one side and out the other without interacting with anything else. (This corresponds to Feynman diagrams in which one initial vertex is connected to one final vertex by a single line that is disconnected from the rest of the diagram.) More generally, if the sets of initial and final particles can each be partitioned into k subsets such that the total momentum in each incoming subset is equal to the total momentum in the corresponding outgoing subset, the main contribution to the total scattering process will be k separate interactions involving the various subsets. (This corresponds to Feynman diagrams with at least k connected components, each component carrying its own momentum-conservation delta-function.)

We assume at the outset that we are only interested in "irreducible" scattering processes in which all the incoming particles participate, that is, for which the Feynman diagrams are connected. We therefore disallow outgoing states in which some subset of the outgoing particles has exactly the same total momentum as some subset of the incoming particles. This is hardly a serious restriction, as the forbidden subset of the joint momentum space of the outgoing particles is a subvariety of positive codimension (and hence measure zero). Under this condition, if $|\text{in}\rangle$ and $|\text{out}\rangle$ are the initial and final states, and p^{in} and p^{out} are their total 4-momenta, the S-matrix element $\langle\text{out}|S|\text{in}\rangle$ has the form

$$S(\text{in} \to \text{out}) = \langle\text{out}|S|\text{in}\rangle = i(2\pi)^4\delta(p_\beta - p_\alpha)M(\text{in} \to \text{out}), \tag{6.65}$$

where $M(\text{in} \to \text{out})$ is free of delta-functions. (The i is conventional.) Here $|\text{in}\rangle$ and $|\text{out}\rangle$ are of the form $a_1^\dagger(\mathbf{p}_1)\cdots a_N^\dagger(\mathbf{p}_N)|0\rangle$ where the subscripts on the a's encode spin states, particle species, and any other relevant parameters. N will always denote the number of particles in such a state, $\mathbf{p} = \sum_1^N \mathbf{p}_j$ will denote its total 3-momentum, $E_\mathbf{p} = \sum\sqrt{|\mathbf{p}_j|^2 + m_j^2}$ its total energy, and $p = (E_\mathbf{p}, \mathbf{p})$ its total 4-momentum; all of these will carry superscripts "in" or "out."

We now replace $\mathbb{R}^3$ by a box $\mathcal{B} = [-\frac{1}{2}L, \frac{1}{2}L]^3$ with volume $V = L^3$. (One may envision L as a length scale that is extremely large in comparison to subatomic particles but perhaps not in comparison to the experimental apparatus: just a region large enough so that all events of interest happen well inside it.) For simplicity (and here again, specifics are unimportant), we impose periodic boundary conditions on the sides of the box, so that the 3-momenta of the particles are restricted to the lattice $[(2\pi/L)\mathbb{Z}]^3$, and the Dirac delta $\delta(\mathbf{p}_1 - \mathbf{p}_2)$ is replaced by the rescaled Kronecker delta

$$\delta_\mathcal{B}(\mathbf{p}_1 - \mathbf{p}_2) = \int_\mathcal{B} e^{i(\mathbf{p}_1-\mathbf{p}_2)\cdot\mathbf{x}}\frac{d^3\mathbf{x}}{(2\pi)^3} = \frac{V}{(2\pi)^3}\delta_{\mathbf{p}_1\mathbf{p}_2}. \tag{6.66}$$

Having cut down space to a finite size, we need also to restrict the time during which the interaction can take place to a large but finite interval $[-\frac{1}{2}T, \frac{1}{2}T]$, since the particles can no longer escape to infinity at large positive and negative times. Here again, in practice T should be only large enough to encompass the interaction in question comfortably; it need not be large on a human or cosmological time scale. The Dirac delta $\delta(E_1 - E_2)$ is then replaced by a smooth bump function,

$$\delta_T(E_1 - E_2) = \int_{-T/2}^{T/2} e^{i(E_1-E_2)t}\frac{dt}{2\pi} = \frac{1}{\pi}\frac{\sin\frac{1}{2}(E_1 - E_2)T}{E_1 - E_2}. \tag{6.67}$$

We must now address the question of the normalization of the initial and final states. In order to obtain a transition probability, we must use state vectors with norm one. In $\mathbb{R}^3$ this is problematic because the states $|\text{in}\rangle$ and $|\text{out}\rangle$ have definite momenta and are therefore not normalizable. But in the box, we use the creation operators $a^\dagger_{\mathcal{B}}(\mathbf{p})$ ($\mathbf{p} \in [(2\pi/L)\mathbb{Z}]^3$) appropriate to the box instead, and then it is easy: the vectors $a^\dagger_{\mathcal{B}}(\mathbf{p})|0\rangle$ always have norm one, and hence so do all states of the form $a^\dagger_{\mathcal{B}1}(\mathbf{p}_1)\cdots a^\dagger_{\mathcal{B}N}(\mathbf{p}_N)|0\rangle$ provided only that no two of the operators $a^\dagger_{\mathcal{B}j}(\mathbf{p}_j)$ are identical (creating particles of the same species in the same spin state with exactly the same momentum).[9] This is an entirely harmless restriction, and we adopt it henceforth. However, there is a rescaling implicit in the replacement of $a^\dagger$ by $a^\dagger_{\mathcal{B}}$, because as we saw in §5.1 (in the calculations leading to (5.11), where we denoted $a^\dagger_{\mathcal{B}}(\mathbf{p})$ by $A^\dagger_{\mathbf{p}}$), it is not $a^\dagger_{\mathcal{B}}$ but $L^{3/2}a^\dagger_{\mathcal{B}} = V^{1/2}a^\dagger_{\mathcal{B}}$ that turns into $a^\dagger$ as the box is removed. We see this again in the rescaling of delta-functions in (6.66): the inner product of $V^{1/2}a^\dagger_{\mathcal{B}}(\mathbf{p}_1)$ and $V^{1/2}a^\dagger_{\mathcal{B}}(\mathbf{p}_2)$ is $V\delta_{\mathbf{p}_1\mathbf{p}_2} = (2\pi)^3\delta_{\mathcal{B}}(\mathbf{p}_1-\mathbf{p}_2)$, which turns into $(2\pi)^3\delta(\mathbf{p}_1-\mathbf{p}_2) = \langle 0|a(\mathbf{p}_1)a^\dagger(\mathbf{p}_2)|0\rangle$ when the box is removed.

We can therefore adapt the S-matrix element (6.65) to the box by replacing the initial and final states $|\text{in}\rangle$ and $|\text{out}\rangle$ with

$$V^{N^{\text{in}}/2}|\text{in}\rangle_{\mathcal{B}} = V^{N^{\text{in}}/2}\prod_1^{N^{\text{in}}} a^\dagger(\mathbf{p}_j^{\text{in}})|0\rangle, \quad V^{N^{\text{out}}/2}|\text{out}\rangle_{\mathcal{B}} = V^{N^{\text{out}}/2}\prod_1^{N^{\text{out}}} a^\dagger(\mathbf{p}_j^{\text{out}})|0\rangle$$

(we are suppressing indices that indicate spin states, particle species, etc.) and the delta-function $\delta(p^{\text{out}} - p^{\text{in}})$ with box delta-functions:

$$\begin{aligned} &S(V^{N^{\text{in}}/2}|\text{in}\rangle_{\mathcal{B}} \to V^{N^{\text{out}}/2}|\text{out}\rangle_{\mathcal{B}}) \\ &\qquad = i(2\pi)^4\delta_{\mathcal{B}}(\mathbf{p}^{\text{out}} - \mathbf{p}^{\text{in}})\delta_T(E^{\text{out}} - E^{\text{in}})M(\text{in}\to\text{out}). \end{aligned}$$

(The factor $M(\text{in}\to\text{out})$, which contains the hard-won information from the quantum field theory, remains the same!) The transition amplitude $S_{\mathcal{B}}(\text{in}\to\text{out})$ for the *normalized* states $u^{\text{in}}_{\mathcal{B}}$ and $u^{\text{out}}_{\mathcal{B}}$ is then

$$S_{\mathcal{B}}(\text{in}\to\text{out}) = iV^{(-N^{\text{in}}-N^{\text{out}})/2}(2\pi)^4\delta_{\mathcal{B}}(\mathbf{p}^{\text{out}} - \mathbf{p}^{\text{in}})\delta_T(E^{\text{out}} - E^{\text{in}})M(\text{in}\to\text{out}),$$

and the transition *probability* is the absolute square of this quantity:

$$P_{\mathcal{B}}(\text{in}\to\text{out}) = V^{-N^{\text{in}}-N^{\text{out}}}(2\pi)^8\delta_{\mathcal{B}}(\mathbf{p}^{\text{out}} - \mathbf{p}^{\text{in}})^2\delta_T(E^{\text{out}} - E^{\text{in}})^2|M(\text{in}\to\text{out})|^2.$$

We can now deal with the squared delta-functions, as they no longer have infinite singularities. Indeed, by (6.66) we have

$$\delta_{\mathcal{B}}(\mathbf{p}_\beta - \mathbf{p}_\alpha)^2 = \delta_{\mathcal{B}}(0)\delta_{\mathcal{B}}(\mathbf{p}_\beta - \mathbf{p}_\alpha) = \frac{V}{(2\pi)^3}\delta_{\mathcal{B}}(p_\beta - p_\alpha).$$

There is no corresponding exact formula for δ_T^2, but we can make an analogous approximation from (6.67):

$$\delta_T(E_\beta - E_\alpha)^2 \approx \delta_T(0)\delta_T(E_\beta - E_\alpha) = \frac{T}{2\pi}\delta_T(E_\beta - E_\alpha).$$

Using this approximation, we obtain

$$P_{\mathcal{B}}(\text{in}\to\text{out}) = V^{-N^{\text{in}}-N^{\text{out}}+1}T(2\pi)^4\delta_{\mathcal{B}}(\mathbf{p}^{\text{out}} - \mathbf{p}^{\text{in}})\delta_T(E^{\text{out}} - E^{\text{in}})|M(\text{in}\to\text{out})|^2.$$

---

[9] If m of the operators are identical, the norm of $u_{\mathcal{B}}$ has a factor of $\sqrt{m!}$.

At this point we can drop the pretense of the box as far as the discreteness of possible momenta is concerned. In practice the probability of finding each outgoing particle with any particular momentum is infinitesimal, so we are interested instead in particles whose momentum lies in a small bit $d^3\mathbf{p}$ of momentum space. The number of points of the lattice of box momenta in this small region is $[V/(2\pi)^3]d^3\mathbf{p}$. Hence, if $\mathbf{p}_1, \ldots, \mathbf{p}_{N^{\text{out}}}$ are the momenta of the outgoing particles (so that $\mathbf{p}^{\text{out}} = \mathbf{p}_1 + \cdots + \mathbf{p}_{N^{\text{out}}}$) and we write

$$d^{3N^{\text{out}}}\mathbf{p}^{\text{out}} = d^3\mathbf{p}_1 \cdots d^3\mathbf{p}_{N^{\text{out}}}$$

for short, the *differential transition probability* for outgoing particles having momenta in the range $d^{3N^{\text{out}}}\mathbf{p}^{\text{out}}$ is

$$\begin{aligned}(6.68)\quad & dP(\text{in} \to \text{out}) \\ & = V^{1-N^{\text{in}}}T(2\pi)^4\delta(\mathbf{p}^{\text{out}} - \mathbf{p}^{\text{in}})\delta_T(E^{\text{out}} - E^{\text{in}})|M(\text{in} \to \text{out})|^2 \frac{d^{3N^{\text{out}}}\mathbf{p}^{\text{out}}}{(2\pi)^{3N^{\text{out}}}}.\end{aligned}$$

Here, having passed to the continuum limit in describing the momentum states, we have done the same for the momentum-conservation factor and replaced $\delta_{\mathcal{B}}$ by δ.

This is the general formula for interpreting the S-matrix elements. What remains is to explain the role of V and T, and for this it is necessary to consider the specific cases of various numbers of initial particles.

For experiments in particle accelerators, by far the most common situation is that there are two initial particles that are made to collide. More precisely, one manufactures a beam of particles of one species that is directed at a target consisting of a bunch of particles of another (or the same) species. (In some experiments, two particle beams are directed at each other; the only difference is in whether one group of particles is at rest in the laboratory frame or not. For the moment we assume that the target is at rest.) In experiments of this sort, the quantity that is usually measured is the *scattering cross section* σ, which is the number of scattering events per unit time, unit volume, unit density of target particles, and unit flux of beam particles. (Here *density* means number of particles per unit volume, and the *flux* of the beam is its density times the speed of its particles.) Since time, volume, density, and flux have dimensions $[t]$, $[l^3]$, $[1/l^3]$, and $[l/tl^3] = [1/tl^2]$, σ has dimensions $[l^2]$, i.e., it is an *area*. One can think of it as the effective cross-sectional area of the target that is scattered by each particle in the beam.

Now, returning to our model, we take the box $\mathcal{B}$ to be the (macroscopic) region where the interaction of the beam and target takes place, and T to be the (correspondingly long) time during which the interaction takes place. Since the wave functions ψ of particles with definite momenta are evenly spread out in space, i.e., $|\psi|^2 = \text{constant}$, $1/V$ can be interpreted as the *density* of each particle, so normalizing a quantity to be "per unit density" means multiplying by V. Thus, from the definition of σ in the preceding paragraph, we see that the differential cross-section for an initial state $|\text{in}\rangle$ to be scattered into a range of states with momenta in the infinitesimal region $d^{3N^{\text{out}}}\mathbf{p}^{\text{out}}$ is

$$d\sigma(\text{in} \to \text{out}) = dP(\text{in} \to \text{out}) \cdot \frac{1}{T} \cdot \frac{1}{V} \cdot V \cdot \frac{V}{|\mathbf{v}|} = \frac{V\, dP(\text{in} \to \text{out})}{T|\mathbf{v}|},$$

where $\mathbf{v}$ is the velocity of the beam. Plugging (6.68) into this, noting that $N^{\text{in}} = 2$, and replacing the smeared-out δ_T by the plain δ, we obtain

$$d\sigma(\text{in} \to \text{out}) = (2\pi)^4 \delta(p^{\text{out}} - p^{\text{in}})|\mathbf{v}|^{-1}|M(\text{in} \to \text{out})|^2 \frac{d^{3N^{\text{out}}}\mathbf{p}^{\text{out}}}{(2\pi)^{3N^{\text{out}}}}. \tag{6.69}$$

Finally the V and T have disappeared, and we have a useful result! Despite the considerable amount of handwaving in its derivation, this is actually the correct formula for computing cross-sections.

Let us say a few words about the Lorentz transformation properties of $d\sigma$, with an eye to removing the hypothesis that the target is stationary. The states $|\text{in}\rangle$ and $|\text{out}\rangle$ particles in definite spin states, so the transformation law for the matrix elements $S(\text{in} \to \text{out})$ or $M(\text{in} \to \text{out})$ involves the transformation of these spin states — in general, a complicated mess. However, the particles are not normally prepared or detected in any particular spin state, so the experimentally measured quantity is obtained by averaging $d\sigma(\text{in} \to \text{out})$ over an orthonormal basis of spin states for the incoming particles and summing it over an orthonormal basis of spin states for the outgoing particles (with the given momenta); we denote this process simply by $\sum_{\text{spins}}$. A close examination of the definitions of the variables that enter into $S(\text{in} \to \text{out})$ and $M(\text{in} \to \text{out})$, which we omit (see Weinberg [**129**], §3.4), shows that the quantity

$$R(\text{in} \to \text{out}) = \sum_{\text{spins}} |M(\text{in} \to \text{out})|^2 \prod E_j^{\text{in}} \prod E_k^{\text{out}} \tag{6.70}$$

is Lorentz-invariant, where the E_j^{in} and E_k^{out} are the energies of the incoming and coutgoing particles, respectively. On the other hand,

$$\frac{d^{3N^{\text{out}}} p^{\text{out}}}{(2\pi)^{3N^{\text{out}}} \prod E_k^{\text{out}}} = \prod \frac{d\mathbf{p}_k^{\text{out}}}{(2\pi)^3 E_k^{\text{out}}}$$

is a Lorentz-invariant measure on the momentum space of the outgoing particles. Therefore, if E_1 and E_2 are the energies of the incoming particles, the quantity

$$\sum_{\text{spins}} d\sigma(\text{in} \to \text{out}) = \frac{(2\pi)^4 \delta(p^{\text{out}} - p^{\text{in}})}{E_1^{\text{in}} E_2^{\text{in}} |\mathbf{v}|} R(\text{in} \to \text{out}) \prod \frac{d\mathbf{p}_k^{\text{out}}}{(2\pi)^3 E_k^{\text{out}}}$$

will be Lorentz-invariant provided we replace the term $E_1 E_2 |\mathbf{v}|$ by the Lorentz-invariant quantity that agrees with it when the second particle is at rest, namely,

$$U = \sqrt{(p_{1\mu} p_2^\mu)^2 - m_1^2 m_2^2},$$

p_1, p_2 and m_1, m_2 being the 4-momenta and masses of the two initial particles. (Indeed, U is clearly Lorentz-invariant; and if $\mathbf{p}_2 = \mathbf{0}$, then $m_2 = E_2$ and $p_{1\mu} p_2^\mu = E_1 E_2$, so $U = E_2 \sqrt{E_1^2 - m_1^2} = E_2 |\mathbf{p}_1| = E_1 E_2 |\mathbf{v}_1|$.)

The quantity $U/E_1 E_2$ can still be interpreted as a "relative velocity" when both incoming particles are in motion provided that they are aimed directly at each other so that $\mathbf{p}_1 \cdot \mathbf{p}_2 = -|\mathbf{p}_1|\,|\mathbf{p}_2|$. Indeed, in this case a short calculation shows that $U = E_1 |\mathbf{p}_2| + E_2 |\mathbf{p}_1|$, so

$$\frac{U}{E_1 E_2} = \frac{|\mathbf{p}_1|}{E_1} + \frac{|\mathbf{p}_2|}{E_2} = |\mathbf{v}_1 - \mathbf{v}_2|,$$

the absolute difference in velocities of the two particles as viewed from the laboratory frame. (It is *not*, however, the speed of one particle as viewed from the rest frame of the other.)

After the case of two initial particles, the most important situation is that of *one* initial particle. The "scattering" process here is the *decay of an unstable particle.* When $N^{\text{in}} = 1$, the V factor in (6.68) disappears, so dividing by T (again, the time during which the decay process occurs) yields the *differential transition rate*

$$d\Gamma(\text{in} \to \text{out}) = \frac{dP(\text{in} \to \text{out})}{T}.$$

If one replaces δ_T by δ (a step that requires some comment — see below), integrates over momenta, and sums over spins and possible decay modes, one obtains a formula for the total decay rate of the particle:

$$\Gamma = \sum_{\text{spins, modes}} \frac{1}{E^{\text{in}}} \int R(\text{in} \to \text{out})(2\pi)^4 \delta(p^{\text{out}} - p^{\text{in}}) \prod \frac{d^3\mathbf{p}_k^{\text{out}}}{(2\pi)^3 E_k^{\text{out}}}, \tag{6.71}$$

where $R(\text{in} \to \text{out})$ is as in (6.70). This expression is Lorentz-invariant except for the energy E^{in} of the initial particle; that factor incorporates the relativistic time dilation, according to which more rapidly moving particles (with respect to the laboratory frame) decay more slowly (with respect to the laboratory clock). The "half-life" of the particle in the usual sense is $(\log 2)/\Gamma_0$, where Γ_0 is the decay rate in the rest frame of the particle.

However, one cannot let $T \to \infty$ with impunity to replace δ_T by δ in (6.71): unstable particles are unstable, after all, and one cannot assume that they come from the infinitely distant past. This is the same problem that we encountered in studying the decay of an excited state in §6.2: for the approximation $\delta_T \approx \delta$ to be valid, T must be large enough so that $1/T$ is much less than the energies E^{in}, E_k^{out} involved in the process, but $dP(\text{in} \to \text{out})/T$ can be interpreted as a transition rate only if T is much less than the mean lifetime of the particle, so there needs to be a gap between these two quantities. The energies involved in typical particle decays correspond to times of 10^{-20} second or less, so this analysis is valid for all but the extremely short-lived particles. (See Peskin and Schroeder [**88**], §7.3, for an analysis of the latter situation.) As with the decay of excited states, the fact that δ_T is not exactly δ manifests itself in the fact that there is an unremovable uncertainty in the experimentally observed energy difference $E^{\text{out}} - E^{\text{in}}$.

Processes with three or more incoming particles are occasionally encountered, and the general formula (6.68) is again the bridge from field theory to experimental or observational results concerning them. However, we shall say no more about them here.

6.9. QED, the Coulomb potential, and the Yukawa potential

It is time to see how to connect our machinery to the real world. To begin with, the description of electromagnetism provided by QED, in terms of emission and absorption of photons by electrons or other Fermions, looks very different from the description given by classical field theory, so we had better be able to relate the latter to the former. For this purpose, let us examine the basic Feynman diagram for the interaction of two electrons (Figure 6.6). We shall calculate the value of this diagram in excruciating detail; later calculations of a similar sort can then be done somewhat more tersely. The momentum-space integral for this diagram is

(6.72)

$$\int \frac{m^2}{4\sqrt{\omega_{\mathbf{p}_1}\omega_{\mathbf{p}_2}\omega_{\mathbf{p}'_1}\omega_{\mathbf{p}'_2}}}\overline{u}(\mathbf{p}'_1, s'_1)(-ie\gamma^\mu)u(\mathbf{p}_1, s_1)\frac{-ig_{\mu\nu}}{q^2}\overline{u}(\mathbf{p}'_2, s'_2)(-ie\gamma^\nu)u(\mathbf{p}_2, s_2)$$
$$\times (2\pi)^4\delta(p'_1 + q - p_1)(2\pi)^4\delta(p'_2 - q - p_2)\frac{d^4q}{(2\pi)^4}.$$

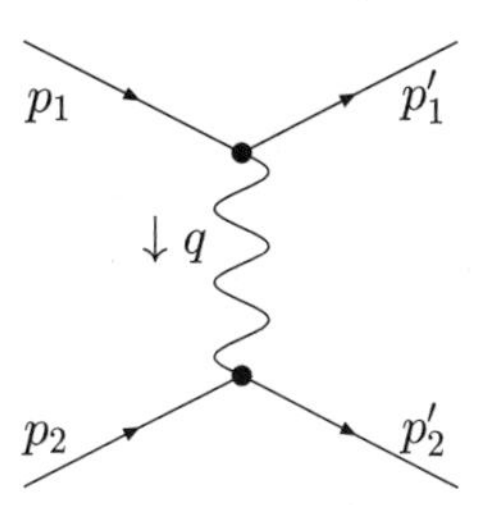

FIGURE 6.6. The basic Feynman diagram for electron-electron scattering.

Some explanations:

i. For the external (on-mass-shell) momenta we have $p = (\omega_{\mathbf{p}}, \mathbf{p})$ as usual.
ii. The internal momentum q is taken to flow from the first particle to the second one. It could equally well be taken to flow the other way; the only change would be in the signs of q in the delta-functions.
iii. The Fourier-transformed photon propagator is the limit of $-ig_{\mu\nu}/(q^2 + i\epsilon)$ as $\epsilon \to 0+$, and we have passed to the limit immediately since the $i\epsilon$ serves no purpose here.
iv. The factors of m and $\omega_{\mathbf{p}}$ come from the $\sqrt{m/2\omega_{\mathbf{p}}}$ in the formula (5.32) for the Dirac field.

The q-integration in (6.72) is easy to perform:

$$\int \frac{-ig_{\mu\nu}}{q^2}(2\pi)^4\delta(p'_1 + q - p_1)(2\pi)^4\delta(p'_2 - q - p_2)\frac{d^4q}{(2\pi)^4}$$
$$= \frac{-ig_{\mu\nu}}{(p'_1 - p_1)^2}(2\pi)^4\delta(p'_1 + p'_2 - p_1 - p_2).$$

(We could equally well write $(p'_2-p_2)^2$ instead of $(p'_1-p_1)^2$.) The overall momentum-conservation delta-function (with its faithful $(2\pi)^4$) can safely remain in the background, so we factor it out and consider only the remaining part, which we denote by iM as in (6.65):

(6.73)

$$iM = \frac{im^2e^2}{4\sqrt{\omega_{\mathbf{p}_1}\omega_{\mathbf{p}_2}\omega_{\mathbf{p}'_1}\omega_{\mathbf{p}'_2}}}\overline{u}(\mathbf{p}'_1, s'_1)\gamma^\mu u(\mathbf{p}_1, s_1)\frac{g_{\mu\nu}}{(p'_1 - p_1)^2}\overline{u}(\mathbf{p}'_2, s'_2)\gamma^\nu u(\mathbf{p}_2, s_2).$$

Since Fermions are involved, we had better check that we have not misplaced a minus sign. The position-space integral corresponding to (6.72) comes from

$$\text{(6.74)}\quad \frac{1}{i^2}\iint \langle 0|a(\mathbf{p}_1', s_1')a(\mathbf{p}_2', s_2')A_\mu(x)\overline{\psi}(x)\gamma^\mu\psi(x)A_\nu(y)\overline{\psi}(y)\gamma^\nu\psi(y) \times a^\dagger(\mathbf{p}_2, s_2)a^\dagger(\mathbf{p}_1, s_1)|0\rangle\, d^4x\, d^4y$$

by pairing up the operators into the contractions

$$\text{(6.75)}\quad \frac{1}{i^2}\iint \overbrace{a(\mathbf{p}_1', s_1')\overline{\psi}(x)}\,\gamma^\mu\,\overbrace{\psi(x)a^\dagger(\mathbf{p}_1, s_1)}\,\overbrace{A_\mu(x)A_\nu(y)} \times \overbrace{a(\mathbf{p}_2', s_2')\overline{\psi}(y)}\,\gamma^\nu\,\overbrace{\psi(y)a^\dagger(\mathbf{p}_2, s_2)}\, d^4x\, d^4y.$$

(Note that the orders of the creation operators for the initial and final states $a^\dagger(\mathbf{p}_2, s_2)a^\dagger(\mathbf{p}_1, s_1)|0\rangle$ and $a^\dagger(\mathbf{p}_2', s_2')a^\dagger(\mathbf{p}_1', s_1')|0\rangle$ are consistent!) One can easily check that the permutation of Fermion operators needed to change the order from that in (6.74) to that in (6.75) is even, so the sign in (6.73) is indeed correct. However, Figure 6.6 is only one of two diagrams that arise from the integral (6.74) and contribute to the basic photon exchange process; the other one is Figure 6.7, with the final particles reversed so that $a(\mathbf{p}_1', s_1')$ is contracted with $\overline{\psi}(y)$ and $a(\mathbf{p}_2', s_2')$ with $\overline{\psi}(x)$. The permutation needed to effect this arrangement is odd, so the value of Figure 6.7 is obtained from (6.73) by switching the $\overline{u}(\mathbf{p}_j', s_j')$'s *and inserting a factor of* -1. This -1 is important because it is the *sum* of the two diagrams that gives the S-matrix element. (Permuting the operators to go from (6.74) to (6.75) may seem to do violence to the matrix algebra, but it does not, as one sees by writing out all the components of the spinor fields explicitly; it is these individual components that are contracted with one another.)

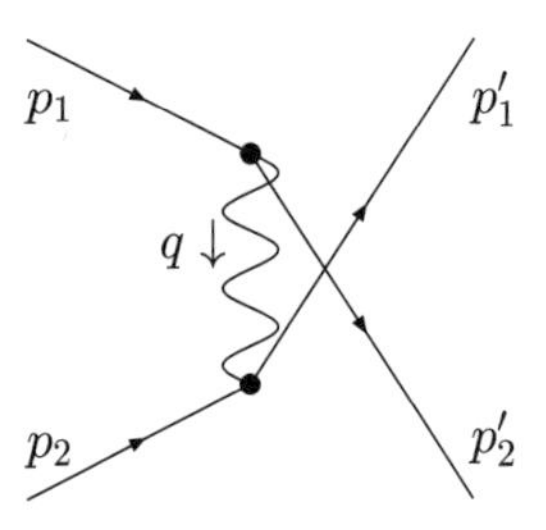

FIGURE 6.7. Electron-electron scattering with outgoing particles interchanged.

By the way, we could equally well have used one of the alternative photon propagators (6.55). The result is exactly the same, because

$$\overline{u}(\mathbf{p}_1', s_1')\gamma^\mu p_{1\mu}' = m\overline{u}(\mathbf{p}_1', s_1'), \qquad p_{1\mu}\gamma^\mu u(\mathbf{p}_1, s_1) = mu(\mathbf{p}_1, s_1),$$

so addition of any term proportional to $(p_1' - p_1)_\mu$ to the $g_{\mu\nu}$ in (6.73) yields two terms that cancel out. Thus our claim that the $q_\mu q_\nu$ term in (6.55) does not contribute to physically meaningful quantities is verified in this simple case.

The fact that the diagrams of both Figures 6.6 and 6.7 contribute to the S-matrix element complicates things a bit, so to make the connection with classical physics as simply as possible we shall consider the scattering of two *different* Fermions with possibly different masses m_1 and m_2 and possibly different charges

e_1 and e_2. Then Figure 6.6, where the subscripts 1 and 2 remain unmixed, is the only relevant one. Its value is still given by (6.73) except that the factor m^2e^2 must be replaced by $m_1m_2e_1e_2$.

Let us examine what happens to (6.73) in the nonrelativistic limit where the external momenta (and hence also the internal momentum $\mathbf{q} = \mathbf{p}_1 - \mathbf{p}_1' = \mathbf{p}_2' - \mathbf{p}_2$) are all very small by replacing all the quantities in (6.73) by their lowest-order terms as functions of the momenta:

$$\omega_{\mathbf{p}_j} \to m_j, \quad \omega_{\mathbf{p}_j'} \to m_j, \quad (p_1' - p_1)^2 \to -|\mathbf{p}_1' - \mathbf{p}_1|^2$$
$$u(\mathbf{p}_j, s) \to u(\mathbf{0}, s), \quad u(\mathbf{p}_j', s_j') \to u(\mathbf{0}, s').$$

Here the $u(\mathbf{0}, s)$, with $s = \pm$, are given by (5.29). We have

$$\overline{u}(\mathbf{0}, s)\gamma^0 u(\mathbf{0}, s') = u(\mathbf{0}, s)^\dagger u(\mathbf{0}, s') = 2\delta_{ss'} \tag{6.76}$$

by (5.29), and for $j = 1, 2, 3$,

$$\overline{u}(\mathbf{0}, s)\gamma^j u(\mathbf{0}, s') = u_R(\mathbf{0}, s)^\dagger \sigma_j u_R(\mathbf{0}, s') - u_L(\mathbf{0}, s)^\dagger \sigma_j u_L(\mathbf{0}, s') = 0 \tag{6.77}$$

where u_R and u_L are as in (4.27) and the σ_j are the Pauli matrices, simply because $u_L(\mathbf{0}, s) = u_R(\mathbf{0}, s)$ for all s. Therefore, since $g_{00} = 1$, in this approximation (6.73) boils down to

$$iM = \frac{-ie_1e_2}{|\mathbf{p}_1' - \mathbf{p}_1|^2}\delta_{s_1s_1'}\delta_{s_2s_2'}. \tag{6.78}$$

The first conclusion is that the spins of the two particles are separately conserved by the interaction. To get to the more interesting electrostatic aspect, concentrate on the first particle and compare (6.78) with the Born approximation (6.25) to the S-matrix element for the scattering of the particle by a potential $V(\mathbf{x})$. Putting aside the energy-conservation delta-function and its 2π as we did for the 4-momentum-conservation delta-function here,[10] we see that (6.78) gives (in the first-order approximation) the amplitude for scattering by the potential V such that $\widehat{V}(\mathbf{q}) = e_1e_2/|\mathbf{q}|^2$. The condition $|\mathbf{q}|^2\widehat{V}(\mathbf{q}) = e_1e_2$ is equivalent to $-\nabla^2 V(\mathbf{x}) = e_1e_2\delta(\mathbf{x})$, so V is just what it should be: the Coulomb potential $V(\mathbf{x}) = e_1e_2/4\pi|\mathbf{x}|$. (If this result isn't familiar, see the calculation at the end of this section and take the limit as $m_\phi \to 0$.) Thus we recover the classical physics in the low-energy limit.

What happens if we replace one of the particles — say, the second one — by its antiparticle? The $a^\dagger(\mathbf{p}_2, s_2)$ and $a(\mathbf{p}_2', s_2')$ in (6.74) must be replaced by $b^\dagger(\mathbf{p}_2, s_2)$ and $b(\mathbf{p}_2', s_2')$, the creation and annihilation operators for the antiparticle. This has the effect that the $u(\mathbf{p}_2, s_2)$ and $\overline{u}(\mathbf{p}_2', s_2')$ in (6.73) are replaced by $v(\mathbf{p}_2, s_2)$ and $\overline{v}(\mathbf{p}_2', s_2')$. In the nonrelativistic approximation, these v's still satisfy (6.76) and (6.77) (this time because $v_L = -v_R$), so it would seem that (6.78) is unchanged. But no: there is an extra minus sign in going from (6.74) to (6.75) because b and $b^\dagger$ contract with ψ and $\overline{\psi}$ rather than the other way around, and this minus sign survives in (6.78). Thus this calculation is consistent with the fact that particles and antiparticles have opposite charges.

Let us carry this line of thought further by considering the corresponding calculations for the Yukawa interaction (6.29): the exchange of a quantum of the

[10] In the center-of-mass frame in which $\mathbf{p}_1 + \mathbf{p}_2 = \mathbf{0}$, the spatial momentum-conservation delta-function disappears on integration over $\mathbf{p}_2$; all that is left is the energy-conservation delta-function.

scalar field ϕ between two distinguishable Fermions (say, a proton and a neutron). The Feynman diagram in Figure 6.6 still serves the purpose, with the wavy line now representing a quantum of the field ϕ. The corresponding quantities (6.73), (6.74), and (6.75) are the same as before except that (i) the field A_μ is replaced by ϕ and the photon propagator $-ig_{\mu\nu}/q^2$ is replaced by the scalar field propagator $-i/(-q^2+m_\phi^2)$, (ii) the e_1 and e_2 are replaced by the coupling constant g for the Yukawa interaction (which we now assume to be the same for both particles, as it is for the proton and neutron), and (iii) the Dirac matrices γ^μ and γ^ν are omitted.[11]

We pass to the nonrelativistic approximation. The analogue of (6.76) is

$$\overline{u}(\mathbf{0},s)u(\mathbf{0},s') = u(\mathbf{0},s)^\dagger u(\mathbf{0},s') = 2\delta_{ss'},$$

as before, because $\gamma_0 u = u$. (There are no terms corresponding to (6.77).) The analogue of (6.78) is therefore

$$iM = \frac{ig^2}{|\mathbf{p}_1' - \mathbf{p}_1|^2 + m_\phi^2}\delta_{s_1s_1'}\delta_{s_2s_2'}, \tag{6.79}$$

which corresponds to the potential V such that $\widehat{V}(\mathbf{q}) = -g^2/(|\mathbf{q}|^2 + m_\phi^2)$. By a calculation that we shall sketch at the end of this section, we find that

$$V(\mathbf{x}) = -\frac{g^2 e^{-m_\phi|\mathbf{x}|}}{4\pi|\mathbf{x}|}. \tag{6.80}$$

This is called the *Yukawa potential.* It resembles the Coulomb potential for $\mathbf{x}$ small but decays rapidly for $\mathbf{x}$ large, and it is negative. Hence it describes an *attractive short-range force* whose range is essentially $1/m_\phi$. The fact that the Yukawa interaction produces an attractive force whereas the electromagnetic interaction between particles of the same charge produces a repulsive force comes out of the opposite signs of q^2 in the photon and scalar field propagators.

What happens if we replace the second particle by its antiparticle in this situation? There is an extra minus sign from the permutation of Fermion operators as with the electromagnetic interaction, but here there is yet another minus sign from the replacement of u's by v's for the second particle, because $\gamma^0 v = -v$ and hence

$$\overline{v}(\mathbf{0},s)v(\mathbf{0},s') = -v(\mathbf{0},s)^\dagger v(\mathbf{0},s') = -2\delta_{ss'}.$$

Thus the force remains attractive! If we also replace the first particle by its antiparticle, we get another pair of minus signs, so (6.79) again remains unchanged. *The Yukawa force is universally attractive* as long as all Fermions in question interact with the scalar field with the same coupling constant g.

Yukawa proposed his interaction in 1935 as a model for the strong force that binds atomic nuclei. The existence of a particle representing the quantum of the ϕ field was one of the main *predictions* of the theory, and Yukawa estimated its mass at around 150 MeV from experimental data about the range of the strong force. The first particle to be discovered that seemed like a reasonable candidate to fit Yukawa's theory was the muon, but it didn't fill the bill for various reasons; Yukawa was finally vindicated with the discovery of pions (with masses around 140 MeV) in 1947. His theory doesn't have many more notable successes, however, because the coupling constant g is too large to allow the effective use of perturbation theory.

---

[11]If one uses the pseudoscalar interaction (6.30) with γ^5 in place of γ^μ instead, the effect is to replace u spinors with v spinors and vice versa, but the final result concerning the Yukawa potential is unchanged.

We conclude this section by sketching the derivation of (6.80), which is a bit delicate because $\widehat{V} \notin L^1$. One can begin by restricting the integration to the ball $|\mathbf{q}| \le R$. Since the result is a radial function of $\mathbf{x}$, it suffices to take $\mathbf{x} = (0,0,a)$ with $a > 0$; then integration in spherical coordinates gives

$$\begin{aligned}\int_{|\mathbf{q}|\le R} \frac{e^{i\mathbf{q}\cdot\mathbf{x}}}{|\mathbf{q}|^2+m_\phi^2}\frac{d^3\mathbf{q}}{(2\pi)^3} &= \int_0^R\int_0^\pi \frac{e^{iar\cos\theta}}{r^2+m_\phi^2} r^2\sin\theta\,\frac{d\theta\,dr}{(2\pi)^2} \\ &= \int_0^R \frac{r(e^{iar}-e^{-iar})}{4\pi^2 ia(r^2+m_\phi^2)}\,dr = \int_{-R}^R \frac{re^{iar}}{4\pi^2 ia(r^2+m_\phi^2)}\,dr.\end{aligned}$$

The limit of this as $R \to \infty$ can be evaluated via the residue theorem to give $e^{-m_\phi a}/4\pi a$. To certify the validity of (6.80) from this, one needs to verify that the convergence take place not just pointwise but in a sense at least as strong as weak convergence of distributions. Alternatively, once one has used this calculation to guess (6.80), it is a completely elementary exercise (using spherical coordinates as above) to check that the Fourier transform of the L^1 function $e^{-m_\phi|\mathbf{x}|}/4\pi|\mathbf{x}|$ is $1/(|\mathbf{q}|^2+m_\phi^2)$.

6.10. Compton scattering

As another illustration of the basic calculations of QED, we sketch the calculation of the cross-section for the scattering of a photon by an electron, known as *Compton scattering*, in the first-order approximation. The initial state $|\mathrm{in}\rangle$ consists of an electron with momentum $\mathbf{p}$ and spin state s and a photon with momentum $\mathbf{k}$ and polarization 4-vector $\epsilon = (0,\boldsymbol{\epsilon})$; the final state $|\mathrm{out}\rangle$ consists of an electron with momentum $\mathbf{p}'$ and spin state s' and a photon with momentum $\mathbf{k}'$ and polarization 4-vector $\epsilon' = (0,\boldsymbol{\epsilon}')$. (Here $\boldsymbol{\epsilon}$ and $\boldsymbol{\epsilon}'$ must be orthogonal to $\mathbf{k}$ and $\mathbf{k}'$, respectively.) We shall assume the electron is initially at rest, so that $\mathbf{p} = \mathbf{0}$ and $p_0 = m$, where m is the mass of the electron. We denote by $\omega = |\mathbf{k}|$ and $\omega' = |\mathbf{k}'|$ the energies of the incoming and outgoing photons, and by p, k, p', k' the 4-momenta of the electrons and photons (i.e., $p = (m,\mathbf{0})$, $k = (\omega,\mathbf{k})$, etc.).

Conservation of energy and momentum means that

$$m+\omega = \sqrt{m^2+|\mathbf{p}'|^2}+\omega', \qquad \mathbf{k} = \mathbf{p}'+\mathbf{k}',$$

so that

$$(m+\omega-\omega')^2 = m^2+|\mathbf{p}'|^2 = m^2+|\mathbf{k}-\mathbf{k}'|^2 = m^2+\omega^2+\omega'^2-2\omega\omega'\cos\theta, \tag{6.81}$$

where θ is the angle between the incoming and outgoing photon momenta. A little algebraic manipulation with this yields the relation between the incoming and outgoing photon energies (or wavelengths):

$$\frac{1}{\omega'} = \frac{1}{\omega}+\frac{1-\cos\theta}{m}. \tag{6.82}$$

which was confirmed in Compton's experiments on the scattering of x-rays by electrons.

(At the end of the nineteenth century the wave theory of light appeared to have won a decisive victory over Newton's corpuscular theory. In 1900 and 1905, Planck and Einstein revived the hypothesis that electromagnetic radiation is quantized to explain the spectrum of black-body radiation and the photoelectric effect, but many physicists remained dubious. Compton's experiments in 1923 clinched the matter: the relation (6.82) is almost obvious if x-rays consist of photons, as we have just seen, but it is hard to account for if one uses a wave model. See Pais [**87**] for a fuller account of this history.)

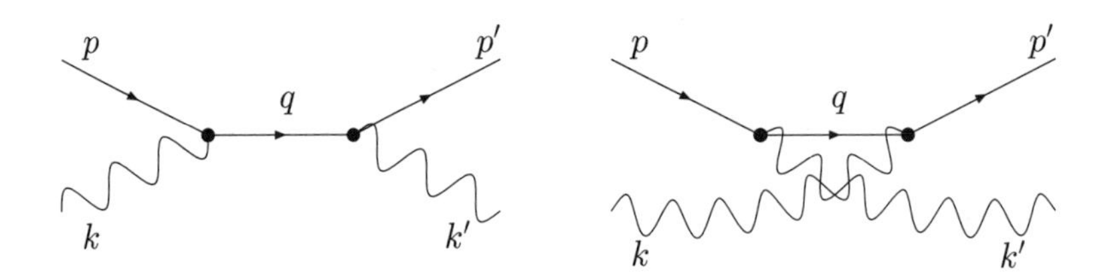

FIGURE 6.8. Feynman diagrams for Compton scattering.

In the first-order approximation, there are two Feynman diagrams that contribute to this process, shown in Figure 6.8. The sum of these diagrams is

$$S = \frac{(-ie)^2(2\pi)^4\sqrt{m}}{4\sqrt{\omega_{\mathbf{p}'}\omega\omega'}} \int \overline{u}(\mathbf{p}', s')\Big[\epsilon'^*_\mu\gamma^\mu \frac{-i(q_\rho\gamma^\rho + m)}{-q^2+m^2}\epsilon_\nu\gamma^\nu\delta(q-p'-k')\delta(q-p-k)$$
$$+ \epsilon_\mu\gamma^\mu \frac{-i(q_\rho\gamma^\rho + m)}{-q^2+m^2}\epsilon'^*_\nu\gamma^\nu\delta(q+k-p')\delta(q+k'-p)\Big]u(\mathbf{0}, s)\, d^4q.$$

(The $(2\pi)^4$ comes from the $(2\pi)^4$'s attached to the delta-functions and the $1/(2\pi)^4$ attached to the d^4q, and we have used the fact that $\omega_{\mathbf{p}} = \omega_{\mathbf{0}} = m$.)

After performing the integration (merely a convolution of delta-functions) and using the facts that $p^2 = m^2$ and $k^2 = k'^2 = 0$ to simplify the resulting denominators $-(p+k)^2 + m^2$ and $-(p-k')^2 + m^2$, we find that

$$S = (2\pi)^4\delta(p' + k' - p - k)iM,$$

where

$$(6.83) \quad M = \frac{-e^2\sqrt{m}}{8\sqrt{\omega_{\mathbf{p}'}\omega\omega'}}$$
$$\times \overline{u}(\mathbf{p}', s')\left[\frac{\epsilon'^*_\mu\gamma^\mu[(p+k)_\rho\gamma^\rho + m]\epsilon_\nu\gamma^\nu}{p_\rho k^\rho} - \frac{\epsilon_\mu\gamma^\mu[(p-k')_\rho\gamma^\rho + m]\epsilon'^*_\nu\gamma^\nu}{p_\rho k'^\rho}\right]u(\mathbf{0}, s).$$

This M is the delta-function-free matrix element that is used to calculate cross-sections. Namely, according to (6.68), the differential cross-section for scattering into the region $d^3\mathbf{p}'\, d^3\mathbf{k}'$ of momentum space is

$$d\sigma = (2\pi)^4\delta(p' + k' - p - k)|M|^2 \frac{d^3\mathbf{p}'\, d^3\mathbf{k}'}{(2\pi)^6}.$$

The factor of $|\mathbf{v}|$ has disappeared because we have assumed that the electron is at rest, so $|\mathbf{v}|$ is just the speed of the incoming photon, namely, 1. Now,

$$(2\pi)^4\delta(p' + k' - p - k)\frac{d^3\mathbf{p}'\, d^3\mathbf{k}'}{(2\pi)^6} = \delta(\mathbf{p}' + \mathbf{k}' - \mathbf{k})\, d^3\mathbf{p}' \cdot \delta(p'_0 + \omega' - m - \omega)\frac{d^3\mathbf{k}'}{(2\pi)^2}.$$

The first delta-function on the right just means that in calculating the cross-section we must set $\mathbf{p}' = \mathbf{k} - \mathbf{k}'$ and drop the $d^3\mathbf{p}'$. Likewise, if we write the element of volume in $\mathbf{k}'$-space in spherical coordinates,

$$d^3\mathbf{k}' = \omega'^2\, d\omega'\, d\Omega,$$

where $d\Omega$ is surface measure on the unit sphere, the second delta-function fixes ω' at the value given by (6.82) and eliminates the $d\omega'$, but an extra factor arises from the general fact that $\delta(f(t)) = \delta(t - t_0)/f'(t_0)$ when f is a smooth function that vanishes at t_0. In our case, by (6.81),

$$f(\omega') = p_0' + \omega' - m - \omega = \sqrt{\omega^2 - 2\omega\omega'\cos\theta + \omega'^2 + m^2} + \omega' - m - \omega,$$

so with $p_0' = m + \omega - \omega'$ and ω' given by (6.82),

$$f'(\omega') = \frac{\omega' - \omega\cos\theta}{p_0'} + 1 = \frac{-\omega\cos\theta + m + \omega}{m + \omega - \omega'} = \frac{m\omega}{\omega'(m + \omega - \omega')}.$$

The upshot is that the differential cross-section for the direction of the outgoing photon to be in the solid angle $d\Omega$ located at an angle θ from the direction of the incoming photon is

$$d\sigma = |M|^2 \frac{(m + \omega - \omega')\omega'^3}{(2\pi)^2 m\omega}\, d\Omega,$$

where ω' is determined by ω according to (6.82).

This is all very well, but the simple notation $|M|^2$ conceals a rather formidable expression (see (6.83)) to calculate. Moreover, $|M|^2$ depends on the spin and helicity states of the incoming and outgoing electrons and photons. Normally one does not wish to specify or measure the electron spin state, so one must average over the two orthonormal states $s = \pm$ for the incoming electron and sum over the two states for the outgoing electron. Plugging in the definition of M and reducing the resulting expression to something manageable takes several pages of tedious algebra involving products of Dirac matrices, which the mathematical tourist would probably prefer to skip. We refer to Weinberg [**129**], §8.7, or Peskin and Schroeder [**88**], §5.5, for the details, with the warning that their normalizations differ from ours. However, the final result is quite simple:

$$\tfrac{1}{2}\sum_{s,s'} d\sigma = \frac{e^4\omega'^2}{64\pi^2 m^2\omega^2}\left[\frac{\omega}{\omega'} + \frac{\omega'}{\omega} - 2 + 4(\boldsymbol{\epsilon}\cdot\boldsymbol{\epsilon}')^2\right] d\Omega.$$

This formula was originally derived in 1929 by Klein and Nishina by more old-fashioned methods.

Furthermore, the polarization $\boldsymbol{\epsilon}$ of the incoming photon is also usually unspecified, so one should average over the two orthonormal possibilities for it. The result is

$$\tag{6.84} \tfrac{1}{4}\sum_{s,s',\epsilon} d\sigma = \frac{e^4\omega'^2}{32\pi^2 m^2\omega^2}\left[\frac{\omega}{\omega'} + \frac{\omega'}{\omega} - 2(\boldsymbol{\epsilon}'\cdot\widehat{\mathbf{k}})^2\right] d\Omega,$$

where $\widehat{\mathbf{k}} = \mathbf{k}/|\mathbf{k}|$. It follows that $d\sigma$ is maximized when $\boldsymbol{\epsilon}'$ is orthogonal to $\mathbf{k}$, and $\boldsymbol{\epsilon}'$ is always orthogonal to $\mathbf{k}'$, so *the outgoing photon is preferentially polarized in the direction orthogonal to the plane of the scattering.*

The cross-section for outgoing photons of arbitrary polarization is obtained by summing (6.84) over two orthonormal values of $\boldsymbol{\epsilon}'$, and the result is

$$\tag{6.85} \tfrac{1}{4}\sum_{s,s',\epsilon,\epsilon'} d\sigma = \frac{e^4\omega'^2}{32\pi^2 m^2\omega^2}\left[\frac{\omega}{\omega'} + \frac{\omega'}{\omega} - 1 + \cos^2\theta\right] d\Omega.$$

(Recall that θ is the angle between the incoming and outgoing photon momenta.) If we assume that the incoming photon energy ω is much less than the mass of the electron (as it is for any photon less energetic than a gamma ray), (6.82) shows that $\omega'/\omega \approx 1$. With this approximation, (6.85) simplifies to

$$\tfrac{1}{4}\sum_{s,s',\epsilon,\epsilon'} d\sigma = \frac{e^4}{32\pi^2 m^2}(1+\cos^2\theta)\,d\Omega,$$

and since

$$\int (1+\cos^2\theta)\,d\Omega = \int_0^{2\pi}\int_0^{\pi}(1+\cos^2\theta)\sin\theta\,d\theta\,d\phi = \frac{16\pi}{3},$$

the *total* cross-section is

$$\sigma_T = \frac{e^4}{6\pi m^2} = \frac{8\pi r_0^2}{3},$$

where r_0 is the Compton radius (1.7).[12] This result was originally derived in the context of *classical* electrodynamics as a formula relating to the scattering of low-energy radiation by a static charge.

6.11. The Gell-Mann–Low and LSZ formulas

In the preceding sections we have taken the most direct path to the integrals corresponding to Feynman diagrams that are used to compute quantities that can be tested against experiment. All of our calculations have been based on free field operators; we have not so much as mentioned operators representing interacting fields.[13] Moreover, the full interacting Hamiltonian was quickly swept under the rug after making a brief appearance at the beginning of the chapter.

From a mathematical point of view, this is all to the good. A precise mathematical construction of the interacting fields that describe actual fundamental physical processes in 4-dimensional space-time is still lacking and may not be feasible without serious modifications to the theory. Similarly, we have no way to define the Hamiltonian H in a mathematically rigorous way as a self-adjoint operator. It was presented as the sum of the free Hamiltonian H_0, which *is* well-defined, and the interaction Hamiltonian H_I; but the latter was presented as the integral of a density consisting of products of fields, which are operator-valued distributions rather than functions. Fortunately, this formal characterization of H_I is sufficient to lead to well-defined calculations in perturbation theory, but we have not really defined H_I as an operator; and if we had, we would still not have shown how to define the appropriate domain to make the sum $H_0 + H_I$ self-adjoint.

The devil now offers us another bargain: If we are willing to accept the *existence* of the interacting fields and the full Hamiltonian associated to them (but without any expectation of finding a mathematically rigorous way to construct them) and to make a few arguments on the level of pure hand-waving about their structure, we can use them to obtain a more complete picture of the theory.

From a practical point of view, it is possible to do a lot of quantum field theory without accepting this bargain. The perturbation theory of this chapter together with the renormalization techniques of the next one suffice to calculate cross sections, anomalous magnetic moments, energy level shifts, and other such items of experimental interest. In this section, however, we shall accept the bargain to the extent of making the connection between scattering processes and Feynman

[12]The subscript T can stand for either "total" or "Thomson."

[13]Hence the carefully chosen title of the chapter.

diagrams, on the one hand, and vacuum expectation values of time-ordered products of interacting fields, on the other. The latter are widely used in the physics literature and are of fundamental importance in the study of rigorous models that satisfy the Wightman axioms. But from a strict mathematical point of view, everything in this section should be taken on the level of heuristics and plausibility arguments. We will add some credibility to main results by rederiving them in an entirely different, still nonrigorous, but intuitively very appealing way in Chapter 8.

To keep the notation manageable, we consider the simple case of the ϕ^4 scalar field theory. The generalization to several fields with arbitrary spin and interactions given by suitable products of field operators is straightforward and will be sketched in due course.

The setup is as follows: We begin with a Hilbert space of states on which we are given a self-adjoint operator H, the full Hamiltonian, whose spectrum is bounded below by some number E_0 that is a simple eigenvalue with eigenvector $|\Omega\rangle$. Furthermore, we are given an operator-valued distribution ϕ on $\mathbb{R}^4$, and as usual, we shall pretend that it is a function for notational purposes. The time dependence of ϕ is "Heisenberg-picture," that is,

$$\phi(t,\mathbf{x}) = e^{iHt}\phi(0,\mathbf{x})e^{-iHt}, \tag{6.86}$$

and H is formally defined in terms of ϕ by

$$H = \int \left[\frac{1}{2}[(\partial_t\phi(t,\mathbf{x}))^2 + |\nabla_{\mathbf{x}}\phi(t,\mathbf{x})|^2 + m^2\phi(t,\mathbf{x})^2] + \frac{\lambda}{4!}\phi(t,\mathbf{x})^4\right] d^3\mathbf{x}. \tag{6.87}$$

(The expression on the right apparently depends on t, but in fact it does not, in view of (6.86) and the fact that H commutes with e^{iHt}.) Associated to ϕ and H is the classical Lagrangian density

$$\mathcal{L} = \frac{1}{2}[(\partial_\mu\phi)(\partial^\mu\phi) - m^2\phi^2] - \frac{\lambda}{4!}\phi^4,$$

and with respect to this Lagrangian there is a canonically conjugate field π, namely, $\pi = \partial\mathcal{L}/\partial(\partial_t\phi) = \partial_t\phi$ just as for the free field. We assume that ϕ and π satisfy the canonical equal-time commutation relations

$$[\phi(t,\mathbf{x}),\pi(t,\mathbf{y})] = i\delta(\mathbf{x}-\mathbf{y}), \qquad [\phi(t,\mathbf{x}),\phi(t,\mathbf{y})] = [\pi(t,\mathbf{x}),\pi(t,\mathbf{y})] = 0. \tag{6.88}$$

We fix a reference time, which may as well be $t_0 = 0$, and expand $\phi(0,\mathbf{x})$ and $\pi(0,\mathbf{x})$ as Fourier integrals,

$$\phi(0,\mathbf{x}) = \int e^{i\mathbf{p}\cdot\mathbf{x}}\widehat{\phi}(\mathbf{p})\frac{d^3\mathbf{p}}{(2\pi)^3}, \qquad \pi(0,\mathbf{x}) = \int e^{i\mathbf{p}\cdot\mathbf{x}}\widehat{\pi}(\mathbf{p})\frac{d^3\mathbf{p}}{(2\pi)^3},$$

where $\widehat{\phi}$ and $\widehat{\pi}$ are operator-valued distributions on $\mathbb{R}^3$. Since $\phi(0,\mathbf{x})$ and $\pi(0,\mathbf{x})$ are Hermitian, we have $\widehat{\phi}(\mathbf{p})^\dagger = \widehat{\phi}(-\mathbf{p})$ and $\widehat{\pi}(\mathbf{p})^\dagger = \widehat{\pi}(-\mathbf{p})$. We next set

$$a(\mathbf{p}) = \frac{\omega_{\mathbf{p}}\widehat{\phi}(\mathbf{p}) + i\widehat{\pi}(\mathbf{p})}{\sqrt{2\omega_{\mathbf{p}}}} \qquad (\omega_{\mathbf{p}} = \sqrt{m^2+|\mathbf{p}|^2}).$$

Then $a^\dagger(-\mathbf{p}) = [\omega_{\mathbf{p}}\widehat{\phi}(\mathbf{p}) - i\widehat{\pi}(\mathbf{p})]/\sqrt{2\omega_{\mathbf{p}}}$, so

$$\widehat{\phi}(\mathbf{p}) = \frac{a(\mathbf{p}) + a^\dagger(-\mathbf{p})}{\sqrt{2\omega_{\mathbf{p}}}}, \qquad \widehat{\pi}(\mathbf{p}) = -i\sqrt{\frac{\omega_{\mathbf{p}}}{2}}\,[a(\mathbf{p}) - a^\dagger(-\mathbf{p})],$$

and hence $\phi(0,\mathbf{x})$ and $\pi(0,\mathbf{x})$ are given by the same integrals (5.11) and (5.19) as for free fields. A short calculation then shows that the commutation relations (6.88) for the fields imply the canonical commutation relations (5.10) for the operators $a(\mathbf{p})$ and $a^\dagger(\mathbf{p})$. In short, the field $\phi(0,\mathbf{x})$ has the same formal structure as the free field studied in Chapter 5; what is different is the time evolution.

We wish to describe things in the "interaction picture," obtained by decomposing the Hamiltonian (6.87), with $t = 0$, into the free part and the interaction part:

$$\begin{aligned} H &= H_0 + H_I, \\ H_0 &= \int \tfrac{1}{2}\left[(\partial_t\phi(0,\mathbf{x}))^2 + |\nabla_{\mathbf{x}}\phi(0,\mathbf{x})|^2 + m^2\phi(0,\mathbf{x})^2\right] d^3\mathbf{x}, \\ H_I &= \frac{\lambda}{4!}\int \phi(0,\mathbf{x})^4\, d^3\mathbf{x}. \end{aligned} \tag{6.89}$$

As in §5.3, H_0 needs to be renormalized by subtracting off an infinite constant, or by Wick ordering, so that the lowest eigenvalue of H_0 is 0. When this is done, the calculations in §5.3 following (5.23) show that H_0 is given in terms of $a(\mathbf{p})$ and $a^\dagger(\mathbf{p})$ by (5.22). We define the *interaction picture field* ϕ_0 by

$$\phi_0(t,\mathbf{x}) = e^{iH_0t}\phi(0,\mathbf{x})e^{-iH_0t}.$$

The quantities on the right are defined in terms of $a(\mathbf{p})$ and $a^\dagger(\mathbf{p})$ just as in the free field case, so ϕ_0 is indeed structurally identical to a free field and is given by (5.12):

$$\phi_0(x) = \int \frac{1}{\sqrt{2\omega_{\mathbf{p}}}}\left(e^{-ip_\mu x^\mu}a(\mathbf{p}) + e^{ip_\mu x^\mu}a^\dagger(\mathbf{p})\right)\frac{d^3\mathbf{p}}{(2\pi)^3}.$$

In the interaction picture, the interaction Hamiltonian is the time-dependent operator $H_I(t)$ defined by

$$H_I(t) = e^{iH_0t}H_Ie^{-iH_0t} = \frac{\lambda}{4!}\int \phi_0(t,\mathbf{x})^4\, d^3\mathbf{x}. \tag{6.90}$$

As in §6.1 and §6.3, we define

$$V(t,t') = e^{iH_0t}e^{-iH(t-t')}e^{-iH_0t'},$$

so that

$$V(t,t')V(t',t'') = V(t,t''), \qquad \phi(t,\mathbf{x}) = V(0,t)\phi_0(t,\mathbf{x})V(t,0), \tag{6.91}$$

and, for $t > t'$, $V(t,t')$ is given perturbation-theoretically by the Dyson series:

$$V(t,t') = \mathcal{T}\exp\left[\frac{1}{i}\int_{t'}^{t} H_I(\tau)\right] d\tau. \tag{6.92}$$

This is the point where the mathematical credibility of the argument falls to the infinitesimal level. One has to face the fact that the Stone-von Neumann theorem is false in infinite dimensions, so there are many inequivalent representations of the canonical commutation relations (6.33) (see Segal [**109**]). We therefore have no right to expect that the field ϕ_0 is just the free field studied in Chapter 5. In fact, a theorem of Haag (see Streater and Wightman [**114**] or Bogolubov et al. [**11**], [**12**]) says in effect that it *cannot* be unless the interaction is trivial. Nonetheless, we shall proceed: after a bit we shall arrive at some results that beg to be interpreted in perturbation theory, in terms of Feynman diagrams, and *on that level* they seem to be perfectly correct. There are some unresolved mysteries here about the radical restructuring of a quantum system that is produced by any nontrivial interaction of fields, but this is not the place to explore them.

The next step is to examine the vacuum states. As mentioned earlier, we denote the vacuum for H (i.e., the unique state whose eigenvalue E_0 is the infimum of the spectrum of H) by $|\Omega\rangle$. The free Hamiltonian H_0 has its own vacuum state, i.e., its eigenstate with eigenvalue 0, which we denote as before by $|0\rangle$. There is no reason for $|\Omega\rangle$ and $|0\rangle$ to coincide, but to extricate ourselves from the wilderness of the preceding paragraph we must take on faith that they have some overlap, i.e., that $\langle\Omega|0\rangle \neq 0$. Granted this, there is a neat device for expressing $|\Omega\rangle$ in terms of $|0\rangle$. Let P be the projection-valued measure associated to H, so that H has the spectral resolution

$$H = \int_{[E_0,\infty)} E\, dP(E) = E_0 P(\{E_0\}) + \int_{(E_0,\infty)} E\, dP(E).$$

Then for any $T \in \mathbb{R}$,

$$e^{-iT(H-E_0)}|0\rangle = |\Omega\rangle\langle\Omega|0\rangle + \int_{(E_0,\infty)} e^{-iT(E-E_0)}\, dP(E)|0\rangle.$$

If we now give T a negative imaginary part, $T \to T(1-i\epsilon)$, and send it to infinity, the second term vanishes and we obtain

$$|\Omega\rangle = \lim_{T\to\infty(1-i\epsilon)} \frac{e^{-iT(H-E_0)}|0\rangle}{\langle\Omega|0\rangle} = \lim_{T\to\infty(1-i\epsilon)} \frac{e^{iE_0T}V(0,-T)|0\rangle}{\langle\Omega|0\rangle}, \tag{6.93}$$

since $e^{iH_0T}|0\rangle = |0\rangle$. At this point ϵ is unrestricted, but in what follows we will need to let it tend to zero, and one should think of it intuitively as being infinitesimal.

We are now ready to express vacuum expectation values of products of interacting fields in terms of quantities defined in terms of free fields. For any $x_1, \ldots, x_n \in \mathbb{R}^4$, let $t_j = (x_j)^0$; then by (6.91) we have

$$\phi(x_1)\cdots\phi(x_n) = V(0,t_1)\phi_0(x_1)V(t_1,t_2)\phi_0(x_2)\cdots V(t_{n-1},t_n)\phi_0(x_n)V(t_n,0),$$

and hence by (6.93),

$$\begin{aligned}\langle\Omega|\phi(x_1)\cdots\phi(x_n)|\Omega\rangle = \lim_{T\to\infty(1-i\epsilon)} \frac{e^{2iE_0T}}{|\langle\Omega|0\rangle|^2}\langle 0|V(T,t_1)\phi_0(x_1)V(t_1,t_2)\phi_0(x_2)\cdots \\ \cdots V(t_{n-1},t_n)\phi_0(x_n)V(t_n,-T)|0\rangle.\end{aligned}$$

We can get rid of the unpleasant numerical factor in front by observing that the same equation holds with $n = 0$:

$$1 = \langle\Omega|\Omega\rangle = \lim_{T\to\infty(1-i\epsilon)} \frac{e^{2iE_0T}}{|\langle\Omega|0\rangle|^2}\langle 0|V(T,-T)|0\rangle.$$

Taking the quotient of these two equalities yields

$$\begin{aligned}&\langle\Omega|\phi(x_1)\cdots\phi(x_n)|\Omega\rangle \\ &= \lim_{T\to\infty(1-i\epsilon)} \frac{\langle 0|V(T,t_1)\phi_0(x_1)V(t_1,t_2)\phi_0(x_2)\cdots V(t_{n-1},t_n)\phi_0(x_n)V(t_n,-T)|0\rangle}{\langle 0|V(T,-T)|0\rangle}.\end{aligned}$$

Finally, suppose that $x_1, \ldots, x_n$ are time-ordered: $t_1 \geq \cdots \geq t_n$. Then, for T sufficiently large, the time parameters in this expression all decrease from left to right (where the ordering refers to the *real* parts of T and $-T$), so we can expand the V's in their Dyson series (6.92). Since $(\exp \int_a^b)(\exp \int_b^c) = \exp \int_a^c$, the integrals in these series combine to yield an elegant final result:

$$(6.94)\quad \langle\Omega|\mathcal{T}[\phi(x_1)\cdots\phi(x_n)]|\Omega\rangle = \lim_{T\to\infty(1-i\epsilon)} \frac{\langle 0|\mathcal{T}\big[\phi_0(x_1)\cdots\phi_0(x_n)\exp(-i\int_{-T}^{T} H_I(\tau)\,d\tau)\big]|0\rangle}{\langle 0|\mathcal{T}\big[\exp(-i\int_{-T}^{T} H_I(\tau)d\tau)\big]|0\rangle}.$$

To be perfectly clear about the meaning of this: the numerator on the right side of (6.94) is the series whose kth term is

(6.95)

$$\left\langle 0\left|\frac{(-i)^k}{k!}\int_{-T}^{T}\cdots\int_{-T}^{T}\mathcal{T}[\phi_0(x_1)\cdots\phi_0(x_n)H_I(\tau_1)\cdots H_I(\tau_k)]\,d\tau_1\cdots d\tau_k\right|0\right\rangle,$$

and similarly for the denominator. These series are to be interpreted perturbation-theoretically: that is, one uses the finite partial sums to obtain approximate results, in a way that we shall explain shortly, without worrying about the convergence of the entire series. The only slightly mysterious thing is the $i\epsilon$, and we shall shed some more light on it in due course.

We considered the simplest self-interacting scalar field theory in deriving (6.94) only in order not to clutter the picture up with complicated notation, but the result is valid in great generality for quantum field theories involving particles of arbitrary spin whose interaction is given by a sum of products of field operators, as our calculations did not depend on the specific properties of the fields or the interaction in any essential way. The $\phi(x_j)$ need not all come from the same field but can be various components of the various different fields in the theory, and $H_I(\tau)$ is the interaction-picture Hamiltonian given by the first equation in (6.90), the expression on the right of the second equation being replaced by the appropriate interaction for the theory in question. One merely has to exercise care, as always, to introduce appropriate minus signs when time-ordering the Fermion fields. (It is always legitimate to assemble all the H_I factors on the right, however, as they each contain even numbers of Fermionic operators.) Let us restate (6.94) with the minor notational change necessary to indicate this generality:

$$(6.96)\quad \langle\Omega|\mathcal{T}[\phi_1(x_1)\cdots\phi_n(x_n)]|\Omega\rangle = \lim_{T\to\infty(1-i\epsilon)} \frac{\langle 0|\mathcal{T}\big[\phi_{01}(x_1)\cdots\phi_{0n}(x_n)\exp(-i\int_{-T}^{T} H_I(\tau)\,d\tau)\big]|0\rangle}{\langle 0|\mathcal{T}\big[\exp(-i\int_{-T}^{T} H_I(\tau)d\tau)\big]|0\rangle}.$$

This result is known as the *Gell-Mann–Low formula*; it was first derived in [**52**].

There are several things that need to be said about the Gell-Mann–Low formula. First, if one writes the interaction Hamiltonian as the integral of a Hamiltonian density, $H_I(t) = \int \mathcal{H}_I(t,\mathbf{x})\,d^3\mathbf{x}$, and then lets $T\to\infty$ (ignoring the $i\epsilon$ for the time being), the typical term (6.95) in the numerator becomes

$$(6.97)\quad I_k(x_1,\ldots,x_n) = \left\langle 0\left|\frac{(-i)^k}{k!}\int_{\mathbb{R}^4}\cdots\int_{\mathbb{R}^4}\mathcal{T}[\phi_{01}(x_1)\cdots\phi_{0n}(x_n)\mathcal{H}_I(y_1)\cdots\mathcal{H}_I(y_k)]\,d^4y_1\cdots d^4y_k\right|0\right\rangle.$$

This is exactly the sort of thing we had to evaluate in computing S-matrix elements in §6.4, except that the creation and annihilation operators for initial and final particles have been replaced by the free-field operators $\phi_{0j}(x_j)$. Hence the value of (6.97) is computed just as before, as an integral of a sum of products of contractions of free-field operators, and it can be represented by a position-space Feynman diagram just as before. The only difference is that the external vertices are labeled by the points x_i and the external lines represent propagators. Conversely, given such a diagram with internal vertices labeled by y_j and external vertices labeled by x_i, one obtains a contribution to (6.97) that is an integral over the internal variables y_j of a product of propagators $-i\Delta_{j_1j_2}(y_{j_1}-y_{j_2})$ and $-i\Delta_k(x_k-y_j)$. (Note for the external lines one indeed has $\Delta_k(x_k - y_j)$ and not $\Delta_k(y_j - x_k)$, because the $\phi(x_k)$ in (6.96) are all on the left.)

It is more convenient to pass to the momentum-space representation, obtained by writing each propagator as a Fourier integral. However, since the external vertices here also involve propagators, one must pass to *their* momentum-space representation too, that is, to consider the Fourier transform

$$\widehat{I}_k(p_1,\dots,p_n) = \int\cdots\int I_k(x_1,\dots,x_n)e^{i\sum p_{j\mu}x_j^\mu}\,d^4x_1\cdots d^4x_n.$$

Writing each propagator in the form

$$-i\Delta(z) = -i\int \widehat{\Delta}(q)e^{-iq_\mu z^\mu}\,\frac{d^4q}{(2\pi)^4} \qquad (z = x_k - y_j \text{ or } y_{j_1} - y_{j_2}),$$

we obtain an integral over all position variables x_i and y_j and all internal momenta q of exponentials times Fourier-transformed propagators. The position variables appear only in the exponentials, so the integration over them reduces to integrals of the form $\int e^{-iq_\mu z^\mu}\,d^4q/(2\pi)^4$, where z is one of the vertices and q is the algebraic sum of the momenta at that vertex; these give the momentum-conservation delta-functions as before. The p_j are the momenta associated to the external lines, which here are arbitrary (not on mass shell), and the Fourier variables q are the momenta associated to the internal lines. Note that according to the remark at the end of the preceding paragraph and the conventions we established in §6.7, the external momenta always flow toward the external vertices.

This is the right place to account for the mysterious $i\epsilon$ in (6.96). It will be convenient to write $e^{-i\epsilon}$ instead of $1 - i\epsilon$; then the position-space integrals of the preceding paragraph are not really $\int e^{-iq_\mu z^\mu}\,d^4q/(2\pi)^4$ but

$$\lim_{T\to\infty e^{-i\epsilon}} \int_{\mathbb{R}^3}\int_{-T}^{T} e^{-ip_0z^0+i\mathbf{p}\cdot\mathbf{z}}\,\frac{dz^0\,d\mathbf{z}}{(2\pi)^4}.$$

If p_0 is real, this is not good: the integrand blows up exponentially at one end or the other, and even when one is doing informal distribution theory this is a disaster. But if we replace p_0 by $e^{i\epsilon}p_0$, the $e^{i\epsilon}$'s cancel and we obtain the purely oscillatory integral whose value is the delta-function we want. This modification of p_0 is one way to produce the correct $i\epsilon$ in the denominator of the Fourier-transformed propagator — we shall explain this in detail in §7.1 — so in fact the $i\epsilon$ fits into the picture perfectly.

We shall say more shortly about the relation between the Gell-Mann–Low formula and the calculation of S-matrix elements, but first we pause to explain the role of the denominator. The denominator in (6.96) is of the same form as the numerator, but without the factors $\phi_{0j}(x_j)$; thus, it is the sum of the values of all the Feynman diagrams with no external vertices. Such diagrams represent fluctuations of the vacuum, and we shall refer to them as *vacuum bubbles.* We encountered one such bubble for the ϕ^4 scalar field theory in Figure 6.4c; a few of them for QED are shown in Figure 6.9.

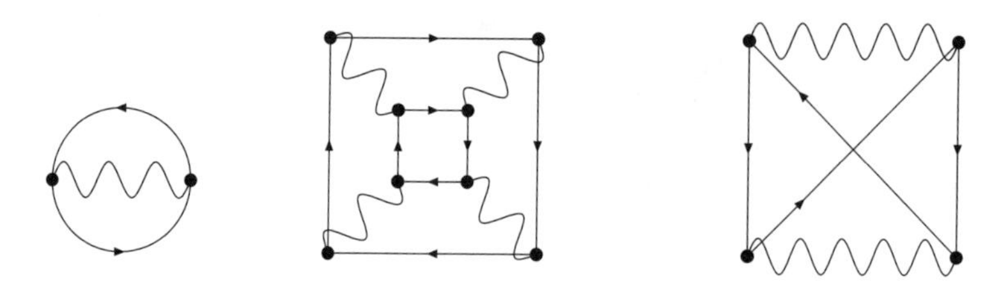

FIGURE 6.9. Some vacuum bubbles in QED.

Let $\{B_i\}_1^\infty$ be a list of all the connected vacuum bubbles for the theory in question. Then every vacuum bubble has the form $B = \sum_{i=1}^\infty n_i B_i$ for some nonnegative integers n_i of which all but finitely many are zero; that is, B is the union of n_1 copies of B_1, n_2 copies of B_2, etc. If V and V_i denote the values of B and B_i, respectively, we have

$$V = \prod_{i=1}^\infty \frac{1}{n_i!} V_i^{n_i}.$$

The $\prod n_i!$ is the symmetry factor of the bubble; it is there because permutation of the n_i copies of B_i does not change B. The denominator of (6.96) is the sum of all these quantities, namely,

$$\left\langle 0 \middle| \mathcal{T}\left[\exp\left(-i\int \mathcal{H}_I(y)\,d^4y\right)\right] \middle| 0 \right\rangle = \sum_{n_1,n_2,\ldots\geq 0,\ n_1+n_2+\cdots<\infty} \prod_{i=1}^\infty \frac{1}{n_i!} V_i^{n_i}.$$

If we provisionally assume that $\sum V_i < \infty$, so that $V_i \to 0$ in particular, we can drop the restriction that $n_1 + n_2 + \cdots < \infty$, since $\prod_{i=1}^\infty V_i^{n_i}/n_i! = 0$ whenever infinitely many n_i are nonzero. In this case we have

$$(6.98)\quad \left\langle 0 \middle| \mathcal{T}\left[\exp\left(-i\int \mathcal{H}_I(y)\,d^4y\right)\right] \middle| 0 \right\rangle = \prod_{i=1}^\infty \sum_{n_i=0}^\infty \frac{1}{n_i!} V_i^{n_i} = \exp\left(\sum_{i=1}^\infty V_i\right).$$

Now, in fact, the values V_i are given by integrals that are generally divergent. However, there are regularization procedures, which will be discussed in the next chapter, for expressing the V_i in a systematic way as limits of finite integrals, and one should replace the V_i by these regularized finite values. Having done this, however, there is still no reason for the series $\sum V_i$ to converge. Rather, both sides of (6.98) should be interpreted in perturbation theory. One thinks of $\mathcal{H}_I$ as containing a coefficient λ; then the V_i also contain various powers of λ, and one drops all terms containing powers of λ higher than N for a fixed but arbitrary N. Then (6.98) is valid with this interpretation, and the extra terms we inserted (with infinitely many $n_i \neq 0$) have no effect because they are of infinite order in λ. (This may seem very murky, but don't give up just yet!)

Now consider the Feynman diagrams contributing to the numerator of (6.96). In general they are disconnected, and many of them contain vacuum bubbles. Indeed, each diagram can be written in the form $D + \sum n_i B_i$ where D is a diagram with no vacuum bubbles (i.e., such that each connected component contains some external vertices). The value of this diagram is the value of D times the value of the bubble $\sum n_i B_i$, so by the preceding calculation, the *sum* of these values, over all possible bubbles, is the value of D times $\exp(\sum V_i)$. Hence, when one divides by the denominator (6.98), the values of all the vacuum bubbles (whatever devices one uses to interpret them) simply cancel out. In short:

The right side of the Gell-Mann–Low formula (6.96) *is the sum of all the position-space integrals represented by Feynman diagrams that contain no vacuum bubbles and that possess n external lines corresponding to the fields $\phi_1, \ldots, \phi_n$, in which the jth external line L_j represents the propagator $-i\Delta_j(x_j - y_j)$. (Here Δ_j is the propagator for ϕ_j and y_j is the variable of integration that labels the vertex at the internal end of L_j).*

It is frequently more convenient to pass to the momentum space representation by taking the Fourier transform. Thus, we set

(6.99)
$$W(p_1, \ldots, p_n) = \int \cdots \int \langle \Omega | \mathcal{T}[\phi_1(x_1) \cdots \phi_n(x_n)] | \Omega \rangle e^{ip_{1\mu}x_1^\mu + \cdots + ip_{n\mu}x_n^\mu} \, d^4x_1 \cdots d^4x_n.$$

In view of the remarks in the second paragraph following (6.96), the preceding result can then be restated as follows:

$W(p_1, \ldots, p_n)$ is the sum of all the momentum-space integrals represented by Feynman diagrams that contain no vacuum bubbles and that possess n external lines corresponding to the fields $\phi_1, \ldots, \phi_n$, in which the jth external line represents the Fourier-transformed propagator $-i\widehat{\Delta}_j(p_j)$ and p_j flows toward the external vertex.

The overall momentum-conservation delta-function that appears in all Feynman diagrams comes out of the translation invariance of the vacuum expectation values. That is, the translation-invariance of the vacuum $|\Omega\rangle$ implies that

$$\langle \Omega | \mathcal{T}[\phi_1(x_1) \cdots \phi_n(x_n)] | \Omega \rangle = \langle \Omega | \mathcal{T}[\phi_1(0)\phi_2(x_2 - x_1) \cdots \phi_n(x_n - x_1)] | \Omega \rangle. \tag{6.100}$$

Substituting this into (6.99) and then making the change of variable $x_j = x'_j + x_1$ turns the integrand of (6.99) into a function whose only dependence on x_1 is a factor $e^{i(\sum p_j)_\mu x_1^\mu}$; integration over x_1 then gives $(2\pi)^4 \delta(\sum p_j)$.

We now return to the relation between vacuum expectation values of products of fields and S-matrix elements, that is, the problem of calculating

$$\langle 0 | a_1(\mathbf{p}_1^{\text{out}}) \cdots a_J(\mathbf{p}_J^{\text{out}}) S a_1^\dagger(\mathbf{p}_1^{\text{in}}) \cdots a_K^\dagger(\mathbf{p}_K^{\text{in}}) | 0 \rangle \tag{6.101}$$

from

$$\langle \Omega | \mathcal{T}[\phi_1(x_1) \cdots \phi_n(x_n)] | \Omega \rangle. \tag{6.102}$$

We shall assume, as we did in §6.8, that no proper subset of $\{\mathbf{p}_1^{\text{in}}, \ldots, \mathbf{p}_K^{\text{in}}\}$ has the same sum as any proper subset of $\{\mathbf{p}_1^{\text{out}}, \ldots, \mathbf{p}_J^{\text{out}}\}$, so that the only Feynman diagrams that contribute to (6.101) are connected ones. (The analysis in the general case is similar but a little more complicated.) The first step is to convert the position dependence of (6.102) into momentum dependence by taking the Fourier transform (6.99). We are going to take $n = J + K$ and the p_j to be, essentially, the on-mass-shell 4-momenta corresponding to the 3-momenta in (6.101), but there has to be a twist because the momenta in (6.99) all flow toward the external vertices, whereas the momentum $\mathbf{p}_k^{\text{in}}$ in (6.101) should flow away from its initial vertex. (Recall (6.43), which shows that an incoming momentum $\mathbf{p}$ contributes a factor $e^{-ip_\mu x^\mu}$ to the position-space integral for the S-matrix element, whereas an ougtoing momentum contributes a factor of $e^{ip_\mu x^\mu}$; the sign is important because $p_0 = \omega_{\mathbf{p}} > 0$ in either case.) Thus, to get the right momentum dependence we must consider

$$W(p_1, \ldots, p_J, -p_{J+1}, \ldots, -p_{J+K}) \qquad (p_l = (\omega_{\mathbf{p}_l}, \mathbf{p}_l)),$$

where

$$\mathbf{p}_j = \mathbf{p}_j^{\text{out}} \text{ for } 1 \leq j \leq J, \qquad \mathbf{p}_{J+k} = \mathbf{p}_k^{\text{in}} \text{ for } 1 \leq k \leq K.$$

(Note that the momentum-conservation delta-function herein is $\delta(\sum_j p_j - \sum_k p_{J+k})$, as it should be.)

The preceding discussion shows that this quantity is the sum of the values of all connected Feynman diagrams with external lines labeled by the momenta $\mathbf{p}_k^{\text{out}}, \mathbf{p}_j^{\text{in}}$ and representing the Fourier-transformed propagators $-i\widehat{\Delta}_j(p_j^{\text{out}})$, $-i\widehat{\Delta}_k(-p_k^{\text{in}})$. On the other hand, (6.101) is the sum of the values of all Feynman diagrams of the same sort in which external lines represent the appropriate coefficient functions $u_k(\mathbf{p}_k^{\text{in}})$ and $u_j^*(\mathbf{p}_j^{\text{out}})$. Thus, to get from (6.99) to (6.101) one has merely to replace the external propagators by coefficient functions. The result is

$$\begin{aligned}
&\langle 0 | a_1(\mathbf{p}_1^{\text{out}}) \cdots a_J(\mathbf{p}_J^{\text{out}}) S a_1^\dagger(\mathbf{p}_1^{\text{in}}) \cdots a_K^\dagger(\mathbf{p}_K^{\text{in}}) | 0 \rangle \\
&= i^{J+K} \prod_{j=1}^{J} \frac{u_j^*(\mathbf{p}_j^{\text{out}})}{\widehat{\Delta}_j(p_j^{\text{out}})} \prod_{k=1}^{K} \frac{u_k(\mathbf{p}_k^{\text{in}})}{\widehat{\Delta}_k(-p_k^{\text{in}})} W(p_1^{\text{out}}, \ldots, p_J^{\text{out}}, -p_1^{\text{in}}, \ldots, -p_K^{\text{in}}),
\end{aligned} \tag{6.103}$$

where W is as in (6.99). This is known as the *Lehmann-Symanzik-Zimmermann reduction formula*, or *LSZ formula* for short.

For a scalar field, the function Δ is even, and it is a fundamental solution for the Klein-Gordon operator $\partial^2 + m^2$, so dividing by $\widehat{\Delta}$ in momentum space is the same as applying the operator $\partial^2 + m^2$ in position space. Thus, for scalar fields the LSZ formula can be restated as

$$\begin{aligned}
&\langle 0 | a_1(\mathbf{p}_1^{\text{out}}) \cdots a_J(\mathbf{p}_J^{\text{out}}) S a_1^\dagger(\mathbf{p}_1^{\text{in}}) \cdots a_K^\dagger(\mathbf{p}_K^{\text{in}}) | 0 \rangle \\
&= i^{J+K} \prod_{j,k} u_j^*(p_j^{\text{out}}) u_k(p_k^{\text{in}}) \int \cdots \int \exp\left[i \sum_j p_{j\mu}^{\text{out}} x^{j\mu} - i \sum_k p_{k\mu}^{\text{in}} x_{J+k}^\mu \right] \times \\
&\quad (\partial_{x_1}^2 + m^2) \cdots (\partial_{x_{J+K}}^2 + m^2) \langle \Omega | \mathcal{T}[\phi_1(x_1) \ldots \phi_{J+K}(x_{J+K})] | \Omega \rangle \, d^4x_1 \cdots d^4x_{J+K}.
\end{aligned}$$

Similarly, the Dirac propagator is the fundamental solution for the Dirac operator $-i\gamma^\mu \partial_\mu + m$, so when spin-$\frac{1}{2}$ fields are involved one obtains a similar formula with $\partial_{x_j}^2 + m^2$ replaced by $-i\gamma^\mu(\partial_{x_j})_\mu + m$ and $\partial_{x_{J+k}}^2 + m^2$ replaced by $i\gamma^\mu(\partial_{x_{J+k}})_\mu + m$.

There is one technicality that must be mentioned here. As we shall discuss in the next chapter, to carry out the evaluation of Feynman diagrams involving loops, it is necessary to renormalize the field operators, that is, to multiply the field ϕ by a constant that is conventionally called $Z^{-1/2}$. (If the theory involves several different fields, each one has its own renormalization factor.) *The fields* ϕ_j *in* (6.99) *and hence in the LSZ formula* (6.103) *must be taken to be renormalized fields.* Otherwise, the appropriate Z factors must be included in the formula.

We shall content ourselves with an informal qualitative explanation of this fact. The Feynman-diagram picture suggests that particles propagate from one vertex to the next without anything happening in between, but in fact they have encounters with virtual particles along the way. As we shall see in §7.6, this effectively turns the $1/(-p^2 + m^2)$ in the propagator into $Z/(-p^2 + (m + \delta m)^2)$ for some constants Z and δm. The δm is a mass shift of the sort we encountered in §6.2, but the Z must be compensated for by renormalizing the field. The point is that the fields in the contractions that give rise to the *external* propagators in (6.102) must also be adjusted in this way, and this gives rise to the extra factors of Z.

We derived the LSZ formula on a diagram-by-diagram basis, that is, in perturbation theory. The physics literature also contains a number of (nonrigorous) nonperturbative derivations of it; for example, Itzykson and Zuber [**65**], §5-1-3, Peskin and Schroeder [**88**], §7.2, and Weinberg [**129**], §10.3. In addition, it is possible to prove the LSZ formula rigorously within the setting of the Wightman axioms and the Haag-Ruelle scattering theory; see Bogolubov et al. [**11**], [**12**], and Araki [**3**].

CHAPTER 7

Renormalization

O God, I could be bounded in a nutshell and count myself a king of infinite space, were it not that I have bad dreams.
—W. Shakespeare, *Hamlet* (II.2)

In this chapter we explore the basic ideas of renormalization in quantum field theory. Our main objective is to make sense out of the divergent integrals represented by Feynman diagrams so as to extract physical information from them. In broad outline, the strategy is as follows:

1. Analyze the Feynman diagrams so as to identify which of them correspond to divergent integrals and to understand just what sorts of divergence are to be expected.
2. Modify the divergent integrals to make them converge; that is, express them as limits of convergent integrals that depend on a parameter ϵ as $\epsilon \to 0$. (There are several ways to do this.) The regularized integrals also depend on the parameters in the Lagrangian: masses, coupling constants, and normalizations of the fields. The key point is that the presence of interactions affects all of these quantities, and a careful examination of the relation of the regularized integrals to quantities that can be measured in the laboratory reveals that the parameters in the Lagrangian are not the true physical ones.
3. Determine the parameters in the Lagrangian as functions of the regularization parameter ϵ, with singularities as $\epsilon \to 0$, in such a way that the divergences all cancel out and yield physically meaningful results in terms of the physical masses and coupling constants.
4. Feynman diagrams are creatures of perturbation theory, so all of the preceding work is to be done to a given finite order in perturbation theory. Each additional order of perturbation theory requires more Feynman diagrams and more readjustments of parameters, and an argument must be made that this can be done consistently to arbitrarily high order.

The first two steps are relatively easy, and we shall study them in a fairly general context. The fourth step presents some interesting and highly nontrivial issues, but it is of limited relevance to practical calculations, which always involve only low-order perturbation theory. We shall give only a brief discussion of it, with references to the literature.

The third step is difficult — it presented an obstacle to the development of quantum field theory for two decades before Feynman, Schwinger, and Tomonaga showed how to deal with it in the case of QED — and the actual evaluation of the integrals for all but the simplest Feynman diagrams is very laborious. We shall restrict attention to two specific examples, QED and the ϕ^4 scalar field theory, and we shall work out the calculations only for the basic one-loop diagrams. Even to do this much requires some rather lengthy computations, and most mathematical tourists will probably not want to go through all of them in detail. The author will sympathize if they wish to skim over the parts where the density of symbols exceeds some (reader-dependent) critical value and proceed to the discussion of the physical consequences. On the other hand, if one wants to feel confident that one really understands how renormalization works, one must be willing to undertake a certain amount of toil.

Tricky though this subject is, it entails the commission of almost no mathematical sins beyond the acceptance of perturbation theory that we agreed to at the beginning of Chapter 6. Only at one point (in §7.6) will we need to step outside of perturbation theory and ask the devil for a bit more assistance in understanding how to proceed.

7.1. Introduction

To warm up, and to make the idea of "subtracting off infinities" seem less mysterious to those who are not aficionados of distribution theory, we begin by discussing a purely mathematical problem that will illustrate some of the ideas of renormalization theory.

Let ϕ be a smooth compactly supported function on $\mathbb{R}$. If $\phi(0) \neq 0$, the integral $\int_{-\infty}^{\infty} |t|^{-1}\phi(t)\,dt$ is unambiguously divergent. The problem is to find ways of extracting a well-defined "finite part" from it — more precisely, to find distributions F on $\mathbb{R}$ that agree with the function $f(t) = |t|^{-1}$ away from the origin. We observe at the outset that if F is one such distribution, then so is $F + G$ where G is any distribution supported at the origin, that is, any linear combination of the delta-function and its derivatives. We can remove most of this ambiguity by stating the defining condition for F in the following stronger form: we require that $tF(t) = \operatorname{sgn} t$ *as distributions*; that is, if $\phi(t) = t\psi(t)$, then $\langle F, \phi\rangle = \int (\operatorname{sgn} t)\psi(t)\,dt$. Since $t\delta^{(n)}(t) = n\delta^{(n-1)}(t)$, this reduces the family of allowable G's to the scalar multiples of δ itself.

There are several ways to solve this problem. Here is one: pick a positive number a and define the distribution F_a by

$$\langle F_a, \phi\rangle = \int_{|t|\le a} \frac{\phi(t) - \phi(0)}{|t|}\,dt + \int_{|t|>a} \frac{\phi(t)}{|t|}\,dt.$$

The integrals on the right are absolutely convergent, the first because $\phi(t) - \phi(0) = O(t)$ and the second because ϕ is compactly supported; moreover, if $\phi(t) = t\psi(t)$, then $\phi(0) = 0$ and we have $\langle F_a, \phi\rangle = \int (\operatorname{sgn} t)\psi(t)\,dt$ as desired. Now, formally we have

$$\langle F_a, \phi\rangle \text{ ``=''} \int_{\mathbb{R}} \frac{\phi(t)}{|t|}\,dt - \phi(0)\int_{|t|\le a} \frac{dt}{|t|},$$

so F_a can be considered as arising from the original function $|t|^{-1}$ by subtraction of $C\delta$ where C is the infinite constant $\int_{-a}^{a} dt/|t|$. Moreover, if $a, b > 0$, F_a and F_b differ by a *finite* multiple of δ:

$$\langle F_b, \phi\rangle - \langle F_a, \phi\rangle = -\phi(0)\left(\int_{-b}^{-a} + \int_{a}^{b}\right)\frac{dt}{|t|} = -2\phi(0)\log\frac{b}{a},$$

so $F_b - F_a = -2\log(b/a)\delta$. To decide which value of a to use, one needs to impose some additional condition on F. The infinite multiple of δ that is subtracted here corresponds to the "counterterms" that are subtracted from Feynman integrals to produce finite results, and the determination of the remaining finite multiple of δ corresponds to matching some finite result to an experimentally determined quantity.

Here is another approach to this problem that is more in the spirit of the dimensional regularization that we shall use for Feynman diagrams. For $\operatorname{Re} z > -1$, let $T^z(t) = |t|^z$. T^z is an analytic distribution-valued function of z in the half-plane $\operatorname{Re} z > -1$, where the function $|t|^z$ is locally integrable. It can be meromorphically continued to the half plane $\operatorname{Re} z > -2$ (and hence, by bootstrapping, to the entire complex plane) by the relation

$$T^z = \frac{d}{dt}\left[(\operatorname{sgn} t)\frac{T^{z+1}}{z+1}\right].$$

There is a simple pole at $z = -1$, and the residue there is 2δ, because

$$\lim_{z\to -1}(z+1)\langle T^z, \phi\rangle = -\lim_{z\to -1}\langle(\operatorname{sgn} t)T^{z+1}, \phi'\rangle = \int_{-\infty}^{0}\phi'(t)\,dt - \int_0^\infty \phi'(t)\,dt = 2\phi(0).$$

We can therefore obtain a distributional version of $|t|^{-1}$ by passing to the limit as $z \to -1$ while subtracting off the singular part of the Laurent expansion. The result is the distribution F_1 of the preceding paragraph, because $\int_{-1}^{1}|t|^z\,dt = 2/(z+1)$:

$$\lim_{z\searrow -1}\left\langle T^z - \frac{2\delta}{z+1}, \phi\right\rangle = \lim_{z\searrow -1}\left[\int_{|t|\le 1}[\phi(t)-\phi(0)]|t|^z\,dt + \int_{|t|>1}|t|^z\,dt\right] = \langle F_1, \phi\rangle.$$

To see a situation where one might want F_a with $a \neq 1$, let us consider the Fourier transform. A classic calculation that is outlined in Exercise 27 of §9.2 of Folland [**47**][1] shows that for $-1 < \operatorname{Re} z < 0$,

$$\widehat{T^z} = \frac{\sqrt{\pi}\,2^{1+z}\Gamma(\frac{1}{2}(1+z))}{\Gamma(-\frac{1}{2}z)}T^{-1-z},$$

and hence

$$\widehat{T^z} - \frac{2\widehat{\delta}}{z+1} = \frac{2}{z+1}\left[\frac{\sqrt{\pi}\,2^{1+z}\Gamma(\frac{1}{2}(3+z))}{\Gamma(-\frac{1}{2}z)}T^{-1-z} - 1\right].$$

Taking the limit as $z \to -1$ via l'Hôpital's rule shows that $\widehat{F_1}$ is the locally integrable function

$$\widehat{F_1}(s) = 2\left[\log 2 + \Gamma'(1) - \frac{\Gamma'(\frac{1}{2})}{\sqrt{\pi}} - \log|s|\right].$$

One might now wish to specify a so that $\widehat{F_a}$ is the function $-2\log|s/c|$ for some given $c > 0$. Since $F_a = F_1 - 2(\log a)\delta$, the solution is

$$a = \frac{2}{c}\exp\left(-\Gamma'(1) - \frac{\Gamma'(\frac{1}{2})}{\sqrt{\pi}}\right).$$

Having refreshed ourselves with this mathematical snack, let us return to physics. The general problem we will be considering is the evaluation of a connected Feynman diagram for some field theory, with

[1] A different convention for the placement of the 2π's is used there.

V internal vertices,
I internal lines,
E external lines (and external vertices),

and hence, by (6.64),

$L = I - V + 1$ independent loops.

We recall that each external line in the diagram is labelled with an external momentum $p_j = (\omega_{\mathbf{p}_j}, \mathbf{p}_j)$, and each internal line is labeled with an internal momentum q_i. The integrand for the momentum-space integral defined by the diagram is the product of Fourier-transformed propagators $-i\widehat{\Delta}_i(q_i)$ (one for each internal line), delta-functions $\delta_v(p,q) = \delta(\Sigma_v)$ for each vertex v (Σ_v being is the algebraic sum of the momenta at v), certain factors coming from the external lines, and factors of 2π, i, and coupling constants. In more detail, the propagators have certain forms depending on the particle species they represent, for example,

$$\widehat{\Delta}_F(q) = \frac{1}{-q^2 + m^2 - i\epsilon} \qquad \text{for a scalar field of mass } m,$$

$$\widehat{\Delta}^{\text{Dirac}}(q) = \frac{q_\mu \gamma^\mu + m}{-q^2 + m^2 - i\epsilon} \qquad \text{for a Dirac field of mass } m,$$

$$\widehat{\Delta}^{\text{Proca}}_{\mu\nu}(q) = \frac{g_{\mu\nu} - (q_\mu q_\nu / m^2)}{q^2 - m^2 + i\epsilon} \qquad \text{for a vector field of mass } m,$$

$$\widehat{\Delta}_{\mu\nu}(q) = \frac{g_{\mu\nu}}{q^2 + i\epsilon} \qquad \text{for a massless vector field.}$$

There is an implicit "$\lim_{\epsilon\to 0+}$" in these formulas; we shall say more about it later, but for now one should think of ϵ as a small but positive number. The integral corresponding to the diagram is then

$$C(p) \int \cdots \int \prod_{\text{vertices}} \delta_v(p,q) \prod_{\text{internal lines}} \widehat{\Delta}_i(q_i)\, d^4 q_i,$$

where $C(p)$ is a factor that incorporates the coefficients from the external lines, coupling constants, factors of 2π, etc.

If there are V vertices, the equations $\Sigma_v = 0$ give V linear equations in the internal and external momenta, of which $V-1$ are independent; the compatibility condition for a solution to exist is the overall conservation of external momenta, $\sum p_j^{\text{out}} = \sum p_k^{\text{in}}$. We use $V-1$ of the delta-functions to dispose of $V-1$ of the integrals and to express $V-1$ of the internal momenta in terms of the rest of them and the external momenta. This having been done, let the remaining internal momenta be (re)labeled as $q_1, \ldots, q_L$ and the external momenta as $(p_1, \ldots, p_E)$; we write $q = (q_1, \ldots, q_L)$ and $p = (p_1, \ldots, p_E)$ for short. Then the integral reduces to

$$C(p)\delta\left(\sum p_j^{\text{out}} - \sum p_k^{\text{in}}\right) \int \cdots \int \prod_{i=1}^{I} \widetilde{\Delta}_i(q,p)\, d^{4L} q. \tag{7.1}$$

Here $\widetilde{\Delta}_i(q,p) = \widehat{\Delta}_i(q_i)$ for $i \leq L$; and for $i > L$, $\widetilde{\Delta}_i(q,p)$ is $\widehat{\Delta}_i(q_i)$ with q_i replaced by its expression in terms of $q_1, \ldots, q_L$ and $p_1, \ldots, p_E$.

The principal difficulty here — what we will be focusing on in the ensuing discussion — is that the integral may diverge because the integrand does not decay fast enough at infinity ("ultraviolet divergence"). When one passes to the limit $\epsilon \to 0+$, it looks as though things will get worse because the factors $-q^2 + m^2$ in the denominator will blow up along the mass shells. This turns out not to be much of a problem as long as the masses m are all positive. We shall say more about this shortly, but for now it may reassure the reader to observe that the function $\lambda(q) = -q^2 + m^2$ vanishes only to first order on the mass shell $q^2 = m^2$, so in suitable local coordinates one is considering integrals similar to $\int_{-1}^{1} f(t)\,dt/(t - i\epsilon)$ where f is smooth, and this has a finite limit as $\epsilon \to 0+$:

$$\lim_{\epsilon\to 0+} \int_{-1}^{1} \frac{f(t)}{t - i\epsilon}\,dt = \text{p.v.} \int_{-1}^{1} \frac{f(t)}{t}\,dt + i\pi f(0).$$

However, if the theory involves massless particles, the singularity of $1/q^2$ at the origin will cause additional headaches ("infrared divergence"). *We therefore assume in this general discussion that the masses of all the field quanta in question are positive, or — as a temporary expedient — that any massless denominators $q^2 + i\epsilon$ have been replaced with $q^2 - \mu^2 + i\epsilon$ where μ is a small positive number.* We shall comment on the removal of this assumption for QED near the end of §7.8.

The analysis of the integrals (7.1) is greatly facilitated by a device known as *Wick rotation*, which is essentially an analytic continuation of the energy variables to imaginary values. To derive this, we need to insert the $i\epsilon$'s in the propagator denominators in a slightly different fashion. To be specific, let us consider the Feynman propagator $\Delta_F(t, \mathbf{x})$. We recall from §6.5 that Δ_F is the distribution whose Fourier transform in $\mathbf{x}$ is the locally (but not globally) integrable function $\mathcal{F}_{\mathbf{x}}\Delta_F(t, \mathbf{q}) = ie^{-i\omega_{\mathbf{q}}|t|}/2\omega_{\mathbf{q}}$ ($\omega_{\mathbf{q}} = \sqrt{|\mathbf{q}|^2 + m^2}$). In §6.5 we derived the formula

$$\widehat{\Delta}_F(q) = \lim_{\epsilon\to 0+} \frac{1}{-\omega^2 + |\mathbf{q}|^2 + m^2 - i\epsilon}$$

for its full Fourier transform by considering the function $t \mapsto ie^{-i\omega_{\mathbf{q}}|t|}/2\omega_{\mathbf{q}}$ as the limit as $\epsilon \to 0+$ of the L^1 function

$$t \mapsto -\frac{\exp[-\sqrt{i\epsilon - \omega_{\mathbf{q}}^2}\,|t|]}{2\sqrt{i\epsilon - \omega_{\mathbf{q}}^2}},$$

where the square root is the one with positive real part. However, we could also consider it as the limit as $\epsilon \to 0+$ of the L^1 function

$$t \mapsto \frac{ie^{-i\epsilon}\exp[-ie^{-i\epsilon}\omega_{\mathbf{q}}\,|t|]}{2\omega_{\mathbf{q}}},$$

and the Fourier transform of this function, evaluated at $q^0 \in \mathbb{R}$, is

$$\begin{aligned}
&\frac{ie^{-i\epsilon}}{2\omega_{\mathbf{q}}}\left[\int_{-\infty}^{0} \exp[(ie^{-i\epsilon}\omega_{\mathbf{q}} - iq^0)t]\,dt + \int_0^{\infty} \exp[(-ie^{-i\epsilon}\omega_{\mathbf{q}} - iq^0)t]\,dt\right]\\
&= \frac{ie^{-i\epsilon}}{2\omega_{\mathbf{q}}}\left[\frac{1}{ie^{-i\epsilon}\omega_{\mathbf{q}} - iq^0} + \frac{1}{ie^{-i\epsilon}\omega_{\mathbf{q}} + iq^0}\right]\\
&= \frac{1}{-(e^{i\epsilon}q^0)^2 + |\mathbf{q}|^2 + m^2}.
\end{aligned}$$

In short, instead of incorporating the $-i\epsilon$ into the expression $-q^2 + m^2$ as an additive term, we can incorporate it as a multiplicative coefficient $e^{i\epsilon}$ of q^0. Either way, the effect is to add a small negative imaginary part. The only significant difference occurs in the case $m = 0$, where the new procedure does nothing to mollify the singularity of $1/q^2$ at the origin — but we have agreed to exclude this case.

The same idea applies to all the other propagators. Thus we may write the propagator denominators in (7.1) as

$$-(e^{i\epsilon}q^0)^2 + |\mathbf{q}|^2 + m^2 \tag{7.2}$$

rather than$-q^2 + m^2 - i\epsilon$. Doing so, of course, destroys the manifest Lorentz covariance, which is recovered only in the limit as $\epsilon \to 0$. Concerning the latter, we have the following rigorous theorem for the case where the integrals with suitably incorporated ϵ's are absolutely convergent.

Let $R(q,p)$ denote the integrand of (7.1), *considered simply as a rational function of $q = (q_1, \ldots, q_L)$ and $p = (p^{\text{in}}, p^{\text{out}})$ (without ϵ's), and let $q^\epsilon = (q_1^\epsilon, \ldots, q_L^\epsilon)$ where $q_l^\epsilon = (e^{i\epsilon}q_l^0, \mathbf{q}_l)$, and likewise for p^ϵ. If the integrals $I_\epsilon(p) = \int R(q^\epsilon, p^\epsilon)\, d^{4L}q^\epsilon$ ($p \in \mathbb{R}^{4E}$) are absolutely convergent for some $\epsilon \in (0, \pi)$, then they are absolutely convergent for every such ϵ. In this case, as $\epsilon \to 0+$, I_ϵ converges in the weak topology of tempered distributions on $\mathbb{R}^{4E}$ to a distribution with the appropriate Lorentz-covariance properties.*

This result is due to Zimmermann [**137**]; see also Manoukian [**79**].

Having gone this far, the temptation is irresistible to let ϵ tend not to 0 but to $\frac{1}{2}\pi$. Indeed, let us agree that, as in the preceding theorem, we shall insert a factor of $e^{i\epsilon}$ not only into the internal energy variables q_l^0 of (7.1) but into the external ones as well. Then (7.1) becomes, in effect, an integral in which the energy variables range not over $\mathbb{R}$ but over $e^{i\epsilon}\mathbb{R}$, and the result of passing to $\epsilon = \frac{1}{2}\pi$ is that the the Lorentz norms $-q^2 = -(q^0)^2 + |\mathbf{q}|^2$ in the denominators in (7.1) are replaced by Euclidean norms $|q|^2 = (q^0)^2 + |\mathbf{q}|^2$. We say that (7.1) has been changed from a *Minkowski-space integral* into a *Euclidean-space integral* by *Wick rotation.*

In summary, Wick rotation transforms an integral of the form (7.1) into an integral of the form

$$\int \cdots \int \prod_{i=1}^{I} \frac{P_i(q,p)}{|f_i(q,p)|^2 + m_i^2}\, d^{4L}q \tag{7.3}$$

where the $P_i(q,p)$ are polynomials, the $f_i(p,q)$ are linear combinations of the q's and p's, and $m_i > 0$. Moreover, transformation properties of the original integral under the Lorentz group turn into transformation properties of the Wick-rotated integral under the rotation group $SO(4)$.

7.2. Power counting

The next order of business is to develop a way of telling whether the integrals (7.1), or rather their Wick-rotated versions (7.3), converge (absolutely) or not. Now, one can consider such integrals in which the variables q_l and p_m belong not to $\mathbb{R}^4$ but to $\mathbb{R}^d$ for any positive integer d, and it will make the general structure clearer (and serve as a warmup for some future manipulations) to do so. Let $\deg_+(R)$ and $\deg_-(R)$ be the degrees of the numerator and denominator of R, and let $\deg(R) = \deg_+(R) - \deg_-(R)$. Then for large $|q|$ we have $|R(q)| \geq C|q|^{\deg(R)}$ except perhaps in some sectors where the numerator grows more slowly than $|q|^{\deg_+(R)}$, so a crude integration in polar coordinates shows that the integral (7.3) (over $(\mathbb{R}^{dL})$ has no hope of converging unless $\deg(R) + dL < 0$.

We define the *superficial degree of divergence superficial degree of divergence* of the integral, or of its associated Feynman diagram, to be

$$D = \deg(R) + dL. \tag{7.4}$$

Thus, the integral is surely divergent if $D \geq 0$.

The graph may also be divergent if $D < 0$, because there may be sectors where the denominator grows more slowly than $|q|^{\deg_-(R)}$ and hence some subintegrations that diverge. (A simple example: The rational function $R(x,y) = 1/(1+x^2)^2$ on $\mathbb{R}^2$ has $\deg(R) = -4$ while $d = 2$, so crude power counting would predict that $\iint R(x,y)\,dx\,dy$ should converge, but of course it does not because R has no decay in y.) Pictorially, the difficulty arises from the fact that superficially convergent diagrams may contain divergent subdiagrams. An example from QED is shown in Figure 7.1.

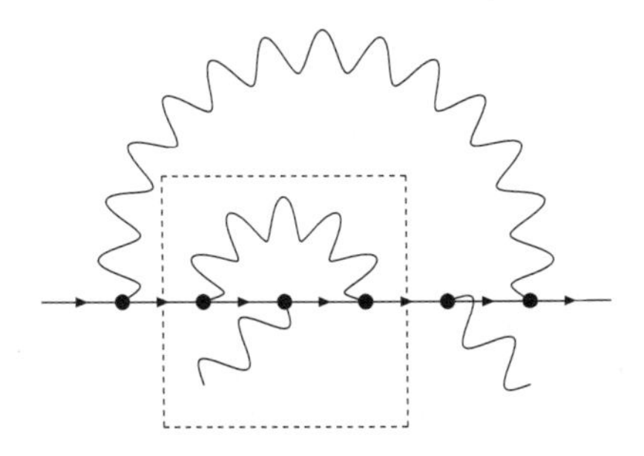

FIGURE 7.1. A superficially convergent QED diagram with a divergent subdiagram (enclosed by dotted lines).

However, the situation is no more complicated than these remarks would indicate. Indeed, we have the following basic result, known as *Weinberg's theorem* (see Weinberg [**127**], Hahn and Zimmerman [**62**], and Manoukian [**79**]; also Zimmermann [**139**] for an extension to the case where massless fields are involved):

Let $R(p,q)$ denote the integrand in (7.3). The integral (7.3) is convergent if and only if the superficial degrees of divergence of all the subintegrals $\int_V R(p,q)\,d^{dk}q$ (including the original integral) are negative, where V is any dk-dimensional affine subspace of $\mathbb{R}^{dL}$ $(k = 1,\dots,L)$ defined by a set of linear equations $\sum_{j=1}^{L} a_i^j q_j = c_i$, $i = 1,\dots,L-k$.

When one restricts the variables q in (7.3) to affine subspaces, the situation that tends to produce the greatest increase in the superficial degree of divergence is where one or more of the linear forms $f_i(q,p)$ is constant on the domain of integration, that is, where the momentum associated to one or more of the propagators is constant. As far as convergence is concerned, making one of the propagators constant in this way is equivalent to cutting the corresponding line in the Feynman diagram. As a result, one has the following diagrammatic version of Weinberg's theorem:

The integral corresponding to a connected Feynman diagram is convergent if and only if the superficial degrees of divergence of the diagram and all its connected subdiagrams are negative.

It is convenient to replace the parameters $\deg(R)$ and L in (7.4) by quantities that can be read off more directly from the diagram. One of these should be the number of external lines, as we are usually interested in processes with a fixed set of incoming and outgoing particles. For this purpose we can use the identity $L = I - V + 1$ together with identities relating the number of vertices to the number of edges, which will depend on the nature of the interactions in the diagram. We shall work this out for some important examples below, but first there is one general point to be addressed.

In general, one expects each internal line of a connected Feynman diagram to contribute a nonconstant factor to the integrand in (7.3). However, if some line has the property that cutting it would disconnect the diagram, its associated momentum is completely determined by the external momenta: up to a factor of ± 1, it is the sum of the external momenta in each of the two subdiagrams that the line connects. In this situation the integral (7.3) is essentially the product of the two integrals associated to these subdiagrams. Hence, to analyze convergence it is enough to consider diagrams that cannot be disconnected by cutting one internal line; such diagrams are called *one-particle irreducible (1PI)*.

We now analyze the superficial degrees of divergence of the 1PI Feynman diagrams for the particular field theories that we listed in §6.3. In this discussion *"vertex" always means "internal vertex."*

For the ϕ^4 scalar field theory, each external line connects to one vertex and each internal line connects to two, whereas each vertex connects to four lines; hence $2I + E = 4V$. Moreover, the degree of each propagator is -2, and there is one propagator for each internal line, so $\deg(R) = -2I$. Therefore,

$$\begin{aligned} D = -2I + dL &= -2I + d(I - V + 1) = (d-2)I - dV + d \\ &= \tfrac{1}{2}(d-2)(4V - E) - dV + d = (d-4)V - \tfrac{1}{2}(d-2)E + d. \end{aligned}$$

In the "real-world" case $d = 4$, we have simply $D = 4 - E$. Thus, only diagrams with at most four external lines are superficially divergent. The superficial degree of divergence is 4 for vacuum bubbles (diagrams with no external lines) and at most 2 otherwise. In dimensions $d = 2$ or 3, things are even better: here there are actually only finitely many superficially divergent diagrams, as the degree of divergence decreases with the number of vertices. However, in dimensions $d \geq 5$, there are diagrams with arbitrarily high degrees of divergence.

For QED, we have to distinguish between electron lines and photon lines; we indicate the former with a subscript e and the latter with a subscript γ. The electron propagator has degree -1 and the photon propagator has degree -2, so $\deg(R) = -I_e - 2I_\gamma$. Moreover, each vertex meets two electron lines and one photon line, so $2I_e + E_e = 2V$ and $2I_\gamma + E_\gamma = V$, and $L = I_e + I_\gamma - V + 1$. Therefore,

$$\begin{aligned} D &= -I_e - 2I_\gamma + d(I_e + I_\gamma - V + 1) = (d-1)I_e + (d-2)I_\gamma - dV + d \\ &= (d-1)(V - \tfrac{1}{2}E_e) + \tfrac{1}{2}(d-2)(V - E_\gamma) - dV + d \\ &= \tfrac{1}{2}(d-4)V - \tfrac{1}{2}(d-1)E_e - \tfrac{1}{2}(d-2)E_\gamma + d. \end{aligned}$$

For $d = 4$ we have $D = 4 - \frac{3}{2}E_e - E_\gamma$, so again D depends only on the number of external lines and is nonnegative only when that number is small; it is 4 for vacuum bubbles and at most 2 otherwise (with equality for $E_e = 0$, $E_\gamma = 2$). For $d < 4$ the situation again improves, and for $d > 4$ it again deteriorates.

The result is exactly the same for the Yukawa interaction, as the propagators for photons and pions both have degree -2. (The Yukawa interaction is simpler than QED in one respect: since pions are massive, it has no infrared divergences.)

Finally, the Fermi model for beta decay involves a four-Fermion interaction. The Feynman diagrams here have the property that (i) every line represents a particle of spin $\frac{1}{2}$, whose propagator has degree -1, and (ii) four lines, one for each particle species, meet at every vertex. Hence, with notation as above, we have $2I + E = 4V$ and $\deg(R) = -I$, from which we obtain

$$D = (2d-3)V - \tfrac{1}{2}(d-1)E + d.$$

Thus there are divergences of arbitrarily high orders in all dimensions $d \geq 2$. It was recognized early on that this presented a serious problem and implied that the Fermi model could be trusted only as a low-energy approximation. Eventually it was supplanted by the Glashow-Salam-Weinberg theory, which we shall discuss in §9.4.

It is not hard to generalize the calculations in the preceding examples to arbitrary field theories. We refer the reader to Weinberg [**129**], §12.1, for a detailed discussion. Here we just state the main result: In a field theory in d dimensions involving several different fields f and several different interactions i, the superficial degree of divergence of a Feynman diagram has the form

$$D = d - \sum_f a_f E_f - \sum_i b_i V_i \tag{7.5}$$

where d is the space-time dimension, E_f is the number of external lines of field type f, V_i is the number of vertices of interaction type i, and the a_f and b_i are coefficients depending on the particular field theory. Specifically, a_f is the number such that the Fourier-transformed propagator $\widehat{\Delta}_f(p)$ for field f has degree $-d + 2a_f$, and b_i is the number such that the coupling constant for the interaction i has dimension $[m^{b_i}]$ (in natural units where $[l] = [t] = [m^{-1}]$). Since the degree of any propagator is at least -2, we have $a_f \geq 0$, with equality possible only when $d = 2$.

From this one sees that there are three possibilities, depending on whether the constants b_i are positive, zero, or negative:

(1) All $b_i > 0$. In this case there are only finitely many diagrams with $D \geq 0$. The theory is said to be *superrenormalizable*.
(2) All $b_i \geq 0$, but some $b_i = 0$. In this case there are infinitely many diagrams with $D \geq 0$, but the degree of divergence is bounded above by d, and superficial divergence can occur only for diagrams with sufficiently few external lines.[2] The theory is said to be *renormalizable*.
(3) Some $b_i < 0$. In this case there are diagrams with arbitrarily high degrees of divergence. The theory is said to be *nonrenormalizable*.

Thus, for example, the ϕ^4 scalar field theory, QED, and Yukawa theory are superrenormalizable in dimensions 2 and 3, renormalizable in dimension 4, but nonrenormalizable in dimensions $d \geq 5$; the Fermi model is nonrenormalizable in any dimension.

It can be shown that the divergences in a renormalizable theory can be removed by renormalizing the field strengths, particle masses, and coupling constants, perhaps after adding a finite number of additional terms to the Lagrangian; we shall

[2]The possible exceptions to this last assertion for $d = 2$ are of no importance.

indicate how this works for ϕ^4 theory and QED later in this chapter. For nonrenormalizable theories one can still remove the divergences one by one but cannot create a consistent theory to all orders of perturbation theory by renormalizing finitely many parameters. For a long time this was taken to mean that nonrenormalizable theories are to be eschewed, but in recent years theorists have adopted a more flexible attitude. Having faced the fact that even renormalizable theories do not come with any guarantee of physical validity in regimes where very high energies are involved, they have become more willing to accept nonrenormalizable theories as useful "effective field theories" in regimes of sufficiently low energy that the contributions of high-order diagrams are suppressed. See Weinberg [**129**], §12.3, and Zee [**136**], §VIII.3, for discussions of this issue.

7.3. Evaluation and regularization of Feynman diagrams

In this section we develop some mathematical techniques for evaluating the integrals arising from Feynman diagrams when they converge and approximating them in a systematic way by convergent integrals when they do not.

Feynman parameters. The first device is a trick, first exploited by Feynman, for converting the product of quadratic functions in the denominator of (7.1) or (7.3) into a single quadratic function raised to a power. It involves integration over the $(n-1)$-simplex Σ_n of points $x \in \mathbb{R}^n$ whose coordinates are nonnegative and add up to 1 (i.e., the convex hull of the standard unit basis vectors). We shall write such integrals as

$$\int_{[0,1]^n} \delta(1 - x_1 - \cdots - x_n) f(x)\, d^n x.$$

In other words, the prescription is to set one of the variables (say, x_n) equal to 1 minus the sum of the others and then integrate over the "base simplex" in $\mathbb{R}^{n-1}$ whose vertices are the origin and the standard unit basis vectors , i.e., the region defined by the conditions $0 \le \sum_1^{n-1} x_j \le 1$, $x_j \ge 0$. (The measure $\delta(1 - \sum x_j)\, dx_1 \cdots dx_n$ is $1/\sqrt{n}$ times the Euclidean surface measure on the simplex Σ_n.) The result we need is as follows:

Suppose $c_1, \ldots, c_n$ $(n \ge 2)$ are complex numbers whose convex hull does not contain the origin. Then

$$\frac{1}{c_1 c_2 \cdots c_n} = (n-1)! \int_{[0,1]^n} \frac{\delta(1 - \sum x_j)}{(\sum c_j x_j)^n}\, d^n x. \tag{7.6}$$

We shall refer to (7.6) as *Feynman's formula*; the variables $x_1, \ldots, x_n$ in it are known as *Feynman parameters.* Its hypothesis is necessary to guarantee the nonsingularity of the integrand on the right.

To prove (7.6) we use induction on n. For $n = 2$, the integral on the right is

$$\int_0^1 \frac{dx}{(c_1 x + c_2(1-x))^2} = \int_0^1 \frac{dx}{((c_1 - c_2)x + c_2)^2} = \frac{1}{c_1 - c_2} \int_{c_2}^{c_1} \frac{dy}{y^2} = \frac{1}{c_1 c_2},$$

as claimed. Moreover, differentiating this formula $n-1$ times with respect to c_1 yields

$$\frac{1}{c_1^n c_2} = \int_0^1 \frac{n x^{n-1}\, dx}{(c_1 x + c_2(1-x))^{n+1}}.$$

We use these formulas (with the variables relabeled) to accomplish the inductive step:

$$\frac{1}{c_1\cdots c_{n+1}} = \frac{1}{(c_1\cdots c_n)c_{n+1}} = \int_{[0,1]^n} \frac{(n-1)!\delta(1-\sum y_j)\,d^n y}{(\sum c_j y_j)^n c_{n+1}}$$
$$= \int_{[0,1]^n}\int_0^1 \frac{n!\delta(1-\sum y_j)z^{n-1}\,d^n y\,dz}{(z\sum c_j y_j + (1-z)c_{n+1})^{n+1}}.$$

Now set $x_j = zy_j$ for $j \le n$ and $z_{n+1} = 1-z$. Then $\delta(1-\sum_1^n y_j) = z\delta(z-\sum_1^n zy_j) = z\delta(1-\sum_1^{n+1} z_j)$ since δ is homogeneous of degree -1, and $z^{n-1}\,d^n y\,dz = z^{-1}d^{n+1}x$ (the change of measure is the *absolute value* of the Jacobian), and we obtain (7.6) with n replaced by $n+1$.

One can apply Feynman's formula either to the integrand of a Feynman integral either in its original form (7.1) or in its Wick rotated form (7.3). For the latter, the result is an integral of the form

$$\text{(7.7)}\qquad \int_{\mathbb{R}^{4L}}\int_{[0,1]^I} \frac{(n-1)!\,P(q,p)}{(\sum_1^I x_i(|f_i(q,p)|^2+m_i^2))^{I+1}}\delta(1-\sum x_i)\,d^I x\,d^{4L}q.$$

If the integral (7.3) is absolutely convergent, so is the integral (7.7), and one can interchange the order of integration. The q-integral can be evaluated explicitly without much difficulty, as we shall explain shortly. One is then left with the x-integral, which, being a proper Riemann integral, can be evaluated numerically if not analytically.

One point should be made here. If one lets one of the masses, say m_1, tend to zero to accomodate a massless particle, the integrand of (7.3) develops a singularity at the origin in q_1-space, and the integral may blow up. The integrand in (7.7) has no singularities as a function of q, but the divergence manifests itself as a singularity with respect to the Feynman parameters near the vertex $(x_1,\ldots,x_n) = (1,0,\ldots,0)$.

The q-integrals can be evaluated by reducing them to integrals of radial functions, and this calculation will also play an important role in the analysis of divergent integrals. To begin with, the expression $\sum_1^I x_i(|f_i(q,p)|^2+m_i^2)$ in (7.7) is a positive quadratic function of the variables q_l, so by a suitable linear change of variables $q_l' = \sum a_l^m q_m + b_l$ it can be reduced to the form $|q'|^2 + c^2(p,x)$. (Of course c^2 also depends on the masses m_i.) The integral (7.7) then takes the form

$$\text{(7.8)}\qquad \int_{[0,1]^I}\left[\int_{\mathbb{R}^{4L}} \frac{P'(q',p,x)}{(|q'|^2+c^2(p,x))^{I+1}}\,d^{4L}q'\right]\delta(1-\sum x_j)\,d^I x,$$

where c^2 is a positive quadratic function of p and x.

We shall show below how to evaluate the inner integral. Once this is done, the integration over the Feynman parameters remains, and then one must Wick-rotate the external momenta back from Euclidean space to Minkowski space. The theorem we quoted earlier guarantees that the latter process converges in the topology of tempered distributions, but in practice one can usually just replace the Euclidean momenta pointwise by the Minkowski momenta in the integrand of the Feynman-parameter integral, using the Wick rotation process to determine the appropriate branch of any fractional powers or logarithms.

If this is the goal, one might wonder why one should bother Wick rotating the external momenta in the first place, as they return to Minkowski space at the end, and branches of multivalued functions can be determined at that point by temporarily replacing p^0 by $e^{i\epsilon}p^0$. The answer is that this step is needed to make the intermediate steps legal: to guarantee that 0 is not in the convex hull of the factors in the original denominator so that Feynman's formula can be applied, and to make the $c^2(p,x)$ in (7.8) positive so that the integrand is nonsingular. However, as a matter of practical computation, it can be omitted, and in physics books it usually is. The common procedure is *first* to introduce the Feynman parameters, interchange the order of integration, and eliminate the p-q cross terms by a linear change of variable in the original Minkowski-space integral, and *then* to Wick-rotate the internal momenta only. In spite of some possible ill-definedness in the intermediate steps, this algorithm works (and one could prove a theorem to that effect).

Reduction to radial functions. We turn to the evaluation of the q-integral in (7.8). Dropping all extraneous parts of the notation and generalizing to arbitrary d-dimensional integrals, we are faced with an integral of the form

$$\int_{\mathbb{R}^d} \frac{P(q)}{(|q|^2+c^2)^n}\, d^d q,$$

where P is a polynomial, necessarily of degree less than $2n-d$ if the original integral is convergent.[3] (One can either take $q=(q_1,\ldots,q_L)$ and $d=4L$ or $q=q_i$ and $d=4$; in the latter case the c^2 depends on the other q_j as well as p and x. Other values of d will appear later!) We think of this as the integral of the polynomial P with respect to the rotation-invariant measure $(|q|^2+c^2)^{-n}d^d q$, and it suffices to consider the case where P is homogenous. We can then use the following general result, in which we use multi-index notation: $\alpha=(\alpha_1,\ldots,\alpha_d)$ will denote a d-tuple of nonnegative integers, and we set

$$q^\alpha = q_1^{\alpha_1}q_2^{\alpha_2}\cdots q_d^{\alpha_d}, \qquad \alpha! = \alpha_1!\alpha_2!\cdots\alpha_d!, \qquad |\alpha| = \alpha_1+\cdots+\alpha_d.$$

Let $\mathcal{P}_k$ be the space of homogeneous polynomials of degree k on $\mathbb{R}^d$, and define the inner product $\prec\cdot,\cdot\succ$ (linear in the second variable, in accordance with physicists' convention) on $\mathcal{P}_k$ by declaring the monomials $q^\alpha/\sqrt{\alpha!}$ ($|\alpha|=k$) to be an orthonormal basis:

$$\prec q^\alpha, q^\beta \succ = \alpha!\delta_{\alpha\beta}. \tag{7.9}$$

Let $d\mu(q)=\rho(|q|)\,dq$ be an $O(d)$-invariant measure on $\mathbb{R}^d$ such that $\mathcal{P}_k\subset L^1(\mu)$. If k is odd, then $\int P\,d\mu=0$ for any $P\in\mathcal{P}_k$. If k is even, then for any $P\in\mathcal{P}_k$,

$$\int P\,d\mu = \frac{\prec r^k, P\succ}{\prec r^k, r^k\succ}\int r^k\,d\mu, \tag{7.10}$$

where $r^k\in\mathcal{P}_k$ is the polynomial defined by $r^k(q)=|q|^k$.

The assertion for k odd is obvious since then every $P\in\mathcal{P}_k$ is odd whereas $d\mu$ is even. To prove the result for k even, we need some preliminary observations. First, we clearly have $\prec q^\alpha, q^\beta\succ = \alpha!\delta_{\alpha\beta} = \partial^\alpha(q^\beta)$, so in general we have $\prec P,Q\succ = P^*(\partial)Q$. Hence, multiplication by r^2 (a map from $\mathcal{P}_{k-2}$ to $\mathcal{P}_k$) is the adjoint of $\nabla^2:\mathcal{P}_k\to\mathcal{P}_{k-2}$, for if $P\in\mathcal{P}_{k-2}$ and $Q\in\mathcal{P}_k$,

$$\prec r^2P, Q\succ = P^*(\partial)\nabla^2 Q = \prec P, \nabla^2 Q\succ .$$

[3] Apologies for the two conflicting uses of the letter d.

It follows that $\mathcal{P}_k = \mathcal{H}_k \oplus r^2\mathcal{P}_{k-2}$, where $\mathcal{H}_k \subset \mathcal{P}_k$ is the space of harmonic polynomials and the sum is orthogonal with respect to the inner product (7.9). By induction, then,

$$\mathcal{P}_k = \mathcal{H}_k \oplus r^2\mathcal{H}_{k-2} \oplus r^4\mathcal{H}_{k-4} \oplus \cdots \oplus r^k\mathcal{H}_0,$$

where the direct sums are again orthogonal with respect to (7.9). Now, the linear functional $P \mapsto \int P\,d\mu$ annihilates all the spaces $r^{2j}\mathcal{H}_{k-2j}$ with $2j < k$, since the mean value of a harmonic polynomial on any sphere about the origin is its value at the origin, which is 0 when the polynomial is homogeneous of degree $k - 2j > 0$. Hence, $\int P\,d\mu = \int \widetilde{P}\,d\mu$ where $\widetilde{P}$ is the orthogonal projection of P onto $r^k\mathcal{H}_0$, the scalar multiples of r^k, and this is just (7.10). This completes the proof.

For $k = 2$, we have $\prec q_iq_j, r^2 \succ = 2\delta_{ij}$ and $\prec r^2, r^2 \succ = 2d$, so (7.10) says that

$$\int q_iq_j\,d\mu = \frac{\delta_{ij}}{d}\int r^2\,d\mu. \tag{7.11}$$

For $k = 4$, we have $r^4 = \sum q_j^4 + 2\sum_{j<k} q_j^2q_k^2$, so $\prec q_j^4, r^4 \succ = 4! = 24$ and $\prec q_j^2q_k^2, r^4 \succ = 2\cdot(2!)^2 = 8$ for $j \neq k$. It follows that $\prec r^4, r^4 \succ = 8d(d+2)$ and hence

$$\int q_j^4\,d\mu = \frac{3}{d(d+2)}\int r^4\,d\mu, \qquad \int q_j^2q_k^2\,d\mu = \frac{1}{d(d+2)}\int r^4\,d\mu,$$

and $\int q^\alpha\,d\mu = 0$ if any $\alpha_j = 1$. These equalities may be combined as follows:

$$\int q_iq_jq_kq_l\,d\mu = \frac{1}{d(d+2)}(\delta_{ij}\delta_{kl} + \delta_{ik}\delta_{jl} + \delta_{il}\delta_{jk})\int r^4\,d\mu. \tag{7.12}$$

Important remark: If we apply these results to a Wick-rotated Feynman integral with a single momentum variable $q \in \mathbb{R}^4$ (or with several $q_1, \dots, q_L$, but doing the integration one q_l at a time) and undo the Wick rotation by replacing q_0 by q_0/i, the δ_{ij}'s in (7.11) and (7.12) turn into g_{ij}'s (the Lorentz metric), and the obvious rotational covariance of these formulas turns into the appropriate Lorentz covariance. The same is true for polynomials of higher degree; we shall not go into detail.

Regularization of divergent integrals. We now turn to the problem of extracting some meaning from the divergent integrals. As we indicated earlier, the first step is to modify, or "regularize," the integrals to make them convergent. There are several ways to do this, of which we shall discuss two. These methods all eventually lead to the same results, but one may be more convenient than another in specific situations; in particular, one may wish to choose a regularization that preserves certain symmetries of the theory.

Perhaps the most straightforward procedure is *Pauli-Villars regularization*, which goes as follows. Let Δ_m be the propagator for one of the particle types occurring in the given diagram, where m is the mass of the particle: one simply replaces Δ_m by $\Delta_m - \Delta_M$ where M is a *large* mass that eventually will be sent to infinity. This gives some extra decay of the momentum space integral: for example, for a scalar particle, after Wick rotation, one has

$$\frac{1}{|q|^2 + m^2} - \frac{1}{|q|^2 + M^2} = \frac{M^2 - m^2}{(|q|^2 + m^2)(|q|^2 + M^2)}.$$

If this decay is not enough, one can replace Δ_m by $\Delta_m - c_1\Delta_{M_1} - c_2\Delta_{M_2}$ where c_1 and c_2 are chosen to satisfy $c_1 + c_2 = 1$ and $c_1 M_1^2 + c_2 M_2^2 = m^2$: the first condition gives $1/|q|^4$ decay, and the second one then gives $1/|q|^6$ decay. Further decay can be obtained by adding more Δ_{M_j} terms with appropriate coefficients. (In position space, the extra decay corresponds to a weakening of the singularities along the light cone.) In any event, one chooses the modified propagator so as to produce a convergent integral, whose value will depend on the M_j. Renormalization is then accomplished by analyzing the behavior as $M_j \to \infty$.

Informally, one can think of Pauli-Villars regularization as a matter of adding some fictitious particles with large masses to the theory; as their masses tend to infinity, they decouple from everything else and disappear from the calculations.

Pauli-Villars regularization is still used for many calculations, but another method has become more popular because it works better for non-Abelian gauge fields: the *dimensional regularization* of 't Hooft and Veltman [**118**]. The idea of dimensional regularization is to perform an analytic continuation in the number d of space-time dimensions. This is not as crazy as it sounds at first! We are not proposing to develop a theory of integration in d dimensions for arbitrary complex d, only to extend certain special kinds of d-dimensional integrals to complex d. Nor are we claiming that there is only one way to do so; all we need is one way that works.

To apply this method to a divergent integral of the form (7.1), one prepares the ground by retracing the steps we have used to evaluate *convergent* integrals:

i. Wick-rotate the momenta.
ii. Convert the denominator from a product of quadratics to a power of a single quadratic by Feynman's formula, obtaining an integral of the form (7.7), and then interchange the momentum space integration with the integration over the Feynman parameters.
iii. Perform a linear change of variable to get rid of the linear terms in the denominator and obtain an integral of the form (7.8).
iv. Reduce the resulting integral to integrals of radial functions on $\mathbb{R}^{4L}$ by using the formula (7.10).

In the present setting, each of these steps consists of *formal* manipulations. They must be regarded simply as parts of a symbolic calculation that will eventually interpret the original divergent integral as a limit of well-defined finite quantities.

It is with the final reduction to integrals of radial functions that we can start doing some honest analysis. Specifically, for a radial function on $\mathbb{R}^d$, say $f(|q|)$, integration in polar coordinates reduces its d-dimensional integral to a one-dimensional one:

$$\int f(|q|)\, d^d q = \Omega_d \int_0^\infty f(r) r^{d-1}\, dr, \tag{7.13}$$

where $\Omega_d = 2\pi^{d/2}/\Gamma(d/2)$ is the area of the unit sphere in $\mathbb{R}^d$. (See, e.g., Folland [**48**].) Now, the expression on the right *does* define an analytic function of d in the domain of those d for which the integral converges. For our purposes $f(r)$ will be of the form $f(r) = r^{2k}/(r^2 + c^2)^n$, so the integral converges provided $0 < d < 2n - k$ and can be evaluated in terms of the gamma function: the substitution $t = (r/c)^2$

yields

$$\begin{aligned}\int_0^\infty \frac{r^{2k+d-1}}{(r^2+c^2)^n}\,dr &= \frac{c^{2k+d-2n}}{2}\int_0^\infty \frac{t^{k+(d/2)-1}\,dt}{(1+t)^{-n}} \\ &= \frac{c^{2k+d-2n}}{2}B(k+\tfrac{1}{2}d,\, n-k-\tfrac{1}{2}d) \\ &= \frac{c^{2k+d-2n}}{2}\frac{\Gamma(k+\frac{1}{2}d)\Gamma(n-k-\frac{1}{2}d)}{\Gamma(n)}.\end{aligned}\tag{7.14}$$

This expression continues analytically to larger values of d except for poles where $n-k-\frac{1}{2}d$ is a nonnegative integer. In our situation, one of these poles will occur at $d=4$, which corresponds to the original divergent integral (when we integrate over the loop momenta one at a time). The "∞" of the original integral then becomes a "$1/(4-d)$" term in a well-defined analytic expression that can be used in further calculations.

The shortcut that we discussed for evaluating convergent integrals after (7.8) can be used here too: that is, one postpones the Wick rotation until after introduction of the Feynman parameters and elimination of the cross terms, and one performs it explicitly only for the internal momenta. When we do our first concrete dimensional regularization in the next section, we shall do it by the more careful procedure, but thenceforth we shall use the shortcut.

Now, as far as convergence or divergence is concerned, the only d-dependence in (7.13) that matters is the r^{d-1}; one could replace the Ω_d by its value $2\pi^2$ at $d=4$ without changing anything essential. On the other hand, the integrals with which we shall be concerned contain other ingredients with a natural dependence on (integer-valued) dimension. For one thing, the momentum space volume element is equipped with a factor of $(2\pi)^4$ in the denominator that arises from Fourier analysis; in d dimensions it would be $(2\pi)^d$. For another, the reductions to radial functions by means of (7.10), such as (7.11), are d-dependent. Finally, certain identities involving Dirac matrices are dimension-dependent. More precisely, for d-dimensional space-time the Dirac matrices are taken to be the generators of the Clifford algebra over the Minkowski space $\mathbb{R}^d$, satisfying $\{\gamma^\mu,\gamma^\nu\}=2g^{\mu\nu}$. It follows that $\gamma_\nu\gamma^\nu=dI$ and hence, for example, that $\gamma_\nu\gamma^\mu\gamma^\nu=2g^{\mu\nu}\gamma_\nu-\gamma_\nu\gamma^\nu\gamma^\mu=(2-d)\gamma^\mu$.

Inclusion or exclusion of these d-dependences in the dimensional regularization algorithm does not affect the pole of the regularized integral at $d=4$ or the residue there, but it does affect the remaining finite part (the constant term in the Laurent series). For the calculations arising from any individual Feynman diagram this ultimately doesn't matter, because if one adds an extra finite part to the integral, one will just end up subtracting it off again in the counterterms (as we shall see). However, if one wants to compare contributions from different diagrams, a little more care is needed to maintain consistency.

The physicists' standard prescription is to include *all* the natural dimension dependences in the dimensional regularization algorithm, so that one obtains correct formulas for d-dimensional space-time whenever d is an integer. Specifically, this means:

i. The d-dependence of formulas such as (7.11) that arise from (7.10) is maintained.

ii. For integrals of radial functions, the d-dependence is given by
(7.15)

$$\int_{\mathbb{R}^4} f(|q|)\frac{d^4q}{(2\pi)^4} \longrightarrow \frac{\Omega_d}{(2\pi)^d}\int_0^\infty f(r)r^{d-1}\,dr = \frac{2}{(4\pi)^{d/2}\Gamma(\frac{1}{2}d)}\int_0^\infty f(r)r^{d-1}\,dr.$$

iii. The d-dependence of contractions of Dirac matrices is maintained:

$$\gamma_\nu\gamma^\nu = dI, \qquad \gamma_\nu\gamma^\mu\gamma^\nu = (2-d)\gamma^\mu, \qquad \gamma_\rho\gamma_\nu\gamma^\mu\gamma^\nu\gamma^\rho = (2-d)^2\gamma^\mu. \tag{7.16}$$

There is an additional dimension dependence that one can take into account, arising from the fact that the action — the integral of the Lagrangian density over space-time — can also be considered in d dimensions. In natural units in which length, time, and reciprocal mass are equivalent, the action is dimensionless, so the Lagrangian density has dimension $[l^{-d}] = [m^d]$, or, as we shall say, "mass dimension d." From this one can deduce the dimensions of the fields and coupling constants. For example, in QED, the fact that the free-field terms $-\frac{1}{4}F^{\mu\nu}F_{\mu\nu}$ and $\overline{\psi}\gamma^\mu\partial_\mu\psi$ have mass dimension d imply that the electromagnetic field A_μ and the electron field ψ have mass dimensions $\frac{1}{2}(d-2)$ and $\frac{1}{2}(d-1)$, respectively. The fact that the interaction term $e\overline{\psi}\gamma^\mu A_\mu\psi$ has mass dimension d then implies that the coupling constant e has mass dimension $\frac{1}{2}(4-d)$. One can make this explicit by setting $e = e_0\mu^{(4-d)/2}$ where e_0 is dimensionless and μ is a parameter with dimensions of mass, which is arbitrary at the outset but can be chosen to correlate with the energy scale at which one is doing physics. This additional d-dependence results in additional terms proportional to $\log\mu$ in the regularized integrals, which are of use in connection with renormalization group techniques. However, as that subject is beyond the scope of this book, we shall not incorporate this refinement in our calculations.

There is one place where factors of 4 turn up that should *not* automatically be turned into d's, namely, calculations involving traces of products of Dirac matrices. They arise because the trace of the 4×4 identity matrix is 4, but this 4 is the dimension of spinor space, not space-time! One could take $\mathrm{tr}(I)$ to be $f(d)$ where, for integer d, $f(d)$ is the dimension of a suitable representation of the Clifford algebra over d-dimensional Minkowski space, but here the common practice is simply to take traces of products of Dirac matrices to be dimension-independent.

For more about various regularization and and renormalization techniques, see the anthology of Velo and Wightman [**124**], particularly the articles by Wightman, Speer, and Lowenstein.

7.4. A one-loop calculation in scalar field theory

In this section we carry out the regularization and renormalization procedure for the simplest possible example: the regularization of a one-loop diagram in the ϕ^4 scalar field theory and the resulting renormalization of the coupling constant.

The Feynman diagram we wish to study is shown in Figure 7.2. It has two incoming particles with 4-momenta p_1, p_2 and two outgoing particles with 4-momenta p_3, p_4. The value of the diagram includes a factor of $(2\pi)^4\delta(p_3+p_4-p_1-p_2)$, which we may ignore after setting

$$p_1 + p_2 = p_3 + p_4 \equiv p,$$

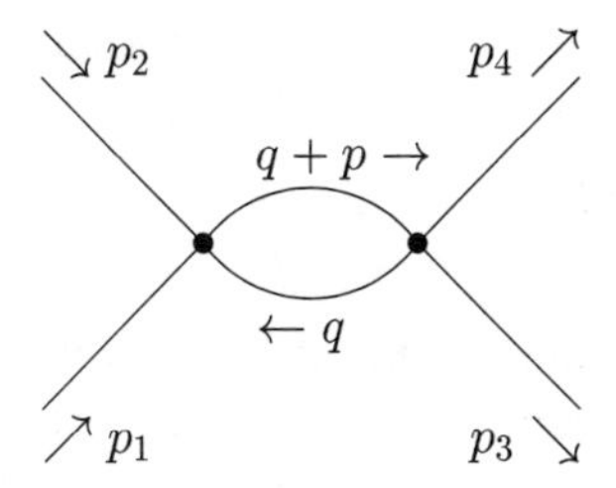

FIGURE 7.2. The basic one-loop diagram in ϕ^4 theory.

as well as factors of $1/\sqrt{2\omega_{\mathbf{p}_j}}$ from the external lines, which play no role in the analysis. After amputating these inert ingredients, the value of the diagram is

$$I(p) = \frac{(-i\lambda)^2}{2} \int \frac{-i}{-q^2 + m^2 - i\epsilon} \cdot \frac{-i}{-(q+p)^2 + m^2 - i\epsilon} \frac{d^4q}{(2\pi)^4}. \tag{7.17}$$

(The 2 in the denominator is the symmetry factor arising from the invariance of the diagram under interchange of the two internal lines.)

This integral has superficial degree of divergence 0; in other words, it is logarithmically divergent. To regularize it, we begin with a Wick rotation: replacing q^0 by iq^0 and p^0 by ip^0 yields the Euclidean-space integral

$$\frac{(-i\lambda)^2}{2} \int \frac{-i}{(|q|^2 + m^2)(|q+p|^2 + m^2)} \frac{d^4q}{(2\pi)^4}.$$

Next, introducing the Feynman parameter x and switching the order of integration gives

$$\frac{(-i\lambda)^2}{2} \int_0^1 \int_{\mathbb{R}^4} \frac{-i}{(|q|^2 + 2xq\cdot p + x|p|^2 + m^2)^2} \frac{d^4q}{(2\pi)^4}\, dx,$$

and then the substitution $k = q + xp$ turns this into

$$\frac{(-i\lambda)^2}{2} \int_0^1 \int_{\mathbb{R}^4} \frac{-i}{(|k|^2 + x(1-x)|p|^2 + m^2)^2} \frac{d^4k}{(2\pi)^4}\, dx. \tag{7.18}$$

Now do the inner integral not in 4 dimensions but in d dimensions with $d < 4$, using (7.14):

$$\begin{aligned}
\int_{\mathbb{R}^4} &\frac{1}{(|k|^2 + x(1-x)|p|^2 + m^2)^2} \frac{d^4k}{(2\pi)^4} \\
&\to \frac{2}{(4\pi)^{d/2}\Gamma(d/2)} \int_0^\infty \frac{r^{d-1}\, dr}{2(r^2 + x(1-x)|p|^2 + m^2)^2} \\
&\qquad = \frac{\Gamma(2-(d/2))}{(4\pi)^{d/2}(m^2 + x(1-x)|p|^2)^{(4-d)/2}}.
\end{aligned}$$

Finally, undo the Wick rotation of p to turn $|p|^2 = -(e^{i\pi/2}p^0)^2 + |\mathbf{p}|^2$ back into $-p^2 = -(p^0)^2 + |\mathbf{p}|^2$. For $p^2 > 4m^2$, the quantity $m^2 - x(1-x)p^2$ is negative for some values of x; in this case the branch of the fractional power is determined by the requirement that it vary continuously as ϵ goes from $\pi/2$ to 0 in $-(e^{i\epsilon}p^0)^2$. For $p^2 \geq 4m^2$, the resulting function of x has singularities, but the x-integral is still absolutely convergent provided d is close to 4. In short, the regularized value of the integral $I(p)$ (a function of d as well as p) is

$$I_d(p) = -i\frac{(-i\lambda)^2}{2}\frac{\Gamma(2-(d/2))}{(4\pi)^{d/2}}\int_0^1 \frac{dx}{(m^2 - x(1-x)p^2)^{(4-d)/2}}.$$

This is a well-defined finite quantity for all p when $3 < d < 4$.

We could also do this computation using the shortcut discussed after (7.8). The reader may verify that if we start with the original integral (7.17), introduce the Feynman parameter x, make the substitution $k = q+xp$, and then Wick-rotate q (all on the level of formal calculation), the result is the integral (7.18) with $|p|^2$ replaced by $-p^2$. The rest of the computation proceeds as before, with p^0 *implicitly* given a positive imaginary part to ensure the convergence of the d-dimensional integral and then returned to the real axis to determine the branch of the fractional power.

It will be convenient to factor out the $(-i\lambda)^2$ and make explicit the fact that $I_d(p)$ depends only on the Lorentz norm of p by setting

$$I_d(p) = (-i\lambda)^2 J_d(p^2),$$
$$J_d(s) = \frac{-i\Gamma(2-(d/2))}{2(4\pi)^{d/2}}\int_0^1 \frac{dx}{(m^2 - sx(1-x))^{(4-d)/2}}. \tag{7.19}$$

For $\epsilon = 4-d$ near 0 we have $x^\epsilon = 1 + \epsilon\log x + O(\epsilon^2)$ and $\Gamma(\epsilon) = \epsilon^{-1} - \gamma + O(\epsilon)$ where γ is the Euler-Mascheroni constant; therefore,

(7.20)

$$\begin{aligned} J_d(s) &= \frac{-i}{32\pi^2}\int_0^1 \left[\frac{2}{4-d} - \gamma + \log(4\pi) - \log(m^2 - sx(1-x)) + O(4-d)\right] dx \\ &= \frac{i}{32\pi^2}\int_0^1 \log(m^2 - sx(1-x))\,dx - \frac{i}{32\pi^2}\left(\frac{2}{4-d} - \gamma + \log(4\pi)\right) + O(4-d). \end{aligned}$$

(The branch of the logarithm is determined as above.) The divergence has now been effectively quarantined as the $2/(4-d)$, and all the p-dependence that survives when $d \to 4$, which presumably contains all the interesting physics, is in the integral.

It should be noted that the constant $-\gamma + \log(4\pi)$ is of no real significance; it is an artifact of our conventions concerning dimensional regularization. For example, we could have kept the original factor of $(2\pi)^4$ instead of turning it into $(2\pi)^d$, which would replace the $\log(4\pi)$ by $-\log\pi$.

Now, what are we to do with the divergent $2/(4-d)$ term? To answer this question we need to consider how the Feynman diagram under consideration enters into a quantity that might actually be measured in a laboratory, namely, the S-matrix element

$$\langle 0|a(\mathbf{p}_3)a(\mathbf{p}_4)Sa^\dagger(\mathbf{p}_1)a^\dagger(\mathbf{p}_2)|0\rangle.$$

We shall calculate this to second order in the coupling constant λ, and part of the game is to throw away all terms containing a factor λ^n with $n > 2$. This cavalier treatment of higher-order error terms may cause some qualms, but it is part of the bargain with the devil that we made in §6.1 that such perturbation-theoretic calculations are to be deemed credible as long as the coupling constant λ is small.

We assume that neither $\mathbf{p}_1$ nor $\mathbf{p}_2$ is equal to $\mathbf{p}_3$ or $\mathbf{p}_4$ to exclude the case of trivial scattering. To second order in λ, then, the Feynman diagrams that contribute to the S-matrix element are the simple one-vertex diagram in Figure 7.3 and the three

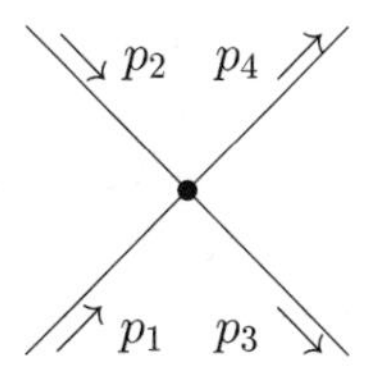

FIGURE 7.3. The basic diagram for two-particle scattering in ϕ^4 theory.

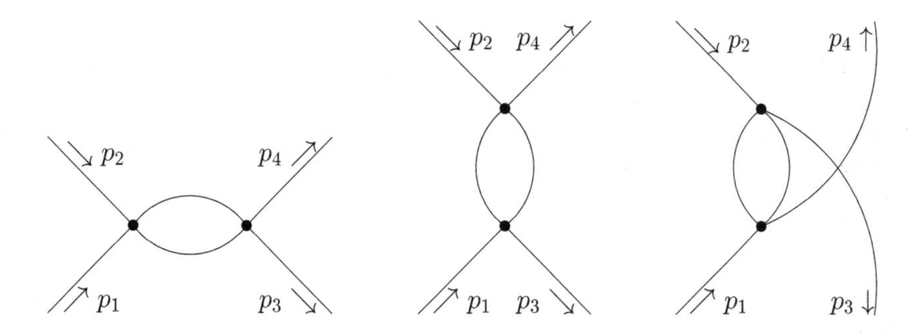

FIGURE 7.4. One-loop diagrams for two-particle scattering in ϕ^4 theory.

one-loop diagrams in Figure 7.4.[4] After setting aside the momentum-conservation delta-function and the factors of $\sqrt{2\omega_{\mathbf{p}_j}}$ from the external lines that are common to all these diagrams, the value of the diagram in Figure 7.3 is $-i\lambda$ and the values of the one-loop diagrams in Figure 7.4 are $I(p_1 + p_2)$, $I(p_1 - p_3)$, and $I(p_1 - p_4)$, respectively. Thus the delta-function-free amplitude $iM = iM_{p_3p_4,p_1p_2}$ as in (6.65), to second order in λ and with external-line factors omitted, is

$$iM = -i\lambda + I(p_1 + p_2) + I(p_1 - p_3) + I(p_1 - p_4).$$

As it stands, this makes no sense because the last three terms are divergent, but on introducing the *Mandelstam variables*

$$s = (p_1 + p_2)^2, \quad t = (p_1 - p_3)^2, \quad u = (p_1 - p_4)^2, \tag{7.21}$$

we can rewrite it as

$$iM = \lim_{d\to 4} \big(-i\lambda + (-i\lambda)^2[J_d(s) + J_d(t) + J_d(u)]\big). \tag{7.22}$$

This limit may be well-defined *if we allow λ to depend on d in such a way that the divergences cancel.*

At first glance this suggestion seems outrageous; isn't λ a coupling constant with physical meaning, not something that can depend on the fictional parameter d? But no, in fact: λ is merely a parameter in the Lagrangian or Hamiltonian. We may call it the "bare coupling constant," but the *physical* coupling constant λ_{phys} is something that must be determined by experiment, and it need not equal λ. Since scalar fields with ϕ^4 interaction do not model any currently observed

[4]We are using the Wick-ordered ϕ^4 interaction here, as explained in §6.6. Without this stipulation there are also diagrams obtained from Figure 7.3 by attaching a one-line loop to one of the external lines. Even if such diagrams are included, they do not contribute to the renormalization of λ that we are about to perform; see Peskin and Schroeder [**88**], §10.2.

physical process, the experiment will have to be a thought experiment, and we have some freedom in deciding what it will be. Here is a reasonable choice: we aim two particles at each other at low velocity and see what happens. That is, we take λ_{phys} to be the magnitude of the scattering amplitude iM in the limit where the 3-momenta of the particles vanish:

$$-i\lambda_{\text{phys}} = iM|_{p_1=p_2=p_3=p_4=(m,\mathbf{0})}. \tag{7.23}$$

Thus, to first order in λ, λ_{phys} is indeed λ, but to second order,

$$\lambda_{\text{phys}} = \lambda\big(1 - i\lambda[J_d(4m^2) + 2J_d(0)]\big). \tag{7.24}$$

This equation can be solved for λ in terms of λ_{phys} and the function J_d of d. One could do so exactly, but it is not worth the trouble because we are only doing perturbation theory to order λ^2; if we go to higher order, additional Feynman diagrams will make additional contributions to this formula. Rather, we perform all calculations modulo terms that are $O(\lambda^3)$. Since $\lambda_{\text{phys}} = \lambda + O(\lambda^2)$, we have $\lambda_{\text{phys}}^2 = \lambda^2 + O(\lambda^3)$ and $O(\lambda_{\text{phys}}^3) = O(\lambda^3)$; thus,

$$\lambda_{\text{phys}} = \lambda[1 - i\lambda_{\text{phys}}(J_d(4m^2) + 2J_d(0))] + O(\lambda^3),$$

and hence

$$\begin{aligned}\lambda &= \lambda_{\text{phys}}[1 - i\lambda_{\text{phys}}(J_d(4m^2) + 2J_d(0))]^{-1} + O(\lambda^3) \\ &= \lambda_{\text{phys}} + i\lambda_{\text{phys}}^2[J_d(4m^2) + 2J_d(0)] + O(\lambda^3).\end{aligned} \tag{7.25}$$

We now substitute (7.25) into (7.22) and drop all $O(\lambda^3)$ terms to obtain

$$iM = \lim_{d\to 4}\big(-i\lambda_{\text{phys}} + (-i\lambda_{\text{phys}})^2[J_d(s) + J_d(t) + J_d(u) - J_d(4m^2) - 2J_d(0)]\big). \tag{7.26}$$

At this point the divergences in the J_d's all cancel, as do the finite constants γ and $\log(4\pi)$. Explicitly, we have

$$\begin{aligned}M = -\lambda_{\text{phys}} - \frac{\lambda_{\text{phys}}^2}{32\pi^2}\int_0^1 \big[&\log(m^2 - sx(1-x)) + \log(m^2 - tx(1-x)) \\ &+ \log(m^2 - ux(1-x)) - \log(m^2 - 4m^2x(1-x)) - 2\log(m^2)\big]\,dx,\end{aligned} \tag{7.27}$$

where s, t, u are given by (7.21).

There is still something a little arbitrary in our definition of coupling constant. What if we had chosen some other "subtraction point" — that is, some values of s, t, u other than $4m^2$, 0, 0 — to define λ_{phys}? *It doesn't matter.* For example, some theorists favor the simple (but unphysical) subtraction point $s = t = u = 0$, which yields $\lambda'_{\text{phys}} = \lambda(1 - 3i\lambda J_d(0))$. To second order, these constants are related by

$$\lambda'_{\text{phys}} = \lambda_{\text{phys}} + i\lambda_{\text{phys}}^2[J_d(4m^2) - J_d(0)],$$

and the quantity $J_d(4m^2) - J_d(0)$ has a finite limit as $d \to 4$. The formula analogous to (7.26) in terms of λ'_{phys} is

$$iM = \lim_{d\to 4}\big(-i\lambda'_{\text{phys}} + (-i\lambda'_{\text{phys}})^2[J_d(s) + J_d(t) + J_d(u) - 3J_d(0)]\big).$$

But on substituting $\lambda_{\text{phys}} + i\lambda_{\text{phys}}^2[J_d(4m^2) - J_d(0)]$ for λ'_{phys} in this formula and discarding terms of order higher than 2, one recovers (7.26). In short, (7.27) *is a well-defined and physically meaningful quantity that can (in principle) be taken to the laboratory and compared with experiment.* If the agreement is pretty good but not excellent, the next step is to go to order λ^3 in hopes of obtaining an improvement.

7.5. Renormalized perturbation theory

Let us review what we have done. We computed the value of the Feynman diagram using the bare coupling constant and regularized the divergence, obtaining a value depending on the parameter $\epsilon = 4 - d$. We then found the relation between the bare coupling constant and the physical coupling constant, which also involves ϵ, and used it to re-express the regularized value in terms of the physical coupling constant. At this point the divergences cancel out and we obtain a meaningful quantity in the limit as $\epsilon \to 0$.

The same procedure can be followed for higher-order diagrams and for other quantum field theories. One starts with a Lagrangian that contains some parameters that are interpreted as bare masses and coupling constants as well as some overall scale factors that are interpreted as field normalizations. One computes the values of Feynman diagrams in terms of the bare parameters, relates the bare parameters to the physical ones, and re-expresses the value of the diagram in terms of the latter, at which point (one hopes that) the divergences cancel. This procedure is called *bare perturbation theory.*

There is another procedure generally known as *renormalized perturbation theory*that is often more convenient, although the two procedures are equivalent; going from one to the other is just a matter of bookkeeping. Its conceptual advantage is that one is perturbing about the fixed physical values of the parameters rather than about the unphysical bare parameters that are going to be sent off to infinity by the renormalization.

To do renormalized perturbation theory, one starts from the realization that the masses and the coupling constants in the Lagrangian $\mathfrak{L}$ may not be the physical ones, and moreover that it may be necessary to rescale the fields — that is, to replace each field ϕ in $\mathfrak{L}$ with a "renormalized field" $\phi_r = C\phi$, where C is a positive constant.[5] (This last phenomenon is not something we have encountered yet. We shall see in the next section how it arises; for now we beg the reader's indulgence.) Let us write $\mathfrak{L}$ as the sum of the Lagrangian for the free fields and the interaction terms:

$$\mathfrak{L} = \mathfrak{L}_{\text{free}} + \mathfrak{L}_{\text{int}}.$$

Moreover, let $\mathfrak{L}'_{\text{free}}$ and $\mathfrak{L}'_{\text{int}}$ be $\mathfrak{L}_{\text{free}}$ and $\mathfrak{L}_{\text{int}}$ with the masses and coupling constants replaced by the physical ones and the fields replaced by the renormalized fields, and let $\mathfrak{L}' = \mathfrak{L}'_{\text{free}} + \mathfrak{L}'_{\text{int}}$. Then we have

$$\mathfrak{L} = \mathfrak{L}' + \mathfrak{L}_{\text{ct}} = \mathfrak{L}'_{\text{free}} + \mathfrak{L}'_{\text{int}} + \mathfrak{L}_{\text{ct}},$$

where $\mathfrak{L}_{\text{ct}}$ is a sum of terms of the same form as the terms in $\mathfrak{L}'$, written in terms of the renormalized fields, whose coefficients involve the differences between those in $\mathfrak{L}$ and those in $\mathfrak{L}'$. These terms are known as *counterterms*, and at this point their coefficients are just some arbitrary constants whose meaning is yet to be specified.

The idea is to calculate S-matrix elements by the Feynman diagram machinery, *taking the interaction to be* $\mathfrak{L}'_{\text{int}} + \mathfrak{L}_{\text{ct}}$. That is, we first calculate Feynman diagrams arising from the interaction $\mathfrak{L}'_{\text{int}}$, as in Chapter 6, obtaining integrals involving the *physical* masses and coupling constants. Some of them diverge, so we regularize

[5] C is conventionally denoted by $Z_j^{-1/2}$ where j is an index to label the field.

them to make them finite but dependent on a regularization parameter. We then add in some new Feynman diagrams arising from the counterterms $\mathcal{L}_{\text{ct}}$ (in conjunction with $\mathcal{L}'_{\text{int}}$), according to some rules that we shall specify below, resulting in some additional terms in the S-matrix element that involve the coefficients of the counterterms. These coefficients are then specified, as functions of the regularization parameter, in such a way that the counterterm contributions cancel the divergences from the original diagrams.

Here is how this works for the scalar field with ϕ^4 interaction. The Lagrangian is

$$\mathcal{L} = \frac{1}{2}(\partial\phi_B)^2 - \frac{1}{2}m_B^2\phi_B^2 - \frac{\lambda_B}{4!}\phi_B^4,$$

where the subscript B indicates that ϕ_B, m_B, and λ_B are the *bare* (unrenormalized) field, mass, and coupling constant. Let m and λ be the *physical* mass and coupling constant, and let $\phi = Z^{-1/2}\phi_B$ be the renormalized field. The Lagrangian written in terms of ϕ is

$$\mathcal{L} = \frac{Z}{2}(\partial\phi)^2 - \frac{Z}{2}m_B^2\phi^2 - \frac{Z^2\lambda_B}{4!}\phi^4,$$

so

$$\mathcal{L}' = \frac{1}{2}(\partial\phi)^2 - m^2\phi^2 - \frac{\lambda}{4!}\phi^4,$$

and

$$\mathcal{L}_{\text{ct}} = \frac{Z-1}{2}(\partial\phi)^2 - \frac{Zm_B^2 - m^2}{2}\phi^2 - \frac{Z^2\lambda_B - \lambda}{4!}\phi^4.$$

It is conventional to write

$$\delta Z = Z - 1, \qquad \delta m^2 = Zm_B^2 - m^2, \qquad \delta\lambda = Z^2\lambda_B - \lambda, \tag{7.28}$$

so that

$$\mathcal{L}_{\text{ct}} = \frac{\delta Z}{2}(\partial\phi)^2 - \frac{\delta m^2}{2}\phi^2 - \frac{\delta\lambda}{4!}\phi^4.$$

The contributions of the counterterms to the momentum-space Feynman diagrams are as follows. The term $(\delta\lambda/4!)\phi^4$, just like the term $(\lambda/4!)\phi^4$ in $\mathcal{L}_0$, corresponds to a vertex at which four lines meet, and it contributes a value of $-i\delta\lambda$ to the corresponding integral. To distinguish such a vertex from the one coming from the ordinary interaction, we indicated it by a crossed circle, as shown in Figure 7.5a. The sum $\frac{1}{2}[(\delta Z)(\partial\phi)^2 - (\delta m^2)\phi^2]$ is considered as a single unit; it corresponds to a vertex at which *two* lines meet, again denoted by a crossed circle, as in Figure 7.5b. As we shall show in the next section, it contributes a factor of $-i(-(\delta Z)p^2 + \delta m^2)$, where p is the momentum of the incoming (or outgoing) line.

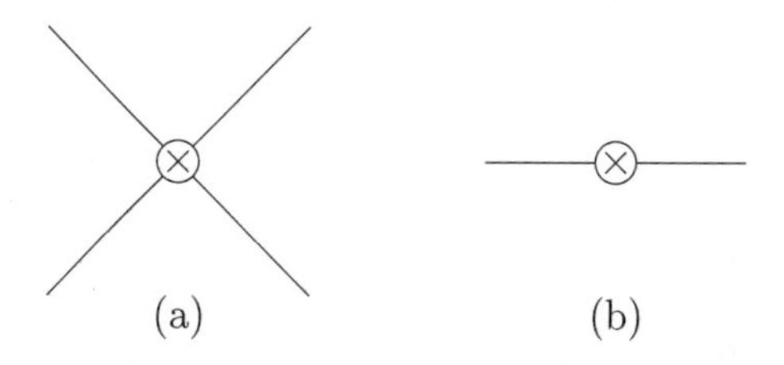

FIGURE 7.5. Counterterm vertices in ϕ^4 theory.

In this approach, the calculation of the two-particle scattering amplitude to second order in perturbation theory that we performed in the preceding section goes as follows. There are five diagrams that contribute: the four that we considered before, and a counterterm diagram of the form in Figure 7.5a to cancel the divergences in the loop diagrams. The first one has the value $-i\lambda$; the last one has the value $-i\delta\lambda$; and the sum of the others (after regularization) has the value $(-i\lambda^2)[J_d(s) + J_d(t) + J_d(u)]$ as in (7.26), except that now λ denotes the *physical* coupling constant. Thus,

$$iM = \lim_{d\to 4} \big(- i\lambda + (-i\lambda)^2[J_d(s) + J_d(t) + J_d(u)] - i\delta\lambda\big).$$

On the other hand, by (7.23),

$$-i\lambda = iM|_{p_1=p_2=p_3=p_4=(m,\mathbf{0})} = \lim_{d\to 4} \big(- i\lambda + (-i\lambda)^2[J_d(4m^2) + 2J_d(0)] - i\delta\lambda\big),$$

so the value of $\delta\lambda$, to order λ^2, is

$$\delta\lambda = i\lambda^2[J_d(4m^2) + J_d(0)].$$

This is consistent with (7.25) and (7.28) if we take $Z = 1$. But indeed $Z = 1$ to zeroth order, as $Z = 1$ is the correct normalization for free fields, and we will see in the next section that there is no first-order correction to Z, so everything is correct to order λ^2.

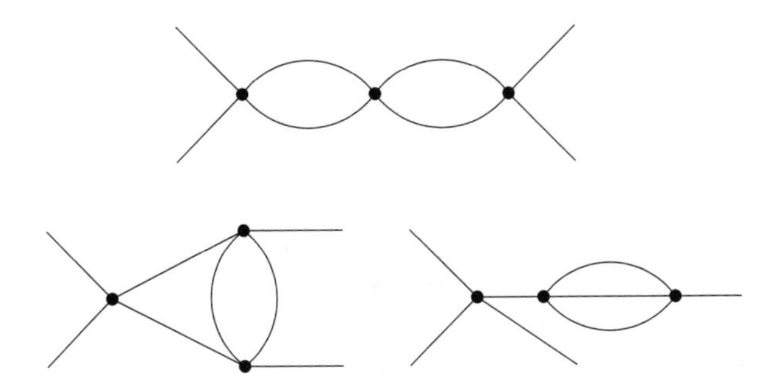

FIGURE 7.6. Third-order diagrams for two-particle scattering in ϕ^4 theory.

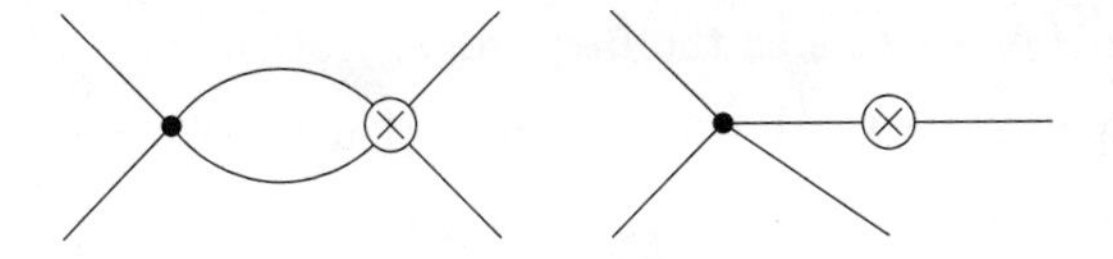

FIGURE 7.7. Counterterm diagrams for two-particle scattering in ϕ^4 theory.

If one proceeds to order λ^3, three more topologically distinct types of "ordinary" diagrams enter the picture, as shown in Figure 7.6. (For each of these, one must consider all the different ways of assigning the incoming and outgoing momenta $p_1, \dots, p_4$ to the four external legs.) There are also two additional types of counterterm diagram, one involving $\delta\lambda$ and one involving $\frac{1}{2}[(\delta Z)(\partial\phi)^2 - (\delta m^2)\phi^2]$, as in Figure 7.7. The first provides an $O(\lambda^3)$ correction to $\delta\lambda$ to cancel the divergences of the first two in Figure 7.6; the second one cancels the divergence of the last one. In fact, this last pair does not contribute to the scattering amplitude at all. It is part of the description of what happens to one of the particles on its journey to or from the scattering event, a problem that we shall study in the next section. (See Peskin and Schroeder [88], §10.5, for the details.)

7.6. Dressing the propagator

The propagation of a free particle from x to y is a simple matter. Diagrammatically, it is represented by a single line whose ends are marked x and y, and its amplitude is just the free propagator $-i\Delta(x-y)$ for the particle species in question. But interacting particles can undergo all sorts of little adventures as they travel. For every diagram that contains a single line from x to y, there will be higher-order diagrams contributing to the same process in which the particle interacts with some virtual particles along the way. Some examples for the propagation of an electron in QED are shown in Figure 7.8. We shall call the propagation amplitude for a particle with all of these virtual interactions taken into account the *dressed propagator* and denote it by $-i\Delta_{\text{dressed}}(x-y)$.

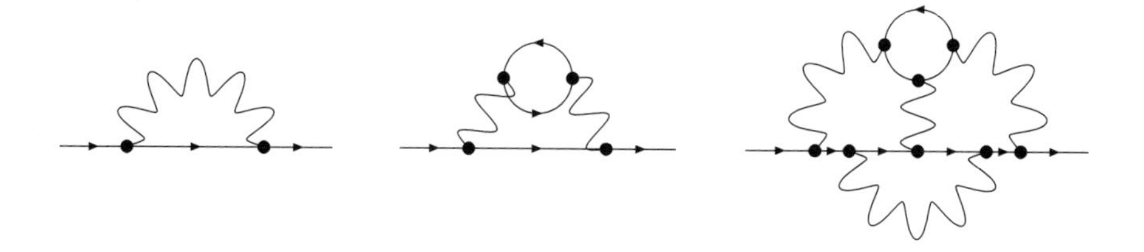

FIGURE 7.8. Some higher-order modifications of the electron propagator.

What should Δ_{dressed} — or, more to the point, its Fourier transform — look like? Virtual particles that are far off mass shell are creatures of perturbative field theory, so one does not expect any concrete general answer to this question. But if a particle is on mass shell, or very nearly so, so that it can be observed, what we should see is just a particle traveling from one point to another; the extra virtual interactions are unobservable. That is, near the mass shell $p^2 = m^2$, $\widehat{\Delta}_{\text{dressed}}(p)$ should look very much like a free-particle propagator; more precisely, it should have the *same singularity structure* as the free propagator there:

$$(7.29)\quad -i\widehat{\Delta}_{\text{dressed}}(p) = -i\Delta(p)\big|_{m=m_{\text{phys}}} + \text{stuff that is nonsingular at } p^2 = m^2_{\text{phys}}.$$

We emphasize the point that the mass in this formula is the physical mass of the particle because a mass renormalization will be necessary. In fact, (7.29) will hold only after all mass and field renormalizations have been performed, and it is the guiding beacon that determines *how* these renormalizations should be performed, just as (7.23) was the guide for coupling constant renormalization in scalar field theory.

To bring this to the level of calculation, consider the set $\mathcal{P}$ of all connected Feynman diagrams for a given theory that contain just two external lines, both of the same particle species. We think of the external lines as representing propagators and pass to the momentum-space representation. Thus the two external lines are both labeled by the momentum p of the particle (not on mass shell) as it enters and leaves the picture, and they correspond to factors $-i\widehat{\Delta}(p)$ in the total propagation amplitude. Each such diagram consists of a finite number of one-particle irreducible pieces linked by simple propagators, which again have the value $-i\widehat{\Delta}(p)$. Let us denote by $-i\Pi(p)$ the sum of all possible nontrivial 1PI subdiagrams with their external lines amputated, with the divergent diagrams regularized in a consistent fashion, as indicated schematically in Figure 7.9. Then the (regularized) sum of all the diagrams in $\mathcal{P}$ is

$$(-i\widehat{\Delta}(p)) + (-i\widehat{\Delta}(p))(-i\Pi(p))(-i\widehat{\Delta}(p)) \\ + (-i\widehat{\Delta}(p))(-i\Pi(p))(-i\widehat{\Delta}(p))(-i\Pi(p))(-i\widehat{\Delta}(p)) + \cdots.$$

This is a geometric series that can be formally summed to give

$$(-i\widehat{\Delta}(p))\sum_0^\infty(-\Pi(p)\widehat{\Delta}(p))^n = -i[\widehat{\Delta}(p)^{-1} + \Pi(p)]^{-1}. \tag{7.30}$$

The geometric series need not converge, of course, but (7.30) is valid in perturbation theory. That is, if one takes $\Pi(p)$ to be the sum of all regularized values of 1PI diagrams up to a given order N in perturbation theory and discards terms of order $> N$ from the geometric series, then the equality in (7.30) holds modulo terms of order $> N$.

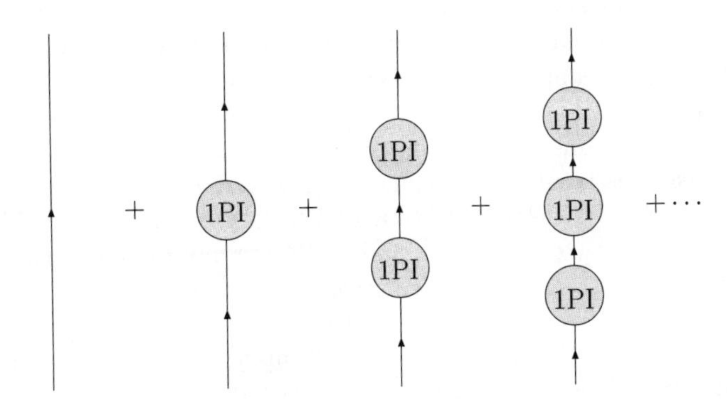

FIGURE 7.9. Scheme of the higher-order modifications to a propagator.

Suppose now that (7.30) can be taken literally, so that the expression on the right is the actual dressed (but unrenormalized) propagator — a supposition that will receive more attention shortly. As we shall see, this expression has almost the form that we were expecting:

$$[\widehat{\Delta}(p)^{-1} + \Pi(p)]^{-1} = Z\widehat{\Delta}(p)\big|_{m\to m'} + \text{stuff that is nonsingular at } p^2 = (m')^2. \tag{7.31}$$

Here Z is a constant, m is the mass occurring in the free propagator $\widehat{\Delta}(p)$, and the notation $m \to m'$ means that m must be shifted to a different value m'. On comparing this with (7.29), we see that (in bare perturbation theory) m' must be identified with the physical mass m_{phys}; this determines the proper mass renormalization. The factor of Z is then disposed of by renormalizing the field ϕ that produces the particle in question — specifically, by replacing ϕ by $Z^{-1/2}\phi$. In renormalized perturbation theory, one recalculates everything with the inclusion of counterterm insertions, which are added to $\Pi(p)$; the counterterms are then adjusted so that the value of m remains m_{phys} and the value of Z is 1.

The algebraic details of these calculations differ from field to field. We shall derive (7.31) and perform the renormalization on a case-by-case basis for scalar field, electron, and photon propagators. However, if one wants to think in terms of interacting field operators as in §6.11, other general arguments are available. Indeed, by the result following (6.99), the sum of all the momentum-space Feynman diagrams of the class $\mathcal{P}$, with the external lines carrying momentum p — that is, the sum on the left side of (7.30) — is the Fourier-transformed vacuum expectation value

$$\int \langle\Omega|\mathcal{T}[\phi(x)\phi(y)]|\Omega\rangle e^{ip_\mu(x-y)^\mu}\,d^4x$$

(which is independent of y by (6.100)), where ϕ here denotes the *interacting* field. One can then fashion nonperturbative arguments to see that this quantity has the form on the right side of (7.31); see Weinberg [**129**], §10.3, or Peskin and Schroeder [**88**], §7.1. Moreover, it is clear from this that the replacement of ϕ by $Z^{-1/2}\phi$ will cancel the Z in that formula.

Now, what about our claim that the expression on the right side of (7.30) really is the (unrenormalized) dressed propagator, or at least that it has the same singularity structure? The perturbation-theoretic validity of (7.30) does not suffice, for it cannot detect the shift in the location of the singularities: all the partial sums of the left side of (7.30) have singularities in the same places as the free propagator $\widehat{\Delta}$. We shall simply accept this as an extension of our bargain with the devil. In Chapter 6 we agreed to use perturbation theory with no rigorous proof that it would work; here we do the same for the nonperturbative validity of (7.30).[6]

We can make this result more plausible, however, by considering an "interaction" of a particularly trivial sort. We start with the free real scalar field of mass m with Lagrangian $\mathcal{L} = \frac{1}{2}[(\partial_\phi)^2 - m^2\phi^2]$ and add some more mass — that is, replace m^2 by $m^2 + \delta m^2$ — but regard the extra term $(\delta m^2)\phi^2$ as an *interaction* to be dealt with by perturbation theory. The set $\mathcal{P}$ of connected Feynman diagrams is very simple in this situation: the interaction Hamiltonian is $\frac{1}{2}(\delta m^2)\phi^2$, so each internal vertex has just two lines attached to it, and its value is $-i$ times the "coupling constant" δm^2. (There is a symmetry factor of 2, like the 4! in ϕ^4 theory, so the $\frac{1}{2}$ drops out.) Thus the dressed propagator is given diagramatically by Figure 7.10 and analytically by

$$\frac{-i}{-p^2 + m^2 - i\epsilon}\sum_0^\infty \left[\frac{(-i\delta m^2)(-i)}{-p^2 + m^2 - i\epsilon}\right]^n = \frac{-i}{-p^2 + m^2 + \delta m^2 - i\epsilon}.$$

The geometric series certainly does not converge for p^2 near m^2; nonetheless, the expression on the right is the true propagator for a scalar particle of mass $\sqrt{m^2 + \delta m^2}$, with singularities at $p^2 = m^2 + \delta m^2$.

A variation on this calculation in which we adjust both the mass and the field strength explains the form of the mass-and-field renormalization counterterm for the scalar field that we asserted in the previous section. We start with the bare scalar field ϕ_B with bare mass m_B, whose propagator is $-i\widehat{\Delta}(p) = -i/(-p^2 + m_B^2)$ (suppressing the infinitesimal $i\epsilon$, which plays no role here). The renormalized field is $\phi = Z^{-1/2}\phi_B$, and since $\overbrace{\phi(x)\phi(y)} = Z^{-1}\overbrace{\phi_B(x)\phi_B(y)}$, its propagator is $-iZ^{-1}\widehat{\Delta}(p)$.

---

[6]Renormalization group techniques, as discussed briefly in §7.12, can be used to put this result on a more solid foundation.

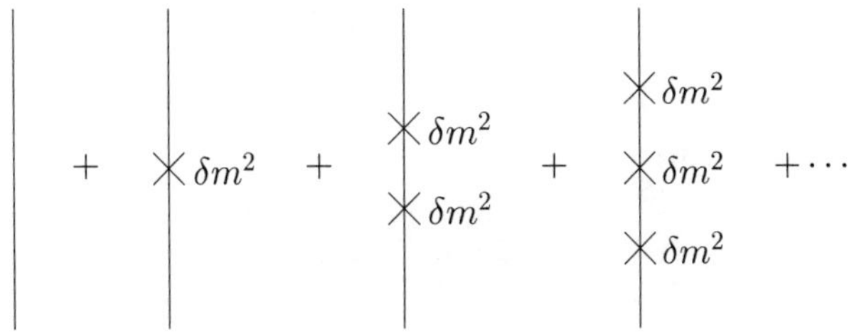

$$+ \quad \delta m^2 \quad + \quad \begin{matrix}\delta m^2 \\ \delta m^2\end{matrix} \quad + \quad \begin{matrix}\delta m^2 \\ \delta m^2 \\ \delta m^2\end{matrix} \quad + \cdots$$

FIGURE 7.10. Added mass as an interaction.

Writing $Z = 1 + \delta Z$ and $Zm_B^2 = m^2 + \delta m^2$ as in (7.28), this becomes

$$\frac{-i}{-p^2 + m^2 - (\delta Z)p^2 + \delta m^2} = \frac{-i}{-p^2 + m^2} \sum_0^\infty \left[(-i)^2 \frac{-(\delta Z)p^2 + \delta m^2}{-p^2 + m^2} \right]^n,$$

and this is the sum of the diagrams in Figure 7.10 if the vertices there are reassigned the asserted value for the counterterm vertex, $-i[-(\delta Z)p^2 + \delta m^2]$.

Let us see how all this works in renormalized perturbation theory for the scalar field with ϕ^4 interaction. At the outset, let us observe that since we are dealing with a scalar field, by Lorentz invariance the sum $\Pi(p)$ of the 1PI insertions in the propagator can only depend on p^2:

$$\Pi(p) = \Pi_0(p^2).$$

The explicit calculation of $\Pi_0(p^2)$ to a given order in perturbation theory is accomplished by evaluation of the appropriate Feynman diagrams; it will not concern us here. The only important thing is that each term in $\Pi_0(p^2)$ contains a power of the coupling constant λ, and for each n there are only finitely many terms containing λ^n.

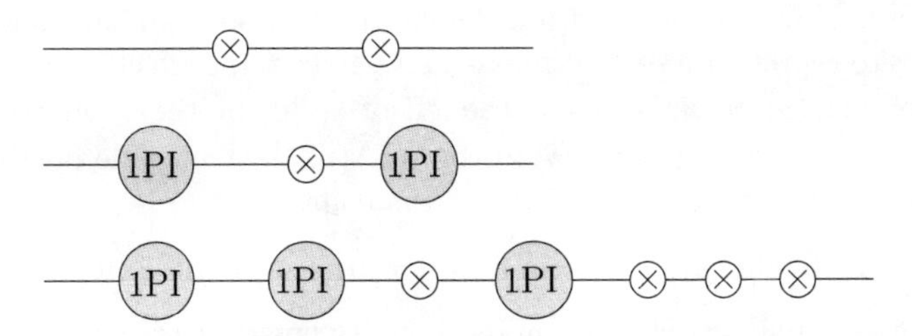

FIGURE 7.11. Modifications to a propagator including counterterms.

To calculate the dressed propagator in renormalized perturbation theory, we start with the free propagator and insert any number of 1PI subdiagrams and counterterm vertices, producing diagrams such as those in Figure 7.11. For any $j, k \geq 0$, there are $\binom{j+k}{j}$ ways to arrange the insertion of j 1PI subdiagrams and k counterterm insertions. After adding up all the possibilities for 1PI subdiagrams, the corresponding contribution to the dressed propagator is the product of j factors of $-i\Pi_0(p^2)$, k factors of $-i[-(\delta Z)p^2 + \delta m^2]$, and $j + k + 1$ factors of the free propagator $-i/(-p^2+m^2-i\epsilon)$. Thus the grand total is

$$\sum_{j,k=0}^{\infty}\binom{j+k}{j}\frac{-i(-1)^{j+k}[\Pi_0(p^2)]^j[(-(\delta Z)p^2+\delta m^2)]^k}{(-p^2+m^2-i\epsilon)^{j+k+1}}$$
$$=\sum_{n=0}^{\infty}\frac{-i[-\Pi_0(p^2)+(\delta Z)p^2-\delta m^2]^n}{(-p^2+m^2-i\epsilon)^{n+1}}$$
$$=\frac{-i}{-p^2+m^2+\Pi_0(p^2)-(\delta Z)p^2+\delta m^2-i\epsilon},$$

with the last equality understood as explained earlier. On performing a Taylor expansion of $\Pi_0(p^2)$ about $p^2=m^2$,

$$\Pi_0(p^2)=\Pi_0(m^2)+(p^2-m^2)\Pi_0'(m^2)+o(p^2-m^2),$$

we can rewrite this last expression as

$$\frac{-i}{(-p^2+m^2)[1-\Pi_0'(p^2)+\delta Z+o(1)]+\Pi_0(m^2)-(\delta Z)m^2+\delta m^2}. \tag{7.32}$$

Thus we obtain the correct pole and residue by setting

$$\delta Z=\Pi_0'(m^2),\qquad \delta m^2=-\Pi_0(m^2)+(\delta Z)m^2=-\Pi_0(m^2)+\Pi_0'(m^2)m^2. \tag{7.33}$$

With this specification, (7.32) has the desired form (7.29). The formulas (7.33) are to be used to calculate δZ and δm^2 order-by-order in perturbation theory, with the adjustments at any given order depending on the previous ones.

There is one more important general point to be made. A Feynman diagram that contributes to an S-matrix element contains external lines as well as internal ones, and the incoming and outgoing particles undergo the same interactions with virtual particles on their way to and from the real interaction as the particles in the middle of the interaction. One might therefore expect a need to renormalize the external coefficient functions (including spinors for Dirac particles and polarization vectors for vector particles) that correspond to the external lines. But in fact, that is not the case. If we go back to the calculation of (7.30) and take one of the two external lines to represent a real on-mass-shell particle, the effect is to replace one of the factors of $-i\widehat{\Delta}(p)$ in the series on the left of (7.30) by the appropriate external coefficient C, that is, to multiply (7.30) by $iC\widehat{\Delta}(p)^{-1}$. The same operation on the renormalized, dressed propagator (7.29) turns it into

$$C+iC\widehat{\Delta}(p)^{-1}(\text{stuff that is nonsingular at } p^2=m^2).$$

But since the external particle is on mass shell, we must set $p^2=m^2$ here, and then the second term vanishes. For the scalar field, for example, $\widehat{\Delta}(p)^{-1}=-p^2+m^2=0$ when $p^2=m^2$; for a Dirac field, $iC\widehat{\Delta}(p)^{-1}|_{p^2=m^2}=-\overline{u}(p)(p_\mu\gamma^\mu-m)=0$ by the Fourier-transformed Dirac equation; and similarly for other types of fields. In all cases, the final result is simply C.

That interactions change the effective mass of a particle is a common phenomenon — we encountered it in §6.2 — and it would not be unexpected even if there were no divergences to be removed. The renormalization of field strength is more mysterious, because the latter (unlike the mass of a particle) is generally not something that can be measured in the laboratory. In fact, by now the reader has probably forgotten how we normalized the field operators in the first place: they were constructed out of creation and annihilation operators, and *those* were normalized to satisfy the canonical (anti)commutation relations (6.33). (Alternatively, the field normalization can be specified by the commutation relation between a field and its canonical conjugate, as in (5.17).) We are assured that this is appropriate when renormalization is not an issue by the success of calculations such as those in §6.9 and §6.10. When it is an issue, the most direct connection of field strength with measurable quantities comes through the propagator; hence our use of (7.29) as a foundation.

There are ways of giving some intuitive content to the renormalization factor Z, but they are fraught with pitfalls. The calculations with vacuum expectation values of interacting fields that we alluded to earlier in this section[7] lead one to believe that Z is a number between 0 and 1, and hence infinitesimal if it is not a well-defined nonzero quantity; but the evaluation of Z in perturbation theory yields $Z = \infty$ — more precisely, in dimensional regularization, its regularized value becomes infinite as $d \to 4$. (This discrepancy is related to the fact that $(1-x)^{-1} = 1 + x$ to first order in x; in effect, we are applying this formula to an x that contains a divergent coefficient.) I asked a well-respected physicist how I could think about this without getting a headache, and his response was, "You just have to get the headache and get over it." It is best to take a ruthlessly pragmatic attitude: we perform the field renormalization as indicated because it works.

7.7. The Ward identities

The next five sections are devoted to a study of renormalization in QED. The messiness of the calculations will be somewhat abated by the adoption of a couple of notational abbreviations. First, if p is a Lorentz 4-vector (usually but not always representing a momentum), we set

$$\not{p} = p_\mu \gamma^\mu,$$

and likewise with p replaced by any other letter. Second, we shall write the Dirac propagator in the form (6.52):

$$-i\widehat{\Delta}_{\text{Dirac}}(p) = \frac{i}{\not{p} - m}.$$

The infinitesimal $i\epsilon$ that should be in there will be incorporated in a Wick rotation when it really matters; otherwise it will be suppressed. We shall also write Dirac spinors as $u(p)$ rather than $u(\mathbf{p}, s)$; here $p = (\omega_{\mathbf{p}}, \mathbf{p})$ as usual, and the dependence on the spin state will be left implicit until it plays a significant role.

Before we proceed to the main order of business, we need to develop a collection of related results known as *Ward identities* or *Ward-Takahashi identities*[8] that play a crucial role in the theory. One of their main applications will be to justify the freedom in choosing the parameter a in the photon propagator (6.55).

We begin with a simple observation. Consider the fundamental interaction diagram of QED (Figure 7.12), where the incoming and outgoing electron lines carry momenta p and q and the photon line carries incoming momentum $k = q - p$ (or outgoing momentum $-k = p - q$). These momenta may or may not be on mass shell, and the lines will represent spinor/vector coefficients or propagators accordingly. If the electron lines are on mass shell, so that they represent Dirac

[7] In particular, the so-called Källén-Lehmann representation.

[8] Ward derived a special case; Takahashi generalized; others then generalized further.

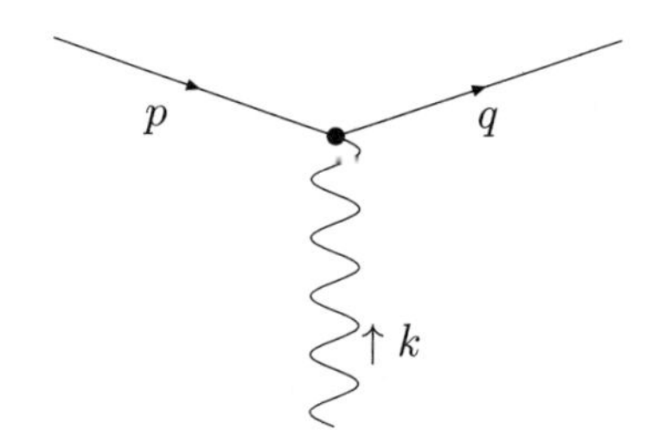

FIGURE 7.12. The fundamental interaction diagram of QED.

spinors $u(p)$ and $\overline{u}(q)$, the value of this diagram with the photon line amputated and the ubiquitous $(2\pi)^4\delta(q-p-k)$ suppressed, is simply $\overline{u}(q)(-ie\gamma^\mu)u(p)$, and the Lorentz product of this with the photon momentum k is

$$\begin{aligned}(7.34)\quad k_\mu\overline{u}(q)(-ie\gamma^\mu)u(p) &= -ie\overline{u}(q)\not{k}u(p)\\ &= -ie\overline{u}(q)(\not{q}-\not{p})u(p) = -ie\overline{u}(q)(m-m)u(p) = 0,\end{aligned}$$

because of the Dirac equation $(\not{p}-m)u(p) = \overline{u}(q)(\not{q}-m) = 0$. On the other hand, if the electron lines are off mass shell and represent propagators, the value of the diagram (modified as before) is $-i^3e(\not{q}-m)^{-1}\gamma^\mu(\not{p}-m)^{-1}$, and the Lorentz product of this with k is

$$\begin{aligned}\frac{i}{\not{q}-m}(-ie\not{k})\frac{i}{\not{p}-m} &= -ie\frac{i}{\not{q}-m}[(\not{q}-m)-(\not{p}-m)]\frac{i}{\not{p}-m}\\ &= e\left[\frac{i}{\not{p}-m}-\frac{i}{\not{q}-m}\right].\end{aligned}\tag{7.35}$$

We can also take the incoming electron on mass shell and the outgoing one off mass shell or vice versa; the same calculations then yield

$$\frac{i}{\not{q}-m}(-ie\not{k})u(p) = eu(p), \qquad \overline{u}(q)(-ie\not{k})\frac{i}{\not{p}-m} = -e\overline{u}(q).\tag{7.36}$$

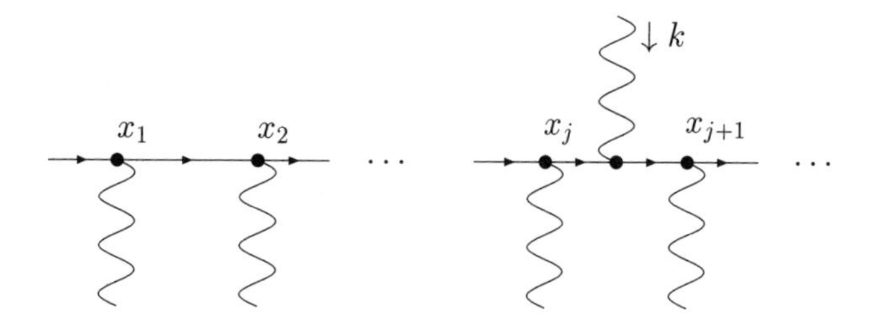

FIGURE 7.13. Inserting a photon line into an electron pathway.

With this in mind, consider an arbitrary connected Feynman diagram D containing a connected sequence $L_0, \ldots, L_J$ of electron lines, where L_0 is incoming, $L_1, \ldots, L_{J-1}$ are internal, L_J is outgoing, and L_{j-1} and L_j meet at a vertex x_j for $1 \leq j \leq J$. Let D_j be the diagram obtained from D by attaching an additional external photon line in the middle of the line L_j, as in Figure 7.13. We are interested in the values of D and D_j as functions of the momenta p, q, and k of the incoming electron, outgoing electron, and added photon, respectively. If the external electron lines represent propagators, for example, the value of D has the form

$$D(p,q)^{\rm prop} = \frac{i}{\not q - m}\gamma^{\nu_J}\frac{i}{\not p_{J-1} - m}\cdots\frac{i}{\not p_1 - m}\gamma^{\nu_1}\frac{i}{\not p - m}\Phi_{\nu_1\cdots\nu_{J-1}},$$

where "prop" stands for "propagators at the ends," $p_1,\ldots,p_{J-1}$ are the momenta on the lines $L_1,\ldots,L_{J-1}$, and the $\Phi_{\nu_1\cdots\nu_J}$ are quantities that depend on everything else in the diagram (including the factors of $-ie$ that should accompany the γ^{ν_j}'s). The corresponding value $D_j^\mu(p,q,k)^{\rm prop}$ of D_j, with the added photon line amputated, is obtained by inserting an extra $-ie\gamma^\mu$ and propagator $i/(\not p_j + \not k - m)$ at the appropriate spot and replacing all the subsequent $\not p_i$ by $\not p_i + \not k$.

Now consider the sum $\sum_{j=0}^J k_\mu D_j^\mu(p,q,k)^{\rm prop}$. We apply (7.35) to the piece of D_j consisting of the added photon line and the line L_j (now broken into two) to express the product of the $(-ie\gamma^\mu)$ and its two adjacent propagators in the term $k_\mu D_j^\mu(p,q,k)^{\rm prop}$ as e times the difference of two propagators. Summing over j yields a telescoping sum: everything cancels out except for one term on each end, with the result that

$$\sum_{j=0}^J k_\mu D_j^\mu(p,q,k)^{\rm prop} = e[D(p,\, q-k)^{\rm prop} - D(p+k,\, q)^{\rm prop}]. \tag{7.37}$$

(Readers who are not convinced by this briefly sketched argument can remedy the situation by performing the calculation for $J=1$ or $J=2$ with their own pencil and paper; or see Peskin and Schroeder [**88**], §7.4, or Zee [**136**], §II.7. The formula (7.35) can be regarded as the case $J=0$, since $q-k=p$ and $p+k=q$ there.)

If we take the external electron lines L_0 and L_J to be on mass shell and hence represent spinors, this calculation goes through without change except that one must use (7.36) instead of (7.35) at the ends. The value $D^\mu(p,q)^{\rm spinor}$ of D with this interpretation is obtained from $D^\mu(p,q)$ by replacing the factors of $i(\not p - m)^{-1}$ and $i(\not q - m)^{-1}$ by $u(p)$ and $\overline{u}(q)$, respectively; likewise for $D_j^\mu(p,q,k)^{\rm spinor}$. But this replacement nullifies the difference between $D(p,\, q-k)$ and $D(p+k,\, q)$, so we have

$$\sum_{j=0}^J k_\mu D_j^\mu(p,q,k)^{\rm spinor} = 0. \tag{7.38}$$

There is one more variation to be played on this theme. Consider a Feynman diagram $\widetilde{D}$ that contains a set $L_1,\ldots,L_J$ of electron lines that form a closed loop (with L_{j-1} and L_j meeting at x_j for $j>1$, and L_J and L_1 meeting at x_1), and again let $\widetilde{D}_j^\mu(k)$ be the value of the diagram obtained from D by attaching a photon line with momentum k to L_j, with that photon line amputated (see Figure 7.14). Let p_j be the momentum on the line L_j in $\widetilde{D}$; thus $p_j = p_1 + \sum_1^{j-1} k_i$ where k_i is the incoming photon momentum at the vertex joining L_i to L_{i+1}. Taking p_1 to be the variable of integration for the loop integral, by (6.59) the value of D has the form

$$-\int \mathrm{tr}\left[\frac{i}{\not p_J - m}\gamma^{\nu_J}\cdots\frac{i}{\not p_1 - m}\gamma^{\nu_1}\Phi_{\nu_1\cdots\nu_J}\right]\frac{d^4p_1}{(2\pi)^4}, \tag{7.39}$$

where the $\Phi_{\nu_1\cdots\nu_J}$ again incorporate all the rest of the diagram. If we apply (7.35) to the diagrams $\widetilde{D}_j$ and sum over j as before, there is again a lot of cancellation,

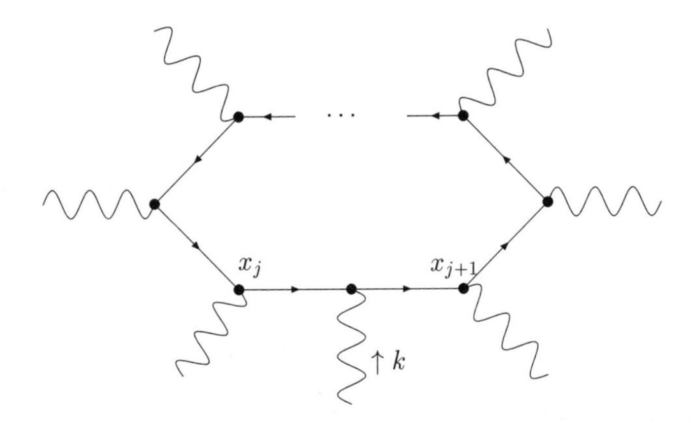

FIGURE 7.14. Inserting a photon line into an electron loop.

and we find that $\sum k_\mu \widetilde{D}_j^\mu(k)$ is equal to the integral (7.39) minus the same integral with $\not{p}_j$ replaced by $\not{p}_j + \not{k}$. The change of variable $p_1 \to p_1 - k$ in the latter shows that these two integrals cancel; thus,

$$\sum_{j=1}^{J} k_\mu \widetilde{D}_j^\mu(k) = 0. \tag{7.40}$$

The last part of this argument needs more buttressing. The integral (7.39) may be divergent, in which case its exact cancellation with the shifted integral is not clear. (Physicists tend to assume that such formal manipulations are innocent of error until proven guilty, but in this case there is a similar calculation in a different situation — the "chiral anomaly" or "axial vector anomaly" in the theory of massless Fermions — that does give an incorrect result.) The resolution of the problem is that the integrals must be regularized, and the cancellation should be valid for the regularized integrals. For this to happen, one needs to take care that the regularization procedure does not distort the structure of the integrals in a way that destroys the cancellation. Dimensional regularization avoids such distortion quite automatically (this is one of its advantages), however, and so (7.40) is indeed valid.

The equalities (7.37), (7.38), and (7.40) are the Feynman-diagram version of the general *Ward identities.* Other versions are obtained by applying these identities to amplitudes M that are given by sums of Feynman diagrams with the property that all of the diagrams D_j or $\widetilde{D}_j$ contribute to M whenever any one of them does. For example, consider a (connected, δ-function-free) S-matrix element M in which one of the initial or final states contains a photon with momentum k and polarization vector $\epsilon_\mu(k)$, and which thus has the form $M = \epsilon_\mu(k)M^\mu(k)$. In any connected Feynman diagram contributing to M, the line for this photon connects to an electron line that is part of either a path from an incoming electron to an outgoing electron, both on mass shell, or an internal loop, and all diagrams obtained by reattaching it at other points of the path or loop also contribute to M. On summing over such diagrams and applying (7.38) and (7.40), we find that

$$k_\mu M^\mu(k) = 0.$$

The same reasoning also applies when the photon with momentum k is internal (so $M = -i\Delta_{\mu\nu}(k)M^{\mu}(k)$ rather than $M = \epsilon_{\mu}(k)M^{\mu}(k)$). *This is the reason why the different choices for the parameter a in the photon propagator* (6.55) *lead to the same result in computing S-matrix elements.* Indeed, consider an S-matrix element M and a diagram D_1 that contributes to M and contains an internal photon line with momentum k. The diagrams D_j obtained from D_1 by reattaching one end of this line at other points on the same electron path or loop also contribute to M. If one applies (7.38) or (7.40) to (appropriate parts of) the D_j's with the photon propagator amputated, one finds on reinserting the propagator that any term in it that is proportional to k_μ contributes nothing to M. Hence one can modify the propagator by inserting any term $ak_\mu k_\nu/k^2$ as in (6.55), and in fact, a need not be a constant but can be a function of k (more precisely, a function of k^2 in order to preserve Lorentz covariance). This observation will be important in the next section.

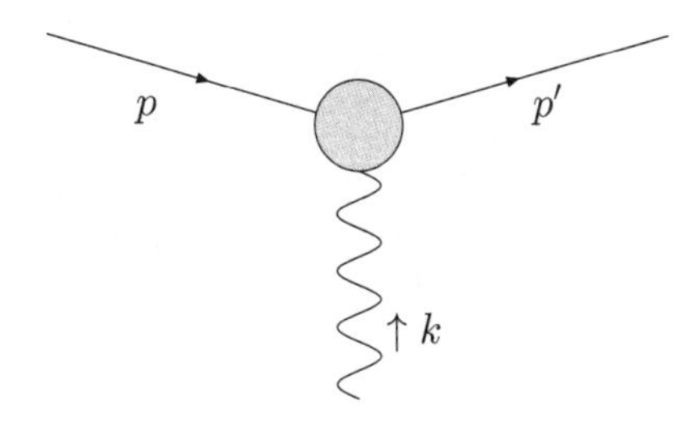

FIGURE 7.15. The fundamental interaction with higher-order corrections.

One can also use (7.37) and (7.40) to obtain results about processes where the external electron lines are off mass shell. The simplest and most important of these is the "dressed vertex function," which is the sum of the basic diagram in Figure 7.12 and all of its higher-order corrections. That is, we consider all Feynman diagrams of the form in Figure 7.15, with one incoming electron with momentum p, one outgoing electron with momentum p', and one incoming photon with momentum $k = p' - p$ (or outgoing photon with momentum $-k$), all of which represent propagators. The virtual processes that modify the basic vertex are of three kinds: those that begin and end on the photon line, which represent corrections to the photon propagator; those that begin and end on an electron line, which represent corrections to the electron propagators; and all the rest — namely, all the one-particle irreducible diagrams of the form in Figure 7.15. In other words, Figure 7.15 can be reexpressed as Figure 7.16, in which the external lines represent *dressed* propagators. The sum of the values of all the 1PI vertex diagrams, with the external legs amputated, is denoted by $-ie\Gamma^{\mu}(p', p)$, where μ is the Lorentz index to be contracted with the photon propagator. Of course $\Gamma^{\mu}(p', p)$, as it stands, is an infinite sum of integrals that generally diverge; it is understood that for actual calculations, one includes only the diagrams up to a fixed finite order in perturbation theory and regularizes all the divergent integrals. (Note that $\Gamma^{\mu}(p', p)$ is a 4×4 matrix and that to zeroth order, i.e., including only the basic vertex diagram (Figure 7.12), $\Gamma^{\mu}(p', p) = \gamma^{\mu}$.)

Let $S(p)$ denote the dressed electron propagator with momentum p, which we shall describe more explicitly in §7.8 and §7.9. The value of the diagram in Figure 7.16 with electron propagators included but the photon propagator still amputated

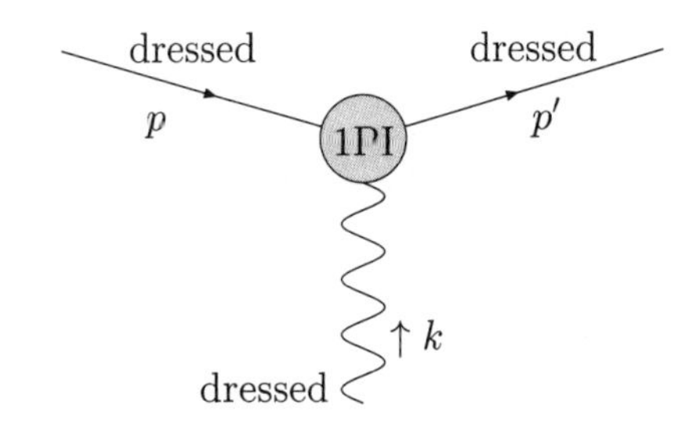

FIGURE 7.16. The fundamental interaction with vertex corrections and dressed propagators.

is then $-ieS(p')\Gamma^\mu(p',p)S(p)$. On applying the Ward identities (7.37) and (7.40) to the various specific diagrams contributing to Figure 7.16, we find that

$$-iek_\mu S(p')\Gamma^\mu(p',p)S(p) = e[S(p) - S(p')]. \tag{7.41}$$

Note that to zeroth order in perturbation theory — that is, considering only Figure 7.12 instead of Figure 7.16 — $S(p) = i(\not{p} - m)^{-1}$ and $\Gamma^\mu(p',p) = \gamma^\mu$, so (7.41) reduces to (7.35). (7.41) is usually restated by multiplying both sides on the left by $S(p')^{-1}$ and on the right by $S(p)^{-1}$:

$$-ik_\mu \Gamma^\mu(p',p) = S(p')^{-1} - S(p)^{-1}. \tag{7.42}$$

This is the Ward-Takahashi identity in the form derived by Takahashi; Ward's original version is obtained by passing to the limit as $q \to 0$, i.e., as $p' \to p$:

$$\Gamma^\mu(p,p) = i\frac{\partial S(p)^{-1}}{\partial p_\mu}. \tag{7.43}$$

The identity (7.42), as well as the analogous results for more complicated processes involving more external electrons, can be interpreted as statements about vacuum expectation values of time-ordered products of interacting fields, as explained in §6.11. From this point of view they can also be derived nonperturbatively, rather than via the diagram-by-diagram calculations we have performed; see Weinberg [**129**], §10.4.

7.8. Renormalization in QED: general structure

The Lagrangian for QED is

$$\mathcal{L} = -\tfrac{1}{4}F_{\mu\nu\mathrm{B}}F^{\mu\nu}_{\mathrm{B}} + i\overline{\psi}_{\mathrm{B}}\gamma^\mu\partial_\mu\psi_{\mathrm{B}} - m_B\overline{\psi}_B\psi_B - e_B\overline{\psi}_{\mathrm{B}}\gamma^\mu A_{\mu\mathrm{B}}\psi_B, \tag{7.44}$$

where the subscripts B are there to remind us that the quantities in question are bare (unrenormalized). We can rewrite it in terms of the renormalized fields

$$A_\mu = Z_3^{-1/2}A_{\mu\mathrm{B}} \quad \text{and} \quad \psi = Z_2^{-1/2}\psi_{\mathrm{B}}$$

and the physical charge e of the electron as follows:

$$\mathcal{L} = \tfrac{1}{4}Z_3F_{\mu\nu}F^{\mu\nu} + iZ_2\overline{\psi}\gamma^\mu\partial_\mu\psi - Z_2m_B\overline{\psi}\psi - Z_1e\overline{\psi}\gamma^\mu A_\mu\psi, \tag{7.45}$$

where

$$Z_1 = Z_2Z_3^{1/2}\frac{e_{\mathrm{B}}}{e}. \tag{7.46}$$

Electron/positron propagator	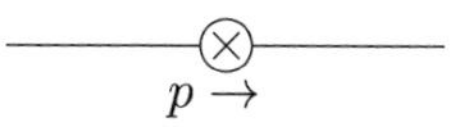 $p \rightarrow$	$i[(\delta Z_2)p_\mu\gamma^\mu - \delta m]$
Photon propagator	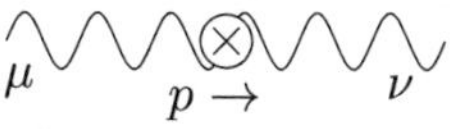 μ $p \rightarrow$ ν	$-i(g^{\mu\nu}p^2 - p^\mu p^\nu)\delta Z_3$
Vertex	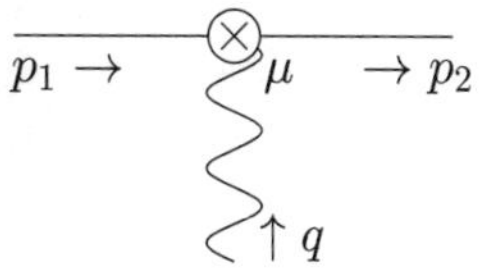 $p_1 \rightarrow$ μ $\rightarrow p_2$ $\uparrow q$	$-ie\gamma^\mu(\delta Z_1)\delta(p_1 + q - p_2)$

TABLE 7.1. Feynman rules for counterterms in QED.

(The assignment of the subscripts to the Z's is firmly fixed by convention.) Introducing also the physical mass m of the electron, we separate the Lagrangian into its "physical" and "counterterm" parts:

$$\begin{aligned}\mathfrak{L} = &-\tfrac{1}{4}F_{\mu\nu}F^{\mu\nu} + i\overline{\psi}\gamma^\mu\partial_\mu\psi - m\overline{\psi}\psi - e\overline{\psi}\gamma^\mu A_\mu\psi \\ &-\tfrac{1}{4}(\delta Z_3)F_{\mu\nu}F^{\mu\nu} + i(\delta Z_2)\overline{\psi}\gamma^\mu\partial_\mu\psi - (\delta m)\overline{\psi}\psi - (\delta Z_1)e\overline{\psi}\gamma^\mu A_\mu\psi,\end{aligned} \tag{7.47}$$

where

$$\delta Z_j = Z_j - 1, \qquad \delta m = Z_2 m_{\mathrm{B}} - m.$$

The Feynman rules for QED in renormalized perturbation theory are given in Table 7.1. As the diagrams would suggest, the three types of counterterm are connected with the radiative corrections to the electron propagator, the photon propagator, and the electron-photon vertex. The form of the counterterms for the electron-photon vertex and the electron propagator are entirely analogous to the corresponding counterterms for the scalar field that we saw in §7.5 and §7.6; the form of the counterterm for the photon propagator requires some explanation that we shall give later in this section. We now examine each of these processes in some detail to see how the renormalization should work. In particular, we shall prove and discuss the very important fact that the renormalization constants Z_1 and Z_2 coincide.

The electron propagator. The discussion of the electron propagator proceeds as in §7.6, with some additional complications from the fact that it is a matrix-valued function. Let $-i\Sigma(p)$ denote the sum of all 1PI diagrams with one incoming and one outgoing electron with momentum p (not on mass shell). (No renormalization is being performed yet. Only ordinary Feynman diagrams without counterterm insertions are included; all divergent integrals are to be regularized; and one sums only the diagrams up to a fixed but arbitrary order in perturbation theory.) As in §7.6, the dressed electron propagator is given diagrammatically by

Figure 7.9 and analytically by

$$S(p) = \frac{i}{\not p - m} + \frac{i}{\not p - m}(-i\Sigma(p))\frac{i}{\not p - m} + \frac{i}{\not p - m}(-i\Sigma(p))\frac{i}{\not p - m}(-i\Sigma(p))\frac{i}{\not p - m} + \cdots,$$

where, for the moment, m is the bare mass of the electron. Now, $\Sigma(p)$ is a 4×4 matrix, and it is a Lorentz scalar (i.e., invariant under Lorentz transformations) just like the free propagator $i(\not p - m)^{-1}$, so it can depend on p only through $\not p$ and p^2. Moreover, $\not p^2 = p^2$ (i.e., $\not p^2 = p^2 I$) so $\Sigma(p)$ is actually function of $\not p$.[9] It therefore commutes with $i(\not p - m)^{-1}$, so we can sum the series (as in §7.6) to obtain

$$S(p) = i(\not p - m - \Sigma(p))^{-1}.$$

The electron mass and field strength now need to be renormalized to make this look as much like the free propagator $i(\not p - m)^{-1}$ as possible.

We therefore redo this calculation in renormalized perturbation theory: add arbitrary numbers of counterterm insertions of the form in the first row of Table 7.1 into the diagrams in Figure 7.9 and take m to be the physical mass of the electron. Just as before, the result is

$$S(p) = i(\not p - m - \Sigma(p) + (\delta Z_2)\not p - \delta m)^{-1}. \tag{7.48}$$

This is the quantity whose pole and residue must be examined.

The physicists' shorthand language for what happens next is: "We choose δZ_2 and δm so that $S(p)$ has a pole at $\not p = m$ with residue i; this is accomplished by using the Taylor expansion of $\Sigma(p)$ about $\not p = m$." Taken literally, the condition $\not p = m$ is nonsense even though m is really mI, for $\mathrm{tr}(\not p) = 0$ whereas $\mathrm{tr}(mI) = 4m$. However, whenever q is a 4-vector such that $q^2 = m^2$, the matrix $\not q - m$ has a 2-dimensional nullspace consisting of the Dirac spinors $u(q)$, so $(\not p - m)^{-1}u(q)$ does have a pole at $p = q$. The condition that "$S(p)$ has a pole at $\not p = m$ with residue i" can likewise be interpreted by applying it to a spinor $u(q)$, but more practically, it means that $S(p) = i[(\not p - m)(1 + E(p))]^{-1}$ where the error term $E(p)$ is itself divisible by $\not p - m$.

The "Taylor expansion of $\Sigma(p)$ about $\not p = m$" is accomplished as follows. Since $\Sigma(p)$ is a function of $\not p$ as we have noted, it has the form $f(p^2)I + g(p^2)\not p$ where f and g are scalar functions. We expand them about $p^2 = m^2$ ($f(p^2) = f(m^2) + f'(m^2)(p^2 - m^2) + \cdots$ and similarly for g), use the identity $(p^2 - m^2) = 2m(\not p - m) + (\not p - m)^2$, and write $\not p = (\not p - m) + m$ to obtain

$$\Sigma(p) = \Sigma(m)I + \Sigma'(m)(\not p - m) + \text{terms divisible by } (\not p - m)^2, \tag{7.49}$$

where

$$\Sigma(m) = f(m^2) + mg(m^2), \qquad \Sigma'(m) = g(m^2) + 2mf'(m^2) + 2m^2 g'(m^2).$$

Substituting (7.49) into (7.48), we find that

$$S(p) = i\big[(\not p - m)[1 - \Sigma'(m) + \delta Z_2 + O(\not p - m)] - \Sigma(m) + (\delta Z_2)m - \delta m\big]^{-1},$$

[9] The implicit assumption here is that $\Sigma(p)$ is an analytic function, so that it can be expanded in a power series in $\not p$ and p^2 and hence in $\not p$ alone. The regularized integrals that contribute to $\Sigma(p)$ are in fact analytic in p; in practice, though, we need to look only at the first couple of terms of the Taylor expansion.

so the renormalization condition (7.29) becomes

$$\delta Z_2 = \Sigma'(m), \qquad \delta m = -\Sigma(m) + (\delta Z_2)m = -\Sigma(m) + m\Sigma'(m). \tag{7.50}$$

We emphasize again that $\Sigma(p)$ contains regularized divergent integrals and is taken to a certain order in perturbation theory; these equations then determine δZ_2 and δm as functions of the regularization parameter and the order. We shall perform the first-order calculation of these quantities in the next section.

The photon propagator. Let $i\Pi^{\mu\nu}(p)$ be the sum of all 1PI diagrams that have just two external photon lines with off-mass-shell momentum p with the external lines amputated (and as usual, taken to some finite order in perturbation theory with all divergent integrals regularized). The general idea is the same as for the corrections to the scalar field propagator and the electron propagator discussed previously. Denoting the bare photon propagator $-ig_{\mu\nu}/p^2$ by $\Delta_{\mu\nu}(p)$, the dressed propagator is given, as with electrons and scalar fields, by

$$\Delta + \Delta(i\Pi)\Delta + \Delta(i\Pi)\Delta(i\Pi)\Delta + \cdots = \Delta(1 - i\Pi\Delta)^{-1} = (\Delta^{-1} - i\Pi)^{-1}, \tag{7.51}$$

where all of the terms are 4×4 spinor matrices labeled with Lorentz indices μ, ν.

We can say more about the form of $\Pi^{\mu\nu}(p)$. It must have the same Lorentz covariance as $(\Delta^{-1})^{\mu\nu} = ip^2 g^{\mu\nu}$ and hence must have the form $\rho(p^2)g^{\mu\nu} + \sigma(p^2)p^\mu p^\nu$ for some scalar functions ρ and σ. Moreover, the Ward identity (7.40) implies that

$$0 = p_\mu \Pi^{\mu\nu} = \rho(p^2)p^\nu + \sigma(p^2)p^2 p^\nu,$$

so that $\rho(p^2) = -p^2\sigma(p^2)$. Thus, setting $\pi_0(p^2) = -\sigma(p^2)$, we have

$$\Pi^{\mu\nu}(p) = (p^2 g^{\mu\nu} - p^\mu p^\nu)\pi_0(p^2). \tag{7.52}$$

It follows that

$$\begin{aligned}(\Delta^{-1} - i\Pi)^{\mu\nu} &= i[p^2 g^{\mu\nu} - (p^2 g^{\mu\nu} - p^\mu p^\nu)\pi_0(p^2)] \\ &= ip^2(1 - \pi_0(p^2)) \left[g^{\mu\nu} + \frac{p^\mu p^\nu}{p^2} \frac{\pi_0(p^2)}{1 - \pi_0(p^2)} \right],\end{aligned}$$

and it is easily verified (by computing their product) that the inverse of this matrix, the dressed propagator, is

$$[(\Delta^{-1} - i\Pi)^{-1}]_{\mu\nu} = \frac{-i}{p^2(1 - \pi_0(p^2))} \left[g_{\mu\nu} - \pi_0(p^2)\frac{p_\mu p_\nu}{p^2} \right].$$

As we showed in the preceding section, however, the Ward identity (7.38) implies that the $p_\mu p_\nu$ term can be dropped for any S-matrix calculations. In short, the effective dressed photon propagator is

$$\frac{-ig_{\mu\nu}}{p^2(1 - \pi_0(p^2))}.$$

We observe, with great relief, that no mass renormalization is necessary: the pole of the propagator still occurs at $p^2 = 0$, so the photon remains massless. However, the factor $1 - \pi_0(p^2)$ necessitates a renormalization of field strength. Thus, we redo this calculation by including the photon-propagator counterterm from Table 7.1, the effect of which is to make the following replacement for Π in (7.51):

$$\Pi^{\mu\nu}(p) \to \Pi^{\mu\nu}(p) - (\delta Z_3)(p^2 g^{\mu\nu} - p^\mu p^\nu) = (p^2 g^{\mu\nu} - p^\mu p^\nu)[\pi_0(p^2) - \delta Z_3].$$

We now see the explanation for the form of the counterterms: it must be the same as the form (7.52) of Π. The dressed propagator is now

$$\frac{-ig_{\mu\nu}}{p^2(1-\pi_0(p^2)+\delta Z_3)},$$

so the renormalization condition, that the residue at the pole $p^2=0$ be $-ig_{\mu\nu}$, is simply

$$\delta Z_3 = \pi_0(0).$$

The renormalized tensor

$$\Pi^{\mu\nu}_{\text{renorm}}(p) = (p^2 g^{\mu\nu} - p^\mu p^\nu)\pi(p^2), \qquad \pi(p^2) = \pi_0(p^2) - \pi_0(0), \tag{7.53}$$

remains finite as the regularization is removed, and it is a quantity of physical significance. We shall calculate it to lowest order and examine its meaning in §7.10.

The vertex function. As explained in the preceding section, $-ie\Gamma^\mu(p',p)$ denotes the sum of all 1PI diagrams containing one incoming electron with momentum p, one outgoing electron with momentum p', and one incoming (or outgoing) photon with momentum $q=p'-p$ (or $-q=p-p'$), with the external lines amputated. We shall call it the *vertex function.* As usual, the divergent integrals in $\Gamma^\mu(p',p)$ are to be regularized and the sum is to be truncated at some finite order of perturbation theory. The renormalization condition for $\Gamma^\mu(p',p)$ comes from considering the interactions of a real, nonvirtual electron, so we assume that the incoming and outgoing electron lines are on mass shell and are labeled by Dirac spinors $u(p)$ and $\overline{u}(p')$, and we are interested in the quantity $\overline{u}(p')\Gamma^\mu(p',p)u(p)$. Some simplifications take place when Γ^μ is thus sandwiched between spinors, and what we shall continue to call Γ^μ below is really the "effective" part of Γ^μ that contributes to $\overline{u}(p')\Gamma^\mu(p',p)u(p)$.

We recall that the Dirac matrices γ^ν generate the whole algebra of 4×4 matrices, so the matrix $\Gamma^\mu(p',p)$ is a linear combination of I, the γ^ν's, and products of 2, 3, or 4 distinct γ^ν's. Moreover, its zeroth-order approximation (coming from the bare vertex) is just γ^μ, so it must transform as a 4-vector under Lorentz transformations (including spatial reflections) just as γ^μ does. From this it follows easily that it must be a linear combination, with coefficients that are Lorentz-invariant scalar functions of p and p', of $p_1^\mu I$, γ^μ, $p_1^\mu \not{p}_2$, $[\gamma^\mu,\not{p}_1]$, $[\not{p}_1,\not{p}_2]p_3^\mu$, and $\not{p}_1\not{p}_2\gamma^\mu$ and similar products with the factors permuted, where p_1, p_2, and p_3 are each either p or p'. But then the Dirac equation

$$\overline{u}(p')(\not{p}'-m) = (\not{p}-m)u(p) = 0$$

together with the anticommutation relations for the γ^ν's implies that, once $\Gamma^\mu(p',p)$ is sandwiched in between $\overline{u}$ and u, the $\not{p}$'s and $\not{p}'$'s can be replaced by m's. Hence, $\Gamma^\mu(p',p)$ can be replaced by a linear combination merely of γ^μ, p^μ, and p'^μ (dropping the I as usual), or equivalently of γ^μ, $(p+p')^\mu$, and $q^\mu=(p'-p)^\mu$:

$$\Gamma^\mu(p',p) = A\gamma^\mu + B(p+p')^\mu + Cq^\mu \qquad (q=p'-p),$$

where A, B, and C, are Lorentz-invariant scalar functions of p and p'. But this means that A, B, and C can depend only on p^2, p'^2, and $p'_\mu p^\mu$, and since $p^2 = p'^2 = m^2$ and $2p'_\mu p^\mu = p^2+p'^2-(p'-p)^2 = 2m^2-q^2$, we conclude that A, B, and C are functions of q^2 alone.

A further simplification comes from the Ward identities (7.38) and (7.40), which imply that $k_\mu\overline{u}(p')\Gamma^\mu(p',p)u(p) = 0$. Since $\overline{u}(p')q_\mu\gamma^\mu u(p) = 0$ by (7.34) and $q_\mu(p+p')^\mu = p'^2 - p^2 = m^2 - m^2 = 0$, this reduces to $Cq^2 = 0$, so that $C = 0$. In short,

$$\Gamma^\mu(p', p) = A(q^2)\gamma^\mu + B(q^2)(p+p')^\mu. \tag{7.54}$$

It is customary to rewrite (7.54) by using the *Gordon identity*:

$$\overline{u}(p')(p+p')^\mu u(p) = \overline{u}(p')(2m\gamma^\mu - i\sigma^{\mu\nu}q_\nu)u(p), \tag{7.55}$$

where

$$\sigma^{\mu\nu} = \tfrac{1}{2}i[\gamma^\mu, \gamma^\nu].[10]$$

To prove this, observe that $\not{p}\gamma^\mu = -\gamma^\mu\not{p} + 2p^\mu$ for any p, so

$$\begin{aligned}
[\gamma^\mu, \gamma^\nu]q_\nu &= (\gamma^\mu\gamma^\nu - \gamma^\nu\gamma^\mu)(p' - p)_\nu \\
&= \gamma^\mu\not{p}' - \not{p}'\gamma^\mu - \gamma^\mu\not{p} + \not{p}\gamma^\mu \\
&= 2p'^\mu - 2\not{p}'\gamma^\mu - 2\gamma^\mu\not{p} + 2p^\mu.
\end{aligned}$$

Since $\overline{u}(p')\not{p}' = m\overline{u}(p')$ and $\not{p}u(p) = mu(p)$, this gives

$$\overline{u}(p')[\gamma^\mu, \gamma^\nu]q_\nu u(p) = \overline{u}(p')[2(p+p')^\mu - 4m\gamma^\mu]u(p)$$

and hence (7.55).

The Gordon identity says that, as long as $\Gamma^\mu(p', p)$ is to be sandwiched between spinors, the $(p+p')^\mu$ in it can be replaced by $2m\gamma^\mu - i\sigma^{\mu\nu}q_\nu$, so we finally obtain

$$\Gamma^\mu(q, p) = F_1(q^2)\gamma^\mu + F_2(q^2)\frac{i\sigma^{\mu\nu}q_\nu}{2m} \qquad (\sigma^{\mu\nu} = \tfrac{1}{2}i[\gamma^\mu, \gamma^\nu]), \tag{7.56}$$

where $F_1 = A + 2mB$, $F_2 = -2mB$. (The factor of $2m$ is adjusted to make F_2, like F_1, dimensionless.) The functions F_1 and F_2 are known as *form factors*.

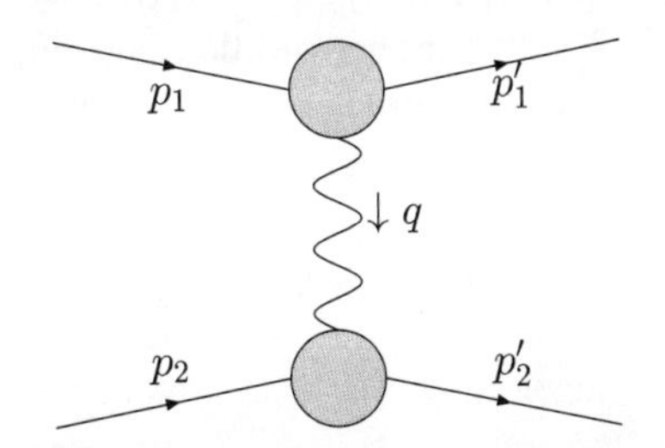

FIGURE 7.17. Diagram for two-electron scattering with vertex corrections.

To understand the physical meaning of this, let us return to the calculation of the scattering of two Fermions that we performed in §6.9 and add in the radiative corrections. That is, diagrammatically we replace the diagram in Figure 6.6 with the diagram in Figure 7.17, and analytically we replace the $\overline{u}\gamma^\mu u$ terms in the calculation with $\overline{u}\Gamma^\mu u$ terms. When we pass to the low-energy limit, we have

$$\Gamma^\mu(p', p) \to \Gamma^\mu(p, p) = F_1(0)\gamma^\mu,$$

so the end result is the insertion of an extra factor of $F_1(0)^2$ in the expression on the right side of (6.78). But classical physics tells us that (6.78) is correct as it stands, so either the charges must be renormalized by a factor of $F_1(0)$ (in bare perturbation theory) or counterterms must be added to F_1 so that $F_1(0) = 1$ (in

[10]This use of the letter σ is a standard convention. It unfortunately invites confusion with the Pauli matrices σ_j, to which the $\sigma^{\mu\nu}$ are of course related.

renormalized perturbation theory). Either way, we have the condition we need to fix the renormalization.

In more detail: in bare perturbation theory, the quantity we have been calling e is really the bare charge e_0,[11] and the renormalization is effected by setting $e_0 = Z_1 e$, where e is the physical charge and

$$Z_1 = \frac{1}{F_1(0)}. \tag{7.57}$$

Let us observe that, to zeroth order — that is, using only the bare vertex diagram — $F_1(0)$ is actually equal to 1. We can therefore write

$$F_1(0) = 1 + \delta F_1(0),$$

where $\delta F_1(0)$ is the sum of all the higher-order corrections. One computes this to a given finite order, regularizing all the divergent integrals involved; then $\delta Z_1 = Z_1 - 1$ is given to that order, as a function of the regularization parameter, by

$$\delta Z_1 = \frac{1}{1+\delta F_1(0)} - 1 = \frac{-\delta F_1(0)}{1+\delta F_1(0)}. \tag{7.58}$$

In renormalized perturbation theory, we recalculate $\Gamma^\mu(p', p)$ with e taken to be the physical charge of the electron and diagrams containing counterterm vertices (as in Table 7.1) included. The result is that

$$F_1(q^2) = 1 + \delta F_1(q^2) + (\text{counterterm contributions}),$$

where $\delta F_1(q^2)$ is the sum of the corrections with ordinary Feynman diagrams. The quantity δZ_1 is calculated, order by order in perturbation theory, so that the counterterm contributions add up to $-\delta F_1(0)$. To first order we have only the diagrams in Figure 7.18: $\delta F_1(0)$ is the value of the middle diagram, which we shall calculate in §7.11, and $\delta Z_1 = -\delta F_1(0)$. This agrees to first order with (7.58), as it should.

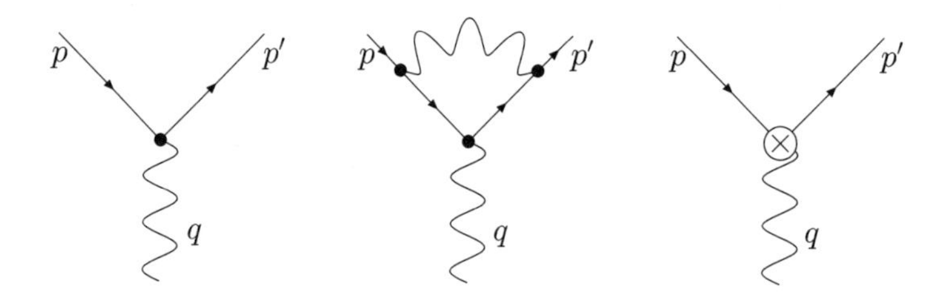

FIGURE 7.18. Diagrams for the first-order calculation of $\Gamma^\mu(p', p)$.

The second form factor $F_2(q^2)$ in (7.56) has played no role in all this, but it has its own story to tell. We shall put it in the spotlight in §7.11.

Let us return for a moment to bare perturbation theory. By (7.56), the defining relation (7.57) for the renormalization constant Z_1 can be written as

$$\Gamma^\mu(p, p) = Z_1^{-1}\gamma^\mu. \tag{7.59}$$

On the other hand, the defining relation for the renormalization constant Z_2 for the electron field is that iZ_2 is the residue of the "pole" of the dressed electron

[11] If the electron and electromagnetic fields have already been renormalized, e_0 is not the e_B in (7.44) but the $Z_2 Z_3^{1/2} e_B$ in (7.45).

propagator "at $\not{p} = m$" *before* renormalization of the electron field; that is, the dressed propagator $S(p)$ satisfies

$$S(p)^{-1} = \frac{1}{iZ_2}(\not{p} - m)[1 + O(\not{p} - m)], \tag{7.60}$$

where m is the physical mass of the electron and $O(\not{p} - m)$ denotes a matrix that is divisible by $\not{p} - m$. Now recall the Ward identity (7.43), which was proved in bare perturbation theory:[12]

$$\Gamma^\mu(p, p) = i\frac{\partial S(p)^{-1}}{\partial p_\mu}.$$

On comparing this with (7.59) and (7.60), we find that

$$Z_1^{-1}\gamma^\mu = Z_2^{-1}\gamma^\mu + O(\not{p} - m).$$

Letting p approach mass shell and applying both sides to a spinor that annihilates $\not{p} - m$, we find that

$$Z_1 = Z_2.$$

In other words, going back to (7.46), we have

$$e = Z_3^{1/2} e_{\rm B}. \tag{7.61}$$

That the renormalization of the electric charge is influenced *only* by the renormalization of the electromagnetic field and *not* that of the electron field is an extremely important result. Indeed, although we have been calling the Fermion field in QED the "electron field," the same theory describes electromagnetic interactions of all other charged spin-$\frac{1}{2}$ particles too, and if the renormalization of charge depended on the particle species in question, one would expect some observable corrections from the differing physical properties of the particles. The fact that it does not is what allows the charges of electrons and protons (for example) to be *exactly* opposite even though electrons and protons are very different beasts.

Another way of looking at this is that the product eA_μ in the QED Lagrangian is invariant under renormalization: $e_{\rm B} A_{\mu{\rm B}} = eA_\mu$. Thus, in the Lagrangian (7.44) and in both the "physical" and "counterterm" parts of the rewritten Lagrangian (7.47), the derivative ∂_μ and the field A_μ couple to the electron field ψ *only* in the combination $\partial_\mu - ieA_\mu$. The universality of this combination is a manifestation of the gauge invariance of QED.

Infrared divergences. Before we proceed, there is one other item to be discussed — or, more honestly, to be swept under the rug. The regularization procedures discussed in §7.3 are designed to handle "ultraviolet" divergences, that is, divergences caused by insufficiently rapid decay at infinity in momentum space. But because the photon is massless, some of the integrals in question also contain an "infrared" divergence, that is, a divergence caused by singularities as certain momenta approach zero. In other words, the contributions of very soft photons to certain scattering processes is apparently infinite. However, an electron in a scattering process may also emit any number of real photons; classically this is the radiation produced by an accelerating charge, known as "Brehmsstrahlung." It turns out that (i) the amplitude for such emissions also diverges in the low-energy limit, but (ii) when the diagrams with real and virtual photons are all added together, the

[12] It can also be derived in renormalized perturbation theory.

divergences cancel. From the empirical point of view, the point is that a real photon that is too soft to be detected by the apparatus at hand cannot be experimentally distinguished from a virtual photon, and the theory provides finite answers if one asks experimentally meaningful questions involving only detectable photons. From the mathematical point of view, the point is that the sharp distinction between virtual and real soft photons is an artifact of the perturbation-theoretic methods we are using, and that although these methods are marvellously effective at sorting out the high-energy components of interactions, they are inefficient at the low-energy end of the scale. Recently some computational methods have been developed that avoid infrared divergences; see Bach et al. [**5**] and the references given there.

In any event, infrared divergences are ultimately spurious, and we shall not worry about them. As a practical matter, one can regularize infrared-divergent integrals by replacing the p^2 in the denominator of photon propagators by $p^2 - \mu^2$ where μ is a small positive number — in effect, by assigning a small mass to the photon; physically meaningful quantities will turn out to be independent of μ. We refer the reader to Weinberg [**129**], Chapter 13, Peskin and Schroeder [**88**], §§6.4–5, and Jauch and Rohrlich [**68**], §16.1, for the detailed calculations. (Weinberg, as usual, does things with a keen eye for generality.)

The divergent one-loop diagrams. We conclude this section by compiling a complete list of the types of superficially divergent 1PI diagrams in QED that contain only one loop. Since there is no question of divergent subdiagrams in a superficially convergent diagram when there is only one loop, all other diagrams with only one loop either are convergent or contain one of the diagrams on this list together with some additional tree structure.

We recall from §7.2 that the superficial degree of divergence of a 1PI diagram in QED is $D = 4 - \frac{3}{2}E_e - E_\gamma$ where E_e and E_γ are the numbers of external electron and photon lines. Thus, a superficially divergent diagram must have $E_e = 0$ and $0 \leq E_\gamma \leq 4$ or $E_e = 2$ and $0 \leq E_\gamma \leq 1$. (E_e must be even, since $E_e + 2I_e = 2V$.) There are no vacuum bubbles ($E_e = E_\gamma = 0$) with only one loop, so there are only six possibilities: $E_e = 0$ with $1 \leq E_\gamma \leq 4$, and $E_e = 2$ with $0 \leq E_\gamma \leq 1$. They are listed in Figure 7.19.

The diagrams (b) and (e) are the lowest-order corrections to the photon and electron propagators, and (f) is the lowest-order correction to the vertex function; we shall analyze them in detail in the next three sections. The other three, however, may be quickly disposed of.

The diagram (a) could be excluded by Wick-ordering the interaction, but it is a fake in any case, and many authors don't even bother to mention it. Its value with the photon line amputated, by (6.59), is

$$-ie\int \frac{\operatorname{tr}[\gamma^\mu(\not{q} + m)]}{-q^2 + m^2 - i\epsilon}\,\frac{d^4q}{(2\pi)^4}.$$

This is the sum of two terms, corresponding to the two terms $\not{q}$ and m in the numerator. The second term vanishes because $\operatorname{tr}\gamma^\mu = 0$; the first is formally divergent but vanishes in any regularization because the integrand is an odd function of q. Hence the value of (a) is zero.

The diagram (c) is also a fake, but for a slightly more subtle reason. For a given set of external photons, there are actually two such diagrams, the one pictured and another with the electron arrows reversed. Any process that includes one of them

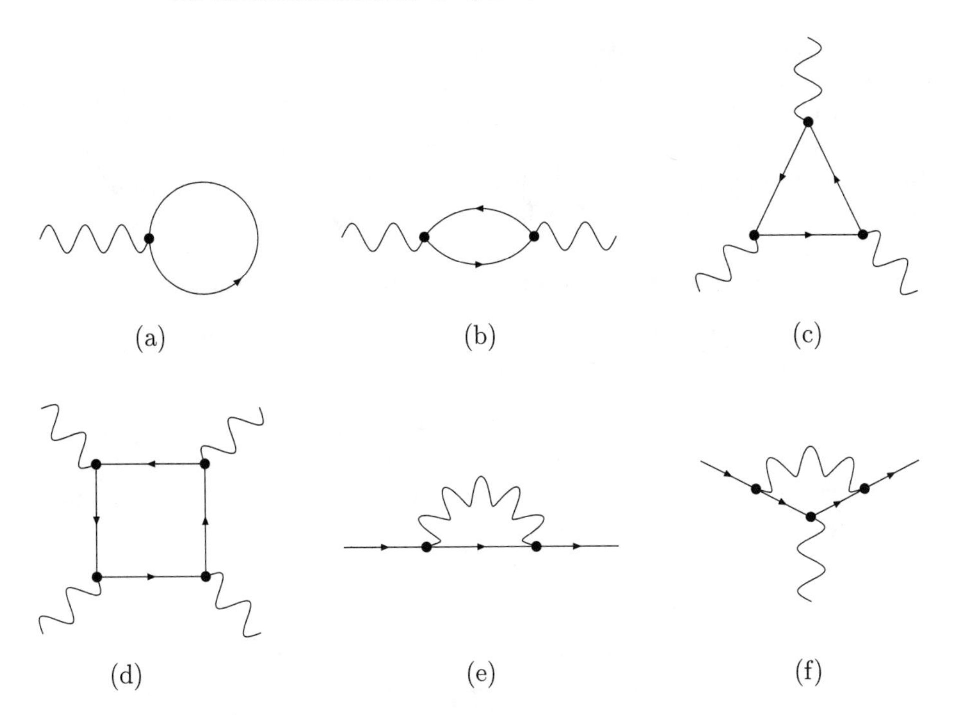

FIGURE 7.19. The six superficially divergent one-loop diagrams.

must include the other one too, and it is a consequence of the charge-conjugation invariance of QED that the values of the two differ by a factor of -1; hence their sum is zero. We omit the details (see Weinberg [**129**], §10.1, or Jauch and Rohrlich [**68**], §8.4) but mention that this is a special case of a more general consequence of charge-conjugation invariance known as *Furry's theorem*: Diagrams containing an electron loop with an odd number of vertices contribute nothing, because the two orientations of the loop cancel each other.

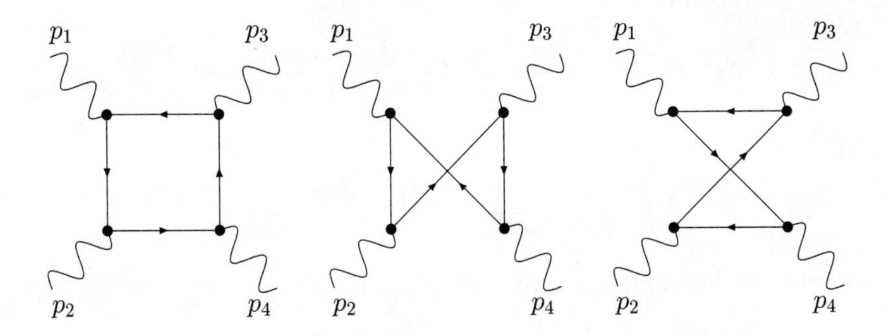

FIGURE 7.20. The one-loop diagrams for photon-photon scattering.

The diagram (d) is is the lowest-order contribution to a real quantum process with no classical analogue, the scattering of light by light. (This effect is hard to observe because it is very weak, i.e., it has an exceedingly small cross section.) Its divergence, however, turns out to be spurious. In fact, when one specifies all the relevant labels on the external lines, there are actually six such diagrams, namely, the ones in Figure 7.20 and the same three with the electron arrows reversed. Here, since the number of vertices is even, the two orientations of the loop give equal rather than opposite results. But when one calculates the sum of the three diagrams shown, one finds that the divergences cancel and one is left with a finite integral; see Jauch and Rohrlich [**68**], §13.1.

7.9. One-loop QED: the electron propagator

In this section we calculate the one-loop correction to the electron propagator, given by the diagram in Figure 7.21. The regularization of the divergent integral and the cancellation of the divergence by the addition of counterterms proceed in much the same manner as the calculations we performed in §7.4 and §7.6 for the scalar field, so we shall be fairly brief. In particular, we shall take the shortcut in computing the regularization that was discussed after (7.8) and before (7.19).

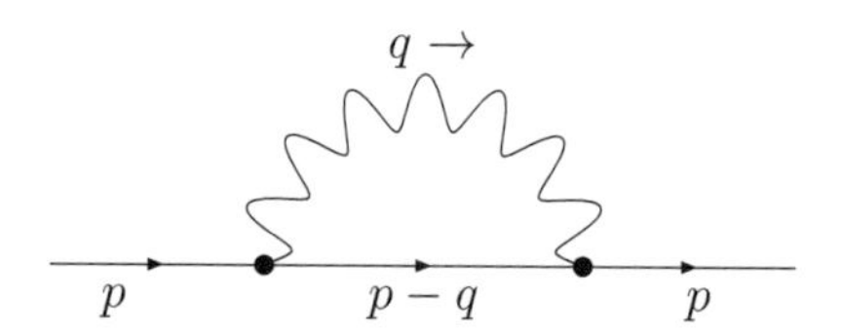

FIGURE 7.21. The one-loop correction to the electron propagator.

We denote the value of the Figure 7.21, with the the external legs amputated, by $-i\Sigma_2(p)$; the subscript 2 indicates that this is the contribution to the 1PI corrections that is second-order in the electric charge e. We have

$$-i\Sigma_2(p) = \int \frac{-ig_{\mu\nu}}{q^2+i\epsilon}(-ie\gamma^\mu)\frac{-i(\not p - \not q + m)}{-(p-q)^2+m^2-i\epsilon}(-ie\gamma^\nu)\frac{d^4q}{(2\pi)^4}. \tag{7.62}$$

Introduction of the Feynman parameter x turns this into

$$\Sigma_2(p) = -ie^2\int_0^1\int_{\mathbb{R}^4}\frac{\gamma_\mu(\not p - \not q + m)\gamma^\mu}{[-x(p-q)^2+xm^2-(1-x)q^2-i\epsilon]^2}\frac{d^4q}{(2\pi)^4}\,dx.$$

Replacing q by $q+xp$ gets rid of the linear term in the denominator:

$$\Sigma_2(p) = -ie^2\int_0^1\int_{\mathbb{R}^4}\frac{\gamma_\mu((1-x)\not p - \not q + m)\gamma^\mu}{[-q^2-x(1-x)p^2+xm^2-i\epsilon]^2}\frac{d^4q}{(2\pi)^4}\,dx.$$

Wick rotation — that is, the substitution of iq^0 for q^0 — then yields

$$\Sigma_2(p) = e^2\int_0^1\int_{\mathbb{R}^4}\frac{\gamma_\mu((1-x)\not p - \not q + m)\gamma^\mu}{[|q|^2-x(1-x)p^2+xm^2]^2}\frac{d^4q}{(2\pi)^4}\,dx.$$

We now perform dimensional regularization. The integral of the term with $\not q$ in the numerator vanishes by symmetry, which reduces the degree of divergence from linear to logarithmic. Moreover, by (7.16), we have

$$\gamma_\mu((1-x)\not p + m)\gamma^\mu = (2-d)(1-x)\not p + dm.$$

Hence, by (7.15), the regularized integral (still denoted by $\Sigma_2(p)$, although it now also depends on d) is

$$\Sigma_2(p) = \frac{2e^2}{(4\pi)^{d/2}} \int_0^1 \int_0^\infty \frac{(2-d)(1-x)\not{p} + dm}{[r^2 - x(1-x)p^2 + xm^2]^2} r^{d-1}\, dr\, dx,$$

so by (7.14),

$$\Sigma_2(p) = \frac{e^2}{(4\pi)^{d/2}} \Gamma(2 - \tfrac{1}{2}d) \int_0^1 \frac{(2-d)(1-x)\not{p} + dm}{[xm^2 + x(1-x)p^2]^{(4-d)/2}}\, dx.$$

Expanding the d-dependent quantities about $d = 4$,

$$\Gamma(2 - \tfrac{1}{2}d) = \frac{2}{4-d} - \gamma + O(4-d), \qquad A^{-(d-4)/2} = 1 - \frac{4-d}{2} \log A + O((4-d)^2),$$

where γ is the Euler-Mascheroni constant, we see that

(7.63)

$$\begin{aligned}
\Sigma_2(p) =& \frac{e^2}{16\pi^2} \left(1 + \frac{4-d}{2} \log 4\pi + \cdots\right) \left(\frac{2}{4-d} - \gamma + \cdots\right) \\
&\times \int_0^1 \big(-2(1-x)\not{p} + 4m + (4-d)(1-x)\not{p} - (4-d)m\big) \\
&\qquad \times \left(1 - \frac{4-d}{2} \log[xm^2 - x(1-x)p^2] + \cdots\right) dx \\
=& \frac{e^2}{16\pi^2} \Bigg[\left(\frac{2}{4-d} - \gamma + \log 4\pi\right)(-\not{p} + 4m) + \not{p} - 2m \\
&\qquad - \int_0^1 [-2(1-x)\not{p} + 4m] \log[xm^2 - x(1-x)p^2]\, dx\Bigg] + O(4-d).
\end{aligned}$$

(We used the fact that $2\int_0^1 (1-x)\, dx = 1$.)

So much for the regularization; now for the renormalization. The "Taylor expansion of $\Sigma(p)$ about $\not{p} = m$" (7.49) is accomplished as follows. Setting

$$C_d = \frac{2}{4-d} - \gamma + \log 4\pi - 1, \qquad f_x(p^2) = \log[xm^2 - x(1-x)p^2],$$

by (7.63) we have

$$\Sigma_2(p) = \frac{e^2}{16\pi^2} \left[C_d(-\not{p} + 4m) + 2m - 2\int_0^1 [-2(1-x)\not{p} + 4m] f_x(p^2)\, dx\right]. \tag{7.64}$$

We have dropped the $O(4-d)$ error term, since eventually we will let $d \to 4$. Now, $\not{p}^2 = p^2 I$, so we have $(p^2 - m^2)I = 2m(\not{p} - m) + (\not{p} - m)^2$ and hence

$$\begin{aligned}
f_x(p^2)I &= [f_x(m^2) + f_x'(m^2)(p^2 - m^2) + o(p^2 - m^2)]I \\
&= f_x(m^2)I + (\not{p} - m)[2m f_x'(m^2) + O(\not{p} - m)].
\end{aligned}$$

Observe also that $f_x(m^2) = \log(x^2 m^2)$ and $f_x'(m^2) = (x-1)/xm^2$. Substituting this into (7.64) and writing $\not{p} = m + (\not{p} - m)$ therein, we obtain

$$\Sigma_2(p) = \Sigma_2(m) + (\not{p} - m)[\Sigma_2'(m)_d + O(\not{p} - m)],$$

where

$$\Sigma_2(m) = \frac{e^2}{16\pi^2}(3C_d + 2)m - \frac{e^2}{8\pi^2} \int_0^1 (1+x) m f_x(m^2)\, dx \tag{7.65}$$

and

(7.66)

$$\Sigma_2'(m) = -\frac{e^2}{16\pi^2}\left[C_d + 2\int_0^1 (1-x)f_x(m^2)\,dx - 4m^2\int_0^1 (1+x)f_x'(m^2)\,dx\right]$$

Unfortunately, since $f_x'(m^2) = (x-1)/xm^2$, the last integral in (7.66) diverges. This is an *infrared* divergence caused by the masslessness of the photon. As a quick fix, we replace the photon denominator $q^2 + i\epsilon$ in (7.62) by $q^2 - \mu^2 + i\epsilon$ where μ is a small positive number. The effect is to add $(1-x)\mu^2$ to the argument of the logarithm in $f_x(p^2)$, so that (7.66) becomes

(7.67) $\Sigma_2'(m)$

$$= -\frac{e^2}{16\pi^2}\left[C_d + \int_0^1 \left([(2-2x)\log[x^2m^2 + (1-x)\mu^2] - \frac{4m^2x(x^2-1)}{m^2x^2 + (1-x)\mu^2}\right)dx\right].$$

We shall say no more about the μ-dependence of $\Sigma_2(p)$, as it is unimportant for the main considerations here.

The required counterterms δZ_2 and δm are

$$\delta Z_2 = \Sigma_2'(m), \qquad \delta m = -\Sigma_2(m) + (\delta Z_2)m = -\Sigma_2(m) + m\Sigma_2'(m)$$

as we found in (7.50), where $\Sigma_2'(m)$ is given by (7.67) and $\Sigma_2(m)$ is given by (7.65) with $f_x(m^2) = \log[x^2m^2 + (1-x)\mu^2]$. With their inclusion, the dressed propagator

$$i(\not{p} - m - \Sigma_2(p) + (\delta Z_2)\not{p} - \delta m)^{-1}$$

has a finite limit as $d \to 4$, with the appropriate pole and residue. (It should be noted that the only part of $\Sigma_2(p)$ that blows up as $d \to 4$, namely, the constant C_d, appears only in the quantities $\Sigma_2(m)$ and $\Sigma_2'(m)$; the $O(\not{p} - m)^2$ remainder is finite.)

A few words about the intuitive interpretation of the mass renormalization: The idea that suggests itself is that the observed, physical mass of an electron (say) is the sum of the "bare mass" that it would possess if the electromagnetic field were turned off and an "electromagnetic mass" due to the presence of the field. There is nothing wrong with this, except that it appears that the "bare mass" must be $-\infty$ and the "electromagnetic mass" must be $\infty + m$. One's faith in the quantum theory of electromagnetism might be shaken by this circumstance were it not for the fact that the same problem occurs in the classical theory. In classical electromagnetism, an electrostatic field $\mathbf{E}$ possesses an energy density proportional to $|\mathbf{E}|^2$. For the inverse-square-law field generated by a point charge at the origin, we have $|\mathbf{E}(\mathbf{x})|^2 = |\mathbf{x}|^{-4}$, so the total electrostatic energy associated to the point charge is a constant times $\int |\mathbf{x}|^{-4}\,d^3\mathbf{x}$, which is infinite. By the equivalence of mass and energy, a point charge carries an infinite "electrostatic mass," so its "bare mass" must also be $-\infty$. Attempts to solve this problem by discarding the notion of a point charge and taking the electron to be a continuous distribution of charge on some suitably small region lead to various other difficulties, mostly because one has to worry about the forces that the different parts of the electron would then exert on each other, and no good resolution has ever been found. (See the Feynman Lectures [**41**], vol. II, Chapter 28, for an extended discussion of this matter.)

In fact, the quantum-mechanical situation is less singular than the classical one in the sense that the integral $\int_{|\mathbf{x}|>\epsilon} |\mathbf{x}|^{-4}\,d^3\mathbf{x}$ diverges linearly in $1/\epsilon$, whereas the integral defining $\Sigma_2(p)$, Wick-rotated into Euclidean space and cut off at $|q| = \Lambda$, diverges only logarithmically in Λ. (Keep in mind that large momenta correspond to small distances.) In any case, the renormalization procedures of QED, mysterious though they may seem, lead to finite, meaningful results more successfully than the classical theory or any of its proposed modifications.

7.10. One-loop QED: the photon propagator and vacuum polarization

We now turn to the evaluation and interpretation of the one-loop correction to the photon propagator shown in Figure 7.22. By (6.59), the value of this diagram, with the the external legs amputated, is

$$(7.68)\quad i\Pi_2^{\mu\nu}(p) = -(-ie)^2 \int \operatorname{tr}\left[\gamma^\mu \frac{i(\not{q}+m)}{-q^2+m^2-i\epsilon}\gamma^\nu \frac{i(\not{q}-\not{p}+m)}{-(q-p)^2+m^2-i\epsilon}\right]\frac{d^4q}{(2\pi)^4}.$$

As in the preceding section, the subscript 2 means that this is the contribution to $\Pi^{\mu\nu}(p)$ of order e^2.

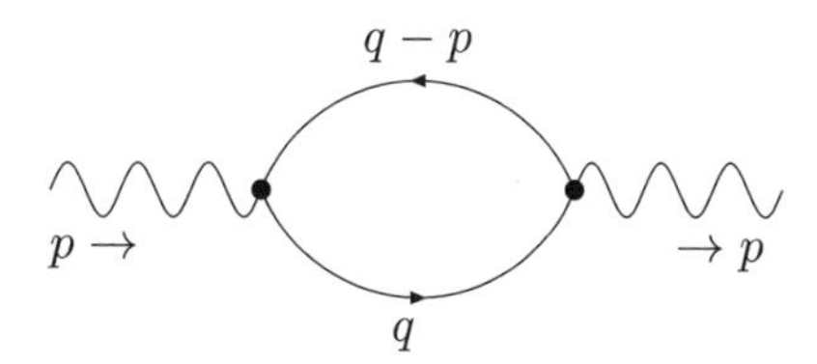

FIGURE 7.22. The one-loop correction to the photon propagator.

By now the regularization procedure should be familiar, so we proceed quickly and focus only on the features that are peculiar to this case. The Feynman formula turns the denominator of the integrand of (7.68) into

$$\int_0^1 \frac{dx}{[(-q^2+m^2-i\epsilon)(1-x)+(-(q-p)^2+m^2-i\epsilon)x]^2} = \int_0^1 \frac{dx}{[-q^2+m^2+2xp_\rho q^\rho m - xp^2 - i\epsilon]^2}.$$

To eliminate the cross term, we substitute $q+xp$ for q, thus obtaining

$$i\Pi_2^{\mu\nu}(p) = -e^2\int_0^1\int_{\mathbb{R}^4} \frac{\operatorname{tr}[\gamma^\mu(\not{q}+x\not{p}+m)\gamma^\nu(\not{q}-(1-x)\not{p}+m)]}{[-q^2+m^2-x(1-x)p^2-i\epsilon]^2}\frac{d^4q}{(2\pi)^4}\,dx.$$

From the formulas for the traces of products of Dirac matrices (which can be found in practically any physics text that deals with them), one finds that the numerator of the integrand is equal to

$$4[(q+xp)^\mu(q-(1-x)p)^\nu + (q+xp)^\nu(q-(1-x)p)^\mu - (q+xp)_\rho(q-(1-x)p)^\rho g^{\mu\nu} + m^2 g^{\mu\nu}]$$
$$= 4[2q^\mu q^\nu - 2x(1-x)p^\mu p^\nu - (-q^2+m^2-x(1-x)p^2)g^{\mu\nu}] + \text{linear terms in } q.$$

The linear terms in q may be dropped, as they are odd and so contribute nothing to the integral; neither do the terms $q^\mu q^\nu$ with $\mu \neq \nu$, as they are odd in q^μ. Next, we perform the Wick rotation in q. The d^4q acquires a factor of i, the q^2 becomes $-|q|^2$, and the q^0q^0 acquires a minus sign. Each of the terms $q^\mu q^\mu$ can be replaced by $(1/d)|q|^2$ by (7.11) (where $d = 4$ at the moment), with the result that $q^\mu q^\nu$ can be replaced by $-(1/d)|q|^2 g^{\mu\nu}$. In short, $i\Pi_2^{\mu\nu}(p)$ is equal to

$$-4ie^2 \int_0^1 \int_{\mathbb{R}^4} \frac{[1-(2/d)]|q|^2 g^{\mu\nu} - 2x(1-x)p^\mu p^\nu + (m^2 - x(1-x)p^2)g^{\mu\nu}}{[|q|^2 + m^2 - x(1-x)p^2]^2} \frac{d^4 q}{(2\pi)^4}\, dx.$$

We now replace this divergent 4-dimensional integral by a d-dimensional one by (7.15):

$$i\Pi_2^{\mu\nu}(p) = \frac{-4ie^2}{(2\pi)^d} \frac{2\pi^{d/2}}{\Gamma(d/2)}$$
$$\times \int_0^1 \int_0^\infty \frac{[1-(2/d)]r^2 g^{\mu\nu} - 2x(1-x)p^\mu p^\nu + (m^2 - x(1-x)p^2)g^{\mu\nu}}{[r^2 + m^2 - x(1-x)p^2]^2} r^{d-1}\, dr\, dx,$$

which converges for $d < 2$. (The expected singularity at $d = 2$, however, turns out to be removable because of the factor of $1 - (2/d)$.) We evaluate it by (7.14):

$$i\Pi_2^{\mu\nu}(p)$$
$$= \frac{-4ie^2}{(4\pi)^{d/2}\Gamma(d/2)} \int_0^1 \left[\left(1-\frac{2}{d}\right)\Gamma\left(1+\frac{d}{2}\right)\Gamma\left(1-\frac{d}{2}\right) g^{\mu\nu}(m^2 - x(1-x)p^2)^{(d/2)-1}\right.$$
$$\left. + \left[(x(1-x)p^2 + m^2)g^{\mu\nu} - 2x(1-x)p^\mu p^\nu\right]\Gamma\left(\frac{d}{2}\right)\Gamma\left(2-\frac{d}{2}\right)(m^2 - x(1-x)p^2)^{(d/2)-2}\right] dx.$$

But

$$\left(1-\frac{2}{d}\right)\Gamma\left(1+\frac{d}{2}\right)\Gamma\left(1-\frac{d}{2}\right) = -\frac{2}{d}\left(1-\frac{d}{2}\right)\Gamma\left(1+\frac{d}{2}\right)\Gamma\left(1-\frac{d}{2}\right)$$
$$= -\Gamma\left(\frac{d}{2}\right)\Gamma\left(2-\frac{d}{2}\right),$$

so the two terms can be combined, and some simplifications result. The upshot is that the regularized $\Pi_2^{\mu\nu}(p)$ has exactly the form (7.52) that it should:

$$\Pi_2^{\mu\nu}(p) = (p^2 g^{\mu\nu} - p^\mu p^\nu)\pi_2(p^2),$$

where

$$\pi_2(p^2) = \frac{-8e^2}{(4\pi)^{d/2}}\Gamma\left(2-\frac{d}{2}\right)\int_0^1 x(1-x)(m^2 - x(1-x)p^2)^{(d/2)-2}\, dx.$$

The factor $\Gamma(2-(d/2))$ blows up at $d = 4$, as expected.

As explained in §7.6, renormalization is accomplished by adding in the counterterm $-(\delta Z_3)(p^2 g^{\mu\nu} - p^\mu p^\nu)$, where

$$\delta Z_3 = \pi_2(0) = \frac{-8e^2}{(4\pi)^{d/2}}\Gamma\left(2-\frac{d}{2}\right) m^{d-4} \int_0^1 x(1-x)\, dx.$$

With this specification of δZ_3, we can take the limit as $d \to 4$ to get the finite renormalized value of $\Pi_2^{\mu\nu}(p)$:

$$\Pi_{2,\text{renorm}}^{\mu\nu}(p) = \pi(p^2)(p^2 g^{\mu\nu} - p^\mu p^\nu), \tag{7.69}$$

where

$$\pi(p^2) = \lim_{d\to 4}[\pi_2(p^2) - \pi_2(0)]$$
$$= \lim_{d\to 4} \frac{-8e^2}{(4\pi)^{d/2}}\Gamma\left(2-\frac{d}{2}\right)\int_0^1 x(1-x)[(m^2 - x(1-x)p^2)^{(d/2)-2} - m^{d-4}]\,dx.$$

(We really should write $\pi_{2,\text{renorm}}(p^2)$ instead of $\pi(p^2)$, but we will use the simpler notation for the rest of this section.) Since

$$\Gamma\left(2-\frac{d}{2}\right) = \left(2-\frac{d}{2}\right)^{-1} + O(1)$$

and

$$(m^2 - x(1-x)p^2)^{(d/2)-2} - m^{d-4} = \left(\frac{d}{2}-2\right)\log\frac{m^2 - x(1-x)p^2}{m^2} + o(d-4),$$

we have

$$\pi(p^2) = \frac{e^2}{2\pi^2}\int_0^1 x(1-x)\log\frac{m^2 - x(1-x)p^2}{m^2}\,dx. \tag{7.70}$$

This is the final result: a finite quantity with physical significance.

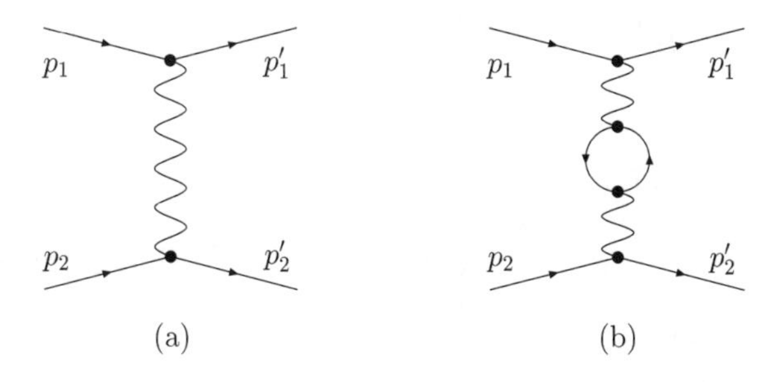

FIGURE 7.23. Fermion-Fermion scattering: the basic diagram and a one-loop correction.

Vacuum polarization. To explain the meaning of $\pi(p^2)$, let us return to the scattering of two distinguishable Fermions with masses m_1 and m_2 and charges e_1 and e_2 that we discussed in §6.9 and consider the contribution of the diagram in Figure 7.23b to the basic process represented by Figure 7.23a. Here the particles have incoming 4-momenta p_1 and p_2, outgoing 4-momenta p_1' and p_2', incoming spinor coefficients $u_1 = u(\mathbf{p}_1, s_1)$ and $u_2 = u(\mathbf{p}_2, s_2)$, and outgoing spinor coefficients $u_1' = u(\mathbf{p}_1', s_1')$ and $u_2' = u(\mathbf{p}_2', s_2')$. We make no assumption about the precise nature of the incoming and outgoing particles, but the virtual particles in the loop in Figure 7.23b are supposed to be an electron-positron pair. (Other virtual particle pairs also contribute to the process, but the more massive they are, the smaller their effect.)

Letting $q = p_1 - p_1' = p_2' - p_2$ be the momentum transferred in the scattering, the contribution of Figure 7.23a to the delta-function-free S-matrix element iM is

$$\frac{-m_1 m_2 e_1 e_2}{4\sqrt{\omega_{\mathbf{p}_1}\omega_{\mathbf{p}_2}\omega_{\mathbf{p}_1'}\omega_{\mathbf{p}_2'}}}\overline{u}_1'\gamma^\mu u_1 \frac{-ig_{\mu\nu}}{q^2}\overline{u}_2'\gamma^\nu u_2,$$

as we saw in §6.9 ((6.73), modified for the case of two distinct particles). The contribution of Figure 7.23b is the same, except that the $-ig_{\mu\nu}/q^2$ is replaced by

$$\frac{-ig_{\mu\rho}}{q^2}[i\Pi^{\rho\sigma}_{2,\text{renorm}}(q^2)]\frac{-ig_{\sigma\nu}}{q^2} = \frac{-ig_{\mu\rho}}{q^2}[i\pi(q^2)(q^2g^{\rho\sigma} - q^\rho q^\sigma)]\frac{-ig_{\sigma\nu}}{q^2}.$$

When sandwiched between the spinors, the $q^\rho q^\sigma$ term drops out by the "baby Ward identity" (7.34), so this boils down to $\pi(q^2)(-ig_{\mu\nu}/q^2)$. Hence the sum of the two diagrams gives

$$iM = \frac{-m_1m_2e_1e_2}{4\sqrt{\omega_{\mathbf{p}_1}\omega_{\mathbf{p}_2}\omega_{\mathbf{p}'_1}\omega_{\mathbf{p}'_2}}}\overline{u}'_1\gamma^\mu u_1[1+\pi(q^2)]\frac{-ig_{\mu\nu}}{q^2}\overline{u}'_2\gamma^\nu u_2.$$

We now pass to the low-energy approximation as in §6.9 to obtain

$$iM = -ie_1e_2\frac{1+\pi(-|\mathbf{q}|^2)}{|\mathbf{q}|^2}\delta_{s_1s'_1}\delta_{s_2s'_2}. \tag{7.71}$$

We henceforth take $s'_1 = s_1$ and $s'_2 = s_2$; the spin variables will play no further role.

As before, (7.71) is the Born approximation to the S-matrix element for scattering by the potential $V(\mathbf{r})$, where

$$\int V(\mathbf{r})e^{-i\mathbf{q}\cdot\mathbf{r}}d^3\mathbf{r} = e_1e_2\frac{1+\pi(-|\mathbf{q}|^2)}{|\mathbf{q}|^2},$$

that is,

$$V(\mathbf{r}) = e_1e_2\int\frac{1+\pi(-|\mathbf{q}|^2)}{|\mathbf{q}|^2}e^{i\mathbf{q}\cdot\mathbf{r}}\frac{d^3\mathbf{q}}{(2\pi)^3}. \tag{7.72}$$

Without the term $\pi(-|\mathbf{q}|^2)$, $V(\mathbf{r})$ would just be the Coulomb potential energy $e_1e_2/4\pi|\mathbf{r}|$ generated by two point charges e_1 and e_2 located at points $\mathbf{a}$ and $\mathbf{b}$ with $\mathbf{a}-\mathbf{b} = \mathbf{r}$. To first order in perturbation theory, i.e., the order to which we have calculated the correction term $\pi(-|\mathbf{q}|^2)$, that term has the effect of replacing the point charges by charge distributions $e_1\rho(\mathbf{x}-\mathbf{a})$ and $e_2\rho(\mathbf{x}-\mathbf{b})$ where

$$\rho(\mathbf{r}) = \delta(\mathbf{r}) + \frac{1}{2}\int\pi(-|\mathbf{q}|^2)e^{i\mathbf{q}\cdot\mathbf{r}}\frac{d^3\mathbf{q}}{(2\pi)^3} = \delta(\mathbf{r}) + \tfrac{1}{2}\check{\pi}(\mathbf{r}), \tag{7.73}$$

$\check{\pi}$ being the inverse Fourier transform of the function $\mathbf{q}\mapsto\pi(-|\mathbf{q}|^2)$. Indeed, the potential energy generated by these distributions is

$$\frac{e_1e_2}{4\pi}\iint\frac{\rho(\mathbf{x}-\mathbf{a})\rho(\mathbf{y}-\mathbf{b})}{|\mathbf{x}-\mathbf{y}|}d^2\mathbf{x}\,d^3\mathbf{y} = \frac{e_1e_2}{4\pi}\iint\frac{\rho(\mathbf{x})\rho(\mathbf{y})}{|\mathbf{x}-\mathbf{y}+\mathbf{a}-\mathbf{b}|}d^3\mathbf{x}\,d^3\mathbf{y},$$

which, on substituting the formula for ρ and setting $\mathbf{r} = \mathbf{a}-\mathbf{b}$ and $r = |\mathbf{r}|$, becomes

$$\frac{e_1e_2}{4\pi r} + \frac{e_1e_2}{4\pi}\int\frac{\check{\pi}(\mathbf{x})}{|\mathbf{x}-\mathbf{r}|}d^3\mathbf{x} + \frac{e_1e_2}{16\pi}\iint\frac{\check{\pi}(\mathbf{x})\check{\pi}(\mathbf{y})}{|\mathbf{x}-\mathbf{y}+\mathbf{r}|}d^3\mathbf{x}\,d^3\mathbf{y}.$$

The sum of the first two terms is $V(\mathbf{r})$ since $1/|\mathbf{q}|^2$ is the Fourier transform of $1/4\pi r$, and the last term is of higher order since it involves two factors of $\check{\pi}$ (which is itself a first-order correction), so it may be dropped.

What does the charge distribution $\rho(\mathbf{r})$ look like? First, observe that

$$\int\rho(\mathbf{r})\,d^3\mathbf{r} = 1 + \tfrac{1}{2}\int\check{\pi}(\mathbf{r})\,d^3\mathbf{r} = 1 + \tfrac{1}{2}\pi(0) = 1,$$

so that the total charge of the distribution $e_j\rho(\mathbf{x})$ is still e_j. (Actually this calculation is suspect since, as we shall shortly see, $\check{\pi}$ is not an L^1 function, but the arguments that follow will suggest how to make rigorous sense of it by suitable manipulations with approximations and limits.) To proceed further, we need to calculate $\check{\pi}(\mathbf{r})$.

As a first step, for $\mathbf{s} \in \mathbb{R}^3$ and $s = |\mathbf{s}|$, let $f(\mathbf{s}) = \log(1 + s^2)$. Then

$$\nabla^2 f(\mathbf{s}) = \left(\frac{d^2}{ds^2} + \frac{2}{s}\frac{d}{ds}\right)\log(1+s^2) = \frac{2}{1+s^2} + \frac{4}{(1+s^2)^2}.$$

This is an L^2 function on $\mathbb{R}^3$, and its inverse Fourier transform (evaluated via spherical coordinates) is

$$\begin{aligned}
(\nabla^2 f)^\vee(\mathbf{r}) &= \frac{1}{(2\pi)^3}\int_0^\infty \left[\frac{2}{1+s^2} + \frac{4}{(1+s^2)^2}\right]\frac{4\pi s}{r}\sin rs\, ds \\
&= \frac{1}{4\pi^2 r}\int_{-\infty}^\infty \left[\frac{2}{1+s^2} + \frac{4}{(1+s^2)^2}\right] s \sin rs\, ds \\
&= \frac{1}{4\pi^2 r}\operatorname{Im}\left\{2\pi i \operatorname{Res}_{s=i} s e^{irs}\left[\frac{2}{1+s^2} + \frac{4}{(1+s^2)^2}\right]\right\} \\
&= \frac{1}{2\pi r}(1+r)e^{-r}.
\end{aligned}$$

On the other hand, $(\nabla^2 f)^\vee(\mathbf{r}) = -r^2 f^\vee(\mathbf{r})$, so $f^\vee$ is a distribution that agrees away from the origin with $-(1/2\pi r^3)(1+r)e^{-r}$. The latter is not integrable at the origin; to make it into a distribution one must renormalize it by adding an infinite multiple of δ as in the example in §7.1. That is, the action of $f^\vee$ on a test function ϕ is actually of the form

$$\langle f^\vee, \phi\rangle = -\frac{1}{2\pi}\int \frac{(1+r)e^{-r}}{r^3}[\phi(\mathbf{r}) - \phi(\mathbf{0})]\, d^3\mathbf{r} + C\phi(\mathbf{0}),$$

where the finite constant C could be determined, for example, by taking ϕ to be a Gaussian and comparing this formula with the formula $\langle f^\vee, \phi\rangle = \langle f, \widehat{\phi}\rangle$. ($f^\vee$ cannot contain any derivatives of the delta-function, for otherwise f would have polynomial growth at infinity.)

Rescaling now shows that the inverse Fourier transform of $f_c(\mathbf{s}) = \log(1+|\mathbf{s}/c|^2)$ agrees with $-(1/4\pi r^3)(1+cr)e^{-cr}$ away from the origin. Combining this result with the formula (7.70) for $\pi(-|\mathbf{q}|^2)$ (with $c = m/\sqrt{x(1-x)}$), we find that

$$\check{\pi}(\mathbf{r}) = -\frac{e^2}{4\pi^3 r^3}\int_0^1 \left(1 + \frac{mr}{\sqrt{\alpha(1-\alpha)}}\right) e^{-mr/\sqrt{x(1-x)}} x(1-x)\, dx + \infty\delta(\mathbf{r}), \tag{7.74}$$

where m and e are the mass and charge of the electron[13] and the coefficient "∞" of $\delta(\mathbf{r})$ is determined (informally speaking) by the requirement that $\int \check{\pi}(\mathbf{r})\, d^3\mathbf{r} = 0$.

In short, the charge distribution $\rho(\mathbf{r})$ consists of an infinite positive "bare" charge at the origin surrounded by a continuous distribution of negative charge, such that the total charge is 1. The density of the negative charge is approximately $e^2/24\pi^3 r^3$ for r small (since $\int_0^1 x(1-x)\, dx = \frac{1}{6}$), and it decays like e^{-2mr} for r large (since the maximum value of $x(1-x)$ is $\frac{1}{4}$). The approximation for r small should not be taken too seriously, since the higher-order corrections affect it

[13] Of course the e inside the integral is $\exp(1)$. Sorry about that.

substantially, but the decay for r large can be trusted. It implies that almost all of the charge in the distribution is located inside the ball about the origin of radius $1/m$, which is about 3.87×10^{-11} cm. (A simple calculation using the estimates $e^{-mr/\sqrt{x(1-x)}} \leq e^{-2mr}$ and $(1/r^3)4\pi r^2\,dr \leq (4\pi/m)\,dr$ shows that the integral of ρ over the complement of this ball is less than .0017 in absolute value; the integral over the complement of a ball k times as large is smaller by a factor of $e^{-2(k-1)}$.) The usual picturesque way of describing this situation is that the bare positive charge at the origin is surrounded by a cloud of virtual electron-positron pairs in which the electrons are attracted to the origin and the positrons are repelled toward infinity — an effect known as *vacuum polarization* — with the result that the bare charge is shielded.

Whether or not one finds the "virtual cloud" picture appealing, the phenomenon that the effective charge of a particle increases at very short distances is an experimental fact. However, short distances are correlated with high energies, and this effect is observable only in very energetic reactions.[14] In scattering processes involving energies of around 90 GeV, available only in the most powerful accelerators in existence today, the effective charge of an electron increases by about 3%. (Not only virtual electron-positron pairs but various other virtual particle-antiparticle pairs contribute to the vacuum polarization to produce this result.) Outside the physics lab this effect is not observed, because the requisite energy density is almost never available in nature. The temperature required for particles to have a mean kinetic energy of m, and hence to be able to probe distances $r \sim 1/m$, is around 6×10^9 K, which is greater by a factor of about 100 than the temperatures in the interiors of even the hottest stars. (Particles with such extremely high energies might be produced in a supernova explosion or in the jets generated in the vicinity of a supermassive black hole.) Thus the familiar observational fact that all electric charges in the "real world" are exact integer multiples of e is safe from contradiction.

The fact that the effective charge of an electron depends on the energy scale has analogues in other quantum field theories, and it suggests a useful shift in point of view. To wit, one has to face the fact that the "coupling constants" of a given theory are not really constants but rather functions of the energy scale. The study of the behavior of these "coupling functions" goes under the general name of *renormalization group analysis.* The name "renormalization group" is rather misleading, at least for mathematicians who might hope to find some interesting group theory at work here; the group in question is simply $\mathbb{R}$, acting as the group of scaling transformations. But renormalization group methods have come to be recognized as an important part of quantum field theory as well as other areas of physics where ideas from quantum field theory have applications; see Zee [**136**], Weinberg [**130**], or Peskin and Schroeder [**88**]. These matters are largely beyond the scope of this book, but we shall say a little more about them in §7.12.

Another place where vacuum polarization produces a measurable effect is in a shift in energy levels of electrons in atoms. Consider, for example, a single electron bound to a nucleus with charge Ze (i.e., atomic number Z). Suppose that in the Dirac model, as discussed in §4.3, it is in a joint eigenstate of energy and total angular momentum, with principal quantum number n (i.e., with n is as in (4.39))

[14]At such high energies, the preceding discussion should be taken as only a very rough guide to reality.

and angular momentum quantum number l (i.e., with total angular momentum $l(l+1)$), and let ψ be its wave function. If the Coulomb potential $-Ze^2/4\pi r$ is replaced by the potential $V(\mathbf{r})$ given by (7.72) (with $e_1 = -e$ and $e_2 = Ze$), by (6.11) the energy level E is shifted by the amount

$$\Delta E = \int \Delta V(\mathbf{r})|\psi(\mathbf{r})|^2 \, d^3\mathbf{r}$$

in the first-order approximation, where ΔV is the correction term,

$$\Delta V(\mathbf{r}) = -Ze^2 \int \frac{\pi(-|\mathbf{q}|^2)}{|\mathbf{q}|^2} e^{i\mathbf{q}\cdot\mathbf{r}} \frac{d^3\mathbf{q}}{(2\pi)^3}.$$

(The perturbation parameter g in the discussion leading to (6.11) is e^2 in the present case, and it is incorporated into the function $\pi(-|\mathbf{q}|^2)$.)

Now, the calculations we performed above easily imply that ΔV is negligibly small (with decay like e^{-2mr}) outside the ball of radius $1/m \sim 10^{-11}$ cm. (One can also approximate ΔV by a direct calculation without using the charge distribution ρ to show that $\Delta V(\mathbf{r}) \approx -Ze^4 e^{-2mr}/4\pi^{1/2} m^{3/2} r^{5/2}$ for $r > 1/m$; see Peskin and Schroeder [**88**], p. 254.) On the other hand, for realistically small Z, the wave function ψ is essentially constant over such a short distance, its characteristic length being the diameter of the atom, which is roughly 100 times bigger. Hence we have

$$\Delta E \approx |\psi(\mathbf{0})|^2 \int \Delta V(\mathbf{r}) \, d^3\mathbf{r}.$$

In this approximation, the only states with nonzero energy shifts are those for which $\psi(\mathbf{0}) \neq 0$, that is, those for which $l = 0$. For such states, explicit calculation of the wave function and its normalization shows that

$$\psi(\mathbf{0}) = \frac{e^3}{8\pi^2}\left(\frac{Zm}{n}\right)^{3/2}.$$

Moreover, the integral of ΔV can be calculated by the Fourier inversion formula and l'Hôpital's rule:

$$\begin{aligned}
\int \Delta V(\mathbf{r}) \, d^3\mathbf{r} &= -Ze^2 \frac{\pi(-|\mathbf{q}|^2)}{|\mathbf{q}|^2}\bigg|_{\mathbf{q}=\mathbf{0}} \\
&= -\frac{Ze^4}{2\pi^2} \frac{d}{ds} \int_0^1 x(1-x) \log \frac{m^2 + x(1-x)s}{m^2} \, dx \bigg|_{s=0} \\
&= -\frac{Ze^4}{2\pi^2 m^2} \int_0^1 x^2(1-x)^2 \, dx = -\frac{Ze^4}{60\pi^2 m^2}.
\end{aligned}$$

In particular, for an electron in the $2S_{1/2}$ state in a hydrogen atom ($Z = 1$, $n = 2$, $l = 0$), we have

$$\Delta E = -\frac{e^6 m^3}{64\pi^2} \cdot \frac{e^4}{60\pi^2 m^2} = -1.122 \times 10^{-7} \text{ eV}.$$

The corresponding spectroscopic shift $\Delta\nu = \Delta E/2\pi\hbar$ is -27.13 MHz. This shift is not directly observable; what is observable is the difference between the shifts in the $2S_{1/2}$ and $2P_{1/2}$ states, which are at exactly the same energy level according to the Dirac model. In our approximation (a good one), the shift for the $2P_{1/2}$ state is zero since $\psi(\mathbf{0}) = 0$, so vacuum polarization contributes -27.13 MHz to the energy level difference between these states. This is known as the *Uehling effect*. In fact, the Uehling effect is a relatively small contribution to the total difference, which is +1057.9 MHz (the Lamb shift), but since theory agrees with experiment to within 0.1 MHz, its existence is decisively confirmed. We shall comment further on the Lamb shift at the end of the next section.

7.11. One-loop QED: the vertex function and magnetic moments

We now come to the evaluation of the diagram in Figure 7.24, which gives the lowest-order radiative correction $-ie\Gamma_2^\mu(p',p)$ in the dressed vertex function $-ie\Gamma^\mu(p',p)$. Here p, p', and k are the momenta of the incoming electron, the outgoing electron, and the internal photon, and $q = p' - p$ is the momentum of the external photon (considered as incoming). With the external photon amputated but the spinors $u(p) = u(\mathbf{p}, s)$ and $u(p') = u(\mathbf{p}', s')$ for the incoming and outgoing electron included, and the $-ie$ factored out, it is

$$\begin{aligned}(7.75)\quad & \overline{u}(p')\Gamma_2^\mu(p',p)u(p) \\ &= \int \overline{u}(p')\frac{-ig_{\nu\rho}}{k^2+i\epsilon}(-ie\gamma^\nu)\frac{i(\not p' - \not k + m)}{(p'-k)^2 - m^2 + i\epsilon} \\ &\qquad \times \gamma^\mu \frac{i(\not p - \not k + m)}{(p-k)^2 - m^2 + i\epsilon}(-ie\gamma^\rho)u(p)\frac{d^4k}{(2\pi)^4}\end{aligned}$$

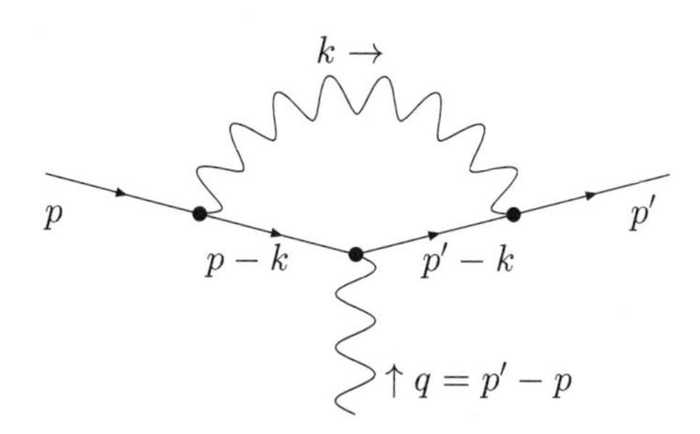

FIGURE 7.24. The one-loop correction to the vertex function.

We proceed to regularize this. We first apply Feynman's formula (7.6) to rewrite the denominator:

$$(7.76)\quad \frac{1}{[k^2+i\epsilon][(p'-k)^2 - m^2 + i\epsilon][(p-k)^2 - m^2 + i\epsilon]} = \int_{[0,1]^3} \frac{2\delta(x+y+z-1)}{D^3}\,dx\,dy\,dz,$$

where

$$D = xk^2 + y[(p'-k)^2 - m^2] + z[(p-k)^2 - m^2] + i\epsilon.$$

Using the fact that $p^2 = {p'}^2 = m^2$, we have

$$D = k^2 - 2q_\mu(yp' + zp)^\mu + i\epsilon.$$

To eliminate the cross term, let $\widetilde{k} = k - yp' - zp$ and use the facts that $2p'_\mu p^\mu = p'^2 + p^2 - (p'-p)^2 = 2m^2 - q^2$ and $y + z = 1 - x$:

$$\begin{aligned} D &= \widetilde{k}^2 - (yp' + zp)^2 + i\epsilon = \widetilde{k}^2 - (y^2 + z^2 + 2yz)m^2 + yzq^2 + i\epsilon \\ &= \widetilde{k}^2 - (1-x)^2 + yzq^2 + i\epsilon. \end{aligned} \tag{7.77}$$

We now attend to the numerator of the integrand in (7.75). We again replace k by $\widetilde{k} + yp' + zp$, then drop all terms that are linear in $\widetilde{k}$ (since they integrate to 0) and replace $\widetilde{k}_\mu \widetilde{k}_\nu$ by $\frac{1}{4} g_{\mu\nu}\widetilde{k}^2$, according to (7.11). Next, we simplify the result further by using the anticommutation relations $\{\gamma^\mu, \not{p}\} = 2p$ to move $\not{p}$'s to the right and likewise $\not{p}'$'s to the left, then using the Dirac equations $\overline{u}(p')\not{p}' = m\overline{u}(p')$, $\not{p}u(p) = mu(p)$ to replace them with m's, and finally using the relation $\gamma_\nu\gamma^\mu\gamma^\nu = -2\gamma^\mu$. We omit the tedious details; the result is that the numerator in (7.75) can be replaced by

$$\begin{aligned} \overline{u}(p')\big[[-\tfrac{1}{2}\widetilde{k}^2 + (1-y)(1-z)q^2 + (1-2x-x^2)m^2]\gamma^\mu \\ - mx(1-x)(p'+p)^\mu + (x-2)(y-z)q^\mu\big]u(p). \end{aligned}$$

Moreover, since D is symmetric in y and z, the q^μ term vanishes on integration over the Feynman parameters and so can be dropped. Finally, we use the Gordon identity (7.55) to rewrite the remaining terms as

$$\begin{aligned} \overline{u}(p')\Big[[-\tfrac{1}{2}\widetilde{k}^2 + (1-y)(1-z)q^2 + (1-4x+x^2)m^2]\gamma^\mu \\ + (2m^2 + 1 - x)\frac{i\sigma^{\mu\nu}q_\nu}{2m}\Big]u(p). \end{aligned} \tag{7.78}$$

When we put together (7.75), (7.76), (7.77), and (7.78), and perform the Wick rotation, we see that $\Gamma^\mu_2(p',p)$ has precisely the correct form according to (7.56):

$$\overline{u}(p')\Gamma^\mu_2(p',p)u(p) = \overline{u}(p')\left[\delta F_1(q^2)\gamma^\mu + \delta F_2(q^2)\frac{i\sigma^{\mu\nu}q_\nu}{2m}\right]u(p).$$

The contributions $\delta F_1(q^2)$ and $\delta F_2(q^2)$ to the form factors are given (after dropping the tildes on the k's) by

$$\delta F_j(q^2) = \int_{[0,1]^3} f_j(x,y,z;q^2)\delta(x+y+z-1)\,dx\,dy\,dz \qquad (j = 1,2), \tag{7.79}$$

where

$$f_1(x,y,z;q^2) = 4e^2 \int_{\mathbb{R}^4} \frac{\frac{1}{2}|k|^2 + (1-y)(1-z)q^2 + (1-4x+x^2)m^2}{[|k|^2 + (1-x)^2m^2 - yzq^2]^3}\frac{d^4k}{(2\pi)^4}, \tag{7.80}$$

$$f_2(x,y,z;q^2) = 8e^2 \int_{\mathbb{R}^4} \frac{m^2x(1-x)}{[k^2 + (1-x)^2m^2 - yzq^2]^3}\frac{d^4k}{(2\pi)^4}. \tag{7.81}$$

These are the integrals we must evaluate. To simplify the notation, we shall set

$$\Xi = \Xi(x,y,z;q^2) = (1-x)m^2 - yzq^2 = (yp' + zp)^2. \tag{7.82}$$

(Note that $\Xi > 0$ since $yp' + zp$ is timelike.)

The integral (7.80) splits into two parts:

$$f_1(x,y,z;k^2) = I_1 + I_2,$$

where

$$I_1 = 2e^2 \int_{\mathbb{R}^4} \frac{|k|^2}{[|k|^2 + \Xi]^3} \frac{d^4k}{(2\pi)^4},$$

$$I_2 = 4e^2 \int_{\mathbb{R}^4} \frac{(1-y)(1-z)q^2 + (1-4\xi + x^2)m^2}{[|k|^2 + \Xi]^3} \frac{d^4k}{(2\pi)^4}.$$

The integral I_1 diverges and must be subjected to dimensional regularization. If we retrace the path from (7.75) to (7.78), working in d dimensions, we find that some -2's coming from $\gamma_\nu\gamma^\mu\gamma^\nu = -2\gamma^\mu$ should be $(2-d)$'s, according to (7.16), and a $\frac{1}{4}$ coming from (7.11) should be $1/d$, with the result that I_1 acquires an extra factor of $(d-2)^2/d$. Hence, by (7.15), the regularized I_1 is

$$\begin{aligned}
&\frac{2(d-2)^2e^2}{d(4\pi)^{d/2}\Gamma(\frac{1}{2}d)} \int_0^\infty \frac{r^{d+1}\,dr}{(r^2+\Xi)^3} \\
=&\frac{2(d-2)^2e^2}{d(4\pi)^{d/2}\Gamma(\frac{1}{2}d)} \frac{\Xi^{(d/2)-2}}{2} \frac{\Gamma(1+\frac{1}{2}d)\Gamma(2-\frac{1}{2}d)}{\Gamma(3)} \\
=&\frac{(d-2)^2e^2}{2(4\pi)^{d/2}} \Gamma(2-\tfrac{1}{2}d)\Xi^{(d/2)-2} \\
=&\frac{e^2}{8\pi^2}\left[\frac{2}{4-d} - \gamma + \log 4\pi - 2 - \log \Xi\right] + O(4-d).
\end{aligned}$$

The integration over the Feynman parameters presents no further problem. Since

$$\int_{[0,1]^3} \delta(x+y+z-1)\,dx\,dy\,dz = \int_0^1 \int_0^{1-x} dy\,dx = \tfrac{1}{2},$$

the regularized part of $\delta F_1(k^2)$ coming from I_1, with the $O(4-d)$ term omitted, is

$$\begin{aligned}
\frac{e^2}{16\pi^2}\Bigg[&\frac{2}{4-d} + \log 4\pi + \gamma - 2 \\
&- 2\int_{[0,1]^3} \log[(1-x)m^2 - yzq^2]\delta(x+y+z-1)\,dx\,dy\,dz\Bigg].
\end{aligned} \tag{7.83}$$

The integral I_2 is convergent:

$$\begin{aligned}
I_2 =&4e^2 \int \frac{(1-y)(1-z)q^2 + (1-4x+x^2)m^2}{(|k|^2+\Xi)^3} \frac{d^4k}{(2\pi)^4} \\
=&\frac{4e^2}{(2\pi)^4}[(1-y)(1-z)q^2 + (1-4x+x^2)m^2]2\pi^2 \int_0^\infty \frac{r^3\,dr}{(r^2+\Xi)^3} \\
=&\frac{e^2}{8\pi^2} \frac{(1-y)(1-z)q^2 + (1-4x+x^2)m^2}{\Xi}.
\end{aligned} \tag{7.84}$$

Unfortunately, this quantity blows up at $x = 1$, $y = z = 0$, so its integral over the Feynman parameters diverges. (This is obvious, for instance, when $q = 0$.) This is another infrared divergence, and as before we tame it by replacing the $k^2 + i\epsilon$ in (7.75) by $k^2 - \mu^2 + i\epsilon$ with μ a small positive number; the effect is to replace Ξ by $\Xi + x\mu^2$. With this modification, adding the integral of (7.84) over the Feynman parameters to (7.83) gives the total contribution to the first form factor:

$$\delta F_1(q^2) = \frac{e^2}{16\pi^2}\left\{\frac{2}{4-d} - \gamma + \log 4\pi - 2 - 2\int_{[0,1]^3}\left[\log[(1-x)^2m^2 - yzq^2 + x\mu^2] - \frac{(1-y)(1-z)q^2 + (1-4x+x^2)m^2}{(1-x)^2m^2 - yzq^2 + x\mu^2}\right]\delta(x+y+z-1)\,dx\,dy\,dz\right\}.$$

The renormalization is therefore accomplished by the addition of the counterterm

$$\delta Z_1 = -\delta F_1(0) = \frac{e^2}{16\pi^2}\left(-\frac{2}{4-d} + \gamma - \log 4\pi + 2 + 2\int_0^1\left[\log[(1-x)^2m^2 + x\mu^2] - \frac{(1-4x+x^2)m^2}{(1-x)^2m^2 + x\mu^2}\right](1-x)\,dx\right). \tag{7.85}$$

We proved at the end of §7.8 that the renormalization constants Z_1 and Z_2 are equal. It is worth verifying directly that the one-loop contributions to these constants are the same. Indeed, if one compares the formulas (7.67) for $\Sigma_2'(m) = \delta Z_2$ and (7.85) for δZ_1 and makes the substitution $x \to 1-x$ in the latter, one finds that

$$\delta Z_2 - \delta Z_1 = \frac{e^2}{16\pi^2}\left(-1 + 2\int_0^1\left[(1-2x)\log[x^2m^2 + (1-x)\mu^2] - \frac{x^2(x-2)m^2}{x^2m^2 + (1-x)\mu^2}\right]dx\right).$$

It is not obvious to the naked eye that this quantity vanishes, but an integration by parts shows that the integral of the logarithmic term is equal to

$$-\int_0^1 (x-x^2)\frac{2xm^2 - \mu^2}{x^2m^2 + (1-x)\mu^2}\,dx = \int_0^1\left[x + \frac{x^2(x-2)m^2}{x^2m^2 + (1-x)\mu^2}\right]dx,$$

from which the result follows. It is a testimony to the power of the Ward identity that it explains this apparently fortuitous outcome.

Having tended to the mathematical difficulties of taming the divergent integrals, let us consider the form factor F_2. The one-loop contribution (7.79), (7.81) is completely finite. By Wick rotation and (7.15), we have

$$f_2(x,y,z;k^2) = \frac{8e^2}{(2\pi)^4}\frac{2\pi^2}{\Gamma(2)}\int_0^\infty \frac{r^3\,dr}{(r^2+\Xi)^3} = \frac{e^2}{4\pi^2\Xi},$$

so

$$\delta F_2(q^2) = \frac{e^2m^2}{4\pi^2}\int_{[0,1]^3}\frac{x(1-x)}{(1-x)^2m^2 - yzq^2}\,\delta(x+y+z-1)\,dx\,dy\,dz. \tag{7.86}$$

The factor of $1-x$ in the numerator tames the singularity at $x = 1$, so this integral (unlike that of (7.84)) is still finite. In particular, we have

$$\delta F_2(0) = \frac{e^2}{4\pi^2}\int_0^1\int_0^{1-x}\frac{x}{1-x}\,dy\,dx = \frac{e^2}{8\pi^2}. \tag{7.87}$$

The anomalous magnetic moment of the electron. To see the physical significance of (7.87), let us consider the scattering of an electron by a weak classical static magnetic field, such as one imposed by a macroscopic magnet in a laboratory. In a little more generality, the setup for describing the scattering of an electron by a weak classical electromagnetic field[15] $A(x)$ is as follows. In diagrammatic terms, we replace the external photon in Figure 7.24 by an interaction with the external field $A(x)$ as indicated in Figure 7.25. In analytic terms, we add an extra term to the QED Hamiltonian that gives the interaction of $A(x)$ with the electron current:

$$H' = \int eA_\mu(x)\overline{\psi}(x)\gamma^\mu\psi(x)\,d^3\mathbf{x}.$$

This Hamiltonian arises from the classical interaction Lagrangian $-A_\mu j^\mu$ (cf. (2.32)) by replacing the classical current j_μ by the quantum current $\overline{\psi}\gamma^m\psi$ as in §4.1.

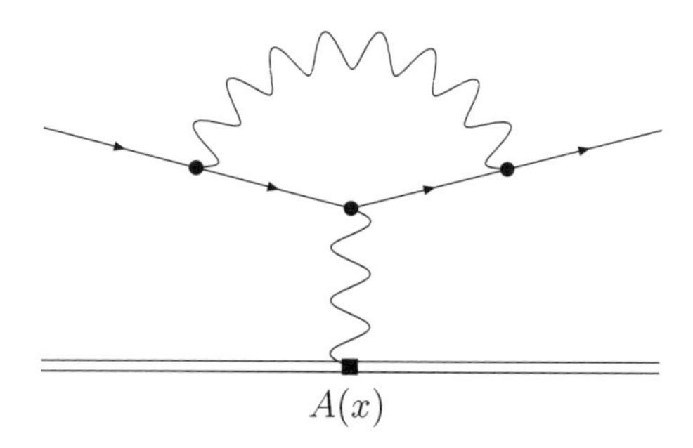

FIGURE 7.25. One-loop correction to the scattering of an electron by an external field.

We are interested in the first-order approximation to the S-matrix element for the scattering by this interaction. Writing $u(p)$ and $a(p)$ for $u(\mathbf{p}, s)$ and $a(\mathbf{p}, s)$ for short, and $q = p' - p$, we have

$$\begin{aligned}
&\left\langle 0 \middle| a(p') \left[-i\int eA_\mu(x)\overline{\psi}(x)\gamma^\mu\psi(x)\,d^4x\right] a(p) \middle| 0 \right\rangle \\
&= \frac{-iem}{2\sqrt{\omega'_{\mathbf{p}}\omega_{\mathbf{p}}}} \int e^{i(p'-p)_\mu x^\mu} A_\mu(x)\overline{u}(p')\gamma^\mu u(p)\,d^4x \\
&= \frac{-iem}{2\sqrt{\omega'_{\mathbf{p}}\omega_{\mathbf{p}}}} \widehat{A}_\mu(q)\overline{u}(p')\gamma^\mu u(p).
\end{aligned}$$

The key point is that we can get a better approximation to this S-matrix element by incorporating the radiative corrections to the vertex in Figure 7.25, that is, by replacing γ^μ by $\Gamma^\mu(p', p)$ (with all renormalizations completed); we denote this improved S-matrix element by S. By (7.56) and the Gordon identity (7.55), we

[15]In quantum mechanical terms, a "weak classical" field is an inexhaustible source and sink for low-energy photons that remains essentially unchanged by the absorption or emission of such a photon.

have

$$\begin{aligned} S &= \frac{-iem}{2\sqrt{\omega'_{\mathbf{p}}\omega_{\mathbf{p}}}}\widehat{A}_\mu(q)\overline{u}(p')\left[F_1(q^2)\gamma^\mu + F_2(q^2)\frac{i\sigma^{\mu\nu}q_\nu}{2m}\right]u(p) \\ &= \frac{-iem}{2\sqrt{\omega'_{\mathbf{p}}\omega_{\mathbf{p}}}}\widehat{A}_\mu(q)\overline{u}(p')\left[F_1(q^2)\frac{(p'+p)^\mu}{2m} + \left(F_1(q^2)+F_2(q^2)\right)\frac{i\sigma^{\mu\nu}q_\nu}{2m}\right]u(p). \end{aligned}$$

We now specialize to the case of a static magnetic field, for which $A(x)$ is independent of x^0 and $A_0(x) = 0$: thus $A_\mu(x)$ becomes $(0, -\mathbf{A}(\mathbf{x}))$ where $\mathbf{A}(\mathbf{x})$ is the vector potential from which the magnetic field $\mathbf{B}$ is given by $\mathbf{B} = \operatorname{curl}\mathbf{A}$. (The minus sign is there in accordance with (2.25).) The 4-dimensional Fourier transform $\widehat{A}_\mu(q)$ thus becomes $-2\pi\delta(q^0)\widehat{\mathbf{A}}(\mathbf{q})$ where $\widehat{\mathbf{A}}$ now denotes the 3-dimensional Fourier transform. We also assume that the momentum transfer q is small (because the external field is supposed to be weak), and thereby discard all terms of order bigger than 1 in q. In particular, since $q^0 = \omega_{\mathbf{p}'} - \omega_{\mathbf{p}} = O(|q|^2)$, we replace q_ν by $(0, -\mathbf{q})$, where the minus sign again comes from the fact that we are dealing with q_ν rather than q^ν. (But there is no minus sign on the $\mathbf{q}$ in $\widehat{\mathbf{A}}(\mathbf{q})$ because of the sign convention for the Fourier transform!) Since $F_1(0) = 1$, the result of these manipulations is that

$$S = \frac{iem}{2\sqrt{\omega'_{\mathbf{p}}\omega_{\mathbf{p}}}}2\pi\delta(p'^0 - p^0)\widehat{A}_j(\mathbf{q})\overline{u}(p')\left[\frac{p'^j + p^j}{2m} - \left(1 + F_2(0)\right)\frac{i\sigma^{jk}q_k}{2m}\right]u(p). \tag{7.88}$$

Here the indices j and k run from 1 to 3, but the summation convention is still in force.

Now, by (1.13), in either the Weyl or Dirac representations we have

$$\sigma^{jk} = \frac{i}{2}[\gamma^j, \gamma^k] = \epsilon^{jkm}\begin{pmatrix}\sigma_m & 0 \\ 0 & \sigma_m\end{pmatrix},$$

where σ_m is a Pauli matrix and ϵ^{jkm} is the sign of the permutation $(123) \to (jkm)$ if j, k, m are distinct and 0 otherwise. Therefore,

$$\begin{aligned} \widehat{\mathbf{A}}_j(\mathbf{q})\sigma^{jk}q_k = (\epsilon^{jkm}\widehat{A}_j(\mathbf{q})q_k)\begin{pmatrix}\sigma_m & 0 \\ 0 & \sigma_m\end{pmatrix} &= -(\mathbf{q}\times\widehat{\mathbf{A}}(\mathbf{q}))^m\begin{pmatrix}\sigma_m & 0 \\ 0 & \sigma_m\end{pmatrix} \\ &= -\frac{1}{i}\begin{pmatrix}\widehat{\mathbf{B}}(\mathbf{q})\cdot\boldsymbol{\sigma} & 0 \\ 0 & \widehat{\mathbf{B}}(\mathbf{q})\cdot\boldsymbol{\sigma}\end{pmatrix}. \end{aligned}$$

Substituting this into (7.88), we obtain

$$\begin{aligned} S = {} & \frac{iem}{2\sqrt{\omega'_{\mathbf{p}}\omega_{\mathbf{p}}}}2\pi\delta(p'^0 - p^0)\overline{u}(p') \\ & \times\left[\frac{p'^j + p^j}{2m}\widehat{A}_j(\mathbf{q}) + \frac{1 + F_2(0)}{2m}\begin{pmatrix}\widehat{\mathbf{B}}(\mathbf{q})\cdot\boldsymbol{\sigma} & 0 \\ 0 & \widehat{\mathbf{B}}(\mathbf{q})\cdot\boldsymbol{\sigma}\end{pmatrix}\right]u(p). \end{aligned}$$

Finally, we write the spinors $u(p)$ and $u(p')$ in the Dirac representation (4.28),

$$u = \begin{pmatrix}u_l \\ u_s\end{pmatrix}, \qquad \overline{u} = \begin{pmatrix}u_l & -u_s\end{pmatrix},$$

and make the nonrelativistic approximation $u_s = 0$ as in §4.3; then

$$S = \frac{iem}{2\sqrt{\omega'_{\mathbf{p}}\omega_{\mathbf{p}}}}2\pi\delta(p'^0 - p^0)u_l(p')^\dagger\left[\frac{p'^j + p^j}{2m}\widehat{A}_j(\mathbf{q}) + \frac{1 + F_2(0)}{2m}\widehat{\mathbf{B}}(\mathbf{q})\cdot\boldsymbol{\sigma}\right]u_l(p).$$

But in the same nonrelativistic approximation, this is the matrix element

$$ie\left\langle \psi_{p'} \left| \frac{1}{2mi}(\nabla \cdot \mathbf{A} + \mathbf{A} \cdot \nabla) + \frac{1 + F_2(0)}{2m}\mathbf{B} \cdot \boldsymbol{\sigma} \right| \psi_p \right\rangle, \tag{7.89}$$

where $\nabla \cdot \mathbf{A}$ denotes the operator $f \mapsto \nabla \cdot (\mathbf{A}f)$ and ψ_p is the Dirac wave function of an electron with definite 4-momentum p:

$$\psi_p(x) = \sqrt{\frac{m}{2\omega_\mathbf{p}}}\, e^{ip_\mu x^\mu} u(p).$$

The normalization factor $\sqrt{m/2\omega_\mathbf{p}}$ is there because $|u(p)| = \sqrt{\omega_\mathbf{p}/m}\,|u(0)|$ by (5.30) and the remarks following (4.43), and $|u(0)| = \sqrt{2}$ by our choice of normalization in (5.29).

In turn, (7.89) is the first-order (Born) approximation to the amplitude for scattering the electron by the "potential"

$$-\frac{e}{2m}[i^{-1}(\nabla \cdot \mathbf{A} + \mathbf{A} \cdot \nabla) + (1 + F_2(0))\mathbf{B} \cdot \boldsymbol{\sigma}].$$

Comparing this with (4.31) and (4.32), we see that, for a constant magnetic field $\mathbf{B}$, these are precisely the orbital angular momentum and spin terms in (4.32) except for the extra $F_2(0)$. The conclusion is that *the magnetic moment of the electron differs from the $(e/2m)\boldsymbol{\sigma}$ predicted by the Dirac equation by the factor* $1 + F_2(0)$, which in the one-loop approximation (7.87) is

$$1 + F_2(0) = 1 + \frac{e^2}{8\pi^2} \approx 1.0011614.$$

The value $e^2/8\pi^2$ for the so-called *anomalous magnetic moment* of the electron was first obtained by Schwinger in 1947, by a *tour de force* of calculation using his own methods (without Feynman diagrams), and experiments at about the same time yielded an experimental value of $.00118 \pm .00003$. This impressive agreement had a profound impact on the physicists of that day and marked the point where quantum electrodynamics became established as a theory worthy of the name. Subsequently, theoreticians have calculated $F_2(0)$ to order e^8, obtaining the value $(1159652192 \pm 74) \times 10^{-12}$; and experimentalists have obtained the empirical value $(1159652188 \pm 38) \times 10^{-12}$.

One can easily believe that it requires experiments of Nobel-prize quality to obtain such a precise experimental value for the anomalous magnetic moment; descriptions can be found in Dehmelt [**20**] and van Dyck [**121**]. The theoretical evaluation of this quantity, however, also requires heroic efforts. One must calculate all the Feynman diagrams representing higher-order corrections to the basic electron-photon vertex with up to four loops. Besides the one-loop diagram we have evaluated above, there are 7 two-loop diagrams, 72 three-loop diagrams, and 891 four-loop diagrams. The two-loop diagrams and some of the three-loop ones can be evaluated analytically (with a lot of work), but the others must be approximated by numerical methods. In addition, one must ascertain that Feynman diagrams containing virtual particles other than electrons and photons do not contribute significantly. (The reason they do not, in a word, is that the other particles that participate in the electromagnetic interaction are all much heavier than electrons, and their contributions are suppressed accordingly.) In the end, one of the limiting factors in obtaining a precise theoretical value is the precision to which the fine structure constant and the mass of the electron are known. More information can be found in Roskies et al. [**101**] and Kinoshita [**72**].

Experiments have also determined the anomalous magnetic moment of the muon to high precision: it is $(1165937 \pm 12) \times 10^{-9}$. Since the muon (as far as we know) differs from the electron only in being about 200 times as massive, the theoretical calculation of this quantity is quite similar to that for the electron, with one significant difference arising from the difference in masses: at two-loop and higher orders, the dominant effects are created by Feynman diagrams containing virtual electrons rather than (or in addition to) virtual muons, and diagrams containing virtual strongly interacting particles must also be taken into account. See Kinoshita and Marciano [**73**] and by Farley and Picasso [**29**] for the theoretical and experimental aspects of this matter.

The Lamb shift. The other classic experimental test of QED — in fact, the principal experimental motivation for the development of QED in the late 1940s — is the Lamb shift, the difference in energy levels between the $2S_{1/2}$ and $2P_{1/2}$ states in the hydrogen atom. This difference is usually stated as a frequency (via the relation $E = 2\pi\hbar\omega$) since it is determined experimentally by spectroscopic methods; the experimental value is

$$\omega_{2S_{1/2}} - \omega_{2P_{1/2}} = 1057.845 \pm .009 \text{ MHz}.$$

The precision of this figure is remarkable (and difficult to improve), for the natural line width for $2S_{1/2}$–$2P_{1/2}$ transitions is about 100 MHz. An account of the experimental methods involved can be found in Pipkin [**89**].

Unlike the anomalous magnetic moment of the electron, which can be neatly explained to a good level of precision by calculation of a single Feynman diagram, the Lamb shift is a relatively complicated phenomenon that involves an interplay of several different effects including all of the QED phenomena we have studied, and a serious account of its theory is unfortunately beyond the scope of this book. In particular, it involves an adaptation of quantum electrodynamics to handle problems involving bound states, which is something we have barely mentioned. Two separate calculations, involving different approximations, are needed to account for the effects of low-energy and high-energy virtual photons (we gave a hint of the low-energy calculation in §6.2), and the results must then be glued together with some care. We refer the reader to Weinberg [**129**], §14.3, and Sakurai [**102**], §2.8 and §4.7, for two accounts of the Lamb shift that both follow the procedure just outlined but are rather different in detail. Also, Sapirstein and Yennie [**104**] give a good account of the theory from a practical (computational) point of view, particularly the ways in which several different physical processes contribute to the Lamb shift and other shifts of energy levels.

7.12. Higher-order renormalization

We have now given a fairly thorough discussion of the renormalization of one-loop Feynman diagrams in QED and ϕ^4 scalar field theory. In this section we give a brief informal account of renormalization in higher orders of perturbation theory. The discussion that follows refers to an arbitrary quantum field theory that is formally renormalizable — that is, a theory for which the set of superficial degrees of divergence of its Feynman diagrams is bounded above — except where specific theories are mentioned.

First, on the practical level, one has the problem of renormalizing and evaluating specific Feynman diagrams in order to obtain higher levels of precision in computing S-matrix elements and related quantities that can be compared with experiment. On the one hand, this is an extremely labor-intensive enterprise. If one goes to nth order in perturbation theory, the number of Feynman diagrams involved as well as the difficulty of extracting the physically meaningful numbers from each of them is a rapidly increasing function of n, as we saw in connection with the magnetic moment problem in the preceding section. (For an account of analytic and numerical techniques used to evaluate Feynman diagrams, see Smirnov [**113**].) On the other hand, there are no additional conceptual hurdles to be overcome here. The divergences of the diagrams are cancelled by adding more counterterms to the Lagrangian, or rather — and this is the essential point — by modifying the coefficients of the counterterms already obtained by adding terms involving higher powers of the coupling constant.

On the theoretical level, the problem is to *prove* that renormalization really can be carried out to all orders of perturbation theory — that is, that the divergence of any diagram can be removed by adding counterterms of the appropriate sort to the Lagrangian, and that this can be done in a consistent way so that the counterterms needed to render one diagram finite do not interfere with those needed for any other. It took a lot of effort to produce a rigorous solution of this problem, and the early literature is full of errors and obscurities. Here is a brief overview.

The analysis of renormalization began with the fundamental paper of Dyson [**25**], who outlined a procedure for performing renormalizations in QED to arbitrary order. However, Dyson's analysis did not adequately address the problem of *overlapping divergences*, that is, diagrams that contain two divergent subdiagrams that share a common internal line. A modification of Dyson's method by Ward seemed to clarify the situation, but later it was found that Ward's method breaks down at order 14. At about the same time, Salam developed a procedure to handle overlapping divergences, but his papers lack mathematical precision. The first attempt to provide a truly mathematically respectable procedure for performing renormalization to arbitrary order for general field theories was made by Bogoliubov and Parasiuk [**10**], but they made some mistakes that were corrected by Hepp [**63**], and the resulting procedure was refined and extended by Zimmermann [**138**]. The theory that finally emerged from these papers is generally known as *BPHZ renormalization.* Manoukian [**79**] contains an account of BPHZ renormalization as well as a more rigorous formulation of Salam's method and a proof that the two are equivalent.

In 1951 Matthews and Salam [**80**] made a famous remark in connection with renormalization theory that has become known as the "Salam Criterion": "The difficulty, as in all this work, is to find a notation which is both concise and intelligible to at least two persons, of whom one may be an author." This began a tradition of expressions of perplexity by workers in this area about the work of their colleagues. Here is Hepp [**63**] speaking of Bogoliubov and Parasiuk: "It is hard to find two theoreticians whose understanding of the essential steps of the proof is isomorphic." Wightman [**133**] on Salam's work: "I will not describe it, if for no other reason than that I have never succeeded in understanding it." And Feldman et al. [**32**] on all of their predecessors: "But we must confess that none of us has yet qualified as that other person who is the guarantor of the [Salam] Criterion."

Accordingly, let the reader who proposes to venture into this jungle beware. The typical mathematical tourist who wants to learn more about the development of renormalization theory up to the mid-1970s would be much better off perusing the volume edited by Velo and Wightman [**124**], which starts off with the delightful survey paper of Wightman just quoted and contains a number of informative articles about various aspects of renormalization theory, including regularization schemes and nonperturbative methods. And there are newer methods, which we shall cite shortly.

Given that renormalization has been accomplished to all orders in perturbation theory, so that the terms in the Dyson series for an S-matrix element or a vacuum expectation value have all been rendered finite, the question remains: does the series converge? In spite of the irrelevance of this question to practical calculations, it is of obvious theoretical interest, for one would then have a rigorous construction of vacuum expectation values of time-ordered products of fields and hence, by the Wightman reconstruction theorem (Streater and Wightman [**114**], Glimm and Jaffe [**54**]), of the interacting fields themselves. But the answer is: Nobody knows for sure whether the series converges, but nobody seriously believes that it does.

The initial optimism that a proof of convergence would be found for QED was destroyed by a 2-page paper of Dyson in 1952 [**26**] in which he gave a nonrigorous but persuasive argument that the series should *not* converge. Here is a sketch of the argument: The Dyson series for any QED process is a power series in the fine structure constant $\alpha = e^2/4\pi$.[16] If the series converges for the physical value $\alpha \approx 1/137$, it will also converge for $-1/137 < \alpha < 0$. So, imagine a universe in which α is negative — in which like charges attract and opposite charges repel. In such a universe there are configurations in which many electrons are clustered together in one region and many positrons are clustered together in another region some distance away. If the clusters are sufficiently large, the Coulomb potential wells for electrons on one side and positrons on the other will be deeper than the rest mass of an electron. At this point it becomes energetically advantageous for electron-positron pairs to be created out of the vacuum, with the electron and positron going to join the respective clusters of their siblings and thus deepening the potential wells even more. As Dyson says, "there will be a rapid creation of more and more particles, an explosive disintegration of the vacuum by spontaneous polarization." Moreover, although there is a high potential barrier between such a pathological state and an "ordinary" state with no large clusters of particles, there is a nonzero probability of any ordinary state turning into a pathological one in a finite time because of tunneling effects. Conclusion: the Dyson series *cannot* converge for any negative value of α, and hence not for any positive value either.

Where does this leave us? It seems entirely possible that there is *no* rigorous nonperturbative mathematical model for QED. However, physicists nowadays are inclined to respond to this bad news with a shrug of the shoulders, because QED, as a stand-alone theory, is physically wrong when one takes sufficiently large energies into account, and its mathematical validity in the high-energy regime is therefore a moot point. More precisely, high-order Feynman diagrams in QED describe electromagnetic processes involving large numbers of virtual particles and hence large

[16]It is a power series in the electron charge e, but the numbers of internal vertices in all Feynman diagrams contributing to the process are congruent mod 2, so it is really a power series in e^2.

virtual energies; but it is physically incorrect to consider such processes without including the other fundamental interactions and their quanta (quarks, gluons, W bosons, neutrinos, ...), and then one's mathematical models must change. If one includes energies all the way up to the Planck scale, then gravity must enter the picture too, but the Planck scale is so far from what is available in our laboratories (or anywhere else in the observable universe) that we really have no idea what new physics might turn up. Perhaps string theory will provide an answer. But meanwhile, the generally accepted point of view is that quantum field theories such as QED that purport to describe real physical processes should all be regarded as "effective" field theories with a finite range of validity.

This attitude may seem like an admission of defeat, but only if one is seduced by the probably unattainable dream of constructing a "theory of everything." On a more realistic level, it is obvious that one needs different tools for different situations. It is a good thing, for example, that in order to calculate the trajectory of a rocket ship one does not have to understand the details of the motion of its constitutent elementary particles. We repeat the recommendation given at the end of §7.2 that the reader consult the discussions of these issues in Weinberg [**129**], §12.3, and Zee [**136**], §VIII.3.

Moreover, a focus on the relationship between the physics on different scales leads to some deep new insights: this is the boon granted to us by renormalization group analysis, to which we alluded briefly in §7.10. In particular, the development of techniques involving the renormalization group has led to new proofs that renormalization can be accomplished consistently to all orders in perturbation theory that are much simpler than the old BPHZ-type arguments. We refer the reader to Feldman et al. [**32**], Hurd [**64**], Rosen and Wright [**99**], Keller and Kopper [**70**], and the references given in these works. The underlying insight is that the process of "subtracting off infinities" is really a matter of subtracting off the irrelevant effects of the (perhaps poorly understood) physics at high energy or short distance scales in order to obtain the meaningful physics at the scales actually studied in the laboratory.

There has also been some remarkable recent work by Alain Connes and his collaborators that sheds more light on this subject. In particular, Connes and Kreimer [**16**], [**17**] have shown how to impose a Hopf algebra structure on the set of Feynman diagrams and interpret renormalization in terms of a factorization problem in this algebra. This thumbnail description only hints at the interesting mathematical structures they have uncovered; the renormalization group also appears naturally in their framework. See also Boutet de Monvel [**13**] for a brief sketch of the Connes-Kreimer work, Kreimer [**75**] for an exposition of related earlier results, and van Suijlekom [**122**] for applications of the Connes-Kreimer machinery to QED, including a derivation of the Ward identities. Finally, ambitious readers should examine the recent book [**18**] of Connes and Marcolli, where they will be taken on an exhausting but exhilarating journey beginning with quantum fields (including Connes-Kreimer theory) and proceeding through noncommutative geometry, Grothendieck motives, and quantum statistical mechanics to number theory and the Riemann hypothesis.

Ultimately, of course, the hope is to find new theories that will provide a more fundamental level of understanding with a more solid mathematical foundation. In particular, there is enough hope that a rigorous model for non-Abelian gauge field theories can be devised that it is one of the Millenium Prize problems; see Jaffe and

Witten [**66**]. But even if such theories are found, it seems likely that perturbative renormalization theory will still be the tool of choice for analyzing many problems, and the structural insight provided by Feynman diagrams will continue to be a conceptual guide for a long time to come.

CHAPTER 8

Functional Integrals

> Thirty-one years ago [in 1948], Dick Feynman told me about his "sum over histories" version of quantum mechanics. "The electron does anything it likes," he said. "It goes in any direction at any speed, forward or backward in time, however it likes, and then you add up the amplitudes and it gives you the wave function." I said to him, "You're crazy." But he wasn't.
> —Freeman Dyson ([**134**], p. 376)

After the very hard work of the last two chapters, the reader is probably ready for some lighter entertainment. The devil, always ready to oblige, arrives with another offer: some new insights into the material we have developed and some tools for extending it further, at the price of working with some mathematically ill-defined integrals over infinite-dimensional spaces. The attractive thing about this offer is that these integrals have an intuitive foundation, lead to meaningful calculations, and are tantalizingly close to being mathematically respectable. They are, in fact, related to some genuine integrals known to probabilists that we shall discuss briefly. Our main objective in this chapter, however, is to describe functional integrals as used by physicists, so the reader must be prepared to exercise a certain amount of suspension of disbelief. There is some solid mathematics here and there, but much of the purported mathematics is fiction. Like good literary fiction, however, it contains a lot of truth.

Functional integrals have proved to be a valuable tool in the development of advanced quantum field theory, and some physicists like to develop the theory from scratch in terms of them. (Zee [**136**] and Ramond [**91**] are good sources for this point of view.) However, in this book we will use them only to rederive some of the results of Chapter 6, the interest being in seeing how to arrive at these results by a completely different route. Accordingly, we shall carry the calculations only far enough to make the connection and leave further developments to the physics texts.

8.1. Functional integrals and quantum mechanics

The cornerstone of the theory of functional integrals is the Gaussian integral on $\mathbb{R}^n$, which we now review. The basic formula is as follows:

Let A be an invertible complex $n \times n$ matrix such that $A = A^T$ and $\operatorname{Re} A$ is positive semidefinite. Then for any $y \in \mathbb{R}^n$,

$$\int e^{-Ax\cdot x/2} e^{iy\cdot x}\, d^n x = \frac{(2\pi)^{n/2}}{\sqrt{\det A}} e^{-A^{-1}y\cdot y/2}. \tag{8.1}$$

Here $\sqrt{\det A} = \prod_1^n \sqrt{\lambda_j}$ *where* $\{\lambda_j\}_1^n$ *are the eigenvalues of* A *and* $\operatorname{Re}\sqrt{\lambda_j} > 0$. *If* $\operatorname{Re} A$ *is singular, the integral is to be interpreted by replacing* A *by* $A + \epsilon I$ *and letting* $\epsilon \to 0+$ *after performing the integration.*

We recall the proof very briefly. First, for the special case $n = 1$ and $A = 1$, the function $f(y) = \int e^{-x^2/2} e^{iyx}\, dx$ is easily seen to satisfy $f'(y) = -yf(y)$ and $f(0) = \sqrt{2\pi}$, whence $f(y) = \sqrt{2\pi} e^{-y^2/2}$. Next, suppose A is real and hence positive definite, and let C be the positive square root of A^{-1}. The substitution $x = Cv$ turns the left side of (8.1) into $(\det C) \int e^{-|v|^2/2} e^{iCy\cdot v}\, d^n v$, which is a product of one-dimensional integrals to which the previous calculation applies. Since $\det C = (\det A)^{-1/2}$ and $|Cy|^2 = A^{-1}y \cdot y$, the result follows. Moreover, both sides of (8.1) are analytic functions of the entries of A in the region $\mathcal{R} = \{A : A = A^T \text{ and } \operatorname{Re} A \text{ is positive definite}\}$ (for the right side this is clear at least on the open dense set of A's with all eigenvalues distinct), and they agree when A is real and positive definite; hence they agree in general. The formula remains valid when $\operatorname{Re} A$ is merely semidefinite because the right side extends continuously to such A and the interpretation via $A + \epsilon I$ makes the left side do so too.

Of course, what (8.1) gives is the Fourier transform of the function $e^{-Ax\cdot x/2}$ — as a Schwartz class function when $\operatorname{Re} A$ is positive definite, and as a tempered distribution in general.

It will be useful to record a rephrasing of (8.1) for Gaussian integrals on $\mathbb{C}^n$. Here and in the sequel, it will be understood, often without explicit mention, that the "$A = \lim_{\epsilon\to 0+}(A + \epsilon I)$" device is to be employed whenever we apply these results to A's whose real part is singular — in particular, to A's that are pure imaginary, the most important case for quantum theory.

Suppose A *is a complex* $n \times n$ *matrix such that* $A = A^T$ *and* $\operatorname{Re} A$ *is positive semidefinite. Then for* $w \in \mathbb{C}^n$,

$$\int e^{-Az\cdot\overline{z}} e^{i(z\cdot\overline{w} + w\cdot\overline{z})}\, d^{2n}z = \frac{\pi^n}{\det A} e^{-A^{-1}w\cdot\overline{w}}. \tag{8.2}$$

Here $a \cdot b = \sum a_j b_j$ *and* $d^{2n}z$ *is the volume element on* $\mathbb{C}^n$.

Thie proof is just a matter of sorting out the notation. We identify $\mathbb{C}^n$ with $\mathbb{R}^{2n}$ by $z = x + ix' \leftrightarrow (x, x')$. Then the condition $A = A^T$ implies that $Az \cdot \overline{z} = Ax \cdot x + Ax' \cdot x'$, and $z \cdot \overline{w} + w \cdot \overline{z}$ is twice the real inner product on $\mathbb{R}^{2n}$. Hence the result follows by applying (8.1) with n replaced by $2n$, A by $2\left(\begin{smallmatrix} A & 0 \\ 0 & A \end{smallmatrix}\right)$, x by (x, x'), and y by $2(y, y')$.

Next, we use the preceding results to evaluate integrals of the form

$$\int P(x) e^{-Ax\cdot x/2}\, d^n x$$

where P is a polynomial. It is enough to consider the case where P is a monomial, and it will be convenient to write monomials as products of linear factors. Thus, given $i_1, \ldots, i_K \in \{1, \ldots, n\}$ (which need not all be distinct), we wish to evaluate

$$\int x_{i_1} \cdots x_{i_K} e^{-Ax\cdot x/2}\, d^n x.$$

Obviously the integral vanishes when K is odd, since the integrand is then odd, so we take $K = 2k$. The key observation is that

$$\int x_{i_1} \cdots x_{i_{2k}} e^{-Ax\cdot x/2}\, d^n x = (-1)^k \frac{\partial^{2k}}{\partial y_{i_1} \cdots \partial y_{i_{2k}}} \int e^{-Ax\cdot x/2} e^{iy\cdot x}\, d^n x \bigg|_{y=0},$$

which by (8.1) is equal to

$$(-1)^k \sqrt{\frac{(2\pi)^n}{\det A}} \frac{\partial^{2k}}{\partial y_{i_1} \cdots \partial y_{i_{2k}}} e^{-A^{-1}y\cdot y} \bigg|_{y=0}.$$

Each of these $2k$ derivatives can do one of two things: (i) contribute a linear factor in y, from differentiating the expression $A^{-1}y \cdot y$, or (ii) differentiate such a linear factor that is already present. The only terms that are nonzero at $y = 0$ are those obtained by using k of the derivatives to bring down linear factors and the other k to reduce the latter to constants, and all such ways contribute equally. The result is:

Suppose A is a complex $n \times n$ matrix such that $A = A^T$ and $\operatorname{Re} A$ is positive semidefinite, and let $A^{-1} = (\alpha_{lm})$. Then for any $i_1, \ldots, i_{2k} \in \{1, \ldots, n\}$,

$$\int x_{i_1} \cdots x_{i_{2k}} e^{-Ax\cdot x/2}\, d^n x = \frac{(2\pi)^{n/2}}{\sqrt{\det A}} \sum \alpha_{i_{j_1} i_{j_2}} \cdots \alpha_{i_{j_{2k-1}} i_{j_{2k}}}, \tag{8.3}$$

where the sum is over all $(2k-1)(2k-3)\cdots 1$ ways of grouping the indices $i_1, \ldots, i_{2k}$ into k unordered pairs.

This ought to look familiar. The combinatorial structure here is exactly the same as in Wick's formula (6.42): the creation and annihilation of linear factors is exactly analogous to the pairing of creation and annihilation operators in the perturbation expansion of the S-matrix. This is no accident. As we shall soon see, a slightly souped-up version of this result plays the role of Wick's theorem in the functional integral formulation of field theory.

With these results in hand, we begin our study of functional integrals in the simple setting of a nonrelativistic particle moving in a potential in $\mathbb{R}^d$, that is, a solution of the Schrödinger equation $i\partial_t \psi = H\psi$ with $H = (-1/2m)\nabla^2 + V(x)$. *Throughout this discussion we omit writing the dimensional superscript on volume elements*; that is, we write dx instead of $d^d x$.

We shall use the informal language of distributions — more specifically, the dialect due to Dirac, as discussed in §3.2. Thus, $|x\rangle$ denotes the (generalized) state whose wave function is the delta-function $f(y) = \delta(y - x)$, $|p\rangle$ denotes the (generalized) state whose wave function is $g(y) = e^{iy\cdot p}$, and we have the completeness relations

$$\int |x\rangle\langle x|\, dx = \int |p\rangle\langle p| \frac{dp}{(2\pi)^d} = I. \tag{8.4}$$

The Schrödinger-picture state whose wave function is ψ will be denoted by $|\psi\rangle$, so that $\psi(x) = \langle x|\psi\rangle$. Furthermore, we set $|xt\rangle = e^{itH}|x\rangle$, so that the wave function at time t is $\psi(t, x) = \langle x|e^{-itH}|\psi\rangle = \langle xt|\psi\rangle$. We then have $\int |xt\rangle\langle xt|\, dx = I$ for any t, so if $t_f > t_i$ (f and i stand for "final" and "initial"),

$$\psi(x, t_f) = \langle xt_f|\psi\rangle = \int \langle xt_f|yt_i\rangle\langle yt_i|\psi\rangle\, dy = \int \langle xt_f|yt_i\rangle \psi(y, t_i)\, dy. \tag{8.5}$$

In other words, the function $K(x, t_f; y, t_i) = \langle x t_f | y t_i \rangle$ is the integral kernel of the time translation operator $e^{-i(t_f - t_i)H}$. Our object is to compute this kernel.

The idea is to break up the interval $[t_i, t_f]$ into N equal subintervals of length $\tau = (t_f - t_i)/N$ with endpoints $t_n = t_i + n\tau$ and apply (8.4) on each subinterval to obtain

(8.6)

$$\langle x_f t_f | x_i t_i \rangle = \int \cdots \int \langle x_f t_f | x_{N-1} t_{N-1} \rangle \cdots \langle x_2 t_2 | x_1 t_1 \rangle \langle x_1 t_1 | x_i t_i \rangle \, dx_1 \cdots dx_{N-1}.$$

We shall calculate the terms $\langle x_{n+1} t_{n+1} | x_n t_n \rangle$ to first order in τ by using the Schrödinger equation and then see what happens as $\tau \to 0$, i.e., as $N \to \infty$. First, since $e^{-i\tau H} = I - i\tau H + o(\tau)$,

$$\begin{aligned} \langle x_{n+1} t_{n+1} | x_n t_n \rangle &= \langle x_{n+1} | e^{-i\tau H} | x_n \rangle \\ &= \langle x_{n+1} | x_n \rangle - i\tau \langle x_{n+1} | H | x_n \rangle + o(\tau) \\ &= \delta(x_{n+1} - x_n) - i\tau \left[\frac{-1}{2m} \langle x_{n+1} | \nabla^2 | x_n \rangle + \langle x_{n+1} | V | x_n \rangle \right] + o(\tau). \end{aligned}$$

Next,

$$\langle x_{n+1} | V | x_n \rangle = V(x_n) \delta(x_{n+1} - x_n),$$

and by (8.4),

$$\begin{aligned} \langle x_{n+1} | \nabla^2 | x_n \rangle &= \iint \langle x_{n+1} | p' \rangle \langle p' | \nabla^2 | p \rangle \langle p | x_n \rangle \frac{dp' \, dp}{(2\pi)^{2d}} \\ &= \iint e^{ip' \cdot x_{n+1}} (-|p|^2) (2\pi)^d \delta(p' - p) e^{-ip \cdot x_n} \frac{dp' \, dp}{(2\pi)^{2d}} \\ &= \int e^{ip \cdot (x_{n+1} - x_n)} |p|^2 \frac{dp}{(2\pi)^d}. \end{aligned}$$

We can also express the delta-functions in the previous two formulas as p-integrals:

$$\delta(x_{n+1} - x_n) = \int e^{ip \cdot (x_{n+1} - x_n)} \frac{dp}{(2\pi)^d}.$$

Putting these all together, we obtain

$$\begin{aligned} \langle x_{n+1} t_{n+1} | x_n t_n \rangle &= \int e^{ip \cdot (x_{n+1} - x_n)} \left[1 - i\tau \left(\frac{|p|^2}{2m} + V(x_n) \right) \right] \frac{dp}{(2\pi)^d} + o(\tau) \\ &= \int \exp \left[i\tau \left(\frac{p \cdot (x_{n+1} - x_n)}{\tau} - \frac{|p|^2}{2m} - V(x_n) \right) \right] \frac{dp}{(2\pi)^d} + o(\tau), \end{aligned}$$

This is a Gaussian integral, so by (8.1) we have

$$\text{(8.7)} \quad \langle x_{n+1} t_{n+1} | x_n t_n \rangle = \left(\frac{m}{2\pi i \tau} \right)^{d/2} \exp i\tau \left[\frac{m}{2} \left| \frac{x_{n+1} - x_n}{\tau} \right|^2 - V(x_n) \right] + o(\tau).$$

This is valid for $n = 1, \ldots, N-2$, and also for $n = 0$ and $n = N-1$ if we set $x_0 = x_i$ and $x_N = x_f$. Plugging it into (8.6) yields

(8.8) $\langle x_f t_f | x_i t_i \rangle$

$$= \left(\frac{m}{2\pi i \tau} \right)^{Nd/2} \int \cdots \int \exp i\tau \left[\sum_0^{N-1} \frac{m}{2} \left| \frac{x_{n+1} - x_n}{\tau} \right|^2 - V(x_n) \right] dx_1 \cdots dx_{N-1} + o(1).$$

So far our calculations have been a bit loose but can easily be tightened up. But now we make the great leap by letting $\tau \to 0$, $N \to \infty$, to obtain, *formally*,

$$(8.9) \qquad \langle x_f t_f | x_i t_i \rangle = (\text{const.}) \int \exp\left[i \int_{t_i}^{t_f} \left[\tfrac{1}{2} m |x'(t)|^2 - V(x(t))\right] dt \right] [dx].$$

Here the constant is the (nonexistent) limit of $(m/2\pi i \tau)^{Nd/2}$, the integration is over all paths $x(t)$ such that $x(t_i) = x_i$ and $x(t_f) = x_f$, and $[dx]$ is the (nonexistent) Lebesgue measure on the space of all such paths.

Leaving aside for the moment the question of how to make real sense out of this, there is clearly something very interesting going on! The quantity $L(x, x') = \frac{1}{2} m |x'|^2 - V(x)$ is the *classical Lagrangian* of the system, so the integrand is $e^{iS(x)}$, where $S(x) = \int_{t_i}^{t_f} L(x, x')\, dt$ is the *classical action* of the path x. Thus we have a way of generating the *quantum* evolution of a nonrelativistic particle from quantities that govern its *classical* physics in the Lagrangian formulation. The nature of the classical-quantum correspondence becomes clearer if we put Planck's constant back in explicitly. As the reader may verify, all that is needed is a factor of $1/\hbar$ in the exponential:

$$(8.10) \qquad \langle x_f t_f | x_i t_i \rangle = (\text{const.}) \int e^{iS(x)/\hbar}\, [dx].$$

On the macroscopic scale where $\hbar$ is very tiny, this integral is highly oscillatory, so that contributions of neighboring paths will tend to cancel out, *except* for those paths x for which the first variation of the action $S(x)$ vanishes so that the nearby paths give constructive rather than destructive interference.[1] In other words, the major contribution to the quantum transition amplitude comes from those paths whose action is stationary. But according to the principle of least action (§2.2), these are precisely the classical trajectories!

The formula for the transition function $\langle x_f t_f | x_i t_i \rangle$ is much less formidable when the time interval $t_f - t_i$ is infinitesimal. In that case it is given by (8.7), which can be restated as

$$(8.11) \qquad \langle y + dy, t + dt | y, t \rangle = (\text{const.}) \exp\left[iL\left(y, \frac{dy}{dt}\right) dt \right].$$

We derived this from the Schrödinger equation, but we can also go the other way by a neat seat-of-the-pants calculation. Indeed, suppose $\psi(x, t)$ is the wave function of the quantum particle, whose evolution is given by

$$\psi(x, t') = \int \langle x t' | y t \rangle \psi(y, t)\, dy.$$

Taking $t' = t + \epsilon$, using (8.11) as a formula for $\langle x t' | y t \rangle$ that is correct to order ϵ (i.e., to within an error that is $o(\epsilon)$), and replacing dt by ϵ and the derivative dy/dt by the finite difference $(x - y)/\epsilon$, we obtain the following, where the approximations

[1] This is just the principle of stationary phase, which has a number of rigorous formulations in the context of finite-dimensional integrals. See, e.g., Erdélyi [**27**].

are valid to order ϵ:

$$\psi(x,\,t+\epsilon) \approx C \int \exp\left[i\epsilon\left[\frac{m}{2}\left(\frac{x-y}{\epsilon}\right)^2 - V(y)\right]\right]\psi(y,t)\,dy$$
$$\approx C \int e^{im(x-y)^2/2\epsilon}[1 - i\epsilon V(y)]\psi(y,t)\,dy.$$

Setting $y = x + \sqrt{\epsilon}\,z$ and continuing to discard terms that are $o(\epsilon)$, we find that

$$\psi(x,\,t+\epsilon) \approx C \int e^{imz^2/2}[1 - i\epsilon V(x+\sqrt{\epsilon}\,z)]\psi(x+\sqrt{\epsilon}\,z,\,t)\,dy$$
$$\approx C \int e^{imz^2/2}[1 - i\epsilon V(x)]\Big[\psi(x,t) + \sqrt{\epsilon}\sum z_j\partial_j\psi(x,t)$$
$$+ \frac{\epsilon}{2}\sum z_j z_k \partial_j\partial_k\psi(x,t)\Big]\epsilon^{d/2}\,dz.$$

The Gaussian integrals

$$\int e^{imz^2/2}\,dz, \qquad \int z_j e^{imz^2/2}\,dz, \qquad \int z_i z_j e^{imz^2/2}\,dz$$

are evaluated in (8.3), and the result is

$$\psi(x,\,t+\epsilon) \approx C\left(\frac{2\pi i\epsilon}{m}\right)^{d/2}\left[\psi(x,t) + i\epsilon\left(\frac{1}{2m}\nabla^2\psi(x,t) - V(x)\psi(x,t)\right)\right].$$

On the other hand, to order ϵ,

$$\psi(x,\,t+\epsilon) \approx \psi(x,t) + \epsilon\partial_t\psi(x,t).$$

Comparison of the terms of order 0 (in ϵ) and then the terms of order 1 in these expressions for $\psi(x,\,t+\epsilon)$ shows first that $C = (m/2\pi i\epsilon)^{d/2}$ and then that

$$\partial_t\psi(x,t) = i\left(\frac{1}{2m}\nabla^2\psi(x,t) - V(x)\psi(x,t)\right),$$

which is the Schrödinger equation.

The historical genesis of all this is of some interest. The initial development of quantum mechanics went by way of Hamiltonian mechanics, as we have sketched in Chapter 3. In 1933 Dirac wrote a short paper [**23**] addressing the question of how the Lagrangian might be connected to quantum mechanics. Based on some calculations with canonical transformations, he concluded that $\langle x+dx, t+dt|x,t\rangle$ "corresponds" to $\exp[iL\,dt/\hbar]$, in that the quantity U defined by $\langle x+dx,\ t+dt|x,t\rangle = e^{iU/\hbar}$ has certain formal similarities to the classical action $L\,dt$. Upon encountering this paper some years later, Feynman wondered what was really meant by "corresponds." On the assumption that the meaning might be "is proportional," Feynman performed the derivation of the Schrödinger equation from (8.11) that we have just seen and concluded that the correspondence really is a proportionality. He then proceeded to redevelop quantum mechanics from scratch, using (8.10) as the starting point. Integrals of the type (8.10) are therefore often called *Feynman path integrals.* Feynman's personal account of this discovery in [**35**] is well worth reading; his path-integral version of quantum mechanics is developed in [**34**] and [**40**].

Let us briefly consider some of the ways of giving rigorous meaning to (8.9).

The first observation is that the easy case of a free particle, $V = 0$, works out perfectly. Here the Schrödinger equation $i\partial_t\psi = -(1/2m)\nabla^2\psi$ is easily solved via the Fourier transform:

$$i\partial_t\widehat{\psi} = \frac{|p|^2}{2m}\widehat{\psi} \quad \Longrightarrow \quad \widehat{\psi}(p, t_0 + \tau) = e^{-i\tau|p|^2/2m}\widehat{\psi}(p, t_0).$$

In other words, $\psi(\cdot, t_0+\tau) = \psi(\cdot, t_0) * K_\tau$ where K_τ is the inverse Fourier transform of $e^{-i\tau p^2/2m}$, which we have calculated in (8.1) to be $(m/2\pi i\tau)^{d/2}e^{im\xi|^2/2\tau}$. In other words, when $V = 0$ the formula (8.7) for the propagator $\langle x_{n+1}t_{n+1}|x_nt_n\rangle$ is *exactly* correct. Hence, the finite-dimensional approximation (8.8) to the Feynman path integral is also an exact formula for $\langle x_ft_f|x_it_i\rangle$ (we are just using the fact that $K_T = K_{T/N} * \cdots * K_{T/N}$ [N factors, $T = t_f - t_i$]), and we are free to let $N \to \infty$.

Now let us put the potential back in. In view of the preceding remarks, the expression on the right side of (8.8) is the integral kernel of the operator

$$\begin{aligned} e^{-i\tau H_0}e^{-i\tau V}e^{-i\tau H_0}e^{-i\tau V}\cdots e^{-i\tau H_0} &= e^{-i\tau H_0}[e^{-i\tau V}e^{-i\tau H_0}]^N \\ &= e^{iTH_0/N}[e^{-iTV/N}e^{-iTH_0/N}]^N, \end{aligned}$$

where $H_0 = (-1/2m)\nabla^2$ and $T = t_f - t_i$. For a wide class of potentials V, this operator converges in the strong operator topology to e^{-iTH}, the desired propagator for the Hamiltonian $H = H_0 + V$. Indeed, we have the *Trotter product formula*:

Suppose A and B are self-adjoint operators such that $A + B$ is essentially self-adjoint on $\mathcal{D}(A)\cap\mathcal{D}(B)$. Then $e^{it(A+B)}$ (the group generated by the closure of $A+B$) is the limit in the strong operator topology of $[e^{itA/n}e^{itB/n}]^n$ as $n \to \infty$.

A proof can be found in Reed and Simon [**93**]. When A and B are bounded (alas, not a case with many important applications), the convergence actually takes place in the norm topology, and the proof is so simple that we present it for the reader's entertainment. Let $C = e^{it(A+B)/n}$ and $D = e^{itA/n}e^{itB/n}$; we wish to show that $\|C^n - D^n\| \to 0$. Now,

$$C^n - D^n = C^{n-1}(C-D) + C^{n-2}(C-D)D + \cdots + C(C-D)D^{n-2} + (C-D)D^{n-1},$$

so since C and D are unitary and hence have norm 1,

$$\|C^n - D^n\| \le \|C - D\|\sum_1^n \|C\|^{n-j}\|D\|^{j-1} = n\|C - D\|.$$

But C and D are both of the form $I + it(A+B)/n + O(1/n^2)$, so $\|C-D\| = O(1/n^2)$ and hence $n\|C - D\| \to 0$.

The Trotter product formula shows that the transition from (8.8) to (8.9) can be made rigorously on the level of operators rather than integral kernels, under suitable conditions on the potential V. These conditions are not very restrictive; for example, they are satisfied for $V \in L^2 + L^\infty$ or for V a polynomial.

Another interesting possibility is to perform a Wick rotation[2] — that is, to replace t by $-it$ — so that e^{-itH} becomes e^{-tH} and the Schrödinger equation becomes the heat equation with a potential term, $\partial_t\psi = (1/2m)\nabla^2\psi - V\psi$. By

[2] In momentum space the Wick rotation is $p^0 \to ip^0$; the corresponding transformation in space-time is $t \to -it$.

the same calculation we did above, the integral kernel $\langle x_f t_f | x_i t_i \rangle$ for the operator $e^{-(t_f - t_i)H}$ is approximated by

$$\langle x_f t_f | x_i t_i \rangle \approx \left(\frac{m}{2\pi\tau}\right)^{Nd/2} \int \cdots \int \exp \tau \left[\sum_0^{N_1} \frac{m}{2} \left| \frac{x_{n+1} - x_n}{\tau} \right|^2 - V(x_n) \right] dx_1 \cdots dx_n$$

($\tau = (t_f - t_i)/N$), which in the limit as $N \to \infty$ looks like

$$(\text{const.}) \int \exp \left[\int_{t_i}^{t_f} \left[-\tfrac{1}{2} m |x'(t)|^2 - V(x(t)) \right] dt \right] [dx],$$

the integral being over the space of all paths x with $x(t_i) = x_i$ and $x(t_f) = x_f$. Now, in this expression the individual ingredients — the constant in front, the Lebesgue measure $[dx]$, and the velocity x' — do not make sense (as there is no reason for the paths to be differentiable), but the *combination* of them does: $(\text{const.}) \exp[-\int_{t_i}^{t_f} m|x'(t)|^2/2\, dt]\, [dx]$ is nothing but Wiener measure on the space of (continuous) paths on the time interval $[t_i, t_f]$.[3] The resulting formula for the solution of the heat equation with potential in terms of a Wiener integral was worked out by Kac after hearing Feynman lecture on his ideas; it is known as the *Feynman-Kac formula*, and it has been a standard tool in stochastic analysis for many years. Expositions of it can be found in Simon [**112**] and in many books that deal with stoachastic processes and differential equations.

One might wonder if one could make sense of (8.9) in a similar way as the integral of $e^{-i\int V(x(t))\, dt}$ with respect to a *complex* measure on path space, or at least as a limit of such integrals. The finite-dimensional version, the "measure"

$$d\mu_{N,i\tau}(x_1, \ldots x_N) = (2\pi i\tau)^{-Nd/2} \exp \left[-\sum_1^N \frac{(x_j - x_{j-1})^2}{2i\tau} \right] dx_1 \ldots dx_n \quad (x_0 = 0)$$

on $\mathbb{R}^{Nd}$, is not a genuine complex measure because its total variation is infinite, but it is the limit in a suitable sense of the complex measure $d\mu_{N,a}$ (defined in the same way, with $i\tau$ replaced by a) as a approaches $i\tau$ in the right half-plane. Unfortunately, as was first pointed out by Cameron [**15**], one cannot let $N \to \infty$ to obtain a corresponding complex measure on path space except when a is real and positive, because the total variation $|\mu_{N,a}|(\mathbb{R}^{Nd})$ is $(|a|/\operatorname{Re} a)^{Nd/2}$ (by an easy computation), and this diverges as $N \to \infty$ as soon as $\operatorname{Im} a \neq 0$.

The idea of making problems in quantum theory more tractable by performing an analytic continuation to turn oscillatory integrands into decaying ones and fake measures on path space into real ones has been quite fruitful, however, and it is the source of many of the rigorous results in the subject. We shall say more about this in §8.5. Here we just mention an idea of Nelson [**85**]: make the mass parameter m in the Schrödinger equation complex. If we replace m by $i\mu$, $\mu > 0$, the Schrödinger equation becomes $\partial_t \psi = (1/2\mu)\nabla^2 \psi - iV\psi$. The Feynman-Kac formula applies to this equation for a large class of potentials V; the resulting solution can be analytically continued to the domain $\operatorname{Re} \mu > 0$, and its limit as $\mu \to m$ exists for a.e. $m > 0$.

[3] More precisely, it is Wiener measure if one fixes the initial position x_i. If one specifies both x_i and x_f, it becomes the "Brownian bridge" measure; other variations are possible. See Folland [**47**] and Simon [**112**] for more about Wiener measure.

8.2. Expectations, functional derivatives, and generating functionals

We now return to the unrefined, unrigorous Feynman integral (8.9) and perform some calculations with it in preparation for the passage from one-particle quantum mechanics to quantum fields. It will be clear that these calculations have a certain formal elegance but not that they really make sense or lead to something useful — at least not at first. The reader is asked to be indulgent for a while; we will eventually come down out of the clouds.

Let $X = (X_1, \ldots, X_d)$ be the position operator for our particle moving in a potential in $\mathbb{R}^d$, and let $X_t = e^{itH}Xe^{-itH}$ be the corresponding Heisenberg-picture operator. Suppose that, instead of computing the transition amplitude $\langle x_f t_f | x_i t_i \rangle$, we wish to compute the matrix element $\langle x_f t_f | g(X_{t_1}) | x_i t_i \rangle$, where $t_i < t_1 < t_f$, g is a (reasonable) function on $\mathbb{R}^d$, and $g(X_{t_1})$ is the operator defined by the spectral functional calculus. (Why would we wish to do this? Again, have patience: this is a warmup for some field-theoretic calculatons.) The key is to insert the eigenfunction expansion for the operators X_{t_1}:

$$\begin{aligned}\langle x_f t_f | g(X_{t_1}) | x_i t_i \rangle &= \int \langle x_f t_f | g(X_{t_1}) | x_{t_1} t_1 \rangle \langle x_{t_1} t_1 | x_i t_i \rangle \, dx_{t_1} \\ &= \int g(x_{t_1}) \langle x_f t_f | x_{t_1} t_1 \rangle \langle x_{t_1} t_1 | x_i t_i \rangle \, dx_{t_1}.\end{aligned}$$

Each of the transition probabilities in this last integral can be expressed as a functional integral as in (8.9):

$$\begin{aligned}&\langle x_f t_f | g(X_{t_1}) | x_i t_i \rangle \\ &\qquad = C \iiint g(x_{t_1}) \exp \Bigg[i \int_{t_i}^{t_1} [\tfrac{1}{2} m x'(t)^2 - V(x(t)] \, dt \\ &\qquad\qquad\qquad + i \int_{t_1}^{t_f} [\tfrac{1}{2} m y'(t)^2 - V(y(t)] \, dt \Bigg] [dx] \, [dy] \, dx_{t_1},\end{aligned}$$

where x runs over all paths on the time interval $[t_i, t_1]$ with $x(t_i) = x_i$ and $x(t_1) = x_{t_1}$, and likewise y runs over all paths on the time interval $[t_1, t_f]$ with $y(t_1) = x_{t_1}$ and $y(t_f) = x_f$. (Here and in the following formulas, C denotes an [ill-defined] normalization constant that may change from instance to instance.) But since we are integrating over the intermediate position x_{t_1} too, this is the same as integrating over all paths on the interval $[t_i, t_f]$ with $x(t_i) = x_i$ and $x(t_f) = x_f$:

$$\langle x_f t_f | g(X_{t_1}) | x_i t_i \rangle = C \int g(x(t_1)) \exp \left[i \int_{t_i}^{t_f} [\tfrac{1}{2} m x'(t)^2 - V(x(t)] \, dt \right] [dx].$$

In other words, inserting $g(X_{t_1})$ into $\langle x_f t_f | x_i t_i \rangle$ is equivalent to inserting $g(x(t_1))$ into the corresponding functional integral.

Let's generalize this: how about $\langle x_f t_f | g_2(X_{t_2}) g_1(X_{t_1}) | x_i t_i \rangle$, where $t_i < t_1 < t_2 < t_f$? The same procedure works. We have

$$\begin{aligned}&\langle x_f t_f | g_2(X_{t_2}) g_1(X_{t_1}) | x_i t_i \rangle \\ &\qquad = C \iint \langle x_f t_f | g_2(X_{t_2}) | x_{t_2} t_2 \rangle \langle x_{t_2} t_2 | g_1(X_{t_1}) | x_{t_1} t_1 \rangle \langle x_{t_1} t_1 | x_i t_i \rangle \, dx_{t_1} \, dx_{t_2}.\end{aligned}$$

The operators $g_j(X_{t_j})$ can now be replaced by their eigenvalues $g_j(x_{t_j})$, after which inner products in this integral can be written as functional integrals that can be combined to yield

$$\langle x_f t_f | g_2(X_{t_2}) g_1(X_{t_1}) | x_i t_i \rangle$$
$$= C \int g_1(x(t_1)) g_2(x(t_2)) \exp\left[i \int_{t_i}^{t_f} [\tfrac{1}{2} m x'(t)^2 - V(x(t)]\, dt \right] [dx].$$

Clearly this procedure extends to any number of factors $g_j(X_{t_j})$. Here is the intriguing point: the product of g_j's on the right is a product of ordinary numerical-valued functions, so the order of the factors is immaterial; but the product of g_j's on the left is a product of noncommuting operators! What must be kept in mind is that we assumed $t_1 < t_2$; the calculation does not work if the time ordering is reversed. In short, we have:

If $t_1, \ldots, t_n \in (t_i, t_f)$, then

$$\left\langle x_f t_f \middle| \mathcal{T} \prod_1^n g_j(X_{t_j}) \middle| x_i t_i \right\rangle \tag{8.12}$$
$$= C \int \prod_1^n g_j(x(t_j)) \exp\left[i \int_{t_i}^{t_f} [\tfrac{1}{2} m x'(t)^2 - V(x(t)]\, dt \right] [dx],$$

where the integration is over all paths x on the interval $[t_i, t_f]$ with $x(t_i) = x_i$ and $x(t_f) = x_f$, and C is a normalization constant that does not depend on $g_1, \ldots, g_n$.

As an application, here's how to derive the perturbation series for a perturbed potential from the functional integral. Suppose the potential V is $V_0 + V_1$, and we regard the Hamiltonian $H_0 = (-1/2m)\nabla^2 + V_0$ as "known" and the extra term V_1 as a perturbation; we wish to calculate the evolution operator e^{-itH} in terms of e^{-itH_0} and V_1. The transition amplitude $\langle x_f t_f | x_i t_i \rangle_H$ for the evolution e^{-itH} is given by

$$\langle x_f t_f | x_i t_i \rangle_H = \int \exp\left[i \int_{t_i}^{t_f} [\tfrac{1}{2} m x'(t)^2 - V(x(t)]\, dt \right] [dx]$$
$$= C \int \exp\left[i \int_{t_i}^{t_f} [\tfrac{1}{2} m x'(t)^2 - V_0(x(t))]\, dt \right] \exp\left[-i \int_{t_i}^{t_f} V_1(x(t))\, dt \right] [dx].$$

Expand the last exponential in its power series and replace the "=" by "∼" to acknowledge that nothing is being claimed about convergence, and use (8.12):

$$\langle x_f t_f | x_i t_i \rangle_H$$
$$\sim \sum_0^\infty C \int \exp\left[i \int_{t_i}^{t_f} [\tfrac{1}{2} m x'(t)^2 - V_0(x(t))]\, dt \right]$$
$$\times \frac{(-i)^n}{n!} \int_{[t_i, t_f]^n} V_1(x(t_1)) \cdots V_1(x(t_n))\, dt_1 \cdots dt_n\, [dx]$$
$$= \sum_0^\infty \frac{(-i)^n}{n!} \left\langle x_f t_f \middle| \int_{[t_i, t_f]^n} \mathcal{T}[V_1(X_{t_1}) \cdots V_1(X_{t_n})] dt_1 \cdots dt_n \middle| x_i t_i \right\rangle_{H_0}.$$

But this is nothing but the Dyson series (6.6), i.e., the transition-amplitude version of the formula

$$e^{-i(t_f - t_i)H} = e^{-i(t_f - t_i)H_0} \mathcal{T} \exp \frac{1}{i} \int_{t_i}^{t_f} V_1(t)\, dt.$$

Returning to (8.12), let us introduce a little notation: we set

$$G(x) = \prod_1^n g_j(x(t_j)), \qquad \mathcal{T}G(X) = \mathcal{T}\prod_1^n g_j(X_{t_j}).$$

It may be that we are interested not in $\langle x_f t_f|\mathcal{T}G(X)|x_i t_i\rangle$ but in $\langle \psi_f|\mathcal{T}G(X)|\psi_i\rangle$ where ψ_f and ψ_i are given initial and final states. This modification is easy to make:

$$\langle \psi_f|\mathcal{T}G(X)|\psi_i\rangle = \iint \langle \psi_f|x_f t_f\rangle\langle x_f t_f|\mathcal{T}G(X)|x_i t_i\rangle\langle x_i t_i|\psi_i\rangle \, dx_f \, dx_i.$$

The $\langle x_f t_f|\mathcal{T}G(X)|x_i t_i\rangle$ on the right can be written as a functional integral over all paths having the specified initial and final values; but we are now integrating over these values too, and the upshot is a functional integral over *all* paths over the time interval $[t_i, t_f]$:

$$\langle \psi_f|\mathcal{T}G(X)|\psi_i\rangle = C\int \psi_f(x(t_f), t_f)^*\psi_i(x(t_i), t_i)G(x)e^{i\int_{t_i}^{t_f} L(x,x')\,dt}\,[dx]. \tag{8.13}$$

In case ψ_i and ψ_f are both the ground state of the Hamiltonian H, we can put this into a more useful form. For simplicity we assume that H has a discrete spectrum, although this restriction is not essential; what is crucial is that the lowest point of the spectrum of H, which we may take to be 0, is a simple eigenvalue. Let $\{|\psi_n\rangle\}_0^\infty$ be an orthonormal basis of eigenstates of H, with eigenvalues $0 = E_0 < E_1 \le E_2 \le \cdots$. Pick $T_+ > t_f$ and $T_- < t_i$, and write

$$\langle x_+ T_+|\mathcal{T}G(X)|x_- T_-\rangle = \iint \langle x_+ T_+|x_f t_f\rangle\langle x_f t_f|\mathcal{T}G(X)|x_i t_i\rangle\langle x_i t_i|x_- T_-\rangle \, dx_f \, dx_i.$$

We have

$$\begin{aligned}\langle x_+ T_+|x_f t_f\rangle &= \sum_m \langle x_+|e^{-iT_+H}|\psi_m\rangle\langle\psi_m|e^{it_fH}|x_f\rangle \\ &= \sum_m \psi_m(x_+)\psi_m(x_f)^* e^{i(t_f-T_+)E_m},\end{aligned}$$

and likewise

$$\langle x_i t_i|x_- T_-\rangle = \sum_n \psi_n(x_i)\psi_n(x_-)^* e^{i(T_- - t_i)E_n},$$

where $\psi_n(x)$ is the Heisenberg-picture (time-independent) wave function of the state $|\psi_n\rangle$. Hence

$$\begin{aligned}&\langle x_+ T_+|\mathcal{T}G(X)|x_- T_-\rangle \\ &= \sum_{m,n}\iint \psi_m(x_+)\psi_n(x_i)\psi_m(x_f)^*\psi_n(x_-)^* e^{i(-T_+E_m + t_fE_m - t_iE_n + T_-E_n)} \\ &\qquad\qquad \times \langle x_f t_f|\mathcal{T}G(X)|x_i t_i\rangle \, dx_f \, dx_i.\end{aligned} \tag{8.14}$$

We wish to isolate the contribution of the ground state $|\psi_0\rangle$. For this purpose, we employ the same device we used in §6.11: we multiply T_+ and T_- by $1 - i\epsilon$ where ϵ is a positive infinitesimal (i.e., a small quantity that ultimately is made to vanish) and then let them tend to $\pm\infty$. The addition of the terms $-\epsilon T_+ E_m$ and

$\epsilon T_- E_n$ to the exponent in (8.14) causes all the terms to vanish in the limit except for $m = n = 0$, with the result that

$$\begin{aligned}\lim_{T_\pm \to \pm\infty(1-i\epsilon)} & \frac{\langle x_+ T_+ | \mathcal{T} G(X) | x_- T_- \rangle}{\psi_0(x_+)\psi_0(x_-)^*} \\ &= \iint \psi_0(x_f)^* \psi_0(x_i) \langle x_f t_f | \mathcal{T} G(X) | x_i t_i \rangle \, dx_f \, dx_i \\ &= \langle \psi_0, t_f | \mathcal{T} G(X) | \psi_0, t_i \rangle.\end{aligned} \tag{8.15}$$

In this last expression we can omit the t_f and t_i, because $|\psi_0, t\rangle = e^{-itH}|\psi_0\rangle = |\psi_0\rangle$. Moreover, the choice of x_+ and x_- is at our disposal; we may assume (by translating the coordinates if necessary) that $\psi_0(0) \neq 0$ and take $x_\pm = 0$.

It is important to note that instead of giving a small negative imaginary part to $T_\pm$, we could equally well replace H by $(1 - i\epsilon)H$, or perform some other operation on H that has the effect of giving the eigenvalues a small negative imaginary part. For example, if H is the Hermite operator $-\nabla^2 + |x|^2$, we could replace it with $-\nabla^2 + (1-i\epsilon)|x|^2$. This amounts to giving the Lagrangian a small *positive* imaginary part, the significance of which will shortly become apparent.

Now, the numerator on the left of (8.15) can be expressed as a functional integral by (8.12) (with t_i, t_f replaced by $T_\pm$), and the denominator is a numerical factor that does not depend on G. We can get rid of the latter by considering the *ratio* of the quantities in (8.15) to the corresponding quantities with $G \equiv 1$. Since $\langle \psi_0 | \psi_0 \rangle = 1$, this gives the final result:

$$\langle \psi_0 | \mathcal{T} G(X) | \psi_0 \rangle = \frac{\int e^{i\int L(x,x')\,dt} G(x)\,[dx]}{\int e^{i\int L(x,x')\,dt}\,[dx]}, \tag{8.16}$$

where the t-integrals are over the whole real line, the x-integrals are over all paths on $\mathbb{R}$ that tend to 0 at $\pm\infty$, and the Lagrangian is taken to contain a positive infinitesimal imaginary part — quite analogously to the implicit convergence-enhancing ϵ in (8.1) when the matrix A is imaginary. The nice thing about (8.16) is that the inconvenient normalization factors on the functional integrals cancel out, being the same for both numerator and denominator.

The formula (8.16) is very reminiscent of the Gell-Mann–Low formula (6.96). The similiarity will turn into an equivalence when we generalize the preceding results to the field-theoretic situation.

Recall that $G(x)$ is a product of polynomials $g_j(x(t_j))$. It is enough to consider monomials $G(x) = x_{i_1}(t_1)x_{i_2}(t_2)\cdots x_{i_J}(t_J)$ (where some t_j's may coincide), corresponding to the operator $\mathcal{T}G(X) = \mathcal{T}[X_{i_1,t_1} X_{i_2,t_2} \cdots X_{i_J,t_J}]$. For these G's there is a very convenient way to encode the preceding information in terms of "functional derivatives" of a suitable generating functional.

Functional derivatives. The functional derivative is one of those handy informal notions that tend to make sticklers for rigor tear their hair. In brief, suppose $\mathcal{X}$ is a space of functions on $\mathbb{R}^n$ (we deliberately decline to be more specific) and $\Phi : \mathcal{X} \to \mathbb{C}$ is a (presumably nonlinear) functional. The *functional derivative* $\delta\Phi(f)/\delta f(x)$ is formally defined as

$$\frac{\delta \Phi(f)}{\delta f(x)} = \lim_{\epsilon \to 0} \frac{\Phi(f + \epsilon\delta_x) - \Phi(f)}{\epsilon},$$

where δ_x is the delta-function with pole at x. Of course, if this definition is to be taken literally, the space $\mathcal{X}$ of "functions" had better contain some generalized functions as well. Alternatively, if $\mathcal{X}$ is a Banach space, one can define the derivative $\Phi'(f)$ to be the element of the dual space $\mathcal{X}^*$ such that $\Phi(f+\delta f) - \Phi(f) = \langle \Phi'(f), \delta f\rangle + o(\|\delta f\|)$. If $\mathcal{X}^*$ can also be identified with a space of functions such that the pairing of $\lambda \in \mathcal{X}^*$ with $f \in \mathcal{X}$ is $\langle \lambda, f\rangle = \int \lambda f$, then $\delta\Phi(f)/\delta f(x)$ is just $[\Phi'(f)](x)$. We shall not worry about the conditions on $\mathcal{X}$ and Φ needed to make sense of this in general, as we need only some simple special cases for which the functional derivative may be defined *ad hoc* and regarded merely as a convenient formalism.

The simplest case is that of a linear functional given by an integral,

$$\Phi(f) = \int f(y)h(y)\,dy,$$

for which we have

$$\frac{\delta\Phi(f)}{\delta f(x)} = h(x).$$

(Pragmatically speaking: think of the integral as a discrete sum $\sum f(y_i)h(y_i)\Delta y_i$ with one of the y_i's equal to x, take the ordinary partial derivative of this sum with respect to $f(x)$, and throw away the Δy_i. The same idea applies to the functionals discussed below.) A slightly less trivial and decidedly more useful case is

$$\Phi(f) = \exp\left[\int f(y)h(y)\,dy\right],$$

for which it is of interest to take functional derivatives of all orders:

$$\frac{\delta^J\Phi(f)}{\delta f(x_1)\,\delta f(x_2)\cdots\delta f(x_J)} = h(x_1)h(x_2)\cdots h(x_j)\Phi(f). \tag{8.17}$$

For a quadratic functional

$$\Phi(f) = \iint f(y)K(y,z)f(z)\,dy\,dz \qquad (K(x,y) = K(y,x)),$$

we have

$$\frac{\delta\Phi(f)}{\delta f(x)} = \int K(x,z)f(z)\,dz + \int f(y)K(y,x)\,dy = 2\int K(x,z)f(z)\,dz,$$

$$\frac{\delta^2\Phi(f)}{\delta f(w)\,\delta f(x)} = 2K(w,x).$$

Here again we are more interested in the exponential,

$$\Phi(f) = \exp\left[\iint f(y)K(y,z)f(z)\,dy\,dz\right],$$

which gives

$$\begin{aligned} \frac{\delta\Phi(f)}{\delta f(x)} &= \left[2\int K(x,z)f(z)\,dz\right]\Phi(f), \\ \frac{\delta^2\Phi(f)}{\delta f(w)\,\delta f(x)} &= \left(2K(w,x) + 4\left[\int K(x,z)f(z)\,dz\right]^2\right)\Phi(f). \end{aligned} \tag{8.18}$$

These formulas will suffice for our needs.

The generating functional. Returning to the calculations involving functional integrals, let us replace the Lagrangian $L(x, x')$ by $L(x, x') + F(t) \cdot x$, where F is a more-or-less arbitrary function of t with values in the same $\mathbb{R}^d$ as the variable x, and consider the functional integral

$$Z[F] = \int e^{i \int (L(x,x') + F(t) \cdot x) \, dt} [dx], \tag{8.19}$$

where the integrand in the exponential is implicitly assumed to carry an infinitesimal positive imaginary part.

We can derive the ground-state expectations $\langle \psi_0 | \mathcal{T} G(X) | \psi_0 \rangle$ of (8.16) by functional differentiation of $Z[F]$ with respect to F. Indeed, by (8.17) we have

$$\frac{\delta^J}{\delta F_{i_1}(t_1) \cdots \delta F_{i_J}(t_J)} \int e^{i \int (L(x,x') + F(t) \cdot x) \, dt} [dx]$$
$$= i^J \int x_{i_1}(t_1) \cdots x_{i_J}(t_J) e^{i \int (L(x,x') + F(t) \cdot x) \, dt} [dx],$$

so

$$\int x_{i_1}(t_1) \cdots x_{i_J}(t_J) e^{i \int L(x,x') \, dt} [dx]$$
$$= (-i)^J \frac{\delta^J}{\delta F_{i_1}(t_1) \cdots \delta F_{i_J}(t_J)} \int e^{i \int (L(x,x') + F(t) \cdot x) \, dt} [dx] \bigg|_{F=0}.$$

For $G(x) = x_{i_1}(t_1) x_{i_2}(t_2) \cdots x_{i_J}(t_J)$, this is the numerator on the right side of (8.16), and the denominator is just $Z[0]$, so we have

$$\langle \psi_0 | \mathcal{T}[X_{i_1,t_1} \cdots X_{i_J,t_J}] | \psi_0 \rangle = (-i)^J \frac{\delta^J}{\delta F_{i_1}(t_1) \cdots F_{i_J}(t_J)} \frac{Z[F]}{Z[0]} \bigg|_{F=0}. \tag{8.20}$$

Since $Z[F]$ can be used to generate all these expectation values, it is called the *generating functional* for them.

Let us say a few words about the physical meaning of F and $Z[F]$. It is appropriate to think of F as an external driving force, because for $L(x, x') = \frac{1}{2} m |x'|^2 - V(x)$, the modified Lagrangian $L(x, x') + F(t) \cdot x$ corresponds to the Newtonian equation $mx'' = -\nabla V(x) + F(t)$ for motion in a potential with an extra force $F(t)$. On the quantum level, the functional $Z[F]$, or rather its normalized version $Z[F]/Z[0]$, can be interpreted as the transition probability for the ground state at time $-\infty$ to return to the ground state at time $+\infty$ in the presence of the external force F, by the obvious modification of (8.16):

$$\langle \psi_0, +\infty | \psi_0, -\infty \rangle^F = \frac{Z[F]}{Z[0]}.$$

In the analogous construction in field theory, which we will consider in the next section, F can be interpreted as an external *field* that provides a source and sink for particles entering and leaving an interaction, and hence it is often referred to as a *source.* The philosophy of systematically using sources in field-theoretic calculations is due to Schwinger and is developed in his book [**108**]. However, the reader need not worry too much about the physical significance of F; for our purposes it is a mathematical device, and it will always be set equal to zero at the end of the calculations, as in (8.20).

8.3. Functional integrals and Boson fields

Scalar fields. We are now ready to see how the functional integral formulation of quantum field theory works. We begin with the easiest case, a free real scalar field $\phi(t, \mathbf{x})$, whose Lagrangian density is

$$\mathcal{L}(\phi, \partial\phi) = \tfrac{1}{2}[(\partial\phi)^2 - m^2\phi^2].$$

All we have to do is to pass from finite to infinite dimensions: the field $\phi(t, \mathbf{x})$ (functions of t parametrized by $\mathbf{x} \in \mathbb{R}^3$) plays the role of the position variables $x_1, \dots, x_d$ (functions of t parametrized by the integer $1, \dots, d$), and the action functional is

$$\int_{t_i}^{t_f} L(\phi, \partial\phi)\, dt = \int_{t_i}^{t_f} \int_{\mathbb{R}^3} \mathcal{L}(\phi, \partial\phi)\, d^3\mathbf{x}\, dt.$$

What we wish to calculate is vacuum expectation values of time-ordered products of field operators, for S-matrix elements can be obtained from them as explained in §6.11. We achieve this simply by adapting the formulas derived in the preceding section for ground-state expectation values of time-ordered products of position operators to the infinite-dimensional situation.

The reader may find this blithe passage from finite to infinite dimensions rather breathtaking, although it is probably no worse a mathematical sin than the use of Feynman path integrals to begin with. It may be reassuring to note that the corresponding passage to infinite dimensions in the theory of real Gaussian integrals is perfectly well-defined and not particularly difficult; we shall sketch the basic ideas in §8.5.

At this point we alert the reader to an upcoming shift of notation. The letter x, which denoted an element of $\mathbb{R}^d$ in the discussion of finite-dimensional theory, will henceforth be used for an element of 4-dimensional space-time, $x = (t, \mathbf{x})$, and we will revert to including the superscript 4 in the volume element d^4x. Moreover, the ground state ψ_0 in the field-theoretic situation is the vacuum state for the free field, and we denote it by $|0\rangle$ as in Chapter 6.

We begin by adding an arbitrary source $F(x) = F(t, \mathbf{x})$ into the Lagrangian:

$$\mathcal{L}^F(\phi, \partial_\mu\phi)(x) = \tfrac{1}{2}[(\partial\phi)^2 - m^2\phi^2](x) + F(x)\phi(x).$$

Thus, the analogue here of the $F(t) \cdot x$ in the finite-dimensional problem is

$$\int F(t, \mathbf{x})\phi(t, \mathbf{x})\, d^3\mathbf{x},$$

and the analogue of the generating functional (8.19) is

$$Z[F] = \int \exp\left[i \int (\tfrac{1}{2}[(\partial\phi)^2 - (m^2 - i\epsilon)\phi^2] + F\phi) d^4x\right] [d\phi]. \tag{8.21}$$

Here we have given the ghostly positive imaginary infinitesimal in the integrand in the exponential a concrete form by replacing m^2 by $m^2 - i\epsilon$; and the integration is over the space of all classical fields vanishing at infinity.

Now, the quantity $\exp\left[i \int \frac{1}{2}[(\partial\phi)^2 - (m^2 - i\epsilon)\phi^2] d^4x\right]$ is a Gaussian function of ϕ, and the integral on the right side of (8.21) is its Fourier transform in the variable F. *We know how to compute such Fourier transforms.* There is just one little step that must be taken: we perform an integration by parts,

$$\int (\partial\phi)^2\, d^4x = \int \partial_\mu\phi\partial^\mu\phi\, d^4x = -\int \phi\partial^2\phi\, d^4x,$$

to rewrite (8.21) as

$$\frac{Z[F]}{Z[0]} = \frac{\int \exp\left[i\int[-\frac{1}{2}\phi(\partial^2+m^2-i\epsilon)\phi+F\phi]\,d^4x\right][d\phi]}{\int \exp\left[i\int[-\frac{1}{2}\phi(\partial^2+m^2-i\epsilon)\phi]\,d^4x\right][d\phi]}. \tag{8.22}$$

By (8.1), generalized to infinite dimensions, the numerator on the right is equal to

$$C\exp\left[\frac{i}{2}\int F(\partial^2+m^2-i\epsilon)^{-1}F\,d^4x\right],$$

where the constant C is formally $(2\pi)^{\infty/2}/\sqrt{\det i(\partial^2+m^2-i\epsilon)}$, and the denominator is the same constant C. The value of C is of no real concern since it cancels out in the fraction.

Behold the miracle: we are left with a perfectly well-defined finite-dimensional integral. Moreover, by virtue of (6.48), the operator $(\partial^2+m^2-i\epsilon)^{-1}$ in it is given by convolution with our old friend the Feynman propagator Δ_F:[4]

$$\begin{aligned}\frac{\int \exp\left[i\int[-\frac{1}{2}\phi(\partial^2+m^2-i\epsilon)\phi+F\phi]\,d^4x\right][d\phi]}{\int \exp\left[i\int[-\frac{1}{2}\phi(\partial^2+m^2-i\epsilon)\phi]\,d^4x\right][d\phi]} &= \exp\left[\frac{i}{2}\int F(\partial^2+m^2-i\epsilon)^{-1}F\,d^4x\right] \\ &= \exp\left[\frac{i}{2}\iint F(x)\Delta_F(x-y)F(y)\,d^4x\,d^4y\right].\end{aligned} \tag{8.23}$$

This can now be functionally differentiated to yield vacuum-to-vacuum expectations for time-ordered products of field operators, by (8.20) (again, generalized to infinite dimensions). Namely, for $x_1,\dots,x_n\in\mathbb{R}^4$,

$$\begin{aligned}\langle 0|\mathcal{T}[\phi_0(x_1)\cdots\phi_0(x_n)]|0\rangle &= (-i)^n\frac{\delta^n}{\delta F(x_1)\cdots\delta F(x_n)}\frac{Z[F]}{Z[0]}\bigg|_{F=0} \\ &= (-i)^n\frac{\delta^n}{\delta F(x_1)\cdots\delta F(x_n)}\exp\left[\frac{i}{2}\iint F(x)\Delta_F(x-y)F(y)\,d^4x\,d^4y\right]\bigg|_{F=0}.\end{aligned} \tag{8.24}$$

Here the subscripts 0 on the left are included to make clear that the ϕ's on the left are free *quantum* fields, i.e., operator-valued functions, as opposed to the ϕ's in (8.23), which are *classical* fields, i.e., numerical-valued functions.

Each differentiation in (8.24) either brings down a factor of the convolution $i\Delta_F * F$ or turns such a factor into $i\Delta_F$, as in (8.18), and all terms with a factor of $i\Delta_F * F$ vanish at $F=0$. Hence, $\langle 0|\mathcal{T}[\phi_0(x_1)\cdots\phi_0(x_n)]|0\rangle$ vanishes if n is odd and is equal to

$$\begin{aligned}&(-i)^n\sum i\Delta_F(x_{i_1}-x_{i_2})i\Delta_F(x_{i_3}-x_{i_4})\cdots i\Delta_F(x_{i_{n-1}}-x_{i_n}) \\ &\quad=\sum[-i\Delta_F(x_{i_1}-x_{i_2})][-i\Delta_F(x_{i_3}-x_{i_4})]\cdots[-i\Delta_F(x_{i_{n-1}}-x_{i_n})]\end{aligned} \tag{8.25}$$

if n is even (since $(-i)^n=(-1)^{n/2}$), where the sum is over all ways of pairing the indices $x_1,\dots,x_n$ up — just as in (8.3). In effect, in view of (6.44) we have recovered the formula (6.42) for vacuum expectation values of free fields.

Well, free fields are boring — but we are now in a position to derive the Feynman diagrams for a self-interacting scalar field. Specifically, let us add a polynomial

[4] Recall that the F in Δ_F is in honor of Feynman and has nothing to do with the source F.

interaction term $-P(\phi)$ to the Lagrangian ($P(\phi) = \lambda\phi^4/4!$ for the standard case). Thus, we take

$$\mathcal{L} = \mathcal{L}_0 - P(\phi) + F\phi, \qquad \mathcal{L}_0 = \tfrac{1}{2}[(\partial\phi)^2 - (m^2 - i\epsilon)\phi^2].$$

The generating functional with source F is

$$Z[F] = \int e^{i\int(\mathcal{L}_0 + F\phi)\,dx} e^{-i\int P(\phi)\,d^4x}[d\phi],$$

and we can obtain vacuum expectation values of time-ordered products of quantum fields by functional differentiation as in (8.24). Here, however, the quantum fields are *interacting*, and the ground state is the *interacting-field vacuum.* We accordingly denote the quantum fields by ϕ_H (the H stands for "Heisenberg-picture") and the vacuum by $|\Omega\rangle$ as in §6.11:

$$(8.26) \qquad \langle\Omega|\mathfrak{T}[\phi_H(x_1)\cdots\phi_H(x_k)]|\Omega\rangle = (-i)^k \frac{\delta^k}{\delta F(x_1)\cdots\delta F(x_k)} \frac{Z[F]}{Z[0]}\bigg|_{F=0}.$$

The integral defining $Z[F]$ is no longer Gaussian, and we cannot evaluate it explicitly as it stands, so we formally expand $e^{i\int P(\phi)\,d^4x}$ in its power series to obtain

$$(8.27) \quad Z[F] = \sum_0^\infty \frac{(-i)^n}{n!} \int\int\cdots\int P(\phi(y_1))\cdots P(\phi(y_n)) e^{i\int(\mathcal{L}_0 + F\phi)\,dx}\, d^4y_1\cdots d^4y_n\,[d\phi].$$

Here the equality must be understood in the sense of perturbation theory: we do not mean to take the whole series in (8.27) seriously, but only to use finitely many terms of it to derive computable results. With this understanding, we combine (8.26) and (8.27) to get

$$\begin{aligned}(8.28) \quad & \langle\Omega|\mathfrak{T}[\phi_H(x_1)\cdots\phi_H(x_k)]|\Omega\rangle \\ &= \frac{1}{Z[0]}\sum_{n=0}^\infty \frac{(-i)^{k+n}}{n!} \int\int\cdots\int \phi(x_1)\cdots\phi(x_k)P(\phi(y_1))\cdots P(\phi(y_n)) \\ &\qquad\qquad \times e^{i\int\mathcal{L}_0\,d^4x}\, d^4y_1\cdots d^4y_n[d\phi].\end{aligned}$$

Now, the integrals on the right are again vacuum expectation values — not of the quantum fields $\phi_H(x_j)$ but of the corresponding *free fields*, with respect to the *free-field vacuum* $|0\rangle$:

$$\begin{aligned}(8.29) \quad & \int\int\cdots\int \phi(x_1)\cdots\phi(x_k)P(\phi(y_1))\cdots P(\phi(y_n))e^{i\int\mathcal{L}_0\,d^4x}\, d^4y_1\cdots d^4y_n\,[d\phi] \\ &= \left\langle 0 \middle| \int\cdots\int \mathfrak{T}[\phi_0(x_1)\cdots\phi_0(x_k)P(\phi_0(y_1))\cdots P(\phi_0(y_n))]\, d^4y_1\cdots d^4y_n \middle| 0 \right\rangle.\end{aligned}$$

Moreover, by (8.27) and (8.29),

$$\begin{aligned} Z[0] &= \sum_0^\infty \frac{(-i)^n}{n!} \int\int\cdots\int P(\phi(y_1))\cdots P(\phi(y_n))e^{i\int(\mathcal{L}_0)\,d^4x}\, d^4y_1\cdots d^4y_n\,[d\phi] \\ &= \sum_0^\infty \frac{(-i)^n}{n!} \left\langle 0 \middle| \int\cdots\int \mathfrak{T}[P(\phi_0(y_1))\cdots P(\phi_0(y_n))]\, d^4y_1\cdots d^4y_n \middle| 0 \right\rangle.\end{aligned}$$

Substituting this and (8.29) into (8.28), we recover the Gell-Mann–Low formula (6.94)!

On the other hand, the integrals on the left of (8.29), with the y-integrations suppressed, are again functional derivatives of the Gaussian integral

$$\int e^{i\int(\mathcal{L}_0+F\phi)\,dx}[d\phi]$$

— each $P(\phi(y_j))$ is obtained by applying $P(-i\delta/\delta F(y_j))$ — and hence they are given by functionally differentiating the finite-dimensional integral on the right of (8.23). The upshot is that these integrals are integrals over $y_1,\dots,y_n$ of products of Feynman propagators $-i\Delta_F(y_i-y_j)$ and $-i\Delta_F(x_i-y_j)$; these are precisely the position-space integrals corresponding to Feynman diagrams, with external as well as internal lines representing propagators. From them one obtains S-matrix elements by replacing the propagators for external lines by suitable coefficients, as we explained in §6.11.

One can also derive the interpretation of Feynman diagrams in terms of S-matrix elements directly from the picture of functional integrals with sources, rather than going through the calculations in §§6.3–7. We refer the reader to Zee [**136**] for a lucid account.

To sum up: *functional integrals have given us a new derivation of the Feynman rules and the Gell-Mann–Low formula for a self-interacting scalar field.* In a subject where complete mathematical rigor is out of reach, it is reassuring to see two quite different paths leading to the same destination. It also offers the hope that functional integrals may be useful at a more advanced level in obtaining results that are inaccessible otherwise; and this indeed turns out to be the case.

The procedure for a charged scalar quantum field ψ of mass M (corresponding to a complex-valued classical field) is similar. The free Lagrangian is $\mathcal{L}_0 = \partial_\mu\psi^*\partial^\mu\psi - M^2\psi^*\psi$. The functional integration must be extended over both ψ and ψ^*, just as the differential form representing area on $\mathbb{C}$ is not dz but $(1/2i)dz\wedge dz^*$. Thus, the generating functional with a source J (a complex-valued function) is

$$Z[J,J^*] = \iint e^{i\int\mathcal{L}_0\,dx}e^{iJ\psi+J^*\psi^*}\,[d\psi][d\psi^*].$$

This is a Gaussian integral that is evaulated by the infinite-dimensional analogue of (8.2). In taking functional derivatives of it, one must treat $J(x)$ and $J^*(x)$ as independent varables, just as one treats z and z^* as independent variables in analyzing nonholomorphic functions on $\mathbb{C}$.

As an example of two interacting fields, consider a neutral scalar field ϕ and a charged scalar field ψ with an interaction Hamiltonian $c\psi^*\psi\phi$. The Lagrangian is

$$\begin{aligned}\mathcal{L} &= \mathcal{L}_\phi + \mathcal{L}_\psi + \mathcal{L}_{\text{int}}\\ &= \tfrac{1}{2}(\partial_\mu\phi\partial^\mu\phi - m^2\phi^2) + (\partial_\mu\psi^*\partial^\mu\psi - M^2\psi^*\psi) - c\psi^*\psi\phi.\end{aligned}$$

The generating functional with a source F for ϕ and a source J for ψ is

$$Z[F,J,J^*] = \iiint e^{i\int\mathcal{L}\,dx}e^{iF\phi+iJ\psi+iJ^*\psi^*}[d\phi][d\psi][d\psi^*]. \tag{8.30}$$

If the interaction term $\mathcal{L}_{\text{int}}$ is omitted from $\mathcal{L}$, this is a Gaussian integral that can be evaulated as before. The interaction is incorporated perturbation-theoretically by writing $e^{i\int\mathcal{L}_{\text{int}}}$ as $\sum(i^n/n!)(\int\mathcal{L}_{\text{int}})^n$, and one obtains vacuum expectation values and S-matrix elements by functional differentiation just as in the case of a single neutral field.

Massless vector fields. The functional integral treatment of massive vector quantum fields is similar to that of scalar fields. For massless vector fields such as electromagnetism, however, gauge invariance requires an extra twist. The action for the free electromagnetic field, as we saw in §2.4, is

$$\begin{aligned} S(A) &= -\tfrac{1}{4}\int F_{\mu\nu}F^{\mu\nu}\,d^4x \qquad (F_{\mu\nu} = \partial_\mu A_\nu - \partial_\nu A_\mu) \\ &= \tfrac{1}{2}\int A_\mu(g^{\mu\nu}\partial^2 - \partial^\mu\partial^\nu)A_\nu\,d^4x, \end{aligned} \tag{8.31}$$

and the analogue of (8.21) with a source J (a vector-valued function) is

$$\int e^{i[S(A)+\int J^\mu A_\mu\,d^4x]}[dA], \tag{8.32}$$

which should equal (up to a normalization constant)

$$\exp\left[\frac{i}{2}\int J^\mu(x)\Delta_{\mu\nu}(x-y)J^\nu(y)\,d^4x\,d^4y\right],$$

where $\Delta_{\mu\nu}$ is a fundamental solution for the operator $-g^{\mu\nu}\partial^2 + \partial^\mu\partial^\nu$. However, as we pointed out in §6.5, this operator has no fundamental solution. The situation is analogous to the finite-dimensional Gaussian integral (8.1) in which the matrix A is not invertible.

The trouble comes from the redundancy in the description of the electromagnetic field. The Lagrangian $-\frac{1}{4}F_{\mu\nu}F^{\mu\nu}$ is invariant under all gauge transformations $A_\mu \mapsto A_\mu + \partial_\mu\chi$, so the integrand in (8.32) is constant along the orbits of the (noncompact, infinite-dimensional) gauge group, just as $e^{-Ax\cdot x}$ is constant along cosets of the nullspace of A when A is singular. The result is that the integral (8.32) diverges badly enough that even physicists are willing to admit that there is a problem. What is needed is a way to factor out the action of the gauge group so that integration is extended only over physically inequivalent fields.

The usual way of doing this is a device due to Faddeev and Popov [**28**].[5] To explain the idea, let us consider a similar but much simpler situation. Let $\{\sigma_t : t \in \mathbb{R}\}$ be a one-parameter group of measure-preserving diffeomorphisms of $\mathbb{R}^n$ whose orbits are (generically) unbounded, and suppose F is a function on $\mathbb{R}^n$ that is invariant under these transformations. We wish to extract a finite and meaningful quantity from the divergent integral $\int F(\mathbf{x})\,d^n\mathbf{x}$. One possibility is to find a hypersurface M that is a cross-section for the orbits (perhaps after judiciously pruning away sets of measure zero) and consider instead the integral $\int_M F(\mathbf{x})\,d\Sigma(\mathbf{x})$ where $d\Sigma$ is surface measure on M. This quantity, however, depends on the choice of M. A related procedure is to find a smooth function h such that $M = h^{-1}(\{0\})$ and consider the integral $\int F(\mathbf{x})\delta(h(\mathbf{x}))\,d^n\mathbf{x}$. This quantity depends on the choices of both M and h, for one must take into account the behavior of the delta-function under a change of variable. (The basic formula is this: if ϕ is a smooth functon on $\mathbb{R}$ and there is a unique t_0 such that $\phi(t_0) = 0$, then $\delta(t - t_0) = \phi'(t_0)\delta(\phi(t))$.)

A better idea is to incorporate the appropriate change-of-measure factor into the integral. Specifically, with M and h as above, let

$$\Delta(\mathbf{x}) = \frac{d}{dt}\big[h(\sigma_t(\mathbf{x}))\big]_{t=t(\mathbf{x})},$$

[5]Short papers that have a big impact are more common in physics than in mathematics, but this sketchy little two-page note is an extreme example.

where $t(\mathbf{x})$ is the unique number such that $h(\sigma_{t(\mathbf{x})}(\mathbf{x})) = 0$, and insert the factor

$$1 = \int \delta(u)\, du = \int \delta(h(\sigma_t(\mathbf{x})))\Delta(\mathbf{x})\, dt$$

into the integral $\int F(\mathbf{x})\, d^n\mathbf{x}$ to obtain (informally speaking)

$$\int F(\mathbf{x})\, d^n\mathbf{x} = \iint F(\mathbf{x})\delta(h(\sigma_t(\mathbf{x})))\Delta(\mathbf{x})\, dt\, d^n\mathbf{x}.$$

Now, F and the measure $d^n\mathbf{x}$ are assumed to be invariant under the transformations σ_t, and it is easily checked that $\Delta(\mathbf{x})$ is too; hence we can make the substitution $\mathbf{x} = \sigma_{-t}(\mathbf{y})$ to obtain

$$\int F(\mathbf{x})\, d^n\mathbf{x} = \left[\int dt\right]\left[\int F(\mathbf{y})\delta(h(\mathbf{y}))\Delta(\mathbf{y})\, d^n\mathbf{y}\right].$$

The divergence has now been isolated as the infinite factor $\int dt$. The remaining $\mathbf{y}$-integral is the quantity we have been seeking: it is finite provided that the restriction of F to M decays suitably at infinity, and one can verify that it is independent of the choice of M and ϕ. (On the informal level, one can observe that $\int dt$ is merely the volume of the transformation group, which does not depend on M, ϕ, or f, so that its removal should yield an invariant quantity too. This, of course, is the reasoning employed in the functional integral situation, where everything is somewhat ill-defined but infinite constants in numerators and denominators cancel out.) The same idea works for multi-parameter groups of transformations; the factor $\Delta(\mathbf{x})$ there is an appropriate Jacobian determinant.

With this as motivation, we are ready to make sense out of the functional integral (8.32). The gauge group here is the additive group G of differentiable functions on $\mathbb{R}^4$ modulo constants, and the gauge transformation corresponding to $\chi \in G$ is $A \mapsto A^\chi = A + \partial\chi$. We begin by choosing a gauge-fixing condition $g(A) = 0$ to obtain a cross-section to the orbits of the gauge group. The Lorentz condition $g(A) = \partial^\mu A_\mu = 0$ is a reasonable choice (the one originally made by Faddeev and Popov), but we shall do something more general with the aim of deriving the whole family of photon propagators that we presented in §6.5. Namely, we take $h(A) = \partial^\mu A_\mu - \omega$ where ω is an arbitrary continuous function on $\mathbb{R}^4$. As before, we write

$$1 = \int_G \delta(f)\, [df] = \int_G \delta(h(A^\chi)) \det \frac{\delta h(A^\chi)}{\delta\chi}\, [d\chi]. \tag{8.33}$$

The integrals here are functional integrals over the space of differentiable functions on $\mathbb{R}^4$; the delta-function represents the point mass at the origin in this space; and since $h(A^\chi) = \partial^\mu(A_\mu + \partial_\mu\chi) - \omega = h(A) + \partial^2\chi$, the functional determinant $\det(\delta h(A^\chi)/\delta\chi)$ — the *Faddeev-Popov determinant* for this situation — is simply $\det(\partial^2)$. In spite of its infiniteness, this determinant has the great virtue of being independent of A. It can therefore be brought outside all integral signs as a constant factor, where it will eventually cancel out when we pass to quotients of integrals.

We insert (8.33) into (8.32), obtaining

$$\begin{aligned} &\int e^{i[S(A)+\int J^\mu A_\mu\, d^4x]}[dA] \\ &\qquad = (\det \partial^2) \iint e^{i[S(A)+\int J^\mu A_\mu\, d^4x]}\delta(h(A^\chi))\, [dA]\, [d\chi]. \end{aligned} \tag{8.34}$$

At this point we must make explicit an assumption on the source J that has been implicitly present all along: J represents an electromagnetic current, so it must satisfy the charge-conservation equation $\partial_\mu J^\mu = 0$; we also assume that J vanishes at infinity to faciliate integration by parts. With this understood, the integral in the exponent is invariant under all gauge transformations, since $\int J^\mu \partial_\mu \chi \, d^4x = -\int (\partial^\mu J_\mu)\chi \, d^4x$. So is the "Lebesgue measure" $[dA]$, since gauge transformations are simply translations in the space of fields. Hence the substitution of $A^{-\chi}$ for A turns the integrand of (8.34) into an expression that is independent of χ, so that

$$(8.35) \quad \int e^{i[S(A)+\int J^\mu A_\mu \, d^4x]} [dA] = (\det \partial^2) \left[\int [d\chi] \right] \int e^{i[S(A)+\int J^\mu A_\mu \, d^4x]} \delta(\partial^\mu A_\mu - \omega) \, [dA].$$

The integral $\int [d\chi]$ (the volume of the gauge group) is another infinite constant that can be ignored, and the integral that is left has some hope of being meaningful; this is the analogue of the finite-dimensional result that we derived above.

To proceed further, we put the arbitrary function ω to use. Since (8.35) is valid for each ω, it remains valid if we take a weighted average over different ω's. We choose the Gaussian weight function $e^{-i\int(\omega^2/2a)\, d^4x}$, where a is a nonzero real number, and obtain

$$\int e^{i[S(A)+\int J^\mu A_\mu \, d^4x]} [dA] = N(a)(\det \partial^2) \left[\int [d\chi] \right] \iint e^{i[S(A)+\int J^\mu A_\mu \, d^4x]} e^{-i\int(\omega^2/2a)\, d^4x} \delta(\partial^\mu A_\mu - \omega) \, [d\omega] \, [dA],$$

where $N(a)$ is another (infinite) normalization factor. Performing the integration over ω and recalling the definition (8.31) of the action $S(A)$, we obtain the final result:

$$(8.36) \quad \int e^{i[S(A)+\int J^\mu A_\mu \, d^4x]} [dA] = C \int e^{i[S(A)-\int((\partial^\mu A_\mu)^2/2a) d^4x + \int J^\mu A_\mu \, d^4x]} \, [dA] = C \int e^{i\int[-(1/2)A_\mu(-g^{\mu\nu}\partial^2 + (1-a^{-1})\partial^\mu\partial^\nu)A_\nu + J^\mu A_\mu] \, d^4x} \, [dA],$$

where C is an infinite but harmless constant.

In short, the net effect of this lengthy calculation is to add the term $-\partial^\mu\partial^\nu/a$ to the differential operator $-g^{\mu\nu}\partial^2 + \partial^\mu\partial^\nu$. The point is that this new operator has a fundamental solution, namely, the distribution $\Delta^{(a)}$ whose Fourier transform is

$$\widehat{\Delta}^{(a)}_{\mu\nu}(p) = \frac{1}{p^2 + i\epsilon} \left[g^{\mu\nu} - (1-a)\frac{p^\mu p^\nu}{p^2} \right],$$

as we discused in §6.5. The functional integral can therefore be evaluated as in the case of scalar fields to yield

$$\int e^{i[S(A)+\int J^\mu A_\mu \, d^4x]} [dA] = C' \exp\left[\frac{i}{2} \int J^\mu(x) \Delta^{(a)}_{\mu\nu}(x-y) J^\nu(y) \, d^4x \, d^4y \right],$$

and one can proceed from there to obtain vacuum expectation values and integrals corresponding to Feynman diagrams. The choice of a remains at one's disposal;

we have generally used $a = 1$ ("Feynman gauge"), but other values of a are advantageous for certain calculations, as we saw in §7.8. One can even take $a = 0$ ("Landau gauge"), which effectively corresponds to the simple choice $g(A) = \partial^\mu A_\mu$ of gauge-fixing function.

8.4. Functional integrals and Fermion fields

To do interesting physics we need Fermions. The functional integral approach can be adapted to them, but only with a rather bizarre twist: the "classical fields" over which functional integration is performed must be taken to have values in a set of "anticommuting numbers," i.e., in a Grassmann algebra. This may seem to make functional integrals for Fermion fields even further removed from honest mathematics than those for Boson fields, but they are not quite as outlandish as they appear at first sight. Just as we led up to Bosonic functional integrals by considering finite-dimensional Gaussian integrals, we shall introduce Fermionic functional integrals by describing the well-defined finite-dimensional analogue.

Grassmann-valued functions: finite dimensions. We start with an n-dimensional real vector space V and consider the complexified Grassmann (exterior) algebra $\mathcal{G} = \mathbb{C} \otimes \bigwedge V$ over V. (The reason for not taking V to be complex to begin with will appear in due course.) We write the exterior product by simple juxtaposition, without the usual $\wedge$. Let $\xi_1, \dots, \xi_n$ be a basis for V. We regard the ξ_j's as elements of $\mathcal{G}$ and hence as "anticommuting variables" of which we can form $\mathcal{G}$-valued functions. The most general such function is

$$f(\xi_1, \dots, \xi_n) = c + \sum_i c_i\xi_i + \sum_{i<j} c_{ij}\xi_i\xi_j + \sum_{i<j<k} c_{ijk}\xi_i\xi_j\xi_k + \cdots + c_{1\cdots n}\xi_1 \cdots \xi_n, \tag{8.37}$$

where the c's are complex constants. Differentiation of such functions (on the left) is defined by the rule

$$\frac{\partial}{\partial \xi_j}\xi_{i_1} \cdots \xi_{i_k} = \begin{cases} 0 & \text{if } j \notin \{i_i, \dots, i_k\}, \\ (-1)^{l-1}\xi_{i_1} \cdots \widehat{\xi_{i_l}} \cdots \xi_{i_k} & \text{if } j = i_l, \end{cases} \tag{8.38}$$

where $\widehat{\xi_{i_l}}$ means that the term ξ_{i_l} in the product is omitted. It is easy to verify that the operators $\partial/\partial\xi_j$ and left multiplication by ξ_j (which we denote simply by ξ_j) satisfy the canonical anticommutation relations

$$\left\{\frac{\partial}{\partial \xi_j}, \xi_k\right\} = \delta_{jk}I, \qquad \left\{\frac{\partial}{\partial \xi_j}, \frac{\partial}{\partial \xi_k}\right\} = \{\xi_j, \xi_k\} = 0.$$

(One can also consider differentiation on the right: $\xi_{i_1} \cdots \xi_{i_k}(\partial/\partial\xi_j)$ is defined by the right side of (8.38) with $(-1)^{l-1}$ replaced by $(-1)^{k-l}$. But we shall have no need of this.)

The main order of business is integration of Grassmann-valued functions. The definition (due to Berezin [**8**]) may seem rather peculiar, so we shall lead up to it gently. The integral is performed one variable at a time: that is, for f of the form (8.37) we are going to define

$$\int d\xi_j \, f(\xi_1, \dots, \xi_n) \tag{8.39}$$

as an analogue of the integral over the whole real line of a function of some real variables,

$$\int_{-\infty}^{\infty} g(x_1, \ldots, x_n)\, dx_j. \tag{8.40}$$

(Writing the $d\xi_j$ on the left is a matter of convention — physicists commonly write ordinary integrals this way too — but it is not without significance, because of the anticommutativity in this situation.) The properties of the real integral (8.40) that we wish the Grassmann integral (8.39) to emulate are as follows, where for notational simplicity we take $j = 1$:

i. $\int g\, dx_1$ is a function of the remaining variables $x_2, \ldots, x_n$ (a complex number if g depends only on x_1).
ii. If $g(x_1, \ldots, x_n) = g_1(x_1) g_2(x_2, \ldots, x_n)$, then

$$\int g\, dx_1 = \left(\int g_1\, dx_1 \right) g_2(x_2, \ldots, x_n).$$

iii. For any function u of the variables $x_2, \ldots, x_n$,

$$\int g(x_1 + u(x_2, \ldots, x_n),\, x_2, \ldots, x_n)\, dx_1 = \int g(x_1, \ldots, x_n)\, dx_1.$$

Again taking $j = 1$, let us see what these conditions mean for (8.39). First, we again require $\int d\xi_1\, f(\xi_1, \ldots)$ to be a function of the remaining variables, or a complex number if there are no remaining variables. Next, any f of the form (8.37) can be written as $f(\xi_1, \ldots) = \alpha + \xi_1 \beta$ where α and β are functions of the remaining variables $\xi_2, \ldots, \xi_n$. (To obtain the analogous formula for $j \neq 1$, $f(\ldots, \xi_j, \ldots) = \alpha + \xi_j \beta$, one must rearrange the orders of the factors in the terms of (8.37), introducing some factors of -1.) Hence, in accordance with (ii) we require that

$$\int d\xi_1 (\alpha + \xi_1 \beta) = \left(\int d\xi_1 \right) \alpha + \left(\int d\xi_1\, \xi_1 \right) \beta,$$

where $\int d\xi_1$ and $\int d\xi_1\, \xi_1$ are complex numbers. Finally, in accordance with (iii) we require that for any function γ of the remaining variables,

$$\int d\xi_1\, (\alpha + (\xi_1 + \gamma)\beta) = \int d\xi_1\, (\alpha + \xi_1 \beta),$$

that is,

$$\left(\int d\xi_1 \right) (\alpha + \gamma\beta) + \left(\int d\xi_1\, \xi_1 \right) \beta = \left(\int d\xi_1 \right) \alpha + \left(\int d\xi_1\, \xi_1 \right) \beta$$

for any γ, which forces $\int d\xi_1$ to be 0. It remains only to normalize the integral by specifying the number $\int d\xi_1\, \xi_1$, which we take to be 1. Thus, going back to a general index j, we have

$$\int d\xi_j\, (\alpha + \xi_j \beta) = \beta \qquad (\alpha,\ \beta \text{ functions of the remaining variables}).$$

A moment's thought then reveals that *integration is the same as differentiation*:

$$\int d\xi_j\, f(\xi_1, \ldots, \xi_n) = \frac{\partial f}{\partial \xi_j}.$$

Multiple integrals are now defined as iterated integrals, with the convention that the integrations are to be performed in order from innermost to outermost:

$$\int\cdots\int d\xi_{i_1}\cdots d\xi_{i_k}\,f(\xi_1,\ldots,\xi_n)=\left[\int d\xi_{i_1}\left[\int d\xi_{i_2}\cdots\left[\int d\xi_{i_k}\,f(\xi_1,\ldots,\xi_n)\right]\right]\right].$$

With this convention, interchanging two integrations changes the result by a factor of -1; in other words, the $d\xi_i$'s should be construed as anticommuting quantities just like the ξ_i's. Thus, for example, $\int\cdots\int d\xi_{i_1}\cdots d\xi_{i_k}\,\xi_{j_1}\cdots\xi_{j_k}$ is the sign of the permutation that takes $\{i_1,\ldots,i_k\}$ to $\{j_k,\ldots,j_1\}$ if the sets $\{i_1,\ldots,i_k\}$ and $\{j_1,\ldots,j_k\}$ are equal, and is 0 otherwise.

It is immediate from the definition that integration commutes with right multiplication by constants. More precisely, if $\{i_1,\ldots,i_k\}$ is a subset of $\{1,\ldots,n\}$ and $\{j_1,\ldots j_{n-k}\}$ is the complementary subset, then

$$\int d\xi_{i_1}\cdots d\xi_{i_k}\,f(\xi_1,\ldots,\xi_n)g(\xi_{j_1},\ldots,\xi_{j_{n-k}})$$
$$=\left[\int d\xi_{i_1}\cdots d\xi_{i_k}\,f(\xi_1,\ldots,\xi_n)\right]g(\xi_{j_1},\ldots,\xi_{j_{n-k}}).$$

Left multiplication is a little less simple; the reader may verify that

$$\int d\xi_{i_1}\cdots d\xi_{i_k}\,g(\xi_{j_1},\ldots,\xi_{j_{n-k}})f(\xi_1,\ldots,\xi_n)$$
$$=g((-1)^k\xi_{j_1},\ldots,(-1)^k\xi_{j_{n-k}})\left[\int d\xi_{i_1}\cdots d\xi_{i_k}\,f(\xi_1,\ldots,\xi_n)\right],$$

which may be restated as

$$\int g(\xi_{j_1},\ldots,\xi_{j_{n-k}})\,d\xi_{i_1}\cdots d\xi_{i_k}\,f(\xi_1,\ldots,\xi_n)$$
$$=g(\xi_{j_1},\ldots,\xi_{j_{n-k}})\left[\int d\xi_{i_1}\cdots d\xi_{i_k}\,f(\xi_1,\ldots,\xi_n)\right]$$

if we make the convention that the $d\xi_i$'s anticommute with the ξ_i's.

In the applications of this machinery to physics, one must sooner or later integrate over *all* the available Grassmann variables so that one ends up with ordinary complex numbers. For future reference we display the result of doing this:
(8.41)

$$f(\xi_1,\ldots,\xi_n)=c_{1\cdots n}\xi_1\cdots\xi_n+\cdots\quad\Longrightarrow\quad\int\cdots\int d\xi_1\cdots d\xi_n f(\xi_1,\ldots,\xi_n)=c_{1\cdots n}.$$

That is, the integration over "the whole space," or *Berezin integral*, simply picks out the coefficient of the highest-order term $\xi_1\cdots\xi_n$.

Suppose now that $\eta_1,\ldots,\eta_n$ is another basis for our vector space V, and let M be the change-of-basis matrix: $\eta_j=\sum_j M_{jk}\xi_k$. Then

$$\eta_1\eta_2\cdots\eta_n=(\det M)\xi_1\xi_2\cdots\xi_n$$

(recall that these products are exterior products!). On the other hand, our rules for integration should apply just as well to the η_j's as to the ξ_j's, so we want

$$\int\cdots\int d\eta_n\cdots d\eta_1\,\eta_1\cdots\eta_n=1=\int\cdots\int d\xi_n\cdots d\xi_1\,\xi_1\cdots\xi_n.$$

To make this consistent, we need

$$d\eta_1 \cdots d\eta_n = (\det M)^{-1} d\xi_1 \cdots d\xi_n \qquad (\eta = M\xi). \tag{8.42}$$

Thus, the formula for the change of the "volume element" under linear transformations is exactly the reverse of the formula in real-variable calculus, which has $(\det M)$ instead of $(\det M)^{-1}$! (A corollary of this is that the correspondence $\xi \to d\xi$ is *not linear*: we have $\eta_j = \sum M_{jk}\xi_k$, the corresponding transformation of differentials is not $d\eta_j = \sum M_{jk} d\xi_k$ but $d\eta_j = \sum (M^{-1})_{jk} d\xi_k$.)

Since Fermions generally have distinct antiparticles, we really want complex Grassmann variables rather than real ones. The transition is easy but requires a bit of care to make things work smoothly. We start with a complex vector space V of complex dimension n and think of it as a real vector space of dimension $2n$ equipped with a complex structure, that is, a linear map $J : V \to V$ such that $J^2 = -I$. Let $\{\xi_1, \ldots, \xi_n\}$ be a basis for V over $\mathbb{C}$; then $\{\xi_1, J\xi_1, \ldots, \xi_n, J\xi_n\}$ is a basis for V over $\mathbb{R}$. We set $\zeta_j = (\xi_j - iJ\xi_j)/\sqrt{2}$ and $\zeta_j^* = (\xi_j + iJ\xi_j)/\sqrt{2}$ to obtain a set of generators for the complex Grassmann algebra $\mathcal{G} = \mathbb{C} \otimes \bigwedge V$ consisting of $\pm i$-eigenvectors of J. (The $\sqrt{2}$'s are there to make the transformation $(\xi_j, J\xi_j) \mapsto (\zeta_j, \zeta_j^*)$ unitary.) The transformation of differentials is given by the inverse map as explained in the preceding paragraph: $d\zeta_j = (d\xi_j + i dJ\xi_j)/\sqrt{2}$ and $d\zeta_j^* = (d\xi_j - i dJ\xi_j)/\sqrt{2}$. With these definitions in hand, we consider elements of $\mathcal{G}$ as "functions of the complex Grassmann variables ζ_j and ζ_j^*" and integrate them just as before. That is, if α and β (resp. α' and β') contain no terms divisible by ζ_j (resp. by ζ_j^*), then

$$\int d\zeta_j(\zeta_j\alpha + \beta) = \alpha, \qquad \int d\zeta_j^*(\zeta_j^*\alpha' + \beta') = \alpha',$$

and so forth.

Next, we develop the finite-dimensional models for the Gaussian integrals of Grassmann variables that will arise from quantum fields. Let ζ_j and ζ_j^* $(j = 1, \ldots, n)$ be complex Grassman variables as in the preceding paragraph, and let A be a complex $n \times n$ matrix. With the notational conventions

$$\zeta^* A\zeta = \sum_{j,k} A_{jk}\zeta_j^*\zeta_k, \qquad \iint [d\zeta^*\, d\zeta] = \iint \cdots \iint d\zeta_n^*\, d\zeta_n \cdots d\zeta_1^*\, d\zeta_1,$$

we claim that

$$\iint [d\zeta^*\, d\zeta] e^{-\zeta^* A\zeta} = \det A. \tag{8.43}$$

(The exponential is defined by the usual power series, which is actually a finite sum since the terms of degree $> 2n$ vanish.) Indeed, the term in $e^{-\zeta^* A\zeta}$ of degree $2n$ is

$$\frac{(-1)^n}{n!} \sum A_{j_1k_1} \cdots A_{j_nk_n} \zeta_{j_1}^*\zeta_{k_1} \cdots \zeta_{j_n}^*\zeta_{k_n} = \frac{1}{n!} \sum A_{j_1k_1} \cdots A_{j_nk_n} \zeta_{j_1}\zeta_{k_1}^* \cdots \zeta_{j_n}\zeta_{k_n}^*,$$

where the sum is over all j's and k's such that $\{j_1, \ldots, j_n\} = \{k_1, \ldots, k_n\} = \{1, \ldots, n\}$. Since the $\zeta_{j_l}\zeta_{k_l}^*$, being of degree 2, all commute with each other, they can be rearranged so that the ζ_j's occur in their canonical order without changing the value. Hence the sum over the j's can be performed immediately to yield

$$\sum A_{1k_1} \cdots A_{nk_n} \zeta_1\zeta_{k_1}^* \cdots \zeta_n\zeta_{k_n}^*.$$

Rearranging the k's to put the ζ_k^*'s in canonical order introduces a factor of $\text{sgn}(\sigma_k)$ where σ_k is the permutation taking $\{k_1, \dots, k_n\}$ to $\{1, \dots, n\}$. (The presence of the ζ_j's does not affect this; when one interchanges two ζ_k^*'s, each of them must pass through the same number of ζ_j's.) Hence we obtain

$$\sum (\operatorname{sgn} \sigma_k) A_{1k_1} \dots A_{1k_n} \zeta_1 \zeta_1^* \cdots \zeta_n \zeta_n^* = (\det A) \zeta_1 \zeta_1^* \cdots \zeta_n \zeta_n^*,$$

and (8.43) follows immediately from this and (8.41).

Now let $\eta_1, \dots \eta_n$ be another set of complex Grassmann variables, independent of $\zeta_1, \dots, \zeta_n$ (i.e., the ζ's, ζ^*'s, η's, and η^*'s together form a basis for a $4n$-dimensional space, and they all anticommute), and assume that the matrix A is invertible and symmetric ($A = A^T$). Setting

$$\eta^* \zeta = \sum \eta_j^* \zeta_j,$$

we have

$$-\zeta^* A \zeta + i(\eta^* \zeta + \zeta^* \eta) = -(\zeta^* - iA^{-1}\eta^*) A (\zeta - iA^{-1}\eta) - \eta^* A^{-1} \eta.$$

(The symmetry of A is needed here so that, for example, $(A^{-1}\eta^*)A\zeta = \eta^*(A^{-1}A)\zeta = \eta^*\zeta$.) Moreover, the quantities $X = (\zeta^* - iA^{-1}\eta^*)A(\zeta - iA^{-1}\eta)$ and $Y = \eta^* A^{-1} \eta$ are of degree 2 and so commute with each other, from which it follows that $e^{-(X+Y)} = e^{-X} e^{-Y}$. In view of (8.43) and the translation-invariance of the integral, we therefore have

(8.44)

$$\iint [d\zeta^*\, d\zeta] e^{-\zeta^* A \zeta + i(\eta^* \zeta + \zeta^* \eta)} = \iint [d\zeta^*\, d\zeta] e^{-(\zeta^* - iA^{-1}\eta^*) A (\zeta - iA^{-1}\eta)} e^{-\eta^* A^{-1} \eta}$$
$$= (\det A) e^{-\eta^* A^{-1} \eta}.$$

This is the "Gaussian Fourier transform" for Grassmann variables. It looks very similar to the corresponding formula (8.2) for complex variables except that the π's are missing and — more significantly — the factor of $\det A$ appears in the numerator instead of the denominator.

One can differentiate (8.44) to obtain integrals of polynomials times Gaussians, just as in (8.3). For example,

(8.45)

$$\iint [d\zeta^*\, d\zeta] \zeta_j^* \zeta_k e^{\zeta^* A \zeta} = \frac{\partial}{\partial \eta_j} \frac{\partial}{\partial \eta_k^*} \iint [d\zeta^*\, d\zeta] e^{\zeta^* A \zeta + i(\eta^* \zeta + \zeta^* \eta)} \Big|_{\eta = \eta^* = 0}$$
$$= (\det A) \frac{\partial}{\partial \eta_j} \frac{\partial}{\partial \eta_k^*} e^{-\eta^* \cdot A^{-1} \eta} \Big|_{\eta = \eta^* = 0} = (\det A)(A^{-1})_{jk}.$$

The verification of this is straightforward if one keeps a couple of things in mind: first, the derivatives $\partial/\partial\eta_j^*$ and $\partial/\partial\eta_k$ anticommute with each $\int d\zeta_l$ and $\int d\zeta_l^*$ and each ζ_l and ζ_l^*; second, the products $\zeta^* A \zeta$, $\eta^* \zeta$, and $\zeta^* \eta$ are of even degree, and their exponentials are sums of terms of even degree, so they all belong to the center of the Grassmann algebra.

We need one more simple extension of these ideas, really just a notational device: the notion of a *Grassmann spinor*. This is just a 4-tuple of complex Grassmann variables, construed as a column vector: $\eta = (\eta_1\ \eta_2\ \eta_3\ \eta_4)^T$. Its Dirac adjoint is $\overline{\eta} = \eta^\dagger \gamma^0 = (\eta_1^*\ \eta_2^*\ \eta_3^*\ \eta_4^*)\gamma^0$, and the corresponding "volume element" is then

$$[d\overline{\eta}\, d\eta] = [d\eta^\dagger\, d\eta] = d\eta_1^*\, d\eta_1 \cdots d\eta_4^*\, d\eta_4.$$

The interchangeability of $\overline{\eta}$ and $\eta^\dagger$ results from the fact that $\det \gamma^0 = 1$.

Fermionic functional integrals. The preceding calculations may look a bit strange and artificial, but at least they make perfectly good sense. But what we need for field theory is the infinite-dimensional analogue in which the discrete index j on ξ_j is replaced by a *continuous* space-time index $x = (t, \mathbf{x})$. That is, the analogue of the generating functional (8.21) for a free Dirac field is

$$\iint [d\overline{\psi}\, d\psi] \exp\left[i \int (\overline{\psi}(i\gamma^\mu \partial_\mu - m)\psi + \overline{\eta}\psi + \overline{\psi}\eta)\, d^4x\right]. \tag{8.46}$$

where the variables ψ and $\overline{\psi}$ and the source fields η and $\overline{\eta}$ are *Grassmann spinor-valued* functions on space-time — that is, 4-tuples of functions on $\mathbb{R}^4$, construed as column vectors in the case of ψ and η and row vectors in the case of $\overline{\psi}$ and $\overline{\eta}$, whose values all *anticommute* with each other.

At first glance, the assertion that a real meaning can be attached to such an expression must strike any mathematician, even one who can cheerfully accept the Bosonic analogue (8.21), as an act of astounding effrontery. We shall sketch a way to make more sense of it below, but for now let us persevere a little longer on a highly informal level. By analogy with (8.44), the value of this "Grassmann functional integral" should be

$$i^\infty [\det(i\gamma_\mu \partial^\mu - m)] \exp\left[i \int \overline{\eta}(-i\gamma_\mu \partial^\mu + m)^{-1} \eta\, d^4x\right]. \tag{8.47}$$

The ghastly factors in front of the exponential give the value of the integral when $\eta = \overline{\eta} = 0$, so they will drop out when we divide by that value as we did in the Bosonic case.[6] Inside the exponential, we at least know what $(-i\gamma_\mu \partial^\mu + m)^{-1}$ means: it is the integral operator whose kernel is the fundamental solution of $-i\gamma_\mu \partial^\mu + m$, i.e., the Dirac propagator $(i\gamma_\mu \partial^\mu + m)\Delta_F$.

Now let us take functional derivatives with respect to various $\eta(x)$ and $\overline{\eta}(x)$ and then set $\eta = \overline{\eta} = 0$. In (8.47), the latter step will give a result of 0 unless each $\delta/\delta\eta(x)$ is paired with some $\delta/\delta\overline{\eta}(y)$, and in any case it will collapse the exponential (which is an element of some immense Grassmann algebra) to the numerical constant 1. Thus, for example,

$$\frac{\delta^2}{\delta\eta(x)\delta\overline{\eta}(y)} e^{-i\int \overline{\eta}(i\gamma_\mu \partial^\mu - m)^{-1}\eta\, dx}\bigg|_{\eta=\overline{\eta}=0} = -i(i\gamma_\mu \partial^\mu + m)\Delta_F(x-y),$$

and similarly for higher derivatives. (One must take a little care here, as the derivatives $\delta/\delta\eta(x)$ and $\delta/\delta\overline{\eta}(y)$ anticommute.) On the other hand, functional differentiation of (8.46) yields functional integrals of products of fields; for example,

$$\frac{\delta^2}{\delta\eta(x)\delta\overline{\eta}(y)} \iint [d\overline{\psi}\, d\psi]\, e^{i\int (\overline{\psi}(i\gamma_\mu \partial^\mu - m)\psi + \overline{\eta}\psi + \overline{\psi}\eta)\, dx}\bigg|_{\eta=\overline{\eta}=0}$$
$$= \iint [d\overline{\psi}\, d\psi]\, \psi(x)\overline{\psi}(y) e^{i\int \overline{\psi}(i\gamma_\mu \partial^\mu - m)\psi\, dx}.$$

(Here again, the derivatives anticommute, as do the components of $\psi(x)$ and $\overline{\psi}(y)$. Both sides of this equation are 4×4 matrices.) One can argue, more or less as we did for Boson fields, that such functional integrals of products of Grassmann-valued

[6]Even though the determinant in (8.47) is ill-defined, the physicists find some significance in the fact that it occurs in the numerator, not in the denominator as in the Bosonic analogue: see Zee [**136**], §II.5.

classical fields are (up to a normalization constant) the vacuum expectation values of the corresponding products of Fermionic quantum fields. For example,

$$(8.48)\qquad \langle 0|\mathcal{T}(\psi_0(x)\overline{\psi}_0(y))|0\rangle = C\iint [d\overline{\psi}\,d\psi]\,\psi(x)\overline{\psi}(y)e^{i\int\overline{\psi}(i\gamma_\mu\partial^\mu - m)\psi\,dx},$$

where the ψ_0's on the left are quantum fields, i.e., operator-valued functions, and the ψ's on the right are "classical fields," i.e., Grassmann-valued functions. (A detailed discussion can be found in Weinberg [**129**], §9.5.) Putting this all together, we recover the fact that the vacuum expectation values of time-ordered products of free spinor fields are given by products of Dirac propagators, just as for Bosons.

We emphasize that that the integral in (8.48) is a *number*, not an element of some Grassmann algebra, because all of the Grassmann-valued variables have been integrated out.

Now one can deal with interactions involving both Fermions and Bosons. For QED, for example, the generating functional is similar to (8.30):

$$(8.49)\qquad Z[J,\eta,\overline{\eta}] = \iiint [dA][d\overline{\psi}\,d\psi]e^{i\int[-F_{\mu\nu}F^{\mu\nu}/4+\overline{\psi}(i\gamma^\mu\partial_\mu-m)\psi-eA_\mu\overline{\psi}\gamma^\mu\psi\,d^4x]}e^{i[J^\mu A_\mu+\overline{\eta}\psi+\eta\overline{\psi}]},$$

where η and $\overline{\eta}$ are spinor-valued functions and J is a vector-valued function such that $\partial_\mu J^\mu = 0$. (We have moved the $[dA]$ to the left merely so that it can keep company with the $[d\overline{\psi}\,d\psi]$.) Without the interaction term $e^{-ie\int A_\mu\overline{\psi}\gamma^\mu\psi\,d^4x}$, this integral can be evaluated in terms of the electron and photon propagators, using the Faddeev-Popov device to tame the integration over A. As with scalar fields, then, one can expand the interaction term in its power series, evaluate the resulting functional integrals by functional differentiation of (8.49), and thereby recover the Feynman calculus for QED. We refer to physics texts for further discussion of these matters; for example, a derivation of the Ward identities through functional integrals can be found in Peskin and Schroeder [**88**], §9.6.

Now let us examine how to make a little more mathematical sense out of these infinite-dimensional Grassmann integrals.

Grassmann-valued functions: infinite dimensions. First, the transition from a finite set of Grassmann-valued variables to a countable discrete set is not hard, although we must take a little care to construct a Grassmann algebra that is big enough to contain everything we need. We start with a separable Hilbert space $\mathcal{H}$ that is equipped with a decomposition $\mathcal{H} = \mathcal{H}_+ \oplus \mathcal{H}_-$ and an antiunitary involution $v \mapsto v^*$ that interchanges $\mathcal{H}_+$ and $\mathcal{H}_-$. We then work with an orthonormal basis $\{\zeta_n\}$ of $\mathcal{H}_+$ and the corresponding basis $\{\zeta_n^*\}$ for $\mathcal{H}_-$; these are the analogues of the ζ's and ζ^*'s in the finite-dimensional case. Let $\bigwedge_0\mathcal{H}$ be the *algebraic* exterior algebra over $\mathcal{H}$, that is, the set of *finite* linear combinations of exterior products of elements of $\mathcal{H}$. $\bigwedge_0\mathcal{H}$ is a pre-Hilbert space, and we could complete it to form the Hilbert exterior algebra over $\mathcal{H}$ (what we have previously called the Fermion Fock space over $\mathcal{H}$), but that does not contain infinite sums such as $\sum\zeta_n^*\zeta_n$, much less their exponentials, that are needed for Gaussian-type integrals. Rather, we consider the full algebraic dual $\bigwedge'\mathcal{H}$ of $\bigwedge_0\mathcal{H}$, which we call the *extended exterior algebra over* $\mathcal{H}$.

$\bigwedge'\mathcal{H}$ still has the structure of a Grassmann algebra. Indeed, each element of $\bigwedge_0\mathcal{H}$ defines a linear functional on $\bigwedge_0\mathcal{H}$ via the inner product, so we can consider

$\bigwedge_0 \mathcal{H}$ as a subspace of $\bigwedge' \mathcal{H}$, and the exterior product extends in a natural way from $\bigwedge_0 \mathcal{H}$ to $\bigwedge' \mathcal{H}$. Namely, if $\phi \in \bigwedge' \mathcal{H}$ and $\mathcal{M}$ is a finite-dimensional subspace of $\mathcal{H}$, let $\phi_{\mathcal{M}}$ be the restriction of ϕ to $\bigwedge \mathcal{M}$, which we consider via the inner product as an element of $\bigwedge \mathcal{M}$ itself. Then the exterior product of $\phi, \psi \in \bigwedge' \mathcal{H}$ is the unique $\chi \in \bigwedge' \mathcal{H}$ such that $\chi_{\mathcal{M}} = \phi_{\mathcal{M}}\psi_{\mathcal{M}}$ for all finite-dimensional $\mathcal{M}$. In other words, the set of finite-dimensional subspaces $\mathcal{M} \subset \mathcal{H}$ is directed in one way by inclusion and in the opposite way by orthogonal projection; $\bigwedge_0 \mathcal{H}$ and $\bigwedge' \mathcal{H}$ are the corresponding inductive and projective limits of the finite-dimensional algebras $\bigwedge \mathcal{M}$.

The Berezin integral is defined on (a subset of) $\bigwedge' \mathcal{H}$ as follows. Let $\mathcal{M}_n$ be the subspace spanned by the basis vectors $\zeta_1, \ldots, \zeta_n, \zeta_1^*, \ldots, \zeta_n^*$. If $\phi \in \bigwedge' \mathcal{H}$, and $\phi_{\mathcal{M}_n}$ is as in the preceding paragraph, we define

$$\iint [d\zeta^* \, d\zeta] \, \phi = \lim_{n \to \infty} \iint \cdots \iint d\zeta_n^* \, d\zeta_n \cdots d\zeta_1^* \, d\zeta_1 \, \phi_{\mathcal{M}_n}, \tag{8.50}$$

provided that the limit exists.

For example, let us abbreviate $\sum \zeta_n^* \zeta_n$ by $\zeta^*\zeta$. This sum is a well-defined element of $\bigwedge' \mathcal{H}$ — more specifically, of the degree-2 subspace with respect to its natural gradation. Hence $e^{-\zeta^*\zeta} = \sum (-\zeta^*\zeta)^n/n!$ is a well-defined element of $\bigwedge' \mathcal{H}$ (the nth term belonging to the degree-$2n$ subspace), and we have

$$\iint [d\zeta^* \, d\zeta] e^{\zeta^*\zeta} = 1, \tag{8.51}$$

as all the finite-dimensional approximants are equal to 1 by (8.43).

The Berezin integral is actually independent of the choice of orthonormal basis ζ_n, ζ_n^*, and its definition can be formulated in a basis-independent fashion with a little more work. (The limit as $n \to \infty$ in (8.50) is replaced by a limit over the directed set of finite-dimensional subspaces of $\mathcal{H}$.) For a lucid and detailed exposition of this theory, including the calculation of a variety of Berezin integrals of Gaussian type, we refer the reader to Robinson [**98**].

The transition from a discrete family of Grassmann-valued variables to a continuous one parametrized by points in space-time is not a matter of redoing this with an uncountable index set but rather of reinterpreting it after adding a little more structure. The ideas that follow are due to Berezin [**8**], but we shall recast them in a more concrete form. To begin with, recall that the underlying Hilbert space $\mathcal{H}$ is equipped with a decomposition $\mathcal{H} = \mathcal{H}_+ \oplus \mathcal{H}_-$ and an anti-unitary involution that interchanges $\mathcal{H}_+$ with $\mathcal{H}_-$. We now take $\mathcal{H}_+$ to be $L^2(\mathbb{R}^4)$. The Grassmann variables parametrized by $x \in \mathbb{R}^4$, however, are going to be the delta-functions $\delta_x(y) = \delta(y - x)$, so we need to enlarge $L^2(\mathbb{R}^4)$ to a space that contains delta-functions. A suitable negative-order Sobolev space will do nicely; say, $\widetilde{\mathcal{H}}_+ = H_{-s}(\mathbb{R}^4)$ (see Folland [**45**] or [**47**]) where s is some fixed number greater than 2. We take $\mathcal{H}_-$ to be another copy of $L^2(\mathbb{R}^4)$, accompanied by its enlargement $\widetilde{\mathcal{H}}_- = H_{-s}(\mathbb{R}^4)$. The anti-involution on $\mathcal{H}$ is given by $(f, g)^* = (g^*, f^*)$, where the stars on the right denote complex conjugation.

We denote the copies of the delta-function δ_x in $\widetilde{\mathcal{H}}_+$ and $\widetilde{\mathcal{H}}_-$ by $\zeta(x)$ and $\zeta^*(x)$. If f is a continuous function with compact support on $\mathbb{R}^4$, $\int f(x)\zeta(x)\,dx = \int f(x)\delta_x\,dx$ is a perfectly good $\widetilde{\mathcal{H}}_+$-valued integral, and its value is f itself, considered as an element of $\mathcal{H}_+$. It is therefore not much of a stretch to write $f = \int f(x)\zeta(x)\,dx$ for any $f \in \mathcal{H}_+$, and likewise $f = \int f(x)\zeta^*(x)\,dx$ for $f \in \mathcal{H}_-$.

(This is just the expansion of f with respect to the "continuous orthonormal basis" of generalized eigenvectors δ_x for the position operators on $\mathbb{R}^4$, as in §3.2. See Gelfand-Vilenkin [??, vol. 4] for a general framework in which such things may be discussed rigorously.)

Now we pass to the exterior algebras $\bigwedge_0 \mathcal{H}$ and $\bigwedge_0 \widetilde{\mathcal{H}}$. In the latter we have the exterior products $\zeta(x)\zeta(y)$, $\zeta(x)\zeta^*(y)$, $\zeta^*(x)\zeta^*(y)$, $\zeta(x)\zeta(y)\zeta(z)$, etc., and these may be used to express elements of the former in terms of integrals as above. For example, the exterior product of $f \in \mathcal{H}_+$ and $g \in \mathcal{H}_+$ is

(8.52)
$$\left(\int f(x)\zeta(x)\,dx\right)\left(\int g(x)\zeta(x)\,dx\right) = \iint f(x)g(y)\zeta(x)\zeta(y)\,dx\,dy$$
$$= \iint \frac{f(x)g(y) - f(y)g(x)}{2}\zeta(x)\zeta(y)\,dx\,dy.$$

This notation makes clear the distinction between f considered as an element of $\mathcal{H}_+$ and f considered as an element of $\mathcal{H}_-$, as well as between the pointwise product of two functions f and g and the anticommutative product in the Grassmann algebra.

The Hilbert-space completion $\bigwedge^2 \mathcal{H}_+$ of $\bigwedge_0^2 \mathcal{H}_+$ can be identified with the space of skew-symmetric functions in $L^2(\mathbb{R}^4 \times \mathbb{R}^4)$; the correspondence is given by

$$F(x,y) \longleftrightarrow \iint F(x,y)\zeta(x)\zeta(y)\,dx\,dy.$$

This space in turn can be identified with the space of *antilinear* Hilbert-Schmidt operators T on $L^2(\mathbb{R}^4)$ that are skew-symmetric in the sense that $\langle Th_1|h_2\rangle = -\langle Th_2|h_1\rangle$; the correspondence is given by

$$F \longleftrightarrow T_F, \quad \text{where} \quad T_F h(x) = \int F(x,y)h(y)^*\,dy.$$

(This is a specialization of a general construction of the tensor product of two Hilbert spaces, as in Folland [**44**].) In particular, the F and T_F corresponding to the exterior product in (8.52) are

$$F(x,y) = \tfrac{1}{2}[f(x)g(y) - g(x)f(y)], \qquad T_F h = \tfrac{1}{2}[\langle h|g\rangle f - \langle h|f\rangle g].$$

More generally, elements of the Hilbert space $\bigwedge^2 \mathcal{H}$ correspond to 2×2 matrices of L^2 functions $(F_{ij})_{i,j=1}^2$ with $F_{ij}(x,y) = -F_{ji}(y,x)$, or to skew-symmetric antilinear Hilbert-Schmidt operators on $\mathcal{H}$. If $F = (F_{ij})$ is such a matrix, the corresponding element of $\bigwedge^2 \mathcal{H}$ is

$$\iint [F_{11}(x,y)\zeta(x)\zeta(y) + F_{12}(x,y)\zeta(x)\zeta^*(y) + F_{21}(x,y)\zeta^*(x)\zeta(y)$$
$$+ F_{22}(x,y)\zeta^*(x)\zeta^*(y)]\,dx\,dy$$
$$= \iint [F_{11}(x,y)\zeta(x)\zeta(y) + 2F_{21}(x,y)\zeta^*(x)\zeta(y) + F_{22}(x,y)\zeta^*(x)\zeta^*(y)]\,dx\,dy,$$

and the corresponding operator T_F is

$$T_F(h_1,h_2) = (T_{F_{11}}h_1 + T_{F_{12}}h_2,\, T_{F_{21}}h_1 + T_{F_{22}}h_2).$$

In particular, the F and T_F corresponding to the exterior product of (f_1,f_2) and (g_1,g_2) (elements of $\mathcal{H}_+ \oplus \mathcal{H}_-$) are given by

$$F_{ij}(x,y) = \tfrac{1}{2}[f_i(x)g_j(y) - f_j(y)g_i(x)], \qquad T_F(h_1,h_2) = (H_1,H_2), \tag{8.53}$$

where

$$
\begin{aligned}
H_1 &= [\langle h_1|g_1\rangle + \langle h_2|g_2\rangle]f_1 - [\langle h_1|f_1\rangle + \langle h_2|f_2\rangle]g_1, \\
H_2 &= [\langle h_1|g_1\rangle + \langle h_2|g_2\rangle]f_2 - [\langle h_1|f_1\rangle + \langle h_2|f_2\rangle]g_2.
\end{aligned} \tag{8.54}
$$

Analogous, but more complicated, formulas hold for products of order > 2.

We can also express many elements of the *extended* exterior algebra $\bigwedge' \mathcal{H}$ as "generalized linear combinations of the continuous family of Grassmann variables $\zeta(x), \zeta^*(x)$" by taking the F's here to be *generalized* functions — say, tempered distributions. Specifically, every bounded antilinear operator on $L^2(\mathbb{R}^4)$ can be expressed as complex conjugation composed with an integral operator with a tempered distribution kernel, by the tempered version of the Schwartz kernel theorem. The correspondence in the preceding paragraph extends to give a correspondence between 2×2 matrices of distributions arising in this fashion with certain degree-2 elements of $\bigwedge' \mathcal{H}$. In particular, let $\{\zeta_n\}$ be an orthonormal basis for $\mathcal{H}_+$, and consider $\zeta^*\zeta = \sum \zeta_n^*\zeta_n \in \bigwedge' \mathcal{H}$. By (8.53) and (8.54) (with $(f_1, f_2) = (0, \zeta_n^*)$ and $(g_1, g_2) = (\zeta_n, 0)$), the corresponding operator is

$$
T(h_1, h_2) = \frac{1}{2}\left[\sum \langle h_1|\zeta_n\rangle \zeta_n^* - \sum \langle h_2|\zeta_n^*\rangle \zeta_n\right] = \frac{1}{2}(-h_2^*, h_1^*),
$$

and hence the corresponding matrix F is

$$
F(x,y) = \frac{1}{2}\begin{pmatrix} 0 & -\delta(x-y) \\ \delta(x-y) & 0 \end{pmatrix}.
$$

In other words,

$$
\sum \zeta_n^*\zeta_n = \iint \delta(x-y)\frac{\zeta^*(x)\zeta(y) - \zeta(x)\zeta^*(y)}{2}\,dx\,dy = \int \zeta^*(x)\zeta(x)\,dx. \tag{8.55}
$$

The "functional integral" in this context is simply the Berezin integral (8.50). That is, one interprets suitable "(generalized) functions of the Grassmann variables $\zeta(x), \zeta^*(x)$" as elements of the extended Grassmann algebra $\bigwedge' \mathcal{H}$ and integrates them by (8.50). For example, in view of (8.55), the basic Gaussian integral (8.51) can be written as

$$
\iint [d\zeta\, d\zeta^*] e^{-\int \zeta^*(x)\zeta(x)\,dx}.
$$

Suitable generalizations of this then lead to physically interesting integrals of the form (8.46) — on the formal level, of course, as the latter integrals have to be taken with some grains of salt even when one understands how to set them up.

8.5. Afterword: Gaussian processes

After the fantasies of the preceding two sections, the reader may feel a need to be brought back to reality with a dose of honest mathematics. We therefore conclude this chapter with a brief discussion of the connection between the Bosonic functional integrals of §8.3 and some genuine integrals with respect to measures on function spaces.

In §8.1 we saw that if one makes a Wick rotation to imaginary time, the functional integral for a particle moving in a potential turns into a well-defined integral with respect to Wiener measure (the Feynman-Kac formula). Similarly, if one starts with the "measure"

$$
C \exp\left[i \int ((\partial\phi)^2 - m^2\phi^2) d^4x\right] \; [d\phi]
$$

that arises from the Lagrangian for a free scalar field, one can replace x^0 by $-ix^0$ to obtain

$$C\exp\left[-\int(|\nabla\phi|^2+m^2\phi^2)\,d^4x\right]\,[d\phi], \tag{8.56}$$

where ∇ means the 4-dimensional Euclidean gradient. This clearly has the appearance of an infinite-dimensional analogue of a real Gaussian measure, and it is not hard to give it a precise mathematical meaning.

Let us give some definitions. A *Gaussian measure* on $\mathbb{R}^n$ (with mean 0) is a Borel probability measure μ on $\mathbb{R}^n$ whose Fourier transform is of the form $\widehat{\mu}(y) = e^{-Ay\cdot y/2}$ where A is a real positive semi-definite matrix. Thus, if A is positive definite, we have $d\mu(x) = [(2\pi)^n \det A]^{-1/2}e^{-A^{-1}x\cdot x/2}\,dx$, whereas if the nullspace $\mathcal{N}(A)$ is nontrivial, we have $d\mu(x) = [(2\pi)^k \det B]^{-1/2}e^{-B^{-1}x\cdot x/2}\,d\sigma(x)$ where σ is surface measure on $\mathcal{N}(A)^\perp$, $B = A|\mathcal{N}(A)^\perp$, and $k = \dim \mathcal{N}(A)^\perp$. The matrix A is called the *covariance* of μ; we observe that $A_{ij} = \int x_i x_j\,d\mu(x)$ (differentiate the formula $\int e^{-ix\cdot y}d\mu(x) = e^{-Ay\cdot y}/2$ with respect to y_i and y_j, then set $y=0$).

A *Gaussian process* (with mean 0) indexed by a set I is a probability space $(\Omega, \mathcal{B}, \mu)$ and a family $\{X_i : i \in I\}$ of random variables on Ω such that for any finite subset $\{i_1, \ldots, i_n\}$ of I, the joint distribution of $X_{i_1}, \ldots, X_{i_n}$ is Gaussian — that is, setting $\mathbf{X} = (X_{i_1}, \ldots, X_{i_n}) : \Omega \to \mathbb{R}^n$, the measure $\mathbf{X}_*\mu(E) = \mu(\mathbf{X}^{-1}(E))$ on $\mathbb{R}^n$ is Gaussian. The matrix $C = (C_{ij})_{i,j\in I}$ defined by $C_{ij} = \int X_i X_j\,d\mu$ is called the *covariance* of the process; the covariance matrices of the Gaussian measures $\mathbf{X}_*\mu$ on $\mathbb{R}^n$ are all submatrices of it.

Here are two examples. (1) Let B be a real positive definite $n\times n$ matrix, let $\Omega = \mathbb{R}^n$, $d\mu(x) = \sqrt{\det B/(2\pi)^n}e^{-Bx\cdot x}\,d^n x$, and $X_i(x) = x_i$. Then $\{X_i : 1 \le i \le n\}$ is a Gaussian process with covariance $C = B^{-1}$. (2) Let μ be Wiener measure on $C([0,\infty))$ and $X_t(\omega) = \omega(t)$ for $t \ge 0$, $\omega \in C([0,\infty))$. Then $\{X_t : t \ge 0\}$ is a Gaussian process with covariance $C_{st} = \min(s,t)$.

Next, let $\mathcal{H}$ be a *real* Hilbert space. A *Gaussian process associated to* $\mathcal{H}$ is a Gaussian process $\{X_f : f \in \mathcal{H}\}$ on a probability space $(\Omega, \mathcal{B}, \mu)$, indexed by $\mathcal{H}$, such that the correspondence $f \mapsto X_f$ is linear and isometric from $\mathcal{H}$ to $L^2(\mu)$. Note that the requirement of isometry means that the covariance is given by the inner product: $C_{fg} = \langle f|g\rangle$. For any $\mathcal{H}$, Gaussian processes associated to $\mathcal{H}$ exist, and they are all isomorphic in the abstract probabilistic sense subject to the technical requirement that the random variables of the form $G(X_{f_1}, \ldots, X_{f_n})$ with G bounded and Borel be dense in $L^2(\mu)$.[7] Hence we speak of *the* Gaussian process associated to $\mathcal{H}$ and refer to its various concrete realizations as *models*. We briefly describe two important models.

Model I: Let $\{e_j\}_{j\in J}$ be an orthonormal basis for $\mathcal{H}$, and let $\Omega = \overline{\mathbb{R}}^J$ where $\overline{\mathbb{R}}$ is the one-point compactification of $\mathbb{R}$. Put the measure $(2\pi)^{-1/2}e^{-x^2/2}\,dx$ on each copy of $\overline{\mathbb{R}}$, let μ be the product measure on the compact Hausdorff space Ω, and let X_{e_i} be the ith coordinate function on Ω (redefined to be 0 at the point at infinity). It is easy to check that the X_{e_i}'s are an orthonormal set in $L^2(\mu)$, so the

[7] To be precise, suppose $(\Omega, \mathcal{B}, \mu, \{X_f\})$ and $(\Omega', \mathcal{B}', \mu', \{X'_f\})$ are two such processes. Let $\mathcal{N}$ be the σ-ideal in $\mathcal{B}$ of sets of μ-measure zero, and let $q : \mathcal{B} \to \mathcal{B}/\mathcal{N}$ be the quotient map; likewise $\mathcal{N}'$ and q'. Then there is a σ-algebra isomorphism $\Phi : \mathcal{B}/\mathcal{N} \to \mathcal{B}'/\mathcal{N}'$ such that $\Phi(q(X_f^{-1}(E))) = q'((X'_f)^{-1}(E))$ for all $f \in \mathcal{H}$ and all Borel $E \subset \mathbb{R}$.

map $e_i \mapsto X_{e_i}$ extends by linearity and continuity to an isometry $f \mapsto X_f$ from $\mathcal{H}$ into $L^2(\mu)$.

Model II: Suppose $\mathcal{H}$ is a space of real-valued functions or distributions on $\mathbb{R}^d$ in which the real-valued Schwartz class $\mathcal{S} = \mathcal{S}_{\mathbb{R}}(\mathbb{R}^d)$ is a dense subspace. We take Ω to be $\mathcal{S}'$, the space of real-valued tempered distributions, and $\mathcal{B}$ to be the σ-algebra generated by the weak-* topology on $\mathcal{S}'$. The map $E(f) = \exp(-\frac{1}{2}\|f\|^2_{\mathcal{H}})$ is a positive definite function on $\mathcal{S}$ (i.e., the matrix $(E(f_i - f_j))$ is positive definite for any finite set $\{f_1, \dots, f_n\} \subset \mathcal{S}$), so by Minlos's extension of Bochner's theorem (see Simon [**112**] for a very nice proof) there is a measure μ on $\mathcal{S}'$ such that

$$\int_{\mathcal{S}'} e^{iT(f)}\, d\mu(T) = \exp(-\tfrac{1}{2}\|f\|^2_{\mathcal{H}}), \qquad f \in \mathcal{S}. \tag{8.57}$$

For $f \in \mathcal{S}$, let X_f be the fth coordinate function on $\mathcal{S}'$, $X_f(T) = T(f)$. Then $\int X_f^2\, d\mu = \|f\|^2_{\mathcal{H}}$, as one sees by replacing f by tf ($t \in \mathbb{R}$) in (8.57) and comparing the t^2 terms in the Taylor expansion of both sides. It follows that the map $f \mapsto X_f$ extends to an isometry from $\mathcal{H}$ into $L^2(\mu)$.

This gives us what we need to make sense of the expression (8.56). Since $\int |\nabla\phi|^2\, dx = -\int \phi\nabla^2\phi\, dx$ (assuming ϕ vanishes at infinity), (8.56) is formally a Gaussian measure with inverse covariance $-\nabla^2 + m^2$. Thus, let us take $\mathcal{H}$ to be the real Sobolev space of order -1 on $\mathbb{R}^4$,

$$\mathcal{H} = \left\{ f \in \mathcal{S}'(\mathbb{R}^4) : f \text{ real valued}, \int \frac{|\widehat{f}(k)|^2}{|k|^2 + m^2} \frac{d^4k}{(2\pi)^4} < \infty \right\},$$

with inner product

$$\langle f|g\rangle_{\mathcal{H}} = \int \frac{\widehat{f}(k)^*\widehat{g}(k)}{|k|^2 + m^2} \frac{d^4k}{(2\pi)^4} = \int f(x)(-\nabla^2 + m^2)^{-1} g(x)\, d^4x.$$

Then the Minlos model of the Gaussian process associated to $\mathcal{H}$ gives a measure μ on $\mathcal{S}'(\mathbb{R}^4)$ with covariance $(-\nabla^2 + m^2)^{-1}$. Formally $d\mu(\phi)$ is precisely of the form (8.56) where C is the (infinite) normalization constant to make μ a probability measure, and the integral in (8.56) makes sense not for all $\phi \in \mathcal{S}'(\mathbb{R}^4)$ but only on a dense subspace, namely, the Sobolev space of order 1. This may still seem a little mysterious if one has not seen such things before, but we are now in a well-explored region of probability theory.

The nth-order moments of the measure μ define a tempered distribution S_n on $\mathbb{R}^{4n}$: for $f_1, \dots, f_n \in \mathcal{S}(\mathbb{R}^4)$,

$$S_n(f_1 \otimes \cdots \otimes f_n) = \int_{C\mathcal{S}'} X_{f_1} \cdots X_{f_n}\, d\mu = \int_{\mathcal{S}'} \phi(f_1) \cdots \phi(f_n)\, d\mu(\phi).$$

In fact, S_n is a locally integrable function, essentially our old friend (8.25) transferred to the Euclidean region:

$$S(x_1, \dots, x_n) = \sum_{\text{pairings}} \Delta_E(x_{j_1} - x_{j_2}) \cdots \Delta_E(x_{j_{2n-1}} - x_{j_{2n}}),$$

where Δ_E is the Euclidean Green's function, i.e., the fundamental solution of $(-\nabla^2 + m^2)^{-1}$. This function has an analytic continuation to a region in $\mathbb{C}^{4n}$ whose boundary includes the noncoincident points of the Minkowski region ($x_j \neq x_k$,

$x_j^0 \in i\mathbb{R}$), and the boundary values of S there give the vacuum expectation values of the free quantum fields, $\langle 0|T(\phi_0(x_1)\cdots\phi_0(x_n))|0\rangle$, according to (8.24) and (8.25). In this way we can recover the field theory from the measure μ.

What we have sketched here is the way to formulate a free field theory in terms of Gaussian processes. The real point, however, is that this is the most fruitful approach to the rigorous construction of *interacting* quantum fields. In brief, the idea is to find a measure μ on $\mathcal{S}'(\mathbb{R}^4)$ whose moments define a suitable set of "Schwinger functions" S (analytic continuations of vacuum expectation values), and then to apply "reconstruction theorems" to recover the field from the Schwinger functions. As we indicated in §5.5, this program has been carried out for some field theories in space-time dimensions 2 and 3, but it remains a major open problem in space-time dimension 4. For the results in dimensions 2 and 3, we refer the reader to the brief introductory articles by Brydges [**14**] and Federbush [**31**] and the more extensive expositions in Glimm and Jaffe [**54**], Simon [**111**], and Rosen [**100**], as well as the references given in these sources. Jaffe and Witten [**66**] gives a good account of the current state of affairs along with recommendations for avenues of future research in dimension 4. Finally, Simon [**112**] is a good reference for applications of rigorous functional integrals to other areas of quantum physics.

CHAPTER 9

Gauge Field Theories

Nature uses only the longest threads to weave her patterns, so each small piece of the fabric reveals the organization of the entire tapestry.
—Richard Feynman ([**36**], p. 34)

This chapter is a very brief introduction to gauge field theories. To do a proper job of developing this subject, one must deal with some difficult issues concerning renormalization and the use of functional integrals, but we shall mention them only briefly. In fact, there is very little actual quantum mechanics in this chapter. We shall work with gauge fields mostly on the pre-quantum level — we say "pre-quantum" rather than "classical" because non-Abelian gauge fields do not correspond to anything in classical physics — with the Lagrangian at center stage. One can learn a remarkable amount about quantum gauge fields simply by first asking what their Lagrangians must look like and then letting those Lagrangians tell their own story.

9.1. Local symmetries and gauge fields

The notion of "gauge invariance" got off to an inauspicious start in 1918 when Hermann Weyl used it in an attempt to build a unified field theory, that is, a unified theory of gravity and electromagnetism. Weyl's idea was simple and attractive. According to general relativity, space-time is modeled by a 4-manifold M equipped with a Lorentz metric g_0, but this presupposes the choice of unit of length (or time). This choice is of no consequence as long as the choice of a unit at one point $p \in M$ determines the choice at every other point; but if one is allowed to choose a unit of length independently at every point, one has not just one Lorentz metric g_0 but a whole conformal class of them — namely, the class $\mathcal{G}$ of all metrics $g = e^{\chi} g_0$ where χ is a (smooth) positive function on M. Equivalently, one has the line bundle L over M whose fiber over $p \in M$ is the set $\{g(p) : g \in \mathcal{G}\}$ of metrics on T_pM arising from $\mathcal{G}$; $\mathcal{G}$ is then the space of sections of L. In order to compare lengths at two different points p and q, one needs to impose a *connection* on L, which allows one to transport an element of L_p along a given curve C from p to q; this amounts to the choice of a mapping Φ from $\mathcal{G}$ to the space of 1-forms on M such that $\Phi(e^{\chi}g) = \Phi(g) - d\chi$. If $\gamma : [0,1] \to M$ is a smooth curve with $\gamma(0) = p$ and $\gamma(1) = q$, the transport of $g_p \in L_p$ from p to q along γ is then $\exp(\int_\gamma \Phi(g))g(q)$, where $g(\cdot)$ is any element of $\mathcal{G}$ such that $g(p) = g_p$.

Weyl suggested that the 1-forms $\Phi(g)$ should be interpreted as electromagnetic potentials; that is, their common exterior derivative $F = d\Phi(g)$ should be the electromagnetic field. Gravitation and electromagnetism would then be tied together by the fact that the choice of potential (a 1-form ϕ such that $d\phi = F$) is tied to the choice of metric $g \in \mathcal{G}$ — that is, to the choice of a length scale or *gauge* at each point — so that a "gauge transformation" $\phi \mapsto \phi - d\chi$ of the potential is accompanied by the literal gauge transformation $g \mapsto e^{\chi} g$. (This is the origin of the various uses of the word "gauge" in gauge field theories!) Weyl's theory made its way into several books on general relativity, but it was a failure as a physical theory for one simple reason: It predicts that if one starts with two identical measuring rods at p and transports them to q along different paths that go through regions with different electromagnetic fields, in general they will no longer have the same length on arrival at q, a phenomenon for which there is not a shred of experimental evidence. (Weyl's mathematics, however, is perfectly sound, and it still has echoes in the study of conformal geometry — usually with Riemannian metrics instead of Lorentzian ones. More details can be found in Folland [**42**].)

Weyl was therefore delighted when quantum mechanics came along a decade later and he found that with the simple substitution of $e^{i\chi}$ for e^{χ}, his "principle of gauge invariance" could be revived as a bond not between electromagnetism and gravity but between electromagnetism and matter fields. This point has already been discussed in §4.2: the Dirac equation $(i\partial_\mu - eA_\mu)\gamma^\mu\psi = m\psi$ is invariant under the simultaneous transformations $\psi \mapsto e^{ie\chi}\psi$, $A_\mu \to A_\mu - \partial_\mu\chi$. More to the point from the perspective of field theory, the Lagrangian $\mathcal{L} = \overline{\psi}(i\partial_\mu - eA_\mu)\gamma^\mu\psi - m\overline{\psi}\psi$ is invariant under the same pair of transformations.

In fact, one can take the principle of gauge invariance as a starting point and reason one's way to the theory of electromagnetism. One starts with the free Dirac Lagrangian, $\mathcal{L}_0 = \overline{\psi}(i\partial_\mu\gamma^\mu - m)\psi$, observing that it is invariant under the transformation $\psi \mapsto e^{i\chi}\psi$ when χ is a constant, and asks how it must be modified in order to be invariant under the transformation $\psi \mapsto e^{i\chi}\psi$ when χ is an arbitrary smooth function. The catch, of course, is that the derivative ∂_μ does not commute with multiplication by $e^{i\chi}$ when χ is nonconstant. One way of looking at this is that if one knows the field ψ only up to a phase factor at each point, there is no way to compare its values at neighboring points in order to form a difference quotient. Rather, one must think of ψ as a section of a vector bundle V (whose fiber is the space of spinors) associated to the principal bundle $P = U(1) \times \mathbb{R}^4$ over the space-time $\mathbb{R}^4$, and replace the ordinary derivative ∂_μ by the covariant derivative D_μ with respect to a connection on P. We shall explain the notion of connection in more detail a little later, but for now it is enough to think of a connection as a map Φ that assigns to each global section σ of P a 1-form $\Phi(\sigma) = A_\mu\, dx^\mu$ on $\mathbb{R}^4$ such that $\Phi(e^{-i\chi}\sigma) = \Phi(\sigma) - d\chi$. The section σ also determines an identification of sections of V with spinor-valued functions ψ on $\mathbb{R}^4$, and when this identification is made, the covariant derivative takes the form $D_\mu = \partial_\mu + iA_\mu$. Replacing σ by $e^{-i\chi}\sigma$ amounts to replacing ψ by $e^{i\chi}\psi$ and A_μ by $A_\mu - \partial_\mu\chi$, so this definition is consistent:

$$[\partial_\mu + i(A_\mu - \partial_\mu\chi)](e^{i\chi}\psi) = e^{i\chi}[\partial_\mu + iA_\mu](\psi).$$

We therefore obtain a gauge-invariant Lagrangian by taking

$$\mathcal{L} = \overline{\psi}(i\partial_\mu - A_\mu - m)\psi,$$

whose physical interpretation is that the matter field ψ is coupled to a "gauge field" A_μ. To complete the picture, we need to add in a free-field Lagrangian for A_μ. The requirement that it be invariant under both Lorentz and gauge transformations narrows down the possibilities enormously; in fact, the only ones that are at most quadratic in the field A_μ and involve derivatives of order at most one are constant multiples of $F_{\mu\nu}F^{\mu\nu}$, where $F_{\mu\nu} = \partial_\mu A_\nu - \partial_\nu A_\mu$ as usual.[1] In particular, gauge invariance forbids a term proportional to $A_\mu A^\mu$ (which Lorentz invariance would allow), so the quanta of the gauge field are massless. Thus, up to adjustment of constant factors (the coupling constant e and the factor of $-\frac{1}{4}$ for $F_{\mu\nu}F^{\mu\nu}$), we have arrived at the standard Lagrangian for electrodynamics.

The construction just outlined can be redone in a much more general setting, with the group $U(1)$ of complex numbers of modulus 1 replaced by a more general Lie group. (The amount of Lie theory we need is quite modest; Hall [**61**] is a good source for more information.) In what follows, we work at the unquantized level, where a "field" is a function (whose values may be scalars, spinors, vectors, ...) on space-time rather than an operator-valued distribution. The ingredients are as follows:

i. an n-tuple $\Phi = (\phi_1, \dots, \phi_n)$ of fields of the same type (scalar, spinor, ...) together with a Lagrangian $\mathfrak{L}_0(\Phi, \partial\Phi)$;
ii. a compact subgroup G of $GL(n, \mathbb{C})$ that acts on the fields (or rather the space of their linear combinations) in the obvious way — $(g \cdot \Phi)_i = \sum_j g_i^j \phi_j$ for $g = (g_i^j) \in G$ — subject to the condition that the Lagrangian be invariant: $\mathfrak{L}_0(g \cdot \Phi,\, g \cdot \partial\Phi) = \mathfrak{L}_0(\Phi, \partial\Phi)$.

We say that G is a group of *global symmetries* of the theory described by the fields Φ. The compactness of G is not a strict requirement at the outset, but it will soon play a role, and it is always satisfied in practice.

We now promote the symmetries from global to *local* ones: that is, we consider not just the transformations $\Phi \mapsto g \cdot \Phi$ but rather $\Phi(x) \mapsto g(x) \cdot \Phi(x)$ where $g(x)$ is an arbitrary smooth G-valued function on space-time. How must the Lagrangian be modified in order to remain invariant under this larger group of transformations? As before, the point is that $\partial_\mu(g(x) \cdot \Phi(x)) = g(x) \cdot \partial_\mu \Phi(x) + \partial_\mu g(x) \cdot \Phi(x)$, and we have to find a way to cancel the second term on the right. Note that $\partial_\mu g(x)$ is an element not of G but of the tangent space to G at $g(x)$; thus $(\partial_\mu g(x))g(x)^{-1}$ belongs to the Lie algebra $\mathfrak{g}$ of G (the tangent space to G at the identity, or equivalently, the space of $n \times n$ matrices X such that $e^X \in G$).

We shall describe the process first in physicists' language and then translate it into the language of differential geometry. In what follows, g will denote a G-valued function on $\mathbb{R}^4$, and we set

$$\Phi^g(x) = g(x) \cdot \Phi(x).$$

As before, the idea is to replace the derivative ∂_μ by a "covariant derivative"

$$D_\mu = \partial_\mu + iA_\mu,$$

where iA_μ is (for each μ) a $\mathfrak{g}$-valued function on $\mathbb{R}^4$. (The factor of i is a standard convention in the physics literature; it is intended to preserve the analogy with electromagnetism, where $G = U(1)$ and hence $\mathfrak{g} = i\mathbb{R}$.) We demand that when we make the transformation $\Phi \mapsto \Phi^g$, A_μ should simultaneously transform to another $\mathfrak{g}$-valued function A_μ^g, and D_μ likewise to $D_\mu^g = \partial_\mu + iA_\mu^g$, in such a way that

[1]If we required invariance only under the restricted Lorentz group $SO^\uparrow(1,3)$, there would be another possibility: $\epsilon^{\mu\nu\rho\sigma}F_{\mu\nu}F_{\rho\sigma}$, where $\epsilon^{\mu\nu\rho\sigma}$ is the sign of the permutation taking (μ, ν, ρ, σ) to $(0, 1, 2, 3)$ if the indices are all distinct and is 0 otherwise.

$D_\mu^g \Phi^g = (D_\mu \Phi)^g$. But

$$(D_\mu\Phi)^g = g\cdot D_\mu\Phi = g\cdot(\partial_\mu\Phi + iA_\mu\cdot\Phi) = \partial_\mu(g\cdot\Phi) - \partial_\mu g\cdot\Phi + i(gA_\mu)\cdot\Phi$$
$$= (\partial_\mu + igA_\mu g^{-1} - (\partial_\mu g)g^{-1})\Phi^g.$$

Hence the desired transformation law for A_μ is

$$A_\mu^g = gA_\mu g^{-1} + i(\partial_\mu g)g^{-1}, \tag{9.1}$$

where the products on the right are simply matrix multiplication.

The modified Lagrangian $\mathfrak{L}_0(\Phi, D\Phi)$ is now invariant under the *gauge transformations* $\Phi \mapsto \Phi^g$, $D \mapsto D^g$. The result is going to be a theory in which the n fields ϕ_j are coupled to $N = \dim(\mathfrak{g})$ vector fields A_μ^k, the components of the $\mathfrak{g}$-valued fields A_μ, which are known as *gauge fields*; the interactions come from the products of the $D_\mu\phi_j$ with each other or with the ϕ_j in $\mathfrak{L}_0(\Phi, D\Phi)$. To complete the picture, we need to add in a free-field Lagrangian for the gauge fields A_μ, which should be invariant under Lorentz transformations and under the gauge transformations (9.1). The simplest way to generate such a quantity is to consider the commutator of two covariant derivatives. We have

$$D_\mu D_\nu\Phi = (\partial_\mu + iA_\mu)(\partial_\nu + iA_\nu)\Phi = \partial_\mu\partial_\nu\Phi + iA_\mu\partial_\nu\Phi + iA_\nu\partial_\mu\Phi + i(\partial_\mu A_\nu)\Phi - A_\mu A_\nu\Phi,$$

so

$$[D_\mu, D_\nu]\Phi = i(\partial_\mu A_\nu - \partial_\nu A_\mu + i[A_\mu, A_\nu])\Phi = iF_{\mu\nu}\Phi$$

where $F_{\mu\nu}$ is the $\mathfrak{g}$-valued function

$$F_{\mu\nu} = \partial_\mu A_\nu - \partial_\nu A_\mu + i[A_\mu, A_\nu]. \tag{9.2}$$

The covariance condition $D_\mu^g D_\nu^g \Phi^g = (D_\mu D_\nu \Phi)^g$, or equivalently $D_\mu^g D_\nu^g(g\cdot\Phi) = g\cdot D_\mu D_\nu\Phi$, implies that $F_{\mu\nu}$ (for each μ and ν) transforms under gauge transformations via the adjoint representation of G on $\mathfrak{g}$:

$$F_{\mu\nu}^g = gF_{\mu\nu}g^{-1}.$$

Therefore, if $\langle\cdot|\cdot\rangle$ is an Ad-invariant inner product on $\mathfrak{g}$, the quantity $-\frac{1}{4}\langle F_{\mu\nu}|F^{\mu\nu}\rangle$ is Lorentz- and gauge-invariant can be used to make a free-field Lagrangian. (This is where the compactness of G comes in: the existence of an Ad-invariant inner product on $\mathfrak{g}$ is equivalent to G being of the form $G = G' \times \mathbb{R}^n$ for some $n \geq 0$, where G' is compact.) The full Lagrangian is then

$$\mathfrak{L}(\Phi, \partial\Phi, A, \partial A) = \mathfrak{L}_0(\Phi, (\partial_\mu + iA_\mu)\Phi)) - \tfrac{1}{4}\langle F_{\mu\nu}|F^{\mu\nu}\rangle. \tag{9.3}$$

Writing things in this way conceals the coupling constants in the choice of Ad-invariant inner product on $\mathfrak{g}$. In the case of electrodynamics, for example, it amounts to substituting A_μ/e for A_μ so that $\partial_\mu + ieA_\mu$ becomes $\partial_\mu + iA_\mu$ and $F_{\mu\nu}F^{\mu\nu}$ becomes $F_{\mu\nu}F^{\mu\nu}/e^2$: that is, the inner product on $\mathfrak{u}(1) = i\mathbb{R}$ is taken to be $\langle is|it\rangle = st/e^2$. If G is simple or $G = U(1)$, there is only one Ad-invariant inner product up to a scalar normalization factor, namely, $\langle X|Y\rangle = -\operatorname{tr}(XY)$. Otherwise, there is one normalization factor for each simple or $U(1)$ factor, so that if $\mathfrak{g} = \bigoplus_1^k \mathfrak{g}^j$, the inner product must have the form

$$\langle X|Y\rangle = -\sum \alpha_j^{-2}\operatorname{tr}(X^jY^j) \qquad \Big(X = \sum X^j,\ Y = \sum Y^j,\quad X^j, Y^j \in \mathfrak{g}^j\Big)$$

for some $\alpha_j > 0$. Alternatively, we can rescale the fields $A_\mu = \sum A_\mu^j$, where A_μ^j is $\mathfrak{g}^j$-valued, by substituting $\alpha_j A_\mu^j$ for A_μ^j; then the inner product on $\mathfrak{g}$ is just $\langle X|Y\rangle = -\operatorname{tr}(XY)$ and the covariant derivative becomes $\partial_\mu + i\sum\alpha_j A_\mu^j$. This is the more common procedure, as it displays α_j explicitly in the coupling between A_μ and Φ.

This construction was first considered by Yang and Mills in 1954 [**135**] in the special case where $G = SU(2)$ and Φ is a pair of Dirac spinor fields. Hence the gauge fields A_μ are often referred to as *Yang-Mills fields*, and the Euler-Lagrange equation for their free action $-\frac{1}{4}\int\langle F_{\mu\nu}|F^{\mu\nu}\rangle\, d^4x$ (sometimes called the *Yang-Mills functional*) is known as the *Yang-Mills equation*.

Two important features of the corresponding quantized field theory can be read off immediately from the form of this Lagrangian. On the one hand, gauge invariance forbids the inclusion of terms that are quadratic in the fields A_μ themselves (without derivatives), such as the Lorentz-invariant $\langle A_\mu|A^\mu\rangle$; hence the quanta of the fields A_μ are *massless* (like photons). On the other hand, if G is non-Abelian, they interact directly with each other (unlike photons), because the $[A_\mu, A_\nu]$ term in (9.2) yields cubic and quartic terms in $\langle F_{\mu\nu}|F^{\mu\nu}\rangle$.

Quantizing Yang-Mills fields is a decidedly nontrivial undertaking; the problems in quantizing the electromagnetic field that we pointed out in §6.5 and §8.3 reappear here in amplified form. A serious discussion of these matters is beyond the scope of this book, and we refer the reader to the physics texts (e.g., Weinberg [**130**], Peskin and Schroeder [**88**], Zee [**136**], and Ramond [**91**]). Here we just give a thumbnail description of a couple of interesting points. It has been found that the most perspicuous way to arrive at Feynman rules for perturbation theory is via the functional integral formalism, much as we did for the electromagnetic field in §8.3. One writes down the functional integral for the gauge field and eliminates the redundancy due to the gauge transformations by introducing a gauge-fixing condition $h(A) = 0$ as in §8.3. However, the Faddeev-Popov determinant $\det(\delta h(A^g)/\delta g)$ (where A^g is given by (9.1)) here turns out to be $\det(\partial^\mu D_\mu)$, where D_μ denotes the covariant derivative for $\mathfrak{g}$-valued fields associated to the adjoint action of G on $\mathfrak{g}$. This depends on the field variables when G is non-Abelian, so it is not just a harmless constant that can be brought outside the integral.

There is a clever stratagem to handle this difficulty: one writes the determinant as a Fermionic Gaussian integral by (8.43), or rather its infinite-dimensional analogue:

$$\det(\partial^\mu D_\mu) = \iint [d\zeta^*\, d\zeta] e^{\zeta^*\partial^\mu D_\mu \zeta},$$

where ζ, ζ^* are Grassmann-valued fields. Inserting this expression for $\det(\partial^\mu D_\mu)$ into the functional integral for the Yang-Mills field amounts to adding an extra field into the theory, called the *Faddeev-Popov ghost*. It is a Fermionic field of spin zero, which violates the spin-statistics theorem, so it cannot have any direct physical significance, but there is no harm in using it as a formal calculational device. One adds Feynman diagrams with ghost lines (in loops only, not as initial or final particles) into the picture, much as we added Feynman diagrams with counterterm vertices for renormalization theory, and the resulting calculations give sensible results. In effect, the contributions from the ghosts cancel out the contributions from the equally unphysical longitudinal modes of the gauge fields.

The unquantized Yang-Mills theory can be recast in a more abstract geometric setting as follows. (See Morgan [**84**] for more details.) The basic ingredients are a manifold M, a Lie group G, a principal G-bundle P over M, a connection on P, and a vector space V equipped with a G-action. From these we can form the associated vector bundle $E = P \times_G V$ over M and a covariant derivative D on sections of E. (That is, for any vector field X on M, $D_X : \Gamma(E) \to \Gamma(E)$ is the covariant derivative of sections of E in the direction X.) We also have the vector bundle $\mathrm{Ad}(P)$ on M with fiber $\mathfrak{g}$, which is the bundle associated to P by the adjoint action of G on $\mathfrak{g}$.

Among several equivalent ways to define the connection on P, the most convenient is to specify a $\mathfrak{g}$-valued 1-form ω on P that transforms under the G-action on P according to the adjoint action of G on $\mathfrak{g}$ and whose value on the vertical vector field $\widetilde{X}$ induced by any $X \in \mathfrak{g}$ is simply X. (The horizontal subspace of T_pP is then the kernel of $\omega(p)$ for any $p \in P$.) The curvature of the connection can be defined either as (i) the $\mathfrak{g}$-valued 2-form $\Omega = d\omega + [\omega, \omega]$ on P (that is, $\Omega(X,Y) = d\omega(X,Y) + [\omega(X), \omega(Y)]$ for any vector fields X, Y on P), or (ii) the $\mathrm{Ad}(P)$-valued 2-form $\widetilde{\Omega}$ on M such that $\pi^*\widetilde{\Omega} = \Omega$, where $\pi : P \to M$ is the projection.[2]

A local trivialization $\tau : U \times G \to P$ of P over the open set $U \subset M$ gives an embedding $\tau(\cdot, 1)$ of U into P; the pullback ω^τ of ω under this embedding is a $\mathfrak{g}$-valued 1-form on U. τ also induces local trivializations of E and $\mathrm{Ad}(P)$. When one uses the former to identify sections of E with V-valued functions on U, the covariant derivative takes the form $D_X f = df(X) + \omega^\tau(X) \cdot f$, where the last term denotes the action of $\mathfrak{g}$ on V induced by the given action of G. Under the trivialization of $\mathrm{Ad}(P)$, the curvature form $\widetilde{\Omega}$ becomes a $\mathfrak{g}$-valued form $\widetilde{\Omega}^\tau$ on U.

The translation from this terminology to the previous one is as follows. We take M to be $\mathbb{R}^4$, V to be the space $\mathbb{C}^n$ in which the fields Φ take their values, and P to be the trivial G-bundle over $\mathbb{R}^4$. The whole point of gauge invariance, however, is that there is no canonical choice of trivialization for P. The gauge field iA_μ in a particular gauge (that is, a particular trivialization τ) is the form ω^τ, and the derived field $iF_{\mu\nu}$ is the curvature $\widetilde{\Omega}^\tau$:

$$\omega^\tau = iA_\mu \, dx^\mu, \qquad \widetilde{\Omega}^\tau = iF_{\mu\nu} \, dx^\mu \wedge dx^\nu.$$

There is an easy and important generalization of the gauge field theory discussed above. To wit, instead of one n-tuple of fields Φ, there can be several of them: $\Phi^1, \dots, \Phi^K$, each with its own free Lagrangian $\mathfrak{L}^k(\Phi^k, \partial\Phi^k)$. One then obtains a theory in which all of the fields ϕ_j^k are coupled to the same gauge fields A_μ by taking the Lagrangian to be

$$\sum_1^K \mathfrak{L}^k(\Phi^k, (\partial + i\alpha_k A)\Phi^k) - \tfrac{1}{4}\langle F_{\mu\nu} | F^{\mu\nu} \rangle.$$

This is what is needed, for example, to do the more general version of QED that includes several different species of charged particles. Allowing α_k to depend on k admits the possibility that different species have different charges; it must be understood that in this case, the gauge transformations of the fields Φ^k take the form $\Phi^k \mapsto e^{i\alpha_k \chi}\Phi^k$.

9.2. A glimpse at quantum chromodynamics

The most important non-Abelian gauge theory with unbroken symmetry is *quantum chromodynamics* (QCD), the theory of strongly interacting particles. Here

[2]Different conventions by different authors may result in extra factors of 2 in these formulas.

$n = 3$ and $G = SU(3)$, and one has *six* 3-tuples of Dirac spinor fields $\Psi^f = (\psi_1^f, \psi_2^f, \psi_3^f)$ whose quanta are called *quarks*, and one $\mathfrak{su}(3)$-valued gauge field A_μ, the components of which are massless vector fields whose quanta are called *gluons*.

There are eighteen quark fields ψ_c^f, and hence eighteen different quarks and eighteen different antiquarks. We shall take the index f to run from 1 to 6 in formulas, but in common parlance f denotes the "flavor" of the quark:

$$f \in \{u, d, s, c, t, b\} = \{\text{up, down, strange, charm, top, bottom}\}.$$

Likewise, the subscript c in ψ_c^f refers to the "color" of the quark, and the set $\{1, 2, 3\}$ is sometimes replaced by $\{\text{red, green, blue}\}$. The antiquarks are said to carry anticolors rather than colors — antired, antigreen, and antiblue, sometimes called cyan, magenta, and yellow; symbolically we may denote them by $\overline{1}$, $\overline{2}$, and $\overline{3}$. Quarks also participate in the electromagnetic and weak interactions: the up, charm, and top quarks carry an electric charge of $\frac{2}{3}|e|$, and the down, strange, and bottom quarks carry an electric charge of $-\frac{1}{3}|e| = \frac{1}{3}e$, where e is the charge of the electron; the antiquarks carry the opposite charges. All transitions between different flavors of quarks are caused by the weak interaction; in pure QCD the different flavors operate completely independently.

There are $8 = \dim(\mathfrak{su}(3))$ different gluons. The gluons carry color-anticolor combinations such as $1\overline{2}$ (red-antigreen); the nine such combinations, subject to the single linear relation $1\overline{1} + 2\overline{2} + 3\overline{3} = 0$, give the eight gluons. Gluons do not participate in the electromagnetic or weak interactions.

Actually, this whole business of color labels is merely an aid to intuition and should not be taken too seriously. It depends on a choice of orthonormal basis for $\mathbb{C}^3$, but nature makes no such choice. It would be more honest just to say that a quark field Ψ takes values in $\mathcal{S} \otimes \mathcal{H}$ where $\mathcal{S}$ is the Dirac spinor space equipped with the action of the Lorentz group and $\mathcal{H}$ is a 3-dimensional Hilbert space equipped with the action of its special unitary group $G = SU(\mathcal{H})$.

The Lagrangian is deceptively simple:

$$\mathcal{L} = \sum_{f=1}^{6} \overline{\Psi}^f([i\partial_\mu - \alpha A_\mu]\gamma^\mu - m^f)\Psi^f - \tfrac{1}{4}\langle F_{\mu\nu}|F^{\mu\nu}\rangle,$$

where the m^f's are the quark masses and α is the strong coupling constant (the same for all flavors of quarks). However, the resulting theory is amazingly rich. The basic interactions are of three types, coming from the $\overline{\psi}A_\mu\gamma^\mu\psi$ terms and the cubic and quartic terms in $\langle F_{\mu\nu}|F^{\mu\nu}\rangle$; they are pictured in Figure 9.1. (Gluons are conventionally denoted by "coiled-spring" lines.) The first diagram in Figure 9.1 can represent, for example, a red quark emitting a red-antigreen gluon and turning into a green quark, and the second can represent a red-antigreen gluon and a green-antiblue gluon combining to yield a red-antiblue gluon.

A fundamental feature of QCD is *color confinement*, which means that all real (non-virtual) particles that can actually be observed are bound states of quarks and gluons that are color-neutral, that is, invariant under the $SU(3)$-action. (This phenomenon is abundantly attested by experiment, but so far it has been proved theoretically only for a lattice approximation to QCD and not for QCD itself.) The observed bound states are of two types: Bosons made of quark-antiquark pairs and Fermions made of three quarks of different colors or three antiquarks of different

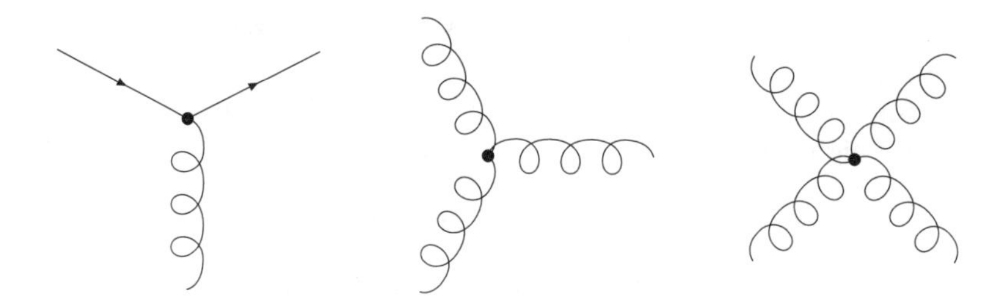

FIGURE 9.1. The basic interactions of QCD.

anticolors.[3] The Bosons can have spin 0 or 1, and the Fermions can have spin $\frac{1}{2}$ or $\frac{3}{2}$, depending on how the spins of the constituent quarks line up.[4] For example, the proton is the spin-$\frac{1}{2}$ bound state of one down and two up quarks, the neutron is the spin-$\frac{1}{2}$ bound state of one up and two down quarks, and the negatively charged pion π^- is the spin-0 bound state of an up antiquark and a down quark; the corresponding spin-$\frac{3}{2}$ and spin-1 particles are called Δ^+, Δ^0, and ρ^-, respectively. There are also color-neutral combinations of gluons, called *glueballs*, which may have been observed although the evidence is not conclusive as of this writing.

Color confinement accounts for the fact that the strong interaction is effectively of very short range even though gluons, like photons, are massless: particles with nontrivial color charges cannot travel alone over long distances.

The other crucial feature of QCD, which is shared with other non-Abelian gauge theories, is *asymptotic freedom*: the property that the effective coupling constant becomes weaker at higher energies, or equivalently at shorter distances, so that perturbation theory works better in the high-energy regime. The demonstration of this property (by 't Hooft, Politzer, and Gross and Wilczek) was a major milestone in the development of QCD and its acceptance as a workable theory of strongly interacting particles. It is quite surprising, as the vacuum polarization by virtual electron-positron pairs that causes the effective coupling in QED to become stronger at higher energies, as we discussed in §7.10, has an analogue in QCD — vacuum polarization by virtual quark-antiquark pairs — that has the same effect. But it turns out that in QCD, this effect is overwhelmed by reverse polarization effects whose existence depends on the non-Abelian nature of the gauge fields. (See Peskin and Schroeder [**88**], §16.7, for an informal discussion of this matter.) Because of asymptotic freedom, QCD has been very successful in accounting for the large variety of strongly interacting particles produced in high-energy collisions and quite successful in making quantitative predictions about their behavior (although the latter are nowhere near as precise as those of QED). However, the situation in the low-energy regime is much less satisfactory. The "strong force" that binds atomic nuclei together is understood to be a by-product of the QCD interaction, but this understanding is not on a very precise quantitative level.

[3]Exercise: Show that the electric charge of such combinations is always an integer multiple of e.

[4]This statement is an oversimplification. Experimental evidence indicates that the spin of a 3-quark Fermion arises in a complicated way from quark spins, gluon spins, and the orbital angular momentum of the quarks, even though on *a priori* grounds it must be exactly a half-integer. See Rith and Schäfer [**97**].

Further information about QCD, at varying levels of completeness and technicality, can be found in Aitcheson and Hey [**2**], Greiner, Schramm, and Stein [**57**], Griffiths [**58**], Peskin and Schroeder [**88**], Weinberg [**130**], and Zee [**136**].

9.3. Broken symmetries

The phrase "broken symmetry" refers to situations in which a symmetry in an underlying problem is lacking in its solutions. For example, suppose the problem is to find the values of $\xi \in \mathbb{R}^n$ that minimize $V(\xi) = |\xi|^4 - 2|\xi|^2$. The function V is invariant under the full orthogonal group $O(n)$, but the minimizing ξ's (those with $|\xi| = 1$) are invariant only under a subgroup of $O(n)$ conjugate to $O(n-1)$. Moreover, if one looks at the graph of V only in a small neighborhood of one of these ξ's, only the smaller symmetry group is evident.

We begin with a result concerning the breaking of a *global* symmetry group in field theory. Suppose $\Phi = (\phi_1, \ldots, \phi_n)$ is an n-tuple of classical real scalar fields — complex fields can be accomodated by breaking them into real and imaginary parts — with a Lagrangian

$$\mathfrak{L} = \tfrac{1}{2}(\partial_\mu \Phi) \cdot (\partial^\mu \Phi) - V(\Phi) = \tfrac{1}{2} \sum (\partial \phi_j)^2 - V(\phi_1, \ldots, \phi_n),$$

where V is a polynomial on $\mathbb{R}^n$ that has a minimum value $V_{\min}$ and is invariant under some compact group $G \subset O(n)$. (The V in the preceding paragraph is a good example.) The Lagrangian $\mathfrak{L}$ is then invariant under G, since the derivative term is invariant under $O(n)$. We are interested in the behavior of the resulting field theory near a field configuration that minimizes the total energy. If such a configuration is unique, it will be invariant under G, but if not, some of the symmetry will probably be broken. The total energy is the sum of the kinetic energy $\frac{1}{2}(\partial_0 \Phi)^2$ and the potential energy $\frac{1}{2}\sum_{i=1}^3 (\partial_i \Phi)^2 + V(\Phi)$, so it is minimized when $\Phi(x)$ is a constant $\xi \in \mathbb{R}^n$ such that $V(\xi) = V_{\min}$. Let

$$M = \{\xi \in \mathbb{R}^n : V(\xi) = V_{\min}\},$$

fix $\xi_0 \in M$, and let

$$H = \{g \in G : g\xi_0 = \xi_0\}.$$

If $H \neq G$, there are broken symmetries: the choice of ξ_0 reduces the symmetry group from G to H.

Let us expand $V(\Phi)$ about $\Phi = \xi_0$. Setting $\check{\Phi} = \Phi - \xi_0$, we have

$$V(\Phi) = V(\check{\Phi} + \xi_0) = V_{\min} + \tfrac{1}{2} \sum_{i,j=1}^{n} \partial_i \partial_j V(\xi_0) \check{\phi}_i \check{\phi}_j + (\text{higher order}).$$

The eigenvalues of the Hessian $(\partial_i \partial_j V(\xi_0))$ are nonnegative since V has a minimum at ξ_0. Let $\eta_1, \ldots, \eta_n \in \mathbb{R}^n$ be an orthonormal eigenbasis for the Hessian, with eigenvalues $m_1^2, \ldots, m_n^2$, and let $\widetilde{\phi}_j = \check{\Phi} \cdot \eta_j$; then

$$V(\Phi) = V_{\min} + \tfrac{1}{2} \sum m_j^2 \widetilde{\phi}_j^2 + (\text{higher order})$$

and hence

$$\mathfrak{L} = -V_{\min} + \tfrac{1}{2} \sum \left[(\partial_\mu \widetilde{\phi}_j)^2 - m_j^2 \widetilde{\phi}_j^2\right] + (\text{higher order}). \tag{9.4}$$

Now, the final point: The nullspace of the Hessian includes (and usually is exactly equal to) the tangent space of M at ξ_0. Indeed, if η is tangent to M at ξ_0, for small

ϵ there is a point $\xi_\epsilon \in M$ such that $|(\xi_0 + \epsilon\eta) - \xi_\epsilon| = O(\epsilon^2)$. Since $V(\xi_\epsilon) = V_{\min} = V(\xi_0)$ and $\nabla V(\xi_0) = 0$, we have

$$\begin{aligned} V(\xi_0 + \epsilon\eta) - V(\xi_0) &= V(\xi_0 + \epsilon\eta) - V(\xi_\epsilon) \\ &= \nabla V(\xi_0 + \epsilon\eta) \cdot (\xi_0 + \epsilon\eta - \xi_\epsilon) + O(|\xi_0 + \epsilon\eta - \xi_\epsilon|^2) \\ &= O(\epsilon) \cdot O(\epsilon^2) + O(\epsilon^4) \\ &= O(\epsilon^3). \end{aligned}$$

It follows that the restriction of the Hessian form to $T_{\xi_0}M$ vanishes. We can therefore choose the basis $\eta_1, \ldots, \eta_n$ so that the η_j with $j \leq \dim(M)$ span $T_{\xi_0}M$ and hence the corresponding m_j are zero.

Now let us interpret this result from a quantum perspective. The "configuration of minimum energy" is the vacuum state $|\Omega\rangle$, and for quantum fields the thing that plays the role of the vector ξ_0 in the preceding argument is the vacuum expectation value $\langle\Omega|\Phi(0)|\Omega\rangle \in \mathbb{R}^n$;[5] the translated field $\check{\Phi}$ is $\Phi - \langle\Omega|\Phi(0)|\Omega\rangle I$. (This, by the way, is where one really needs the assumption that the ϕ_j's are scalar fields. The vacuum expectation value of a field of higher spin always vanishes — it must transform under the Lorentz group according to the representation of appropriate spin, but at the same time it must be Lorentz-invariant since the vacuum is — and so $H = \{g \in G : g \cdot 0 = 0\} = G$.) When the Lagrangian is written in the form (9.4) in terms of the translated and rotated fields $\widetilde{\phi}_j$, it appears as the Lagrangian for a set of interacting scalar fields of masses $m_1, \ldots, m_n$. (The constant $V_{\min}$ does not affect the physics and can be discarded; we could have assumed that $V_{\min} = 0$ from the beginning.) The number of massless particles is at least $\dim(M)$, and since M includes the G-orbit of ξ_0, which can be identified with G/H, this number is at least $\dim(G) - \dim(H)$, the "number of broken symmetries."

In short, our classical argument has led us, on the heuristic level, to a result known as *Goldstone's theorem.* In its general form it says: *Suppose the Lagrangian* $\mathcal{L}$ *for a set of* n *Hermitian scalar fields* $\Phi = (\phi_1, \ldots, \phi_n)$ *is invariant under a group* $G \subset O(n)$ *(acting on* Φ *through its canonical action on* $\mathbb{R}^n$*), but the vacuum expectation value* $\langle\Omega|\Phi|\Omega\rangle \in \mathbb{R}^n$ *is invariant only under a proper subgroup* $H \subset G$*. Then there are at least* $\dim(G) - \dim(H)$ *massless particles in the theory.* These massless particles are known as *Goldstone Bosons* or *Nambu-Goldstone Bosons.* Their existence in particular models was first pointed out by Nambu and Goldstone; the general result is due to Goldstone, Salam, and Weinberg [**56**]. For the quantum-mechanical derivation of the theorem we refer the reader to this paper or to Weinberg [**130**], §19.2, or Peskin and Schroeder [**88**], §11.1.

Since the set of massless, spinless particles in the real world is apparently empty, one may wonder what the point of all of this is. In fact, the attitude in the paper [**56**] just cited is that the existence of Goldstone Bosons in theories with broken symmetry is a reason to reject such theories. However, one might envision a theory with an *approximate* symmtery group that is broken to yield a particle that is *approximately* massless, and a few years after the appearance of [**56**] people realized that the pion can be considered as the Goldstone Boson for a broken approximate symmetry of strongly interacting particles. (See Weinberg [**130**], §19.4, or Zee [**136**], §IV.2.) For our purposes, however, the important point is that if one promotes the symmetry group G to a gauge group — that is, incorporates not just the global

[5] $\langle\Omega|\Phi(x)|\Omega\rangle$ is independent of x, because the vacuum is translation-invariant.

symmetries $\Phi \mapsto g\Phi$ but also the local ones $\Phi(x) \mapsto g(x)\Phi(x)$ into the theory — the outcome is rather different.

Before we study that situation, however, a point needs to be made that pertains to both global and local broken symmetries. The description of the physics in terms of particles of certain masses, as in Goldstone's theorem, pertains only to field configurations that are not too far from the ground state or vacuum and whose energy is therefore not too much greater than $V_{\min}$; it is only for these that the Taylor expansion (9.4) is a reliable guide. At high energies where one can, so to speak, leap over the bumps in the potential V, the nature of the physics changes. In the applications of gauge field theories to elementary particle physics, such as the one discussed in the next section, the low-energy regime generally includes all energies that are available in the laboratory (or the observatory) with room to spare. However, these ideas also have applications in condensed matter physics — in particular, superconductivity — and there the "phase transition" from low to high energy is a directly observable phenomenon. (See Weinberg [**130**], §21.6, and Zee [**136**], Parts V and VI.)

To begin the study of broken local symmetries, let us return to the classical fields Φ considered above, with their energy-minimizing configuration ξ_0, and make the Lagrangian gauge-invariant. That is, as in §9.1, we introduce a $\mathfrak{g}$-valued vector field iA_μ and replace ∂_μ by the covariant derivative $D_\mu = \partial_\mu + iA_\mu$, yielding the Lagrangian

$$\mathcal{L} = \tfrac{1}{2}(D_\mu\Phi)\cdot(D^\mu\Phi) - V(\Phi) - \tfrac{1}{4}\langle F_{\mu\nu}|F^{\mu\nu}\rangle, \tag{9.5}$$

where the coupling constants are implicit in the inner product in the last term. This Lagrangian is invariant under the transformations

$$\Phi \mapsto g\Phi, \qquad A_\mu \mapsto gA_\mu g^{-1} + i(\partial_\mu g)g^{-1}$$

for arbitrary (smooth) G-valued functions g on $\mathbb{R}^4$. We now add the modest extra assumption that G acts transitively on M, so that $M = \{g\xi_0 : g \in G\} \cong G/H$ and $T_{\xi_0}M = \{Y\xi_0 : Y \in \mathfrak{g}\}$. The key point is the following lemma:

For any $\eta \in \mathbb{R}^n$ there is a $g(\eta) \in G$ such that $g(\eta)\eta \cdot Y\xi_0 = 0$ for all $Y \in \mathfrak{g}$. Moreover, there is a neighborhood U of ξ_0 such that $\gamma(\eta)$ can be chosen to depend smoothly on η for $\eta \in U$.

To prove the first assertion, consider the real-valued function $F_\eta(g) = \eta\cdot g\xi_0$ on G. Since G is compact, this function has a maximum at some g_0, where $dF_\eta|_{g_0} = 0$. But for $Y \in \mathfrak{g}$ (considered as a left-invariant vector field on G) we have

$$0 = dF_\eta\Big|_{g_0}(Y) = \frac{d}{dt}F_\eta(g_0e^{tY})\Big|_{t=0} = \eta\cdot g_0Y\xi_0 = g_0^{-1}\eta\cdot Y\xi_0,$$

since $G \subset O(n)$. Thus we can take $g(\eta) = g_0^{-1}$. This choice is not unique: in particular, for any $h \in H$ we have $h\xi_0 = \xi_0$, hence $F_\eta(gh) = F_\eta(g)$, so $g(\eta)$ can be replaced by $hg(\eta)$.

For $\eta = \xi_0$, since $|g\xi_0| = |\xi_0|$ we have $\eta\cdot g\xi_0 \leq |\xi_0|^2$ with equality if and only if $g\xi_0 = \xi_0$, i.e., $g \in H$, so the maximizers of F_{ξ_0} are precisely the elements of H. Let V be a complementary subspace to $\mathfrak{h}$ in $\mathfrak{g}$. It is easily seen by via the implicit function theorem that for η near ξ_0 there is a unique $g_0(\eta)$ near the identity in $\exp(V)$, depending smoothly on η, that maximizes F_η, and then $g(\eta) = g_0(\eta)^{-1}$ is the desired solution for the second assertion. We leave the details to the reader. (For the case we need in the next section, we will be able to produce an explicit solution that is valid on the largest possible neighborhood U of ξ_0, namely, $U = \mathbb{R}^n \setminus \{0\}$. However, $g(\eta)$ clearly cannot depend continuously on η in a neighborhood of 0.)

Now, recall that we are interested in fields that are close to ξ_0; more precisely, we restrict attention to smooth fields that take values in the neighborhood U of ξ_0 specified in the lemma. If Φ^0 is any such field, the lemma guarantees that that there is a smooth G-valued function g such that the gauge-transformed field $\Phi(x) = g(x)\Phi^0(x)$ satisfies

$$\Phi(x) \cdot Y\xi_0 = 0 \qquad (x \in \mathbb{R}^4,\ Y \in \mathfrak{g}). \tag{9.6}$$

We henceforth assume that Φ satisfies (9.6), and as before we set $\check{\Phi} = \Phi - \xi_0$ and $\widetilde{\phi}_j = \check{\Phi} \cdot \eta_j$ where $\{\eta_j\}_1^n$ is a basis for $\mathbb{R}^n$ such that $\{\eta_j\}_1^m$ is a basis for $T_{\xi_0}M$. Then $\eta_j = Y_j\xi_0$ for $j \le m$, where $Y_j \in \mathfrak{g}$. Since $\xi_0 \cdot Y\xi_0 = 0$ for $Y \in \mathfrak{g}$ (as $\mathfrak{g} \subset \mathfrak{so}(n)$ and $\mathfrak{so}(n)$ consists of skew-symmetric matrices), we find that *the Goldstone fields $\widetilde{\phi}_j(x)$ $(j = 1, \dots, m)$ vanish identically.* The gauge invariance of the theory therefore implies that *there are no physical Goldstone Bosons.*

Moreover, we have

$$D_\mu\Phi = (\partial_\mu + iA_\mu)(\check{\Phi} + \xi_0) = D_\mu\check{\Phi} + iA_\mu\xi_0,$$

so

$$(D_\mu\Phi)\cdot(D^\mu\Phi) = (D_\mu\check{\Phi})\cdot(D^\mu\check{\Phi}) + 2i(\partial_\mu\check{\Phi})\cdot A^\mu\xi_0 - (A_\mu\xi_0)\cdot(A^\mu\xi_0).$$

The term $(\partial_\mu\check{\Phi})\cdot(A^\mu\xi_0) = (\partial_\mu\Phi)\cdot(A^\mu\xi_0)$ vanishes by (9.6) because A^μ is $\mathfrak{g}$-valued, and the term $(A_\mu\xi_0)\cdot(A^\mu\xi_0)$ represents *masses* for the fields that are the components of A_μ. To make this more explicit, let $Y_1, \dots, Y_N$ be an orthonormal basis for $\mathfrak{g}$ (with respect to the given Ad-invariant inner product) such that $Y_1\xi_0, \dots, Y_m\xi_0$ are a basis for $T_{\xi_0}M$ and $Y_{m+1}, \dots, Y_N$ are a basis for $\mathfrak{h}$. Then $Y_j\xi_0 = 0$ for $j > m$, and the matrix $((Y_i\xi_0)\cdot(Y_j\xi_0))_{i,j=1}^m$ is positive definite, so we can subject $Y_1, \dots, Y_m$ to an orthogonal transformation to make it diagonal: $(Y_i\xi_0)\cdot(Y_j\xi_0) = M_j^2\delta_{ij}$ for some positive constants M_j.

With Φ satisfying (9.6) and A_μ expanded in terms of the basis $\{Y_j\}$ as $A_\mu = \sum_j A_\mu^j Y_j$, the Lagrangian (9.5) now takes the form

$$\mathfrak{L} = -\tfrac{1}{2}\sum_{j=m+1}^{n}(D\widetilde{\phi}_j)^2 - \tfrac{1}{2}\sum_{j=m+1}^{n} m_j^2\widetilde{\phi}_j^2 - \tfrac{1}{2}\sum_{j=1}^{m} M_j^2 A_\mu^j A^{\mu j} - V(\Phi) - \tfrac{1}{4}\langle F_{\mu\nu}|F^{\mu\nu}\rangle. \tag{9.7}$$

(Recall that $N = \dim(G)$ is the number of symmetries and $m = \dim M$ is the number of broken ones.) The m Goldstone fields have disappeared, and we are left with $n-m$ massive scalar fields $\widetilde{\phi}_j$, m massive gauge fields A_μ^j $(j = 1, \dots, m)$, and $N-m$ massless gauge fields A_μ^j $(j = m+1, \dots, N)$. This phenomenon is picturesquely described by saying that "the gauge fields $A_\mu^1, \dots, A_\mu^m$ eat the Goldstone Bosons and become massive." It was explored by several people, including Peter Higgs, in the mid-1960s, and is now generally known as the *Higgs mechanism.*[6]

It is worth noting that in the conversion from m massless scalar fields and m massless gauge fields to m massive gauge fields there is a "conservation of degrees of freedom," because massive vector fields have longitudinal modes of vibration whereas massless ones do not. In effect, the Goldstone fields turn into the longitudinal modes of the massive gauge fields.

[6]The use of the term "Higgs mechanism" by Deligne and Freed in [**21**], p. 185, is incorrect.

We refer the reader to the physics texts for the development of these matters on the level of quantum fields except to comment on one essential point. As we showed in §6.5, the Fourier-transformed propagator for a quantum vector field of mass M is

$$-i\Delta_{\mu\nu}(p) = -i\frac{g_{\mu\nu} - p_\mu p_\nu/M^2}{p^2 - M^2 + i\epsilon}. \tag{9.8}$$

Since this has no decay as $p \to \infty$ in sectors where p_μ and p_ν are both large, there are clearly going to be serious divergence problems when one inserts this into the integrals defined by Feynman diagrams. Indeed, since the gauge fields can interact with each other, one can easily construct Feynman diagrams with many loops but few lines whose propagators decay at infinity, and such diagrams can have arbitrarily high degrees of divergence. It therefore appears that a theory with massive gauge fields will necessarily be nonrenormalizable. But, in fact, gauge invariance comes to the rescue again. Just as we found that the photon propagator $-ig_{\mu\nu}/(p^2 + i\epsilon)$ is only one of a family of admissible propagators whose interplay is an useful feature of the theory, it turns out that (9.8), which results from the gauge-fixing condition (9.6), is only one of a family of admissible propagators for the massive gauge fields associated to different gauge-fixing conditions, and some of them have better decay at infinity. This discovery, due to 't Hooft, was a crucial ingredient in the development of the modern "standard model" for elementary particles. Specifically, by introducing suitable gauge-fixing delta-functions into the functional integrals for the gauge fields, as we did for the photon field in §8.3 (but with the additional complications described in §9.1), one can derive the propagators

$$-i\Delta^{(a)}_{\mu\nu}(p) = \frac{-i}{p^2 - M^2 + i\epsilon}\left[g_{\mu\nu} - \frac{(1-a)p_\mu p_\nu}{p^2 - aM^2}\right],$$

where a is an arbitrary real parameter. These propagators, after Wick rotation, have the healthy $1/|p|^2$ decay at infinity, and (9.8) can be recovered as the limiting case $a \to \infty$.

9.4. The electroweak theory

In this section we discuss a strikingly successful application of the ideas of the preceding section, the unified model for electromagnetic and weak interactions or "electroweak" theory due (independently) to Salam and Weinberg. To prepare the ground, we give a little background on the weak interaction. The story of the development of the theoretical and experimental understanding of the weak interaction is a fascinating one that is too long to tell here in detail, but I cannot resist devoting a page or two to a brief outline. A much more comprehensive account can be found in Franklin [**50**].

The weak interaction: a historical sketch. The first manifestation of the weak interaction to be observed was *beta decay*, the form of radioactivity in which an atomic nucleus emits an electron and increases its electric charge by one unit. In the early 1900s when this process was first studied, an atomic nucleus of atomic number Z and atomic weight n was generally thought to be composed of n protons and $n-Z$ electrons, and beta decay was understood as the expulsion of an electron. But there were difficulties: beta decay did not seem to respect conservation of energy, and there were anomalies in the spins of some nuclei. (For example, the N^{14} nucleus was observed to have spin 1, which could not be explained if it consisted of 14 protons and 7 electrons that all have spin $\frac{1}{2}$.) In 1930 Pauli proposed a "desperate way out" of both difficulties: nuclei should contain not only protons and electrons but light neutral particles of spin $\frac{1}{2}$ that he called "neutrons," one of which would be emitted along with the electron in beta decay. A little over a year later, Chadwick isolated the particle that we now call the neutron, and subsequent experiments led to the conclusion that nuclei consist of protons and neutrons rather than protons and electrons. This accounted correctly for nuclear spin but not for the missing energy in beta decay, which is much less than the rest mass of a neutron, so a second neutral particle was postulated, and Fermi dubbed it the "neutrino." Nowadays, having developed an appreciation for the law of lepton conservation, we call it an antineutrino instead.

By the mid-1930s, beta decay was generally understood to be the process $n \to p+e+\overline{\nu}$, although the distinction between ν and $\overline{\nu}$ was not yet clear, and the neutrino remained a shadowy device for making the books balance until its existence was confirmed by observation of the "reverse beta decay" $\overline{\nu}+p \to n+\overline{e}$, where $\overline{e}$ denotes a positron, in the mid-1950s. The first good theoretical model for its mechanism was provided in 1934 by Fermi [**33**], who proposed a theory in which proton, neutron, electron, and neutrino fields (denoted by ψ_p, ψ_n, ψ_e, and ψ_ν, respectively) interact via a Hamiltonian of the form

$$\mathcal{H}_{\text{int}} = (\text{const.})(A + A^\dagger), \qquad A = (\overline{\psi}_e\gamma^\mu\psi_\nu)(\overline{\psi}_p\gamma_\mu\psi_n). \tag{9.9}$$

This was one of the early successes of the field-theoretic description of elementary particles. It was recognized right away, however, that several variants of this interaction that still maintained Lorentz invariance were possible: one could replace the A in (9.9) by

$$\begin{array}{cc} (\overline{\psi}_e\psi_\nu)(\overline{\psi}_p\psi_n), & (\overline{\psi}_e\gamma^\mu\gamma^\nu\psi_\nu)(\overline{\psi}_p\gamma_\mu\gamma_\nu\psi_n), \\ (\overline{\psi}_e\gamma^\mu\gamma^5\psi_\nu)(\overline{\psi}_p\gamma_\mu\gamma^5\psi_n), & (\overline{\psi}_e\gamma^5\psi_\nu)(\overline{\psi}_p\gamma^5\psi_n), \end{array} \tag{9.10}$$

or combinations of these. (The interaction (9.9) was called V for "vector"; the ones corresponding to (9.10) were called S, T, A, and P for "scalar," "tensor," "axial-vector," and "pseudoscalar," respectively.) The picture was enriched by the discovery of the muon and its decay process $\mu \to e + \nu_\mu + \overline{\nu}_\mu$,[7] which could be modeled by the interactions (9.9) or (9.10) with n and p replaced by μ and ν_μ. The history of weak-interaction physics in the quarter-century after Fermi's paper is a tangled tale involving an accumulation of experimental evidence (some of it faulty) that suggested first one hypothesis on the correct form of the interaction and then another.

The breakthrough came in 1956, when Lee and Yang proposed the shocking idea that the weak interaction is not parity-invariant. Experimental confirmation was quickly obtained by three groups of experimenters, and within a couple of years Feynman, Gell-Mann, and the team of Marshak and Sudarshan independently found the correct form for any weak interaction of four spin-$\frac{1}{2}$ Fermion fields $\psi_1, \ldots, \psi_4$:

$$\mathcal{H}_{\text{int}} = \sqrt{8}\, G_F(A + A^\dagger), \qquad A = (\overline{P\psi_1}\gamma^\mu P\psi_2)(\overline{P\psi_3}\gamma_\mu P\psi_4), \qquad P = \tfrac{1}{2}(1-\gamma^5), \tag{9.11}$$

where the fields must be paired up correctly for each specific process. The coupling constant is called $\sqrt{8}\, G_F$ for historical reasons, G_F being the constant in Fermi's

[7] The distinction between muon-neutrinos and electron-neutrinos was not yet established, however.

original paper. It was supposed to be *universal* — that is, the same for beta decay, muon decay, and all other such interactions — with the experimentally determined value of 3.25×10^{-5} GeV^{-2} in natural units where $\hbar = c = 1$. This turns out to be wrong, for a rather subtle reason that we shall explain near the end of this section, but it is approximately correct for beta decay and muon decay. Feynman and Gell-Mann were persuaded to write up their results in a joint paper [**39**] which, as one might expect, is a gem. Feynman's personal account [**38**] of his discovery is also very entertaining and instructive.

Let us examine the meaning of (9.11). We use the Weyl representation (4.12) of the Dirac matrices, in which $\gamma^5 = \left(\begin{smallmatrix} -I & 0 \\ 0 & I \end{smallmatrix}\right)$ and hence $P = \left(\begin{smallmatrix} I & 0 \\ 0 & 0 \end{smallmatrix}\right)$. Thus P and $I - P = \frac{1}{2}(1 + \gamma^5)$ are just the projections onto the first and second factors of the decomposition $\mathbb{C}^4 = \mathbb{C}^2 \times \mathbb{C}^2$. (Note that $\overline{P\psi} = \overline{\psi}(I - P)$ and $(I - P)\gamma^\mu = \gamma^\mu P$, so that $\overline{P\psi_i}\gamma^\mu P\psi_j = \overline{\psi}_i\gamma^\mu P\psi_j$. It follows that the interaction (9.11) is a combination of the V and A interactions of (9.9) and (9.10).) As in §4.3, for any spinor ψ we call $\psi_L = P\psi$ and $\psi_R = (I - P)\psi$ the *left-handed* and *right-handed* components of ψ; these may be regarded as 2-component (Pauli) spinors. The point of (9.11) is that *the weak interaction involves only the left-handed components of the Fermion fields.* In making this statement, however, one must be careful to distinguish between particles and antiparticles, as the antiparticle of a left-handed particle is right-handed. It is implicit in this theory that one can distinguish "matter" from "antimatter" in a coherent way so that all the fields ψ_i in (9.11) are *matter* fields; then the interaction represented by (9.11) involves annihilation of ψ_2 and ψ_4 particles or creation of their antiparticles; vice versa for ψ_1 and ψ_3.

There is now persuasive (although indirect) evidence that neutrinos have a small but nonzero mass, but for most purposes it is a good approximation to assume that they are massless. In that case they can be described from the outset by two-component fields with left- or right-handed helicity, as we pointed out in §4.4. The theory of weak interactions just outlined then entails that only left-handed neutrinos and right-handed antineutrinos participate in the weak interaction, and since they apparently don't interact in any other way (except with gravity), one loses nothing by assuming that *all neutrinos are left-handed and all antineutrinos are right-handed.*

In any case, the theory of weak interactions based on (9.11) was able to account for a large and diverse body of experimental data in terms of tree-level calculations. But trouble loomed as soon as one tried to include Feynman diagrams with loops: as we pointed out in §7.2, simple dimension counting shows that the interaction (9.11) is not renormalizable. Moreover, it was natural to hope that the weak interaction could be explained in terms of a new field whose quanta ("intermediate vector Bosons") would mediate the interaction, like photons for electromagnetism or pions for the Yukawa model of the strong nuclear force. A model of this sort was proposed by Glashow [**53**] in 1961, but it included symmetry-breaking terms in the Lagrangian in order to make the intermediate vector Bosons massive. In 1967 Salam [**103**] and Weinberg [**128**][8] independently developed an improved theory of the weak interaction combined with electromagnetism that maintained the symmetry of the Lagrangian and used the Higgs mechanism to generate the required masses. They conjectured that it would be renormalizable, and the proof was provided in 1971 by 't Hooft [**117**].

[8]A three-page paper worthy of a Nobel prize!

The rest of this section is devoted to an outline of the Salam-Weinberg electroweak theory. In it, neutrinos are assumed to be *massless*. We shall indicate at the end how the theory can be modified to accommodate nonzero neutrino masses.

Construction of the Salam-Weinberg model. To keep the notation reasonably compact, we first describe a simplified version of the Salam-Weinberg model whose only Fermions are electrons and electron-type neutrinos (and their antiparticles). Later we shall show how to incorporate more species of Fermions. Moreover, we shall employ a shorthand notation in which the field of a Fermion f is denoted simply by f rather than ψ_f. There will not be a confusion in the use of e to denote both electric charge and the electron field, because from the outset we shall separate the electron field into its left-handed and right-handed components $e_L = \frac{1}{2}(1-\gamma^5)\psi_e$ and $e_R = \frac{1}{2}(1+\gamma^5)\psi_e$, reserving e itself for the *absolute* charge of the electron. Thus, the Fermionic ingredients for our model are left-handed electrons e_L, right-handed electrons e_R, and (necessarily left-handed) electron-type neutrinos ν, all of which are described by four-component spinor fields of which two components vanish.

We begin by writing down a Lagrangian for free *massless* electrons and neutrinos:

$$\mathcal{L}_0 = i\overline{\nu}\gamma^\mu\partial_\mu\nu + i\overline{e}_L\gamma^\mu\partial_\mu e_L + i\overline{e}_R\gamma^\mu\partial_\mu e_R. \tag{9.12}$$

This Lagrangian is invariant under the group $U(3)$, acting in the natural way on the triple (e_L, ν, e_R) of fields, and our gauge group will be based on a carefully chosen subgroup of $U(3)$.

The key idea is that the action of $SU(2)$ on the first two components implements a symmetry between left-handed electrons and neutrinos under the weak interaction, which is manifested in the older theory by the fact that e_L and ν are always paired up in (9.11) as $\overline{\nu}\gamma^\mu e_L$ or $\overline{e}_L\gamma^\mu\nu$. An analogous idea had already been used in hadron physics, where protons and neutrons behave essentially identically under the strong interaction and can be considered as two states of one particle, the "nucleon." There one considers the action of $SU(2)$ on the pair of fields (p,n) as an analogue of the action of $SU(2)$ on two-component spinors; the proton $(p,0)$ and neutron $(0,n)$ then appear as the $\pm\frac{1}{2}$-eigenstates of the Pauli matrix $\frac{1}{2}\sigma_3 \in i\mathfrak{su}(2)$, just like the "$z$-spin up" and "$z$-spin down" states of a spin-$\frac{1}{2}$ particle. One says that the "isospin" (short for "isotopic spin") of the proton — or more precisely, its third component — is $\frac{1}{2}$, and the isospin of the neutron is $-\frac{1}{2}$. The analogous operator

$$I_3 = \begin{pmatrix} \frac{1}{2} & 0 & 0 \\ 0 & -\frac{1}{2} & 0 \\ 0 & 0 & 0 \end{pmatrix}$$

on the triple of fields (ν, e_L, e_R) is called "weak isospin." Thus, we have (with a minor abuse of notation, writing $I_3(\xi) = \lambda$ when ξ is an eigenvector of I_3 with eigenvalue λ) $I_3(\nu) = \frac{1}{2}$, $I_3(e_L) = -\frac{1}{2}$, $I_3(e_R) = 0$.

The other operator on the triple of fields that has obviously important eigenvalues is the electric charge Q, measured in units of the absolute charge of the electron: $Q(e_L) = Q(e_R) = -1$, $Q(\nu) = 0$. However — and it took some cleverness to realize this — the right thing to look at in this situation is not Q but the "weak hypercharge"

$$Y = 2(Q - I_3) = \begin{pmatrix} -1 & 0 & 0 \\ 0 & -1 & 0 \\ 0 & 0 & -2 \end{pmatrix}.$$

(Again, the name is derived from an analogous operator in hadron physics.) This operator generates an action of $U(1)$ on the triple of fields, $e^{i\theta} \mapsto e^{i\theta Y}$, that commutes with the $SU(2)$ action of the preceding paragraph. Putting them together, then, we have an action Π of $SU(2) \times U(1)$ on the 3-dimensional field space:

$$\Pi\left(\begin{pmatrix} a & b \\ c & d \end{pmatrix}, e^{i\theta}\right) = \begin{pmatrix} e^{-i\theta}a & e^{-i\theta}b & 0 \\ e^{-i\theta}c & e^{-i\theta}d & 0 \\ 0 & 0 & e^{-2i\theta} \end{pmatrix}.$$

This is the symmetry group on which the Salam-Weinberg theory is based. Note that the matrices I_3 and Y come from the corresponding Lie algebra action: $I_3 = i^{-1}d\Pi(A, 0)$ and $Y = i^{-1}d\Pi(0, B)$ where $A = \frac{i}{2}\sigma_3 \in \mathfrak{su}(2)$ and $B = i \in i\mathbb{R} = \mathfrak{u}(1)$.

We now promote this group of global symmetries to a group of local symmetries, i.e., a gauge group. This entails the introduction of an $\mathfrak{su}(2) \times \mathfrak{u}(1)$-valued gauge field, which we denote by $(\frac{1}{2i}\mathbf{W}_\mu \cdot \boldsymbol{\sigma},\, \frac{1}{2i}X_\mu)$. Here $\mathbf{W}_\mu = (W_{1\mu}, W_{2\mu}, W_{3\mu})$ where $W_{j\mu}$ are real-valued, as is X_μ, and $\boldsymbol{\sigma}$ denotes the triple of Pauli matrices. (The factors of $\frac{1}{2}$ are conventional. $\frac{1}{2i}\boldsymbol{\sigma}$ is the basis of $\mathfrak{su}(2)$ that corresponds to the standard basis of $\mathfrak{so}(3)$, and the $\frac{1}{2i}$ on X_μ is then dictated by consistency.) The fields $\mathbf{W}_\mu$ and X_μ are normalized as in the paragraph following (9.3) so that the coupling constants appear in the covariant derivative rather than in the inner product. There are two of them: g for the $\mathfrak{su}(2)$ field and g' for the $\mathfrak{u}(1)$ field. The covariant derivative acting on the pair (ν, e_L) is thus

$$D_\mu \begin{pmatrix} \nu \\ e_L \end{pmatrix} = (\partial_\mu + \tfrac{1}{2i}g\mathbf{W}_\mu \cdot \boldsymbol{\sigma} - \tfrac{1}{2i}g'X_\mu)\begin{pmatrix} \nu \\ e_L \end{pmatrix},$$

where $\mathbf{W}_\mu \cdot \boldsymbol{\sigma}$ acts on the column vector by matrix multiplication and ∂_μ and X_μ acts componentwise, and the covariant derivative acting on e_R is

$$D_\mu e_R = (\partial_\mu - \tfrac{1}{i}g'X_\mu)e_R.$$

(The coefficients -1 and -2 multiplying $\frac{1}{2i}g'X_\mu$ in these formulas are the relevant eigenvalues of the hypercharge Y.) Replacing the ordinary derivatives in the Lagrangian (9.12) by covariant derivatives yields the gauge-invariant Lagrangian

$$\text{(9.13)}\quad \mathcal{L}_{e\nu} = \begin{pmatrix} \overline{\nu} & \overline{e}_L \end{pmatrix} \gamma^\mu (i\partial_\mu + \tfrac{1}{2}g\mathbf{W}_\mu \cdot \boldsymbol{\sigma} - \tfrac{1}{2}g'X_\mu)\begin{pmatrix} \nu \\ e_L \end{pmatrix} + \overline{e}_R\gamma^\mu(i\partial_\mu - g'X_\mu)e_R,$$

to which we must add the pure gauge field Lagrangian

$$\text{(9.14)}\qquad \mathcal{L}_F = -\tfrac{1}{4}\langle F_{\mu\nu} | F^{\mu\nu}\rangle,$$

where $F_{\mu\nu}$ is the $\mathbb{C}^3 \times \mathbb{C}$-valued function

$$F_{\mu\nu} = \big(\partial_\mu \mathbf{W}_\nu - \partial_\nu \mathbf{W}_\mu + g\mathbf{W}_\mu \times \mathbf{W}_\nu,\, \partial_\mu X_\nu - \partial_\nu X_\mu\big)$$

and the inner product is the standard inner product on $\mathbb{C}^4$.

At this point we have a theory of massless electrons, neutrinos, and gauge fields, so we need to add something to break the symmetry. For this purpose we introduce a pair of complex scalar fields (or a 4-tuple of real scalar fields)

$$\Phi = \begin{pmatrix} \phi^+ \\ \phi^0 \end{pmatrix} = \begin{pmatrix} \phi_1^+ + i\phi_2^+ \\ \phi_1^0 + i\phi_2^0 \end{pmatrix}$$

equipped with an action of the symmetry group $SU(2) \times U(1)$: $SU(2)$ acts in the canonical way on the two-dimensional vector Φ, and $U(1)$ acts as scalar multiples of the identity, that is,

$$\left(\begin{pmatrix} a & b \\ c & d \end{pmatrix}, e^{i\theta} \right) \mapsto \begin{pmatrix} e^{i\theta}a & e^{i\theta}b \\ e^{i\theta}c & e^{i\theta}d \end{pmatrix}.$$

We then have $I_3(\phi^+) = \frac{1}{2}$, $I_3(\phi^0) = -\frac{1}{2}$, and $Y(\phi^+) = Y(\phi^0) = 1$, where I_3 and Y are defined in terms of the corresponding Lie algebra action as before. It follows that the electric charge $Q = I_3 + \frac{1}{2}Y$ is $+1$ for ϕ^+ and 0 for ϕ^0 (hence the superscripts). We endow Φ with a gauge-invariant Lagrangian that will induce symmetry breaking:

$$\mathcal{L}_\Phi = \tfrac{1}{2}(D_\mu\Phi)^\dagger(D^\mu\Phi) + \kappa\Phi^\dagger\Phi - \lambda(\Phi^\dagger\Phi)^2, \tag{9.15}$$

where κ and λ are positive constants and the covariant derivative here is

$$D_\mu = \partial_\mu + \tfrac{1}{2i}g\mathbf{W}_\mu \cdot \boldsymbol{\sigma} + \tfrac{1}{2i}g'X_\mu.$$

(Normally one would omit the factor of $\frac{1}{2}$ on the derivative term in (9.15) since Φ is complex-valued; we include it so that the real scalar field that remains after the Goldstone components have been removed is correctly normalized.) The final ingredient is an interaction Lagrangian for Φ and the lepton fields:

$$\mathcal{L}_{e\nu\Phi} = -G\begin{pmatrix} \overline{\nu} & \overline{e}_L \end{pmatrix}\begin{pmatrix} \phi^+ \\ \phi^0 \end{pmatrix} e_R - G\overline{e}_R\begin{pmatrix} \phi^{+*} & \phi^{0*} \end{pmatrix}\begin{pmatrix} \nu \\ e_L \end{pmatrix}, \tag{9.16}$$

where G is another coupling constant. The total Lagrangian is the sum of (9.13)–(9.16):

$$\mathcal{L} = \mathcal{L}_{e\nu} + \mathcal{L}_F + \mathcal{L}_\Phi + \mathcal{L}_{e\nu\Phi}. \tag{9.17}$$

This completes the construction of the Salam-Weinberg model.

Interpretation of the Salam-Weinberg model. First of all, the potential $V(\Phi) = -\kappa\Phi^\dagger\Phi + \lambda(\Phi^\dagger\Phi)^2$ in the Lagrangian (9.15) is minimized when $\Phi^\dagger\Phi \equiv |\phi^+|^2 + |\phi^0|^2 = \kappa/2\lambda$. Among the constant fields satisfying this equation, we take the one corresponding to the physical vacuum to be

$$\phi^+ = 0, \quad \phi^0 = a, \qquad \text{where} \quad a = \sqrt{\frac{\kappa}{2\lambda}}. \tag{9.18}$$

(Recall that this means that when the fields are quantized, the vacuum expectation values of ϕ^+ and ϕ^0 are 0 and a, respectively.) The subgroup H of $SU(2) \times U(1)$ that fixes $\binom{0}{a}$ is

$$H = \left\{ \left(\begin{pmatrix} e^{i\theta} & 0 \\ 0 & e^{-i\theta} \end{pmatrix}, e^{i\theta} \right) : \theta \in \mathbb{R} \right\}.$$

As explained in §9.3, we can perform a gauge transformation to put any field with values near $\binom{0}{a}$ into the form

$$\Phi = \begin{pmatrix} 0 \\ a + \phi \end{pmatrix} \tag{9.19}$$

where ϕ is real-valued (Hermitian, after quantization). In fact, there is a canonical choice of gauge transformation $g(x)$ here that works for any Φ such that $\Phi(x)$ never vanishes: namely, the one such that $g(x) \in SU(2)$ for all x, because for any $(z_1, z_2) \neq (0,0) \in \mathbb{C}^2$ there is a unique $g \in SU(2)$ such that $g(z_1, z_2) = (0, y)$ with $y > 0$. The field ϕ is called the *Higgs field*; its quanta are particles of spin 0 known as *Higgs Bosons.*

Henceforth all our calculations will be in the gauge such that (9.19) is valid. We then have

$$D_\mu \Phi = \begin{pmatrix} 0 \\ \partial_\mu \phi \end{pmatrix} + \left[\frac{g}{2i} \begin{pmatrix} W_m^3 & W_\mu^1 - iW_\mu^2 \\ W_\mu^1 + iW_\mu^2 & -W_\mu^3 \end{pmatrix} + \frac{g'}{2i} X_\mu \right] \begin{pmatrix} 0 \\ a + \phi \end{pmatrix}$$
$$= \frac{1}{2i} \begin{pmatrix} ag(W_\mu^1 - iW_\mu^2) + g(W_\mu^1 - iW_\mu^2)\phi \\ 2i\partial_\mu \phi + a(-gW_\mu^3 + g'X_\mu) + (-gW_\mu^3 + g'X_\mu)\phi \end{pmatrix},$$

so

$$D_\mu \Phi^\dagger D^\mu \Phi$$
$$= \partial_\mu \phi \partial^\mu \phi - \frac{a^2 g^2}{4} (W_\mu^1 + iW_\mu^2)(W^{1\mu} - iW^{2\mu}) - \frac{a^2}{4} (gW_\mu^3 - g'X_\mu)(gW^{3\mu} - g'X^\mu) + \cdots,$$

where the dots, here and in the following formulas, denote terms involving products of three or four fields. The quadratic terms are mass terms, and the field combinations with definite masses and charges[9] (which therefore correspond to particles) are

$$W = \frac{W^1 + iW^2}{\sqrt{2}}, \qquad Z = \frac{gW^3 - g'X}{\sqrt{g^2 + g'^2}}, \qquad A = \frac{g'W^3 + gX}{\sqrt{g^2 + g'^2}}. \tag{9.20}$$

Note that the single complex field W has replaced the two real fields W^1 and W^2; the quanta of W are particles with distinct antiparticles, whereas the Z and A quanta are their own antiparticles. We now have

$$D_\mu \Phi^\dagger D^\mu \Phi = \partial_\mu \phi \partial^\mu \phi - \frac{a^2 g^2}{4} W_\mu^* W^\mu - \frac{a^2}{8} (g^2 + g'^2) Z_\mu Z^\mu + \cdots.$$

Moreover,

$$\kappa \Phi^\dagger \Phi - \lambda (\Phi^\dagger \Phi)^2 = \kappa (a + \phi)^2 - \lambda (a + \phi)^4 = \frac{\kappa^2}{4\lambda} - 2\kappa \phi^2 + \cdots,$$

since $a^2 = \kappa / 2\lambda$. Thus, after discarding the constant $\kappa^2 / 4\lambda$, we have

$$\mathfrak{L}_\Phi = \tfrac{1}{2} \partial_\mu \phi \partial^\mu \phi - \frac{a^2 g^2}{4} W_\mu^* W^\mu - \frac{a^2}{8} (g^2 + g'^2) Z_\mu Z^\mu - 2\kappa \phi^2 + \cdots.$$

The Higgs mechanism has done its job: The gauge fields W and Z have acquired masses, and the Higgs field is also massive:

$$m_W = \frac{ag}{2}, \qquad m_Z = \frac{a\sqrt{g^2 + g'^2}}{2}, \qquad m_\phi = 4\kappa. \tag{9.21}$$

(Recall that for complex fields the mass coefficient is m^2 rather than $\frac{1}{2} m^2$.) Only the last gauge field A remains massless.

Next, we observe that under the condition (9.19) the Lagrangian (9.16) becomes

$$\mathfrak{L}_{e\nu\Phi} = -Ga(\bar{e}_L e_R + \bar{e}_R e_L) - G(\phi \bar{e}_L e_R + \phi \bar{e}_R e_L).$$

---

[9] We shall identify the charges of these fields shortly.

Note that the neutrino field has dropped out. The second pair of terms represents the interaction of the electron field with the Higgs field, but the first pair is a mass term for electrons. Indeed, the mass of the electron should appear in the Lagrangian as $-m\overline{\psi}_e\psi_e$ where ψ_e is the full electron field. We have $\psi_e = e_L + e_R$, $\overline{\psi}_e\psi_e = \psi_e^\dagger\gamma^0\psi_e$, and $\gamma^0 = \left(\begin{smallmatrix} 0 & I \\ I & 0 \end{smallmatrix}\right)$; it follows easily that $\overline{\psi}_e\psi_e = \overline{e}_L e_R + \overline{e}_R e_L$. In short, the presence the field Φ imparts a mass to the electron:

$$m_e = aG. \tag{9.22}$$

Finally, we take a close look at the Lagrangian (9.13):

$$\begin{aligned}\mathcal{L}_{e\nu} = (\overline{\nu} \quad \overline{e}_L)\,\gamma^\mu \left[i\partial_\mu + \frac{1}{2}\begin{pmatrix} gW_\mu^3 - g'X_\mu & g(W_\mu^1 - iW_\mu^2) \\ g(W_\mu^1 + iW_\mu^2) & -gW_\mu^3 - g'X_\mu \end{pmatrix}\right]\begin{pmatrix} \nu \\ e_L \end{pmatrix} \\ + \overline{e}_R\gamma^\mu(i\partial_\mu - g'X_\mu)e_R.\end{aligned} \tag{9.23}$$

In terms of the fields W, Z, and A, the 2×2 matrix in (9.23) is

$$\begin{pmatrix} (g^2 + g'^2)^{1/2}Z_\mu & \sqrt{2}\,gW_\mu^* \\ \sqrt{2}\,gW_\mu & (g^2 + g'^2)^{-1/2}((g'^2 - g^2)Z_\mu - 2gg'A_\mu) \end{pmatrix},$$

whereas

$$g'X_\mu = (g^2 + g'^2)^{-1/2}(gg'A_\mu - g'^2 Z_\mu).$$

Substituting these formulas into (9.23), we find that

$$\begin{aligned}\mathcal{L}_{e\nu} =& i[\overline{\nu}\gamma^\mu\partial_\mu\nu + \overline{e}_L\gamma^\mu\partial_\nu e_L + \overline{e}_R\gamma^\mu\partial_\mu e_R] \\ &+ \frac{g}{\sqrt{2}}[\overline{\nu}\gamma^\mu W_\mu^* e_L + \overline{e}_L\gamma^\mu W_\mu\nu] \\ &- \frac{gg'}{\sqrt{g^2 + g'^2}}[\overline{e}_L\gamma^\mu A_\mu e_L + \overline{e}_R\gamma^\mu A_\mu e_R] \\ &+ \frac{1}{2\sqrt{g^2 + g'^2}}[(g^2 + g'^2)\overline{\nu}\gamma^\mu Z_\mu\nu + (g'^2 - g^2)\overline{e}_L\gamma^\mu Z_\mu e_L + 2g'^2\overline{e}_R\gamma^\mu Z_\mu e_R],\end{aligned} \tag{9.24}$$

where we have collected the terms into the groups that involve ∂, W, A, and Z. The first group is just the usual free-field terms for the lepton fields, but each of the other three has some new information to impart.

First, in the language of Feynman diagrams, the first term on the second line of (9.24) represents a vertex with an incoming electron or outgoing positron, an outgoing W-particle or incoming W-antiparticle, and an outgoing neutrino or incoming antineutrino. Charge conservation therefore dictates that the W-particle have the same charge as the electron. It is therefore usually denoted by W^-, and its antiparticle is denoted by W^+. W^- and W^+ are the field quanta that mediate the familiar forms of the weak interaction such as beta decay and muon decay, as we shall see shortly. For the same reason, the interactions on the third and fourth lines show that the quanta of the A and Z fields have no electric charge.

Second, we observe that the massless, neutral field A_μ does not couple to the neutrino field at all, and it couples to e_L and e_R in exactly the same way, namely, the coupling for electromagnetism. We may therefore take A_μ to represent the electromagnetic field, and the coupling constant $-gg'/\sqrt{g^2 + g'^2}$ is then the charge of the electron:

$$\frac{gg'}{\sqrt{g^2 + g'^2}} = e. \tag{9.25}$$

Finally, the last line of (9.24), involving the neutral vector Boson Z, represents the "weak neutral current" interaction. It produces effects such as electron-neutrino scattering, the basic diagram for which is shown in Figure 9.2a. Such effects were the most striking new prediction of the Salam-Weinberg theory and the earlier version of Glashow, and their observation in experiments beginning in 1973 were one of the main confirmations of the theory.

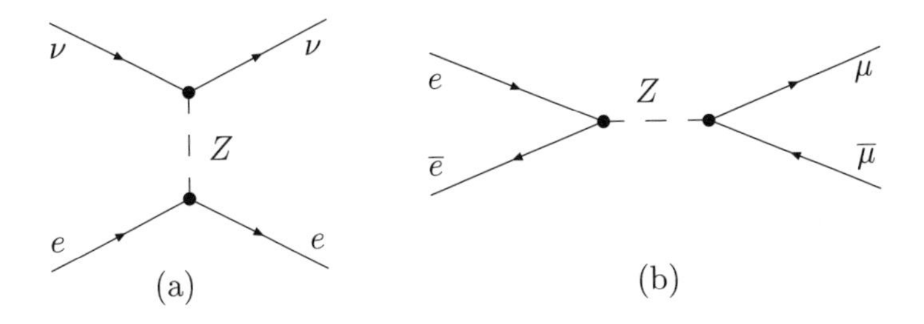

FIGURE 9.2. Two weak neutral current processes.

To complete the picture of the weak interaction, we need to introduce more Fermion species. Adding more leptons — muons and tauons and their associated neutrinos — is easy. One first adds subscript e's to all the neutrino fields ν in the preceding calculations and to the coupling constant G in (9.16). To incorporate muons, one then adds two more terms $\mathfrak{L}_{\mu\nu_\mu}$ and $\mathfrak{L}_{\mu\nu_\mu\Phi}$to the Lagrangian (9.17), which are just like the $\mathfrak{L}_{e\nu_e}$ and $\mathfrak{L}_{e\nu_e\Phi}$ already present except that the electron and electron-neutrino fields are replaced by muon and muon-neutrino fields, and a new coupling constant $G_\mu = G_e m_\mu / m_e$ is used for the $\mu - \Phi$ coupling. Similarly for tauons. The three generations of leptons then interact with each other through exchange of $W^\pm$ and Z particles.

Two such interactions are depicted in Figure 9.2b and Figure 9.3. Figure 9.3 shows the basic process for muon decay. Figure 9.2b shows an interaction by which an electron and a positron can collide and turn into a muon and an antimuon. Such transmutations are observed in particle accelerators. At moderate energies, they occur more often through the intermediation of a photon rather than a Z particle, but the Z contribution becomes particularly strong, resulting in an enhanced cross-section, when the total energy of the electron-positron pair is close to the mass m_Z. This phenomenon has provided a very precise experimental determination of m_Z.

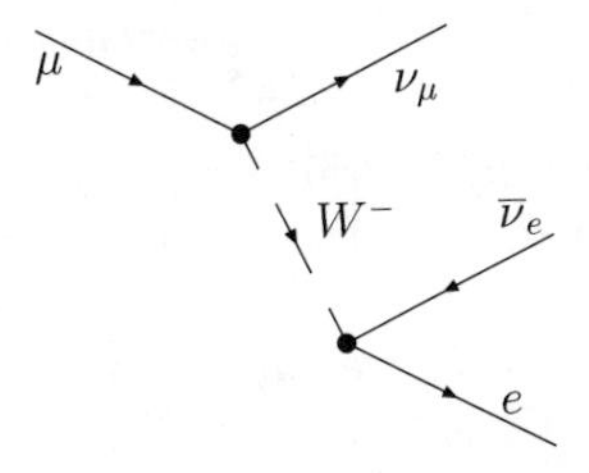

FIGURE 9.3. The basic process for muon decay.

Before putting hadrons into the picture, let us examine what we have developed a little more closely. The preceding formulas are commonly rewritten in terms of the *Weinberg angle*

$$\theta_W = \arctan \frac{g'}{g}.$$

Specifically, (9.20) says that

$$Z = (\cos\theta_W)W^3 - (\sin\theta_W)X, \qquad A = (\sin\theta_W)W^3 + (\cos\theta_W)X,$$

(9.25) says that

$$e = g\sin\theta_W = g'\cos\theta_W, \tag{9.26}$$

and (9.21) therefore says that

$$m_W = \frac{ea}{\sin\theta_W}, \qquad m_Z = \frac{m_W}{\cos\theta_W} = \frac{ea}{\sin\theta_W \cos\theta_W}. \tag{9.27}$$

Now, what are all these quantities?

Since the lepton-W coupling is given by the third line of (9.24), and the propagator of the W^- is of the form (6.53), the diagram in Figure 9.3 corresponds to the S-matrix element

$$(-i)^3 \frac{g^2}{2} \overline{u}(e_L)\gamma^\rho u(\nu_e) \frac{g_{\rho\sigma} - q_\rho q_\sigma/m_W^2}{q^2 - m_W^2} \overline{u}(\nu_\mu)\gamma^\sigma u(\mu_L)$$

for muon decay. Here $u(\mu_L)$ is the spinor of the incoming muon, projected into the left-handed eigenspace (actually a function of the *momentum* of the muon), and likewise for the other u's; q is the momentum transferred by the W^-; and there is one factor of $-i$ for each vertex and another one from the propagator. In the low-energy regime where $|q|$ is negligible in comparison with m_W, this reduces to

$$-i\frac{g^2}{2m_W^2}\overline{u}(e_L)\gamma^\rho u(\nu_e)\overline{u}(\nu_\mu)\gamma_\rho u(\mu_L).$$

But this is exactly the matrix element given by the old theory (9.11), except that $g^2/2m_W^2$ here is replaced by $\sqrt{8}\,G_F$ there, so the coupling constant g is related to the Fermi constant G_F by

$$G_F = \frac{g^2}{\sqrt{32}\,m_W^2}. \tag{9.28}$$

Precise experiments with muons have yielded the value $G_F = 1.16637\times 10^{-5}\ \text{GeV}^{-2}$, specifically for muon decay, so this relation together with (9.21) determines the vacuum expectation value a:

$$a = \frac{2m_W}{g} = \frac{1}{2^{1/4}G_F^{1/2}} = 246.2\ \text{GeV}.$$

Moreover, experiments with weak neutral current processes have yielded the value 0.231 for $\sin^2\theta_W$, or 0.448 for θ_W itself, so we can now read off the masses of the W and Z particles from (9.27): with $e = \sqrt{4\pi/137}$ we get $m_W = 77.6$ GeV and $m_Z = 88.4$ GeV. However, these figures are somewhat inaccurate, because there are various higher-order corrections to be taken into account. The biggest of these is something we have pointed out in §7.10: the fine structure constant $e^2/4\pi$ needs to be taken not as the usual low-energy value $1/137$ but as the effective value at the energy scale ~ 90 GeV, which turns out to be about $1/129$. Taking $e = \sqrt{4\pi/129}$ yields the improved values $m_W = 79.9$ GeV and $m_Z = 91.1$ GeV. The experimental values (not obtained until the 1980s, when particle accelerators were built that were powerful enough to create particles with such energies) are $m_W = 80.4$ GeV and $m_Z = 91.2$ GeV.

A word about coupling strength: since $g = 2m_W/a$, the W-Fermion coupling constant $g/\sqrt{2}$ in the third line of (9.24) is about .46, which is comparable to but larger than the (low-energy) electromagnetic coupling $e \approx .30$. As (9.28) makes clear, the observed weakness of the weak interaction in low-energy processes such as particle decays is due not to the intrinsic smallness of its coupling but to the relative immensity of m_W.

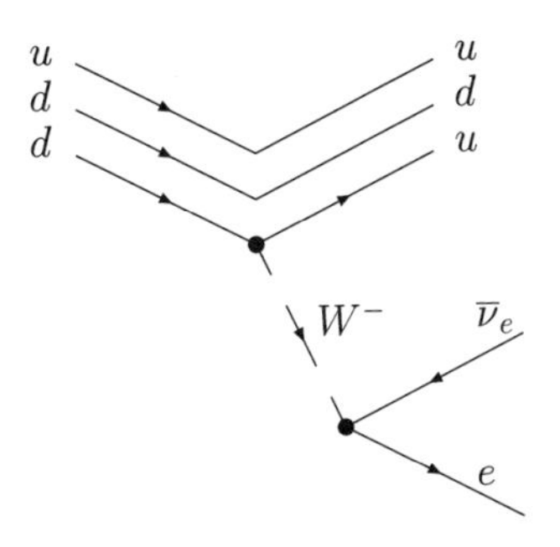

FIGURE 9.4. The basic process for neutron decay.

Our model so far does not include the original manifestation of the weak interaction, beta decay, and it is time to do something about that. Beta decay $n \to p + e + \overline{\nu}_e$ involves the transmutation of one of the neutron's down quarks d into an up quark u; the basic Feynman diagram is shown in Figure 9.4. To accomodate this in the electroweak theory, one can add another piece $\mathfrak{L}_{ud}$ to the Lagrangian, which looks almost like (9.13) but with u and d replacing e and ν $(= \nu_e)$:

$$\begin{aligned}(9.29)\quad \mathfrak{L}_{ud} = (\overline{u}_L \quad \overline{d}_L)\, \gamma^\mu (i\partial_\mu + \tfrac{1}{2} g \mathbf{W}_\mu \cdot \boldsymbol{\sigma} + \tfrac{1}{6} g' X_\mu) \begin{pmatrix} u_L \\ d_L \end{pmatrix} \\ + \overline{u}_R \gamma^\mu (i\partial_\mu + \tfrac{2}{3} g' X_\mu) u_R + \overline{d}_R \gamma^\mu (i\partial_\mu - \tfrac{1}{3} g' X_\mu) d_R.\end{aligned}$$

There are two changes from (9.13): an extra term for the right-handed component of d, and different coefficients for the X_μ terms due to the fact that the electric charges of the quarks are $Q(u) = \frac{2}{3}$ and $Q(d) = -\frac{1}{3}$ and hence the hypercharges are $Y(u_L) = Y(d_L) = \frac{1}{3}$, $Y(u_R) = \frac{4}{3}$, and $Y(d_L) = -\frac{1}{3}$.

However, this is not quite right. The coupling constants for neutron decay and muon decay were originally thought to be the same, but more precise experiments revealed that the former is about 3% smaller, so the g of (9.13) does not seem appropriate for (9.29). Moreover, a theory of hadronic weak processes must include decays of particles containing strange quarks s, for which the coupling is weaker yet. There is a common resolution of these difficulties: one builds the new Lagrangian not out of u and d but rather out of u and d', where d' is a linear combination of d and s:

$$d' = (\cos\theta_C) d + (\sin\theta_C) s.$$

θ_C is called the *Cabibbo angle* after the physicist who first suggested this idea (in a pre quark model); it is experimentally determined to be about .23, so that $\cos\theta_C \approx .97$ (3% less than one!) and $\sin\theta_C \approx .23$.

This improved model worked well for beta decay and for some strange particle decays but gave incorrect results for others. To solve that problem, Glashow, Iliopoulos, and Maiani proposed in 1970 that there should be a fourth quark c ("charm") and a new piece $\mathfrak{L}_{cs'}$ of the Lagrangian involving c and

$$s' = -(\sin\theta_C)d + (\cos\theta_C)s.$$

Decisive experimental confirmation was obtained in 1974 with the discovery of the J or ψ meson[10] which consists of a bound state of c and its antiparticle $\bar{c}$. Since then, a third generation of quarks (t and b) has entered the picture, and the (presumably final) version of the theory involves three new pieces of the Lagrangian of the form (9.29) (with the original g and g') built from the three doublets (u, d''), (c, s''), and (t, b''), where d'', s'', and b'' are linear combinations of d, s, and b whose coefficients form a 3×3 unitary matrix, the *Cabibbo-Kobayashi-Maskawa matrix*. (The entries of this matrix are known only from experiment, and only to a rather crude approximation.) To complete the picture, one needs to add more terms to the Lagrangian to give the interaction of the quark fields with the Higgs field in order to generate the quark masses. We omit the details, which are a bit involved; see Weinberg [**130**], §21.3, for a terse account.

For further information on the applications of the electroweak theory to experimental particle physics, see Griffiths [**58**], Aitcheson and Hey [**2**], Weinberg [**130**], and Peskin and Schroeder [**88**].

To conclude, we say a few words about neutrino masses. There is now persuasive observational evidence for "neutrino oscillations," that is, the ability of neutrinos of the three different species (electron, muon, and tauon) to change into one another, a phenomenon that would be impossible if they were truly massless. The theoretical explanation for it relies on the assumption that the mass eigenstates of the triple of neutrino fields $(\nu_e, \nu_\mu, \nu_\tau)$ are not ν_e, ν_μ, and ν_τ themselves but certain linear combinations of them. (As we saw earlier, the same is true of the electroweak gauge fields, where the mass eigenstates are A and Z rather than W^3 and X.) A neutrino created in muon decay, for example, begins as a pure ν_μ, which is a superposition of particles of three different masses. As the neutrino propagates, these components travel with slightly different speeds, creating interference patterns that yield a nonzero probability that it will behave as a ν_e or ν_τ when it arrives at a detector. Neutrino rest masses are estimated to be on the order of 1 eV, although they have not been measured directly: all observed neutrinos travel with a speed that is experimentally indistinguishable from that of light, so their rest masses $m = \sqrt{p^2}$ are negligible in comparison with their observed masses p^0.

So where does this leave the electroweak theory? The main point to be made is that it needs only modification, not outright replacement. The weak interaction still involves only left-handed neutrinos, so for many purposes one can simply add a subscript L to the ν's in the theory and proceed as before. To complete the picture, one possibility is to incorporate right-handed neutrinos into the theory in much the same way as right-handed electrons by adding suitable terms to the Lagrangian:

---

[10]This particle was discovered independently by two groups of experimentalists, who gave it two different names.

kinetic terms $i\overline{\nu}_R\gamma^\mu\partial_\mu\nu_R$ in (9.13) (with no X since $Y(\nu_R) = 0$) and extra terms in the Fermion-Higgs interaction (9.16) to impart masses to the neutrinos. The right-handed neutrinos are then "sterile": they do not interact with anything except the Higgs field and gravity. However, there are also other models that employ a different kind of mass term in the Lagrangian, the so-called "Majorana masses." The explanation of these ideas is beyond the scope of this book; more information can be found in Ramond [**92**] and the references given there. In any case, the correct theory of massive neutrinos is, as of this writing, still a matter of conjecture; the question awaits further experimental input.

The combination of the electroweak theory with QCD is the *standard model* which (with some modification to accomodate massive neutrinos) has come to be accepted as the correct description of elementary particle interactions with gravity excluded — as far as it goes. However, some aspects of it (strong interactions at low energies, for one thing) are still incompletely understood, and it leaves some fundamental questions unanswered. Much effort has been devoted to attempts to incorporate the standard model into a more complete theory of fundamental interactions — "grand unified theories" that embed the electroweak theory and QCD into a single gauge field theory, supersymmetry, string theory — but the physical validity of these theories remains a matter of speculation.

Mathematical visitors may wish to return for a closer look at these regions of physical theory, as they offer much of mathematical interest. But the subject of this tourist guide is honest physics that can be checked against experiment, so this is a good place to stop.

Bibliography

[1] R. Abraham and J. E. Marsden, *Foundations of Mechanics* (2nd ed.), Benjamin-Cummings, Reading, MA, 1978.

[2] I. A. R. Aitcheson and A. J. G. Hey, *Gauge Theories in Particle Physics*, Adam Hilger Ltd., Bristol, UK, 1982.

[3] H. Araki, *Mathematical Theory of Quantum Fields*, Oxford University Press, Oxford, 2000.

[4] V. I. Arnold, *Mathematical Methods of Classical Mechanics* (2nd ed.), Springer, New York, 1989.

[5] V. Bach, J. Fröhlich, and A. Pizzo, An infrared-finite algorithm for Rayleigh scattering amplitudes, and Bohr's frequency condition, *Comm. Math. Phys.* **274** (2007), 457–486.

[6] V. Bargmann, On unitary ray representations of continuous groups, *Annals of Math.* **59** (1954), 1–46.

[7] V. Bargmann, Note on Wigner's theorem on symmetry operations, *J. Math. Phys.* **5** (1964), 862–868.

[8] F. A. Berezin, *The Method of Second Quantization*, Academic Press, New York, 1966.

[9] H. A. Bethe, The electromagnetic shift of energy levels, *Phys. Rev.* **72** (1949), 339–341; reprinted in [**107**], pp. 139–141.

[10] N. N. Bogoliubow and O. S. Parasiuk, Über die Multiplikation der Kausalfunktion in der Quantentheorie der Felder, *Acta Math.* **97** (1957), 227–266.

[11] N. N. Bogolubov, A. A. Logunov, and I. T. Todorov, *Introduction to Axiomatic Quantum Field Theory*, W. A. Benjamin, Reading, MA, 1975.

[12] N. N. Bogolubov, A. A. Logunov, A. I. Oksak, and I. T. Todorov, *General Principles of Quantum Field Theory*, Kluwer, Dordrecht, 1990.

[13] L. Boutet de Monvel, Algèbre de Hopf des diagrammes de Feynman, renormalisation et factorisation de Wiener-Hopf (d'après A. Connes et D. Kreimer), Séminaire Bourbaki 2001–2002, *Astérisque* **290** (2003), 149–165.

[14] D. C. Brydges, What is a quantum field theory?, *Bull. Amer. Math. Soc.* (N.S.) **8** (1983), 31–40.

[15] R. H. Cameron, A family of integrals serving to connect the Wiener and Feynman integrals, *J. Math. and Phys.* **39** (1960), 126–140.

[16] A. Connes and D. Kreimer, Renormalization in quantum field theory and the Riemann-Hilbert problem I: the Hopf algebra structure of graphs and the main theorem, *Comm. Math. Phys.* **210** (2000), 249–273.

[17] A. Connes and D. Kreimer, Renormalization in quantum field theory and the Riemann-Hilbert problem II: the β-function, diffeomorphisms, and the renormalization group, *Comm. Math. Phys.* **216** (2001), 215–241.

[18] A. Connes and M. Marcolli, *Noncommutative Geometry, Quantum Fields and Motives*, American Mathematical Society, Providence. RI, 2008.

[19] F. H. Croom, *Basic Concepts of Algebraic Topology*, Springer, New York, 1978.

[20] H. Dehmelt, Experiments with an isolated subatomic particle at rest, *Rev. Mod. Phys.* **62** (1990), 525–530.

[21] P. Deligne et al. (eds.), *Quantum Fields and Strings: A Course for Mathematicians* (2 vols.), American Mathematical Society, Providence, RI, 1999.

[22] P. A. M. Dirac, *The Principles of Quantum Mechanics* (1st ed.), Clarendon Press, Oxford, 1930 (plus three later editions).

[23] P. A. M. Dirac, The Lagrangian in quantum mechanics, *Phys. Zeit. Sowjetunion* **3** (1933), 64–72; reprinted in [**107**], pp. 312–320.

[24] N. Dunford and J. T. Schwartz, *Linear Operators, Part II: Spectral Theory*, Wiley-Interscience, New York, 1963.

[25] F. J. Dyson, The S matrix in quantum electrodynamics, *Phys. Rev.* **75** (1949), 1736–1755; reprinted in [**107**], pp. 292–311, and in Dyson's *Selected Papers*, American Mathematical Society, Providence, RI. 1996, pp. 227–246.

[26] F. J. Dyson, Divergence of perturbation theory in quantum electrodynamics, *Phys. Rev.* **85** (1952), 631–632; reprinted in Dyson's *Selected Papers*, American Mathematical Society, Providence, RI. 1996, pp. 255–256.

[27] A. Erdélyi, *Asymptotic Expansions*, Dover, New York, 1956.

[28] L. D. Faddeev and V. N. Popov, Feynman diagrams for the Yang-Mills field, *Phys. Lett.* **25B** (1967), 29–30.

[29] F. J. M. Farley and E. Picasso, The muon $g-2$ experiments, pp. 479–559 in [**71**].

[30] G. Farmelo (ed.), *It Must Be Beautiful: Great Equations of Modern Science*, Granta, London, 2002.

[31] P. Federbush, Quantum field theory in ninety minutes, *Bull. Amer. Math. Soc.* (N.S.) **17** (1987), 93–103.

[32] J. S. Feldman, T. R. Hurd, L. Rosen, and J. D. Wright, *QED: A Proof of Renormalizability*, Springer, Berlin, 1988.

[33] E. Fermi, Versuch einer Theorie der β-Strahlen I, *Zeit. für Phys.* **88** (1934), 161–177.

[34] R. P. Feynman, Space-time approach to non-relativistic quantum mechanics, *Rev. Mod. Phys.* **20** (1948), 367–387; reprinted in [**107**], pp. 321–341, and in Feynman's *Selected Papers*, World Scientific, Singapore, 2000, pp. 177–197.

[35] R. P. Feynman, The development of the space-time view of quantum electrodynamics (Nobel lecture), *Science* **153** (1966), 699–708; reprinted in Feynman's *Selected Papers*, World Scientific, Singapore, 2000, pp. 9–32.

[36] R. P. Feynman, *The Character of Physical Law*, MIT Press, Cambridge, MA, 1967.

[37] R. P. Feynman, *QED: The Strange Theory of Light and Matter*, Princeton University Press, Princeton, NJ, 1985.

[38] R. P. Feynman, *"Surely You're Joking, Mr. Feynman": Adventures of a Curious Character*, W. W. Norton, New York, 1985.

[39] R. P. Feynman and M. Gell-Mann, Theory of the Fermi interaction, *Phys. Rev.* **109** (1958), 193–198; reprinted in Feynman's *Selected Papers*, World Scientific, Singapore, 2000, pp. 417–422.

[40] R. P. Feynman and A. R. Hibbs, *Quantum Mechanics and Path Integrals*, McGraw-Hill, New York, 1965.

[41] R. P. Feynman, R. B. Leighton, and M. Sands, *The Feynman Lectures on Physics* (3 vols.), Addison-Wesley, Reading, MA, 1963–5.

[42] G. B. Folland, Weyl manifolds, *J. Diff. Geom.* **4** (1970), 145–153.

[43] G. B. Folland, *Harmonic Analysis in Phase Space*, Princeton University Press, Princeton, NJ, 1989.

[44] G. B. Folland, *A Course in Abstract Harmonic Analysis*, CRC Press, Boca Raton, FL, 1995.

[45] G. B. Folland, *Introduction to Partial Differential Equations* (2nd ed.), Princeton University Press, Princeton, NJ, 1995.

[46] G. B. Folland, Fundamental solutions for the wave operator, *Expos. Math.* **15** (1997), 25–52.

[47] G. B. Folland, *Real Analysis* (2nd ed.), John Wiley, New York, 1999.

[48] G. B. Folland, How to integrate a polynomial over a sphere, *Amer. Math. Monthly* **108** (2001) 446–448.

[49] G. B. Folland and A. Sitaram, The uncertainty principle: a mathematical survey, *J. Fourier Anal. Appl.* **3** (1997), 207–238.

[50] A. Franklin, *Are There Really Neutrinos?*, Westview Press, Boulder, CO, 2004.

[51] I. M. Gelfand and N. Ya. Vilenkin, *Generalized Functions, Volume 4: Applications of Harmonic Analysis*, Academic Press, New York, 1964.

[52] M. Gell-Mann and F. Low, Bound states in quantum field theory, *Phys. Rev.* **84** (1951), 350–354.

[53] S. L. Glashow, Partial-symmetries of weak interactions, *Nucl. Phys.* **22** (1961), 579–588.

[54] J. Glimm and A. Jaffe, *Quantum Physics: A Functional Integral Point of View* (2nd ed.), Springer, New York, 1987.

[55] H. Goldstein, *Classical Mechanics* (2nd ed.), Addison-Wesley, Reading, MA, 1980.

[56] J. Goldstone, A. Salam, and S. Weinberg, Broken symmetries, *Phys. Rev.* **127** (1962), 965–970.

[57] W. Greiner, S. Schramm, and E. Stein, *Quantum Chromodynamics* (2nd ed.), Springer, Berlin, 2002.

[58] D. Griffiths, *Introduction to Elementary Particles*, John Wiley, New York, 1987.

[59] R. Haag, *Local Quantum Physics: Fields, Particles, Algebras* (2nd ed.), Springer, Berlin, 1996.

[60] G. A. Hagedorn, Semiclassical quantum mechanics I: the $\hbar \to 0$ limit for coherent states, *Comm. Math. Phys.* **71** (1980), 77–93.

[61] B. C. Hall, *Lie Groups, Lie Algebras, and Representations*, Springer, New York, 2003.

[62] Y. Hahn and W. Zimmermann, An elementary proof of Dyson's power counting theorem, *Comm. Math. Phys.* **10** (1968), 330–342.

[63] K. Hepp, Proof of the Bogoliubov-Parasiuk theorem on renormalization, *Comm. Math. Phys.* **2** (1966), 301–326.

[64] T. R. Hurd, A renormalization group proof of perturbative renormalizability, *Comm. Math. Phys.* **124** (1989), 153–168.

[65] C. Itzykson and J.-B. Zuber, *Quantum Field Theory*, McGraw-Hill, New York, 1980.

[66] A. Jaffe and E. Witten, Quantum Yang-Mills theory, in J. Carlson, A. Jaffe, and A. Wiles (eds.), *The Millennium Prize Problems*, American Mathematical Society, Providence, RI, 2006, pp. 129–152.

[67] J. M. Jauch, *Foundations of Quantum Mechanics*, Addison-Wesley, Reading, MA, 1968.

[68] J. M. Jauch and F. Rohrlich, *The Theory of Photons and Electrons* (2nd ed.), Springer, Berlin, 1976.

[69] R. Jost, *The General Theory of Quantized Fields*, American Mathematical Society, Providence, RI, 1965.

[70] G. Keller and C. Kopper, Renormalizability proof for QED based on flow equations, *Comm. Math. Phys.* **176** (1996), 193–226.

[71] T. Kinoshita (ed.), *Quantum Electrodynamics*, World Scientific, Singapore, 1990.

[72] T. Kinoshita, Theory of the Anomalous magnetic moment of the electron—numerical approach, pp. 218–321 in [**71**].

[73] T. Kinoshita and W. J. Marciano, Theory of the muon anomalous magnetic moment, pp. 419–478 in [**71**].

[74] A. A. Kirillov, *Lectures on the Orbit Method*, American Mathematical Society, Providence, RI, 2004.

[75] D. Kreimer, *Knots and Feynman Diagrams*, Cambrige University Press, Cambridge, 2000.

[76] L. D. Landau and E. M. Lifshitz, *Quantum Mechanics (Non-relativistic Theory)* (3rd ed.), Pergamon, Oxford, 1977.

[77] E. H. Lieb, The stability of matter: from atoms to stars, *Bull. Amer. Math. Soc.* (N.S.) **22** (1990), 1–49.

[78] G. W. Mackey, *Mathematical Foundations of Quantum Mechanics*, Benjamin, New York, 1963; reprinted by Dover Books, New York, 2004.

[79] E. B. Manoukian, *Renormalization*, Academic Press, New York, 1983.

[80] P. T. Matthews and A. Salam, The renormalization of meson theories, *Rev. Mod. Phys.* **23** (1951), 311-314.

[81] R. D. Mattuck, *A Guide to Feynman Diagrams in the Many-body Problem* (2nd ed.), McGraw-Hill, New York, 1976; reprinted by Dover Books, New York, 1992.

[82] A. Messiah, *Quantum Mechanics* (2 vols.), Wiley-Interscience, New York, 1961–2; reprinted by Dover Books, New York, 1999.

[83] R. Montgomery, Review of *Symmetry in Mechanics* by S. F. Singer, *Amer. Math. Monthly* **110** (2003), 348–353.

[84] J. W. Morgan, An introduction to gauge theory, in R. Friedman and J. W. Morgan (eds.), *Gauge Theory and the Topology of Four-Manifolds*, American Mathematical Society, Providence, RI, 1998, pp. 51–143.

[85] E. Nelson, Feynman integrals and the Schrödinger equation, *J. Math. Phys.* **5** (1964), 332–343.

[86] E. Nelson, Quantum fields and Markoff fields, in D. C. Spencer (ed.), *Partial Differential Equations* (Proc, Symp. Pure Math., vol. XXIII), American Mathematical Society, Providence, RI, 1973, pp. 413–420.

[87] A. Pais, Einstein on particles, fields, and the quantum theory, pp. 197–251 in [**134**].

[88] M. E. Peskin and D. V. Schroeder, *An Introduction to Quantum Field Theory*, Perseus Books, Cambridge, MA, 1995.

[89] F. M. Pipkin, Lamb shift measurements, pp. 696–773 in [**71**].

[90] E. M. Purcell, *Electricity and Magnetism* (2nd ed.), McGraw-Hill, New York, 1985.

[91] P. Ramond, *Field Theory: A Modern Primer* (2nd ed.), Addison-Wesley, Redwood City, CA, 1989.

[92] P. Ramond, *Journeys Beyond the Standard Model*, Westview Press, Boulder, CO, 2004.

[93] M. Reed and B. Simon, *Methods of Modern Mathematical Physics I: Functional Analysis*, Academic Press, New York, 1972.

[94] M. Reed and B. Simon, *Methods of Modern Mathematical Physics II: Fourier Analysis, Self-Adjointness*, Academic Press, New York, 1975.

[95] M. Reed and B. Simon, *Methods of Modern Mathematical Physics III: Scattering Theory*, Academic Press, New York, 1979.

[96] M. Reed and B. Simon, *Methods of Modern Mathematical Physics IV: Analysis of Operators*, Academic Press, New York, 1978.

[97] K. Rith and A. Schäfer, The mystery of nuclear spin, *Scientific American*, July 1999, 58–63.

[98] P. L. Robinson, The Berezin calculus, *Publ. RIMS Kyoto Univ.* **35** (1999), 123–194.

[99] L. Rosen and J. D. Wright, Dimensional regularization and renormalization of QED, *Comm. Math. Phys.* **134** (1990), 433–466.

[100] L. Rosen, Barry Simon's contributions to quantum field theory, in F. Gesztesy et al. (eds.), *Spectral Theory and Mathematical Physics: A Festschrift in Honor of Barry Simon's 60th Birthday*, Amreican Mathematical Society, Providence, RI, 2007, pp. 69–101.

[101] R. Z. Roskies, E. Remiddi, and M. J. Levine, Analytic evaluation of sixth-order contributions to the electron's g factor, pp. 162–217 in [**71**].

[102] J. J. Sakurai, *Advanced Quantum Mechanics*, Addison-Wesley, Reading, MA, 1967.

[103] A. Salam, Weak and electromagnetic interactions, pp. 369–377 in N. Svartholm (ed.), *Elementary Particle Theory — Relativistic Groups and Analyticity* John Wiley, New York, 1968; also pp. 244–254 in Salam's *Selected Papers*, World Scientific, Singapore, 1994.

[104] J. R. Sapirstein and D. R. Yennie, Theory of hydrogenic bound states, pp. 560–674 in [**71**].

[105] S. S. Schweber, *QED and the Men Who Made It: Dyson, Feynman, Schwinger, and Tomonaga*, Princeton University Press, Princeton, NJ, 1994.

[106] S. S. Schweber, Quantum Field Theory: From QED to the standard model, in M. J. Nye (ed.), *The Cambridge History of Science, Volume 5: The Modern Physical and Mathematical Sciences*, Cambridge University Press, Cambridge, UK, 2003, pp. 375–393.

[107] J. Schwinger (ed.), *Quantum Electrodynamics*, Dover, New York, 1958.

[108] J. Schwinger, *Particles, Sources, and Fields* (2nd ed.), Addison-Weley, Redwood City, CA, 1988.

[109] I. E. Segal, Distributions in Hilbert space and canonical systems of operators, *Trans. Amer. Math. Soc.* **88** (1958), 12–41.

[110] D. J. Simms, *Lie Groups and Quantum Mechanics*, Lecture Notes in Math. v. 52, Springer, Berlin, 1968.

[111] B. Simon, *The $P(\phi)_2$ Euclidean (Quantum) Field Theory*, Princeton University Press, Princeton, NJ, 1974.

[112] B. Simon, *Functional Integration and Quantum Physics* (2nd ed.), American Mathematical Society, Providence, RI, 2005.

[113] V. A. Smirnov, *Feynman Integral Calculus*, Springer, Berlin, 2006.

[114] R. F. Streater and A. S. Wightman, *PCT, Spin, Statistics, and All That* (2nd revised printing), Benjamin, New York, 1978; reprinted by Princeton University Press, Princeton, NJ, 2000.

[115] M. E. Taylor, *Partial Differential Equations I*, Springer, New York, 1996.

[116] B. Thaller, *The Dirac Equation*, Springer, Berlin, 1992.

[117] G. 't Hooft, Renormalizable Lagrangians for massive Yang-Mills fields, *Nucl. Phys. B* **35** (1971), 167–188.

[118] G. 't Hooft and M. Veltman, Regularization and renormalization of gauge fields, *Nucl. Phys. B* **44** (1972), 189–213.

[119] R. Ticciati, *Quantum Field Theory for Mathematicians*, Cambridge University Press, Cambridge, UK, 1999.

[120] F. Treves, *Basic Linear Partial Differential Equations*, Academic Press, New York, 1975.
[121] R. S. van Dyck, Anomalous magnetic moment of single electrons and positrons: experiment, pp. 322–388 in [**71**].
[122] W. D. van Suijlekom, The Hopf algebra of Feynman graphs in quantum electrodynamics, *Letters Math. Phys.* **77** (2006), 265–281.
[123] V. S. Varadarajan, *The Geometry of Quantum Theory* (2 vols.), Van Nostrand Reinhold, New York, 1968–70.
[124] G. Velo and A. S. Wightman, *Renormalization Theory*, D. Reidel, Dordrecht, 1976.
[125] D. A. Vogan, Representations of reductive Lie groups, in *Proceedings of the International Congress of Mathematicians 1986*, American Mathematical Society, Providence, RI, 1987, pp. 245–266.
[126] D. A. Vogan, Review of *Lectures on the Orbit Method* by A. A. Kirillov, *Bull. Amer. Math. Soc.* (N.S.) **42** (2005), 535–544.
[127] S. Weinberg, High-energy behavior in quantum field theory, *Phys. Rev.* **118** (1960), 838–849.
[128] S. Weinberg, A model of leptons, *Phys. Rev. Lett.* **19** (1967), 1264–1266.
[129] S. Weinberg, *The Quantum Theory of Fields I: Foundations*, Cambridge University Press, Cambridge, UK, 1995.
[130] S. Weinberg, *The Quantum Theory of Fields II: Modern Applications*, Cambridge University Press, Cambridge, UK, 1996.
[131] S. Weinberg, *The Quantum Theory of Fields III: Supersymmetry*, Cambridge University Press, Cambridge, UK, 2000.
[132] E. P. Wigner, On unitary representations of the inhomogeneous Lorentz group, *Annals of Math.* **40** (1939), 149–204.
[133] A. S. Wightman, Orientation, pp. 1–24 in [**124**].
[134] H. Woolf (ed.), *Some Strangeness in the Proportion: A Centennial Symposium to Celebrate the Achievements of Albert Einstein*, Addison-Wesley, Reading, MA, 1980.
[135] C. N. Yang and R. L. Mills, Conservation of isotopic spin and isotopic gauge invariance, *Phys. Rev.* **96** (1954), 191–195.
[136] A. Zee, *Quantum Field Theory in a Nutshell*, Princeton University Press, Princeton, NJ, 2003.
[137] W. Zimmermann, The power counting theorem for Minkowski metric, *Comm. Math. Phys.* **11** (1968), 1–8.
[138] W. Zimmermann, Convergence of Bogoliubov's method of renormalization in momentum space, *Comm. Math. Phys.* **15** (1969), 208–234.
[139] W. Zimmerman, The power counting theorem for Feynman integrals with massless propagators, pp. 171–184 in [**124**].

Index

Titles in This Series

149 **Gerald B. Folland,** Quantum field theory: A tourist guide for mathematicians, 2008

148 **Patrick Dehornoy with Ivan Dynnikov, Dale Rolfsen, and Bert Wiest,** Ordering braids, 2008

147 **David J. Benson and Stephen D. Smith,** Classifying spaces of sporadic groups, 2008

146 **Murray Marshall,** Positive polynomials and sums of squares, 2008

145 **Tuna Altinel, Alexandre V. Borovik, and Gregory Cherlin,** Simple groups of finite Morley rank, 2008

144 **Bennett Chow, Sun-Chin Chu, David Glickenstein, Christine Guenther, James Isenberg, Tom Ivey, Dan Knopf, Peng Lu, Feng Luo, and Lei Ni,** The Ricci flow: Techniques and applications, Part II: Analytic aspects, 2008

143 **Alexander Molev,** Yangians and classical Lie algebras, 2007

142 **Joseph A. Wolf,** Harmonic analysis on commutative spaces, 2007

141 **Vladimir Maz′ya and Gunther Schmidt,** Approximate approximations, 2007

140 **Elisabetta Barletta, Sorin Dragomir, and Krishan L. Duggal,** Foliations in Cauchy-Riemann geometry, 2007

139 **Michael Tsfasman, Serge Vlăduţ, and Dmitry Nogin,** Algebraic geometric codes: Basic notions, 2007

138 **Kehe Zhu,** Operator theory in function spaces, 2007

137 **Mikhail G. Katz,** Systolic geometry and topology, 2007

136 **Jean-Michel Coron,** Control and nonlinearity, 2007

135 **Bennett Chow, Sun-Chin Chu, David Glickenstein, Christine Guenther, James Isenberg, Tom Ivey, Dan Knopf, Peng Lu, Feng Luo, and Lei Ni,** The Ricci flow: Techniques and applications, Part I: Geometric aspects, 2007

134 **Dana P. Williams,** Crossed products of C^*-algebras, 2007

133 **Andrew Knightly and Charles Li,** Traces of Hecke operators, 2006

132 **J. P. May and J. Sigurdsson,** Parametrized homotopy theory, 2006

131 **Jin Feng and Thomas G. Kurtz,** Large deviations for stochastic processes, 2006

130 **Qing Han and Jia-Xing Hong,** Isometric embedding of Riemannian manifolds in Euclidean spaces, 2006

129 **William M. Singer,** Steenrod squares in spectral sequences, 2006

128 **Athanassios S. Fokas, Alexander R. Its, Andrei A. Kapaev, and Victor Yu. Novokshenov,** Painlevé transcendents, 2006

127 **Nikolai Chernov and Roberto Markarian,** Chaotic billiards, 2006

126 **Sen-Zhong Huang,** Gradient inequalities, 2006

125 **Joseph A. Cima, Alec L. Matheson, and William T. Ross,** The Cauchy Transform, 2006

124 **Ido Efrat, Editor,** Valuations, orderings, and Milnor K-Theory, 2006

123 **Barbara Fantechi, Lothar Göttsche, Luc Illusie, Steven L. Kleiman, Nitin Nitsure, and Angelo Vistoli,** Fundamental algebraic geometry: Grothendieck's FGA explained, 2005

122 **Antonio Giambruno and Mikhail Zaicev, Editors,** Polynomial identities and asymptotic methods, 2005

121 **Anton Zettl,** Sturm-Liouville theory, 2005

120 **Barry Simon,** Trace ideals and their applications, 2005

119 **Tian Ma and Shouhong Wang,** Geometric theory of incompressible flows with applications to fluid dynamics, 2005

118 **Alexandru Buium,** Arithmetic differential equations, 2005

117 **Volodymyr Nekrashevych,** Self-similar groups, 2005

TITLES IN THIS SERIES

116 **Alexander Koldobsky,** Fourier analysis in convex geometry, 2005

115 **Carlos Julio Moreno,** Advanced analytic number theory: L-functions, 2005

114 **Gregory F. Lawler,** Conformally invariant processes in the plane, 2005

113 **William G. Dwyer, Philip S. Hirschhorn, Daniel M. Kan, and Jeffrey H. Smith,** Homotopy limit functors on model categories and homotopical categories, 2004

112 **Michael Aschbacher and Stephen D. Smith,** The classification of quasithin groups II. Main theorems: The classification of simple QTKE-groups, 2004

111 **Michael Aschbacher and Stephen D. Smith,** The classification of quasithin groups I. Structure of strongly quasithin K-groups, 2004

110 **Bennett Chow and Dan Knopf,** The Ricci flow: An introduction, 2004

109 **Goro Shimura,** Arithmetic and analytic theories of quadratic forms and Clifford groups, 2004

108 **Michael Farber,** Topology of closed one-forms, 2004

107 **Jens Carsten Jantzen,** Representations of algebraic groups, 2003

106 **Hiroyuki Yoshida,** Absolute CM-periods, 2003

105 **Charalambos D. Aliprantis and Owen Burkinshaw,** Locally solid Riesz spaces with applications to economics, second edition, 2003

104 **Graham Everest, Alf van der Poorten, Igor Shparlinski, and Thomas Ward,** Recurrence sequences, 2003

103 **Octav Cornea, Gregory Lupton, John Oprea, and Daniel Tanré,** Lusternik-Schnirelmann category, 2003

102 **Linda Rass and John Radcliffe,** Spatial deterministic epidemics, 2003

101 **Eli Glasner,** Ergodic theory via joinings, 2003

100 **Peter Duren and Alexander Schuster,** Bergman spaces, 2004

99 **Philip S. Hirschhorn,** Model categories and their localizations, 2003

98 **Victor Guillemin, Viktor Ginzburg, and Yael Karshon,** Moment maps, cobordisms, and Hamiltonian group actions, 2002

97 **V. A. Vassiliev,** Applied Picard-Lefschetz theory, 2002

96 **Martin Markl, Steve Shnider, and Jim Stasheff,** Operads in algebra, topology and physics, 2002

95 **Seiichi Kamada,** Braid and knot theory in dimension four, 2002

94 **Mara D. Neusel and Larry Smith,** Invariant theory of finite groups, 2002

93 **Nikolai K. Nikolski,** Operators, functions, and systems: An easy reading. Volume 2: Model operators and systems, 2002

92 **Nikolai K. Nikolski,** Operators, functions, and systems: An easy reading. Volume 1: Hardy, Hankel, and Toeplitz, 2002

91 **Richard Montgomery,** A tour of subriemannian geometries, their geodesics and applications, 2002

90 **Christian Gérard and Izabella Łaba,** Multiparticle quantum scattering in constant magnetic fields, 2002

89 **Michel Ledoux,** The concentration of measure phenomenon, 2001

88 **Edward Frenkel and David Ben-Zvi,** Vertex algebras and algebraic curves, second edition, 2004

87 **Bruno Poizat,** Stable groups, 2001

86 **Stanley N. Burris,** Number theoretic density and logical limit laws, 2001

For a complete list of titles in this series, visit the AMS Bookstore at **www.ams.org/bookstore/**.